Lecture Notes in Earth Sciences   50

Editors:
S. Bhattacharji, Brooklyn
G. M. Friedman, Brooklyn and Troy
H. J. Neugebauer, Bonn
A. Seilacher, Tuebingen

Reiner Rummel Fernando Sansò (Eds.)

Satellite Altimetry in Geodesy and Oceanography

Springer-Verlag Berlin Heidelberg GmbH

Editors

Prof. Dr. Reiner Rummel
Faculty of Geodesy, Delft University of Technology
Thijsseweg 11, NL-2629 JA Delft, The Netherlands

Prof. Dr. Fernando Sansò
Politecnico di Milano, D.I.I.A.R.
Piazza Leonardo da Vinci 32, I-20133 Milano, Italy

"For all Lecture Notes in Earth Sciences published till now please see final pages of the book"

ISBN 978-3-540-56818-6   ISBN 978-3-540-47758-7 (eBook)
DOI 10.1007/978-3-540-47758-7

Typesetting: Camera ready by author

32/3140-543210 - Printed on acid-free paper

FOREWORD

The International Summer School of Theoretical Geodesy is an Italian tradition, initiated by Antonio Marussi and carried on by me in recent years, with the very fundamental help of Reiner Rummel.

So it has been more than 4 years ago, with the course on "Theory of Satellite Geodesy and Gravity Field Determination", and even more this year with the course on "Satellite Altimetry in Geodesy and Oceanography".

The success of the school is absolutely due to the high scientific level and the bright personality of the teachers, who have nicely complemented and integrated each other, and the merit of putting them together is entirely of Reiner Rummel, who has been the true scientific organizer of the school.

The subject has been chosen years ago with the perspective of giving support to a research which was likely to grow, due to the foreseen (now flying) new satellite missions; the interdisciplinarity of the subject was a challenge that has been turned into a success by the spectacular competence of the teachers.

Also the practical organization of the school has worked very well, thanks first of all to the hospitality of the International Centre of Theoretical Physics, of its Deputy Director, Prof. Luciano Bertocchi, and its staff, in particular Miss Daniela Giombi; but the real soul of the organization has been Miss Elena Raguzzoni to whom go the sincere thanks of all the attendees as well as my personal congratulations.

Special thanks are also due to those organizations that have given us a real possibility to organize this course by supporting it financially, primarily creating fellowships for many students: so we want to acknowledge the support of IUGG and IAG (also thanks to their secretaries general, Dr. George Balmino and Dr. Claude Boucher) as well as the Italian firm Alenia.

Special thanks are also due to the Department of Geodesy of the University of Delft and to the Department of Environmental Engineering of the Politecnico of Milano, which have supplied a decisive support, substituting at the last moment other organizations, that traditionally helped the courses of this School, but this time did not.

Fernando Sansò

TABLE OF CONTENTS

Quantifying Time-Varying Oceanographic Signals with Altimetry
V. Zlotnicki

Principle of Satellite Altimetry and Elimination of Radial Orbit Errors
R. Rummel

Orbit Choice and the Theory of Radial Orbit Error for Altimetry
G. Balmino

Theory of Geodetic B.V.P.s Applied to the Analysis of Altimetric Data
F. Sansò

Use of Altimeter Data in Estimating Global Gravity Models
R.H. Rapp

SEMINARS

The Direct Estimation of the Potential Coefficients by Biorthogonal Sequences
M.A. Brovelli and F. Migliaccio

Frozen Orbits and their Application in Satellite Altimetry
E.J.O. Schrama

Integration of Gravity and Altimeter Data by Optimal Estimation Techniques
P. Knudsen

Comparing the UK Fine Resolution Antarctic Model (FRAM) with 3-years of Geosat Altimeter Data
R.C.V. Feron

INTRODUCTION

The International Summer School of Theoretical Geodesy on **Satellite Altimetry in Geodesy and Oceanography** was held in Trieste (Italy) from May 25 to June 6, 1992. It was organized by Prof. R. Rummel of the Delft University of Technology and by Prof. F. Sansò of the Politecnico di Milano and was attended by 63 participants and 7 lecturers from 17 countries. The School was hosted by the International Centre of Theoretical Physics of Trieste.

Satellite altimetry provides a lot of data that require more and more sophisticated models in order to be interpreted and exploited. One of the main problems related to the practical treatment of the data can be summarized as follows: oceanographers would like to ask geodesists to compute precise orbits and a precise geoid in order to put into evidence the Sea Surface Topography that can be interpreted as an oceanographic signal related to currents and to several physical parameters; on the other hand, geodesists would like to ask oceanographers to a-priori determine the Sea Surface Topography, in order to be able to extract from the altimeter data the geoid and the orbit errors to be used in the gravity field modelling.

The solution to this dilemma can only be found in a cooperative frame. An integrated model to be used for a single-step treatment of altimetry is probably far to be defined, so at present geodesists and oceanographers must cooperate to obtain step-wise and iterative modelling of the gravity field and of the oceanographic phenomena.

This is precisely the reason why the school on Satellite Altimetry was organized on an interdisciplinary basis.

The programme of the school was quite intense: 48 hours were devoted to the main lectures and about 8 hours to the seminars.

Prof. <u>C. Wunsch</u> lectured on the <u>Physics of the Ocean Circulation</u> giving a clear introduction to physical oceanography with a particular attention to the scaling problem and geostrophic currents; these lectures were intentionally dedicated to the non-specialists.

Prof. G. Balmino lectured on the Orbit Choice and the Theory of Radial Orbit Error for Altimetry: the fundamental theory of orbits was followed by more specialized topics related to satellite altimetry.

Prof. R. Rummel treated the Principles of Satellite Altimetry and Elimination of Radial Orbit Errors introducing the audience to the general principles of the main topic of the school, and also treating more technical problems like regional and global cross-over adjustments, analysis of the theoretical and numerical rank deficiency of the adjustment and analysis of the separability of the Sea Surface Topography and the geoid.

Prof. F. Sansò treated the Theory of Geodetic B.V.P.s Applied to Altimetric Data. The excursus from the basic definitions of some relevant functionals (like the gravity anomaly) to the analysis of the Altimetry-Gravimetry B.V.P.s using the tools of functionals analysis clearly showed the connection between the theoretical studies and the construction of numerical procedures for the treatment of real data.

Prof. D.E. Cartwright lectured on the Theory of Ocean Tides with Application to Altimetry. Tides are a quite evident phenomenon and their study has therefore a long story, but it is still a very interesting field of active research and their presence in altimetric signals has allowed to test some nowadays classical theory like that of Shwidesky or to improve our knowledge of these phenomena.

Prof. V. Zlotnicki treated the topic of Measuring Oceanographic Phenomena with Altimetric Data; the basic behaviour of the ocean was also described starting from the geostrophic motion and introducing more and more advanced tools to describe different kind of waves. This has achieved also using mimic expressions which helped the intuitive understanding of the matter by the non-specialists.

Finally Prof. R.H. Rapp described the Use of Altimeter Data in Estimating Global Gravity Models. Both the enormous usefulness of the altimeter data and the problems connected to their practical treatment were illustrated, with particular care of the finite dimensional treatment in the form of giant least squares and with deep discussion of the weighting problem.

A short list of the seminars presented is as follows:
- The ERS 1 altimeter; principles and performance errors (G. Levrini from Alenia)
- Processing of GEOS-3 and SEASAT altimetric data on the basis of algorithm of preliminary multipoles analysis (A.N. Marchenko)
- Sea surface height and geoid separation in shallow water with examples in the Nordic Sea (M. Metzner)

- The direct estimation of potential coefficients by biorthogonal sequences (M. Brovelli, F. Migliaccio)
- Comparing the UK fine resolution antarctic model (FRAM) with 3 years of Geosat altimeter data (R. Feron)
- Integrating gravimetric and altimetric data by optimal estimation techniques (P. Knudsen)
- Geophysical structures from altimetry (R. Haagmans)
- ERS 1 orbital equations (R.C.A. Zandbergen)
- Frozen orbits (E. Schrama)
- The ERS 1 altimeter (B. Greco from ESRIN).

The already mentioned interdisciplinary character of the school, together with the very high scientific and didactic level of the lectures, was a reason of great satisfaction for all the participants and the main motive of the success of the school.
It is also interesting to look at the distribution of the participants between the various disciplinary fields. The statistic presented hereafter was elaborated by R. Rummel on the basis of a short questionnaire:

Field of interest   % of interest

Geodesy in general 9.1
Gravimetry 21.8
Satellite geod. in general 10.9
Orbit computation 16.4
Satellite altimetry 25.4
Oceanography 16.4

The participants enjoyed not only the scientific contents of the lectures but also the friendly atmosphere of the school and the rich social programme carefully organized by the secretary of the school Mrs. Elena Raguzzoni.

                                        Battista Benciolini

Lecturers
Balmino, Cartwright, Rapp, Rummel, Sansò, Wunsch, Zlotnicki

Participants
Albertella, Andersen, Barbosa, Barzaghi, Benciolini, Betti, Blanc, Blomenhofer, Bonini, Bonnefond, Brovelli, Carpino, Cascioli, Crippa, De Munck, Denker, Devoti, Drottning, Ekholm, Erlandsen, Farrelly, Fenoglio, Feron, Francis, Fuerst, Haagmans, Hernandez, Huang, Keller, Klees, Knudsen, Kobrle, Le Traon, Luceri, Lyszkowicz, Marchenko, Martinez, Metzner, Migliaccio, Moliterni, Noreus, Otero, Rodrigues, Sacerdote, Samuel, Scharroo, Schrama, Schuh, Sciarretta, Sevilla, Sguerso, Sideris, Sneeuw, Solheim, Solvsteen, Sona, Spoecker, Stum, Toncic, Van Leeuwen, Wang, Wiejak, Zandbergen

LECTURES

Physics of the Ocean Circulation

Carl Wunsch

Department of Earth, Atmospheric and Planetary Sciences
Massachusetts Institute of Technology
Cambridge MA 02139, U.S.A.

1 The Ocean Circulation

The intersections of geodesy and physical oceanography take place at the seasurface, and there is a long history of cross-fertilization, and conflict, between the two subjects. Geodesists have long used the concept of the "mean sealevel" as a useful reference point in land surveys. Tide gauges are operated by organizations with both geodetic and oceanographic missions and the interpretation of the records requires a knowledge of both subjects. The concept of a resting fluid ocean usefully defines the geoid in a nearly operational sense. Discussion of the interpretation of the slope of the actual seasurface relative to the geoid led to a notable debate between oceanographers on the one hand (Sturges, 1974, Balazs and Douglas, 1979) and the geodetic community on the other, and was resolved only comparatively recently.

But it must be admitted that the intersections have tended in the past to be somewhat peripheral to both subjects, with a geodesist having to know little or nothing about oceanography, apart from the tides, and vice-versa. The advent of accurate and precise satellite altimeters has drastically changed this situation. That the actual seasurface deviates slightly from the geoid only becomes of geodetic concern when one attempts to estimate the geoid, and gravity with an accuracy and precision at the level of those deviations. The possibility of distinguishing the mean seasurface from the geoid opens to the oceanographer the revolutionary possibility of having a global tool for determining the ocean circulation with a coverage in space and time which is completely impossible by other means.

It is my purpose in these lectures to touch on the major elements of physical oceanography which a geodesist should at least know about, both so she can be aware of the significance of her own results to oceanographers, and so that the uncertainties of oceanography which influence her results can also be appreciated. I will however, leave the subject of tides wholly to Dr. Cartwright. I have no illusions that in a week of lectures that anyone can digest major chunks of physical oceanography; but perhaps we can make a useful start. The dynamical discussion is almost a cartoon: a sketch of a large and interesting subject, but which the reader will have to pursue in the references provided. In general, I have tried to focus on the central issues which would concern a geodesist attempting to use altimetric data for mainly geodetic purposes, who needs to understand the "contamination" of the measurements by the ocean circulation, but who might take an interest in using the same measurements to help understand the circulation itself.

Why should one worry about the ocean circulation? There are many answers to that question, the central one for a physical oceanographer being that it is simply an immensely fascinating and challenging problem in large scale fluid dynamics. At a more practical level, the ocean circulation is a central element in any understanding of the earth's climate, how it was different in the past and how it might be different in the future. It is a sometimes decisive element in the waxing and waning of fisheries, is inextricably tied to the questions of sealevel rise and its influence on the shorefront, and has important military connotations (now fortunately much reduced in interest). Lambeck (1988) has an interesting discussion of many of the geophysical problems that connect physical oceanography and geodesy.

Basic Physical Elements

The mean oceanic depth is about 3700 m (Pickard and Emery, 1982), which makes it a very thin skin at the surface of the earth (recalling that the mean earth radius is about 6.3×10^6m). This thinness is an essential element in understanding the physics governing the flow of the fluid.

A zero order understanding for the magnitudes of the influence of the water motion on the apparent shape of the earth may be obtained from simple consideration of a rotating, nearly spherical earth covered by a thin shell of constant density fluid. The apparent equipotentials of a sphere rotating with radian angular velocity Ω (see figure 1) are given by

$$-\frac{GM_E}{r} + \Omega^2 r^2 \cos^2\phi = C \ (= constant) \tag{1}$$

where ϕ is the latitude, G is the gravitational constant and M_E is the mass of the earth. Eq (2) defines $r_0 = r(\phi, C)$. Suppose a zonal motion, $u(\phi)$, is set up on the sphere (which for present purposes is covered by water - i.e. no continents in the way). Then the effective rotation rate changes from Ω to $\Omega + u$, and the potential is displaced radially to

$$-\frac{GM_E}{r} + \left(\Omega + \frac{u}{r}\right)^2 r^2 \cos\phi^2 = C \ (= constant) \tag{2}$$

Writing the displaced radius as $r = r_0(1 + \varepsilon(\phi))$,

$$-\frac{GM_E}{r_0(1+\varepsilon)} + \left(\Omega^2 + \frac{2\Omega u}{r_0(1+\varepsilon)} + \frac{u^2}{r_0^2(1+\varepsilon)^2}\right) r_0^2(1+\varepsilon)^2 \cos^2\phi = C \tag{3}$$

and expanding to lowest order, produces

$$\varepsilon \cong \frac{2\Omega u \cos^2\phi}{g'} \tag{4}$$

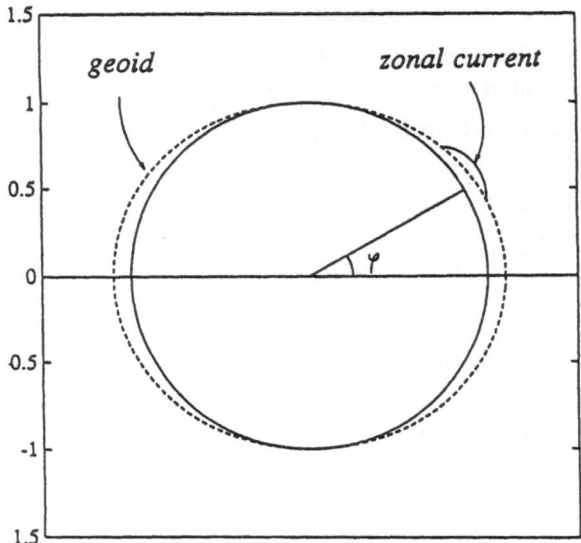

1. Schematic of the deflection of the sea surface relative to the geoid owing to currents on a rotating earth. The effect of a water velocity may be understood as a centrifugal force in a fixed frame. In a reference frame rotating with the earth, it manifests itself as a Coriolis force.

where g' is apparent local gravity. Putting in numbers, $u \cong 1m/s$ produces $\varepsilon \cong 20cm$ which suggests that water movement can shift a gravitational equipotential by some tens of centimeters - and which is the size of the signal we will be seeking. The centrifugal force exerted by such a zonal motion can be resolved into components radially and tangentially. Since the radial component in a thin shell does nothing but slightly perturb ordinary gravity, we can basically ignore it. The centrifugal force in the latitudinal direction is readily derived from (4) and is proportional to $2\Omega \sin \phi u$, vanishing, as it must, at the equator. This force, when written into the equations of fluid flow in a coordinate frame rotating with the earth is usually relabelled as the "Coriolis force". Characteristically it is proportional to 2Ω and the sine of the latitude.

This formulation is too crude for general purposes (we would have to consider the centrifugal potential of a flow at an arbitrary direction), but is the basis for understanding the slightly more complicated Coriolis terms appearing later on.

The other central, zero-order physical characteristics governing the large scale circulation of the ocean are that it is a fluid, mainly water, but containing about 3.5% by weight of salt, that the earth is rotating, and that it exchanges momentum, heat and moisture with the atmosphere.

The presence of salt has several consequences. The first consequence derives from its contribution to the density of seawater. The second consequence may be less obvious, but is more profound. A fluid containing dissolved salt is a conductor; conductors prevent electromagnetic radiation from penetrating themselves. The ocean is thus in practice virtually opaque to electro-magnetic radiation over most of the radio spectrum (except at ultra-low frequencies). We cannot see into it, we cannot send messages through it, and we cannot probe it with any part of the useful electromagnetic spectrum. This fact becomes central when we turn to discussions of the observation technologies available to oceanographers, with consequences for the importance of altimetry.

What does the ocean look like? It is characteristic of fluids that they exhibit structures and movement on all scales from the molecular on up to the size of the container in which they reside. Thus the ocean exhibits motions visible to the eye from millimeters (small ripples), on up to structures visible to astronauts: hundreds to perhaps a thousand kilometers across. Notice that these things just mentioned are ocean surface phenomena. With our eyes, we cannot penetrate more than a few meters into the ocean interior, because of the opacity. The motions of chief interest in these lectures range up to the size of the ocean basin and the globe itself, and which loosely speaking, comprise the oceanic general circulation. Again, speaking roughly, these comprise motions on horizontal scales of 30 km. and larger, changing in time over a few days and longer.

It is a matter of time, space, and my own expertise, that causes me to limit the discussion mainly, but not wholly, to these scales, thus leaving out such fascinating phenomena as internal waves, surface waves, shallow water wave motions, etc. Not coincidentally, the chief application of altimetry is on the large scales. But it remains essential to recognize that in a fluid, motions on scales of no apparent immediate interest can have important consequences for the fluid motion as a whole. Thus for example, internal waves, which typically occur on spatial scales of meters to tens of kilometers, are believed to play an important role in mixing the ocean (e.g. Munk, 1981), and are therefore an essential element of understanding of the ocean circulation on much larger scales. Certain large scale wave motions, comprising such phenomena as Rossby, Kelvin, and Poincaré waves etc. are also essential to the ocean circulation. Many of these motions make an appearance in tidal and tropical phenomena as well as the general circulation. I will necessarily leave their discussion to other lecturers, and to the references.

To provide some background feeling for the basic physical characteristics, let us consider what oceanographers use to describe what is going on out there. Since about 1860, the chief tool that scientists have had for understanding the ocean has been the ship. In the mid-to-late nineteenth century, a non-electronic age, very clever people had learned how to measure the temperature and salt content of the water at depth from a ship stopped in mid-ocean (the reader might want to think about ways to do that - using nothing electrical). Today, although made with sophisticated electronic equipment, such measurements remain a central piece of information about the ocean.

Consider figure 2, a "hydrographic section" obtained from a ship which steamed from near Newfoundland to South America along a nominal longitude of 53° W in 1985 (Knapp and Stommel, 1985; Fukumori, et al., 1991). The ship stopped at intervals to make measurements and produced these sections of temperature and salinity. Some simple features are worth noting. First, the ocean is stratified, roughly horizontally, in both temperature and salt ("salinity" is the salt content per unit mass although technically defined in somewhat different units). The ocean is generally a good deal warmer at the surface than at depth. Second, there is a strong, if slightly less clear-cut, tendency for it to be saltiest at the surface. The temperature change from around 25°C to less than 2.4°C near bottom is remarkably large, and occurs mainly in a region of most rapid change near 800m depth, called the "thermocline", or "main thermocline" to distinguish it from a seasonal effect very near the surface. Notice that the thermocline coincides with a "halocline" of rapid salinity change.

One of the central goals of physical oceanography is to explain the presence, structure, depth, etc. of these various "clines". Notice the region near the northern end of the section where the

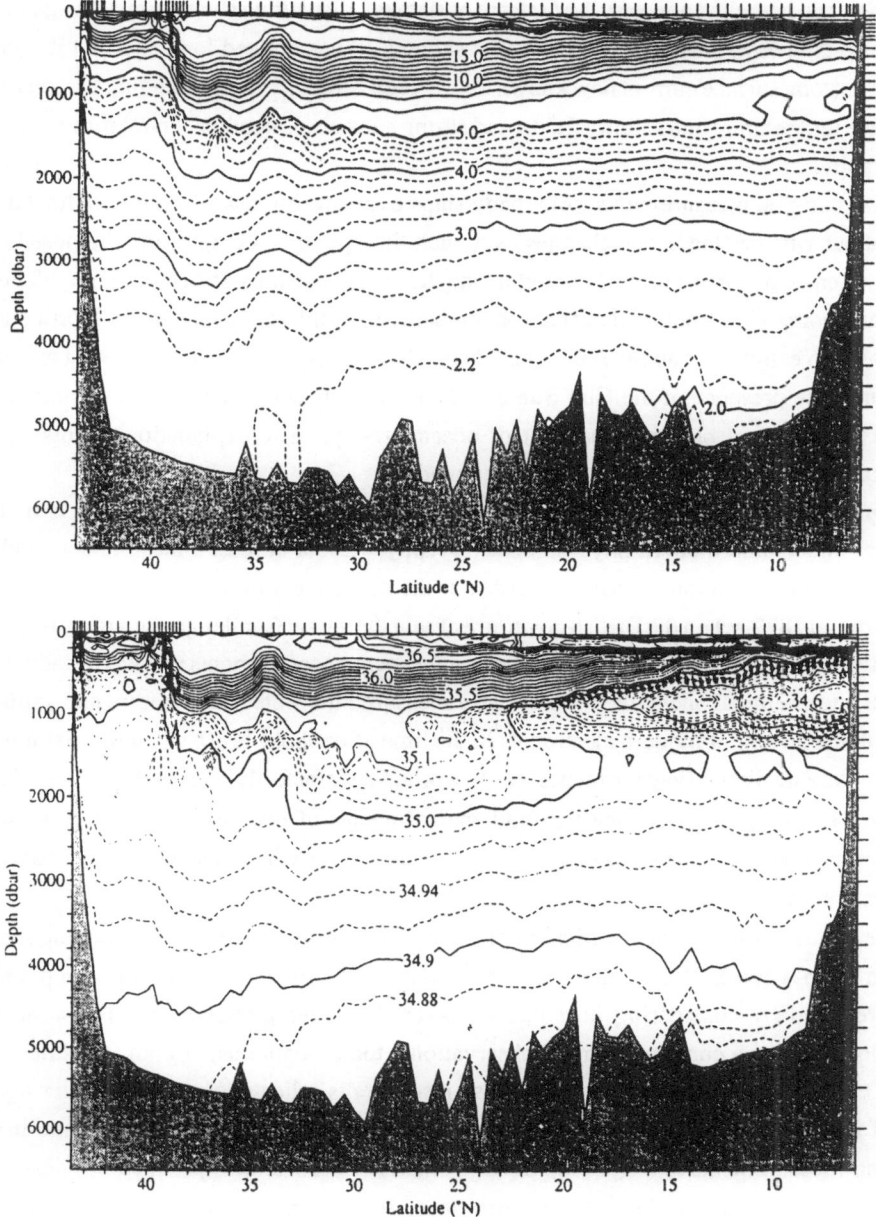

2a, b. Typical modern temperature and salinity sections showing the temperature (a) and salinity (b) stratifications of the ocean along 53°W in the North Atlantic. Strong horizontal gradients near the surface to the north are a reflection of the presence of the Gulf Stream - a powerful current required to balance the pressure forces exerted by the stratification differences on the two sides of the gradient.

thermocline and halocline appear to be far from horizontal, displaying a very rapid change in depth. This feature is always present near where it is shown; it was noticed long ago that it coincided with a very strong surface current now known as the "Gulf Stream". Understanding the relationship between the temperature and salinity fields, and strong currents, is another central focus of physical oceanography.

If we compile measurements from many ships over several years, and ignore for now, any worries about whether the ocean changes over such time periods, i.e. ignoring the possibility of aliasing, we can make maps of the horizontal temperature and salinity structure. Figure 3 displays an estimate (somewhat smoothed out) of the temperature at the seasurface in the North Atlantic, and 4 shows an estimate at about 700 meters depth. The strong gradients on the west are another indication of the presence of the Gulf Stream. That such currents occur on the western sides of all oceans is a central observational fact of oceanography. The explanation of this "western intensification" is something we need to understand.

Figures 2-4 are typical of what historically could be measured from ships. It is really not possible to understand modern physical oceanography without at least a basic understanding of the problems of observation. What is regarded as the modern era of oceanography is commonly dated to around 1872, when the British government sent the vessel Challenger on a round-the-world voyage, along a track depicted in figure 5. Among the many measurements that were made (largely biological and chemical ones), the science party measured the water temperature and salinity.

With the advent of modern electronics in the time after about 1960, when transistor devices became widely available (vacuum tubes never worked satisfactorily at sea) the devices became electronic. But that didn't change the fundamental means of observing the ocean: going to a particular place with a ship, and lowering devices to the depths where one wanted an observation. Why does one do it this way?

The central difficulty arises because of the opacity of the sea to electromagnetic radiation (light and radio waves). Why is this such a formidable difficulty? The question can perhaps be best appreciated by considering how one observes the sister global fluid - the atmosphere. Meteorologists have a complex array of observational tools - balloons, aircraft, satellites, rocket profilers etc. Consider that every one of these instruments relies one way or another upon the ability of light and radio waves to propagate through the atmosphere: the balloons and profilers return their measurements to the ground through small radios; satellites produce wonderful cloud pictures and measurements of the temperature structure of the atmosphere by the signals carried

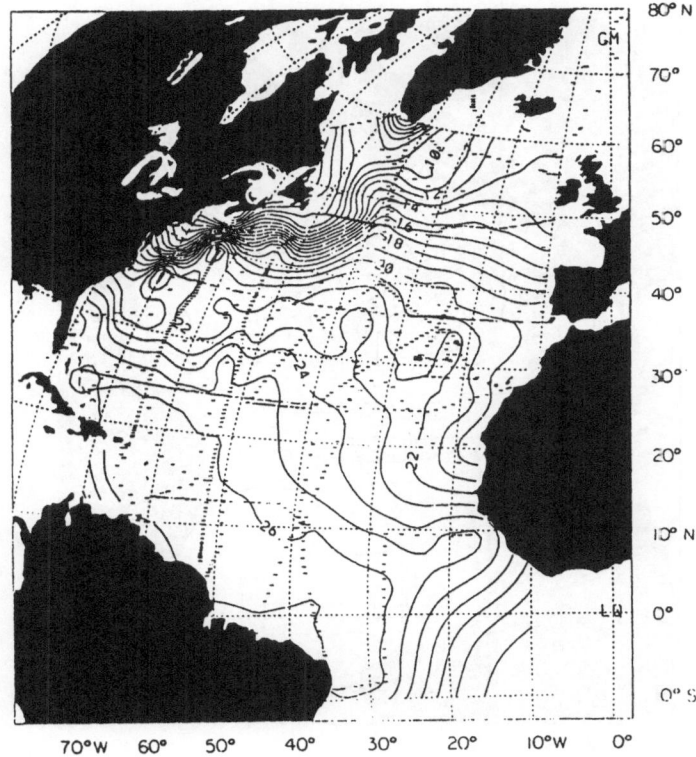

3. Temperature maps at the sea surface constructed by objective mapping ("collocation" to the geodetic community) using modern data obtained over several years in the early 1980's (Fukumori, et al., 1991). Strong spatial gradient near US coast is the Gulf Stream, but which is mapped much more broadly than it exists in practice owing to the smearing effects of the objective analysis.

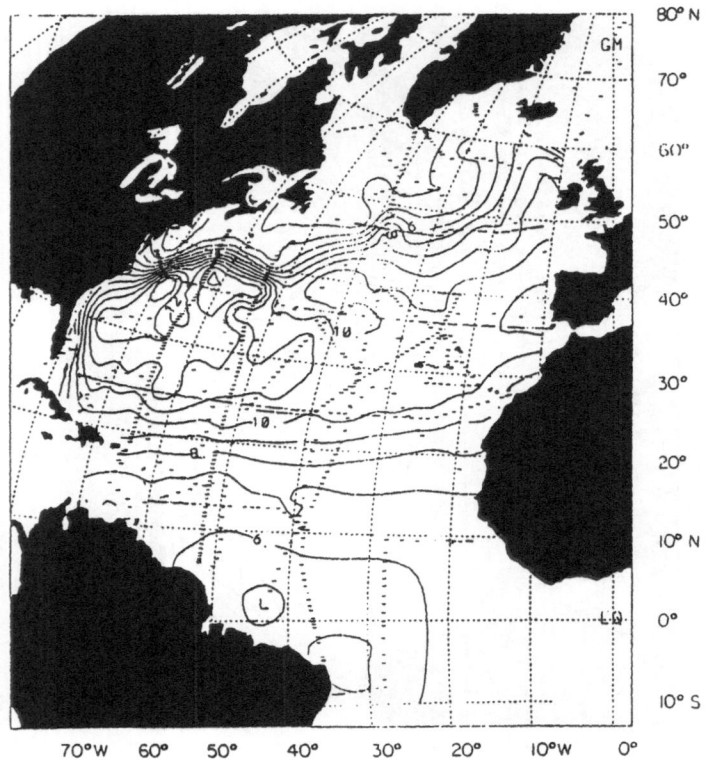

4. Same as fig. 3, except at about 700 m depth .

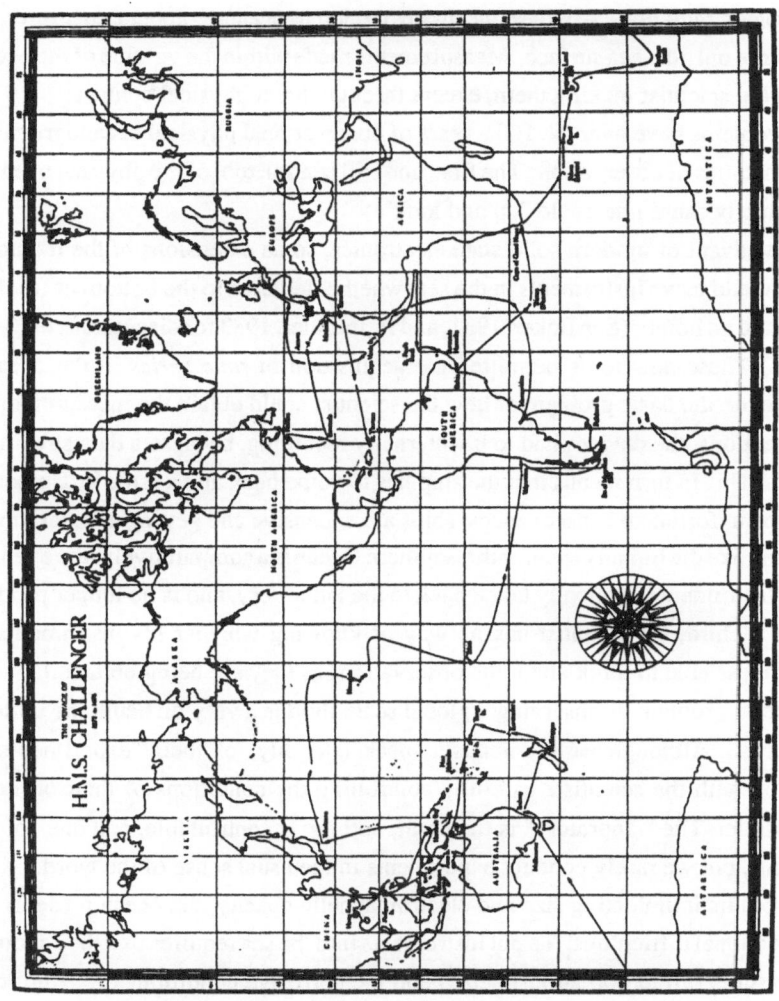

5. Track of the British vessel Challenger in the early 1870's (from Deacon, 1979). Several years were required to go around the world. Although modern ships are somewhat faster, their ability to cover the enormous area of the world ocean is not very different, relative to the rate at which the circulation changes, than it was in the nineteenth century.

by electromagnetic radiation. All these means are denied to oceanographers: measurements from space can observe only the sea surface. Measurements made within the volume of the ocean cannot be returned to the scientist making them, except through direct physical contact.

These problems have over the 100+ years of observational physical oceanography, dictated very specific methods of observation. The first, and still central, tool of the physical oceanographer has been the ship because one could "go and look".

With the advent of modern solid state electronics, some extensions of the method became available: one could leave instruments in the sea, whether tethered to the bottom or freely drifting, and the ship could go home. (See Baker, 1981, and Heinmiller, 1983 for discussions of the technical developments.) These new tools permitted the acquisition of *time series* in the ocean. There remained however, the basic problem of how the scientist could obtain the measurements? With some rare exceptions, the devices had to be internally recording, because a data stream could not be transmitted. This in turn means that the ship has to come back and get the instrument and this in turn has some unfortunate consequences: ships are expensive and getting a research vessel into remote locations like the Indian Ocean or the Southern Ocean is a comparatively rare event. Second, the recovered instrument is typically taken back to the laboratory, and is no longer producing any measurements. Third, the scientist has no way of knowing whether his instruments are even working, much less able to think about the observations as they are being obtained.

The general problem of observing a global scale fluid is a very difficult one under the best of circumstances. Although earth scientists speak (sloppily) of their "experiments", normal experimentation with the scientists carefully controlling the conditions of their observations is usually impossible. The "laboratory" is the earth; nothing is controllable, and one "observes" or runs expeditions, but we rarely conduct experiments in the usual sense of the word.

Apart from its intimidating size and electromagnetic opacity, the ocean presents some forbidding observational difficulties. To put instruments into the sea requires that they should survive enormous pressures (up to 600 atmospheres) and the corrosive conditions of a fluid containing 3.5% by weight of salt. In contrast, measurements in space or in the atmosphere encounter neither such extreme corrosive media, nor pressures exceeding one atmosphere.

But about 20 years ago, partially as a result of measurements with a multitude of new techniques, it became clear that the ocean was intensely turbulent, and that it made no sense to regard it as a steady, unchanging system. Measurements made from a ship could not be made as fast as the ocean was changing. Consider for example, that a modern oceanographic vessel moves no faster than a rather sluggish bicyclist (10-14 knots). By the time a ship setting sail across the Pacific Ocean to make measurements has reached the other side at this slow pace, plus whatever time is spent in stopping to observe, two or more months have passed, and the system will have changed in many ways during that period of time. Furthermore, only one line across the ocean will have been measured, and the oceanographers will have little or no idea what was happening even 50 km from where they made their observations. In addition, the nature of the oceanic variability and turbulence is on a spatial scale which is so small (about 50 km for major changes), that even the clever new generation of measurement devices could never be produced in numbers adequate to observe the changing fluid. The problem is a combination of financial and human costs. Measurements at sea, whether from shipboard or from self-contained instruments, are very expensive. (The daily charges for the R/V Knorr are now US$18,500 per day, not including the costs of the scientific party.) It seems unlikely that one could find the money or human resources to greatly increase the number of research vessels at sea, or to produce and maintain the many thousands of in situ devices which are required.

The upshot of all this is that measurements of the ocean are comparatively rare, and expensive, and is one of the reasons that satellite altimetry has attracted so much attention. But before turning to that subject, we need to explore somewhat further what we do know of the ocean circulation.

The Steady Circulation Idea

Because observations are so scarce and expensive, and because they have historically been so grossly inadequate to define an ever-changing turbulent fluid, oceanographers resorted to making a plausible and surprisingly useful assumption: that on "large-enough" scales, the ocean could be treated as though the flow were steady. If this assumption is correct, one can then use data obtained over many years, and decades, as though obtained simultaneously. The huge advantage of being able to lump together measurements over long time spans will be obvious. To a large extent, this idea works quite well.

Consider figures 6 showing various property maps of the ocean. One sees giant property "tongues", which are stable features over many decades (at least), and which strongly suggest for example, that the fluid somehow emanates from up the property gradient. Maps such as these are important elements in what we know about the ocean. But great care is required in using them: despite their suggestiveness, these maps simply tell us "what is there", i.e. the standing crop. They tell us little, directly, about how the fluid got there, whether it is still moving, and if so how fast.

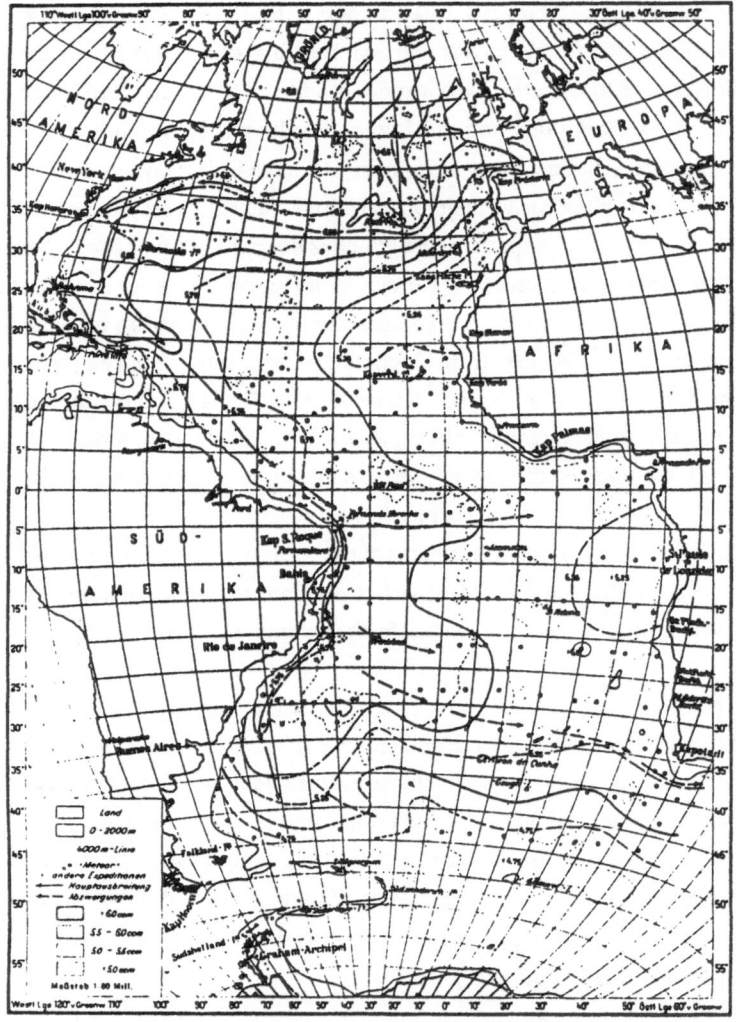

6a. Wüst's (1935) chart of the oxygen distribution at mid-depths in the North Atlantic. Notice the strong "tongues". The arrows Wüst drew show his inferences about the way the water must "spread" from sources to sinks. The timescale over which this movement must take place (days?, millenia?) is not specified.

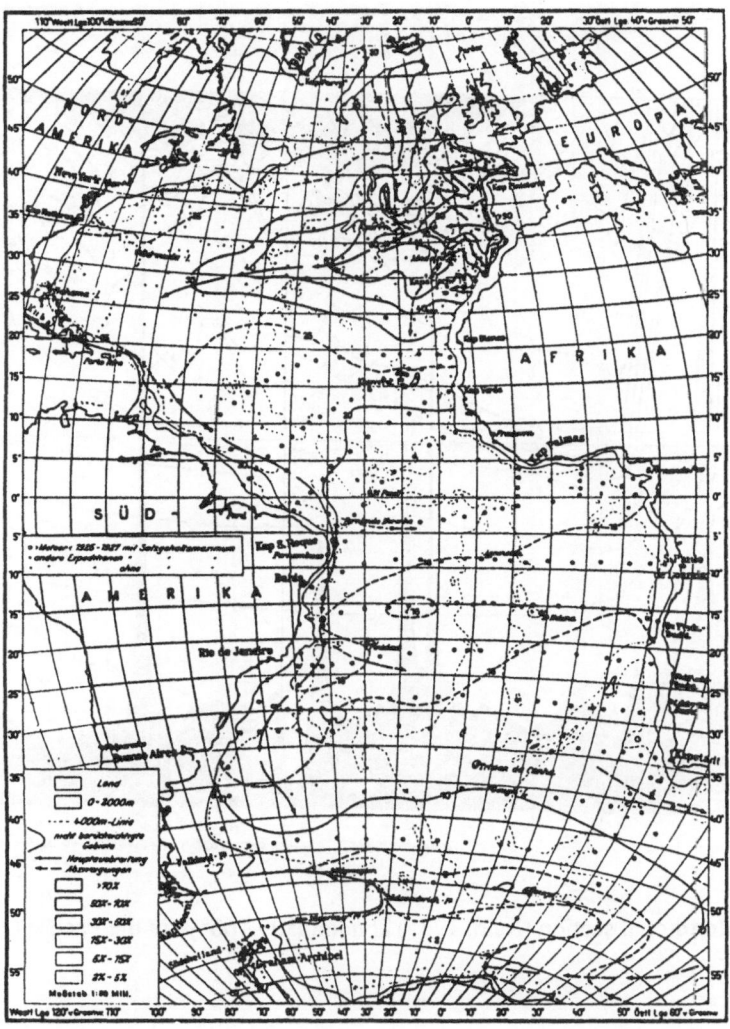

6b. Similar to 6b, except for salinity.

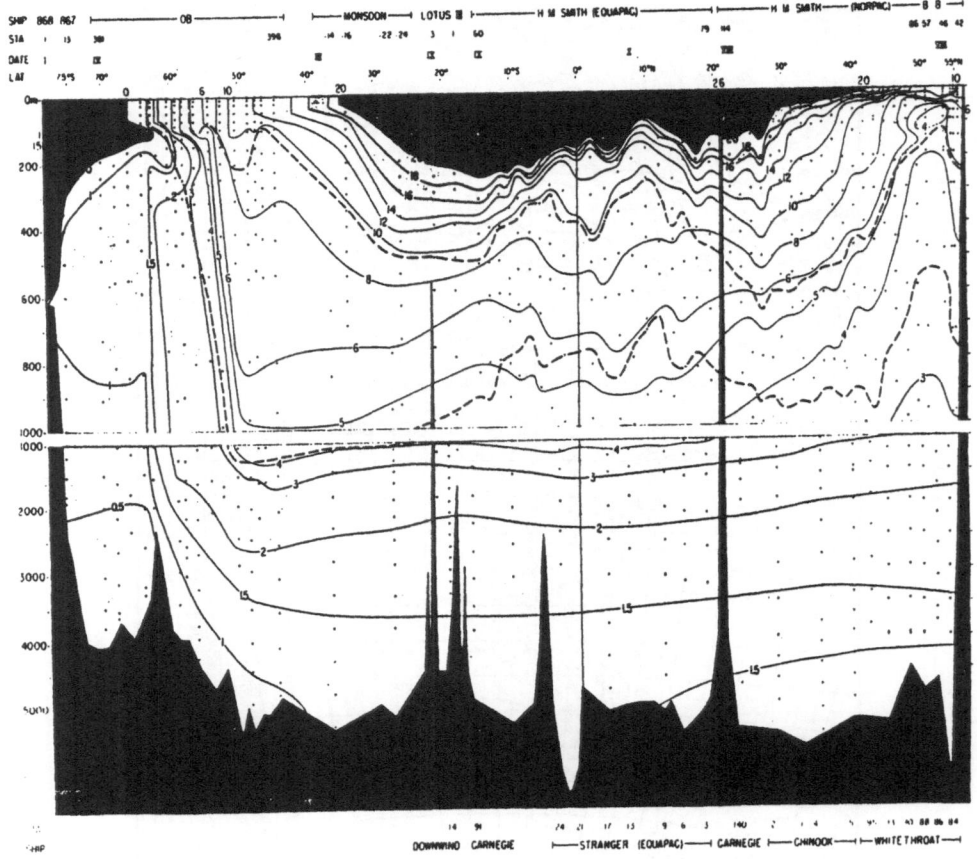

6c. Meridional section (Reid, 1965) at nominally 160°W in the Pacific Ocean.

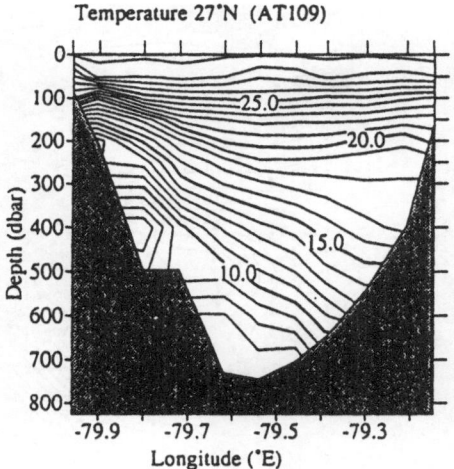

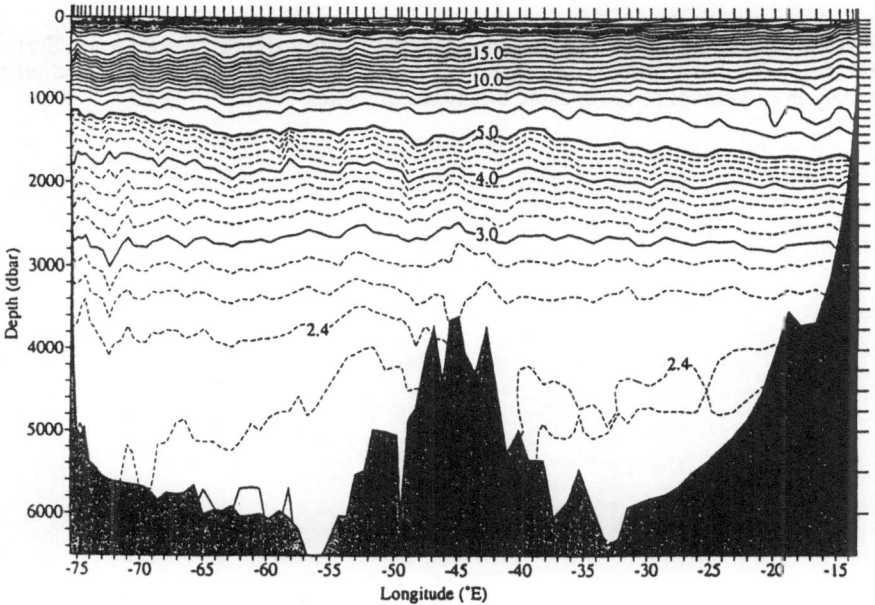

6d. Zonal temperature section across the Gulf Stream and the subtropical North Atlantic at 25°N (Roemmich and Wunsch, 1985).

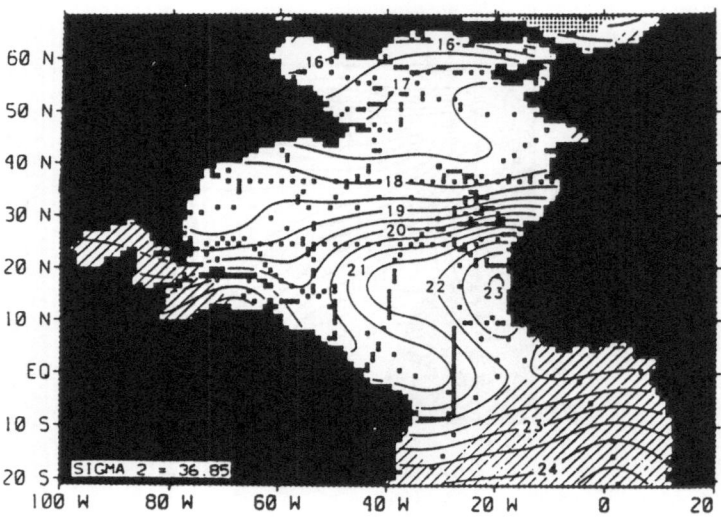

6e. The nitrate distribution at about 1600 m depth in the North Atlantic (Kawase and Sarmiento, 1986). Nitrates and other nutrients are useful tracers of the ocean circulation, but their use is complicated by involvement in the oceanic biochemical cycles.

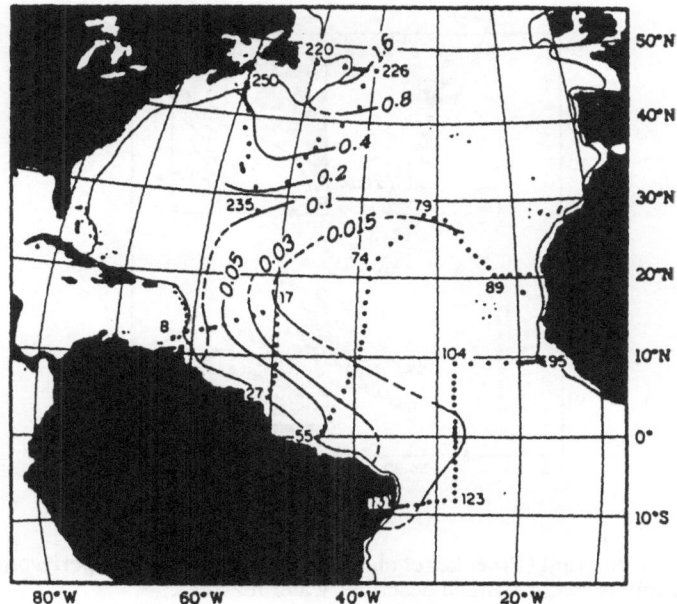

6f. Fluorocarbon concentration (Weiss, et al., 1985) at mid-depths in the North Atlantic. Fluorocarbons do not occur naturally in the ocean, and their presence at great depths in the ocean a decade or two after their introduction into the atmosphere is an indication of the surprising rapidity with which the ocean transfers information from the surface to the abyss.

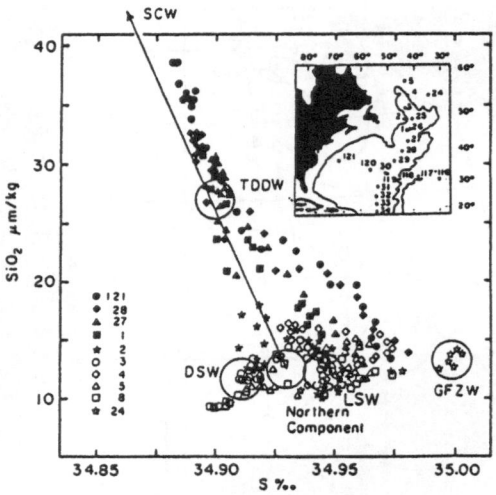

6g. A silica-salinity diagram (Broecker et al., 1976) typical of the "property-property" diagrams used by oceanographers to discuss and describe "water masses".

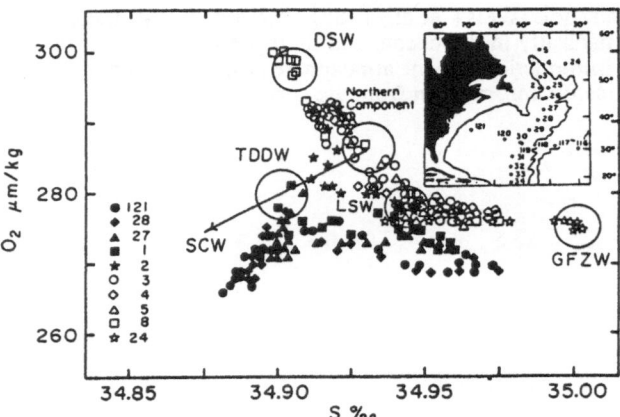

6h. Same as 6g, except for oxygen-salinity.

Consider how one can use measurements from ships to determine how the fluid ocean moves. To go any further, we need to write down the equations of motion. These equations are nothing but Newton's laws of motion for a fluid and are usually known as the Navier-Stokes equations. Because the earth is nearly spherical[1], it proves most convenient to write the equations in spherical coordinates using a coordinate system tied to ordinary latitude and longitude. The chief effect of this latter choice is that we must account for the earth's rotation, because the equations refer to a non-inertial frame. Deriving these equations would take too much time for a short lecture series. I will merely write them down and the reader for whom they are unfamiliar should probably accept them as being sensible postulates whose consequences we will explore.

A useful set of equations, on a spherical earth rotating at angular rate Ω are *approximately* of form (e.g. Phillips, 1966; Veronis, 1981; Pedlosky, 1987; Gill, 1982)

$$\frac{du}{dt} - \frac{uv\tan\phi}{a} + \frac{uw}{a} - 2\Omega\sin\phi v = -\frac{1}{a\cos\phi\,\rho\partial\lambda}\frac{\partial p}{} + F_u \tag{5}$$

$$\frac{dv}{dt} + \frac{u^2\tan\phi}{a} + \frac{vw}{a} + 2\Omega\sin\phi u = -\frac{1}{a\,\rho\partial\phi}\frac{\partial p}{} + F_v \tag{6}$$

$$\frac{dw}{dt} - \frac{u^2+v^2}{a} = -\frac{\partial p}{\rho\partial z} - g \tag{7}$$

$$\frac{1}{a\cos\phi}\left(\frac{\partial u}{\partial\lambda} + \frac{\partial(v\cos\phi)}{\partial\phi}\right) + \frac{\partial w}{\partial z} = 0 \tag{8}$$

$$\frac{d\rho}{dt} = 0 \tag{9a}$$

where

$$\frac{d}{dt} \equiv \frac{\partial}{\partial t} + \frac{u}{a\cos\phi}\frac{\partial}{\partial\lambda} + \frac{v}{a}\frac{\partial}{\partial\phi} + w\frac{\partial}{\partial z}$$

The latitude is ϕ, longitude is λ, the radius is a, and the radial position is $r = a + z$ (see Fig.7). u, v, w are the velocities in the λ, ϕ, r directions respectively and p is the pressure. These equations have been simplified by neglecting molecular viscosity and ignoring non-sphericity. Terms F_u, F_v represent complex, turbulent mechanisms by which the large scales we will discuss are broken

---

[1] In lecturing to geodesists, I feel compelled to state that I recognize that the earth is not really a sphere. But to my knowledge, there are no ocean dynamics which are affected by the discrepancy. The kinematic effects of the non-sphericity on the use of altimeters are quite important, but that is a very separate issue.

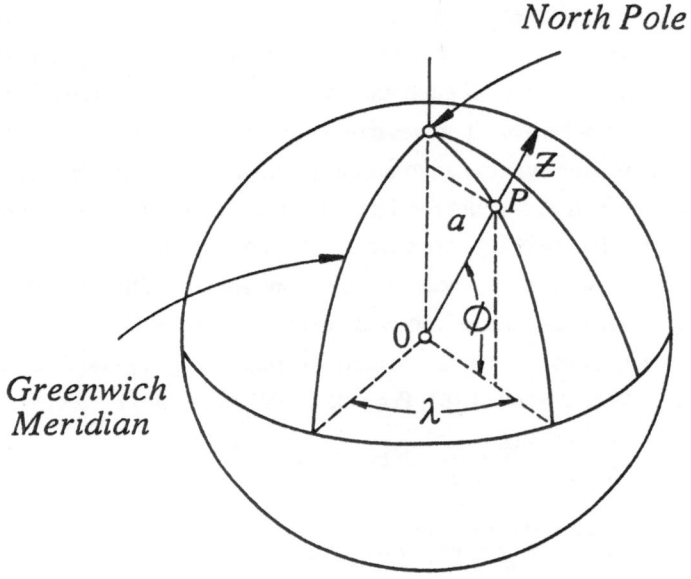

7. Spherical coordinate system for equations of motion of large scale ocean circulation.

down into smaller scales in a process which will be regarded as essentially dissipative (but in practice is not that simple).

Eq. (8) is a statement that the fluid is effectively incompressible, and (9a) asserts that density is also conserved. These two last equations can be combined, if one wishes, into

$$\frac{1}{a\cos\phi}\left(\frac{\partial(\rho u)}{\partial\lambda}+\frac{\partial(\rho v\cos\phi)}{\partial\phi}\right)+\frac{\partial(\rho w)}{\partial z}=0 \tag{9b}$$

where the term $\partial\rho/\partial t$ has been omitted under the assumption that there is no time variation.

To the set (5-9), it is often useful to append a statement of conservation of a general tracer,

$$\frac{d(\rho C)}{dt}=q \tag{10}$$

where C represents any one of a variety of fields, such as oxygen, one of the nutrients, tritium, etc. q represents any internal sources or sinks for the tracer, and so "conservation" is something of a misnomer, but rarely causes confusion. Eq (9b) is a special case with $C=1$, $q=0$.

An equation of state of seawater is also needed. For present purposes, an adequate statement is that, given the temperature, θ, and salt content per unit mass, S, of a parcel of fluid, its density can be computed as a function of the pressure from an empirically measured equation of state,

$$\rho(\rho,\phi,\lambda,t)=\rho(\theta(\rho,\phi,\lambda,t),S(\rho,\phi,\lambda,t),p) \tag{11}$$

(a modern version of the form of this equation may be found in Millero et al. (1980). Various computer implementations exist.) Concentration evolution equations like (10) may be written separately for θ, S.

The Geostrophic Approximation

Despite the fact that these equations have already been simplified, they describe an enormous variety of possible fluid flows; they are a set of non-linear partial differential equations in four dimensions. Consequently their general solution is impossible. Further approximations are necessary to proceed.

Observation suggests that for oceanic fluid flows which evolve on time scales longer than about one day, over spatial scales exceeding about 10 kms., and away from the immediate vicinity of the seasurface and the equator, the dominant terms in the horizontal momentum equations (5,6) are the second term on the left, and the first term on the right, i.e, "geostrophic balance",

$$-2\Omega\rho\sin\phi v \approx -\frac{\partial p}{a\cos\phi\partial\lambda} \tag{12}$$

$$2\Omega\rho\sin\phi u \approx -\frac{\partial p}{a\partial\phi} \tag{13}$$

The third momentum equation (7) is also accurately given by hydrostatic balance,

$$0 \approx -g\rho - \frac{\partial p}{\partial z} \tag{14}$$

Notice that the system is now *linear*, and has many fewer derivatives than before. The latter property means that we have reduced the order of the differentials involved and have consequently thrown out much physics. In particular, the time derivatives are missing, and the balance in (12-14) is essentially a static one - meaning that these equations, useful as they will prove to be, cannot describe how such a field is established, nor how it might decay or be sustained against dissipation.

The local Cartesian approximation to (8-10) is adequate for many purposes (it can be derived systematically from full the spherical coordinate equations; see Phillips, 1963; Veronis, 1981), and is Rossby's "beta-plane". Local geostrophic, hydrostatic balance for steady flow in Cartesian coordinates may be written

$$-\rho f v = -\frac{\partial p}{\partial x} \tag{15}$$

$$\rho f u = -\frac{\partial p}{\partial y} \tag{16}$$

$$0 = -\frac{\partial p}{\partial z} - g\rho \tag{17}$$

$f = 2\Omega\sin\phi$ is the "Coriolis parameter" and is usually expanded locally as $f = f_0 + \beta y$ to account for its dependence on latitude. The continuity equation is

$$\frac{\partial u}{\partial x} + \frac{\partial v}{\partial y} + \frac{\partial w}{\partial z} = 0 \tag{18}$$

Conservation of density is

$$u\frac{\partial \rho}{\partial x} + v\frac{\partial \rho}{\partial y} + w\frac{\partial \rho}{\partial z} = 0 \tag{19a}$$

or if combined with (18) is,

$$\frac{\partial(\rho u)}{\partial x}+\frac{\partial(\rho v)}{\partial y}+\frac{\partial(\rho w)}{\partial z}=0 \tag{19b}$$

The equation of state is simply expressed in terms of the local coordinate system. The concentration conservation equation becomes

$$u\frac{\partial(\rho C)}{\partial x}+v\frac{\partial(\rho C)}{\partial y}+w\frac{\partial(\rho C)}{\partial z}=q(x,y,z,t) \tag{19c}$$

or if combined with (18), is

$$\frac{\partial(\rho Cu)}{\partial x}+\frac{\partial(\rho Cv)}{\partial y}+\frac{\partial(\rho Cw)}{\partial z}=q(x,y,z,t) \tag{15b}$$

Consider a very special ocean that has no stratification, i.e. $\rho = constant$, and where f is treated as constant; then cross-differentiation of (15) and (16) to eliminate the pressure, and use of (18) produces the well-known result

$$\frac{\partial u}{\partial x}+\frac{\partial v}{\partial y}=0$$

from which it follows,

$$\frac{\partial w}{\partial z}=0 \tag{20}$$

$$\frac{\partial}{\partial z}(u,v)=0 \tag{21}$$

i.e., *purely geostrophic motion of a homogeneous fluid is horizontally non-divergent* and the flow field is independent of z (this result is sometimes known as the Taylor-Proudman theorem). A tendency for the oceanic flows to behave like this is evident, locally, in the abyss, where the stratification tends to quite weak and where small topographic features intrude into the flow (manifested by a tendency for the disturbed flow fields to be depth independent).

On the other hand if the fluid is arbitrarily stratified, and if the variation of f is accounted for, we obtain instead

$$\beta v = f\frac{\partial w}{\partial z} \tag{22}$$

the "linear vorticity equation", which is central to understanding much of what we believe about the ocean in regions where equations like (15-19) are thought to be approximately valid. Among other things, (22) shows that if there is to be any meridional motion at all, there *must* be a vertical velocity.

The "thermal wind equations" are obtained from (15-17) by eliminating the pressure in terms of the density field, to give

$$-f\frac{\partial(\rho v)}{\partial z} = g\frac{\partial \rho}{\partial x} \tag{23}$$

$$f\frac{\partial(\rho u)}{\partial z} = g\frac{\partial \rho}{\partial y} \tag{24}$$

and which can be integrated in the vertical to produce

$$\rho u(x,y,z,t) = \frac{g}{f}\int_{z_0}^{z}\frac{\partial \rho}{\partial y}dz + c(x,y,t,z_0) \equiv \rho(u_R + c) \tag{25}$$

$$\rho v(x,y,z,t) = -\frac{g}{f}\int_{z_0}^{z}\frac{\partial \rho}{\partial x}dz + b(x,y,t,z_0) \equiv \rho(v_R + b) \tag{26}$$

(The time, t, has been placed into the argument as a reminder that this balance can be a very good one even in a time evolving field. Normally we will omit it.) The depth, z_0, is usually called the "reference depth", and is arbitrary: the two constants of integration, $b(x,y,t,z_0)$, $c(x,y,t,z_0)$ will change accordingly if z_0 is shifted, but normally we will suppress the explicit dependence upon it. u_R, v_R are called the "relative velocities", and the integration constants b, c are the "reference level velocities".

The two thermal wind equations have been central to physical oceanography because for the reasons already described, the only routinely measurable quantities obtainable from ships were temperature and salinity. Apart from the integration constants, the geostrophic flow is then calculable from shipboard measurements alone. The procedure is known as the "dynamic method". Although the integration constants, i.e. the reference level velocities, are mathematically trivial, the inability to determine them has plagued oceanography for a hundred years, and has been one of the major obstacles to understanding the ocean circulation.

Suppose for the moment that we have somehow determined the reference level velocity, and therefore know the absolute geostrophic velocity u, v. In an altimetric/geodetic context, we are very interested in establishing what such a flow does to the seasurface, getting beyond the rough order of magnitude value given in eq (4). The pressure in the ocean is nearly hydrostatic , and thus we can integrate (17) vertically from the seasurface, $\eta(x, y)$, which produces a pressure field

$$p(x,y,z,t) = \int_z^{\eta(x,y,z,t)} g\rho(x,y,z,t)dz \tag{27}$$

where a time dependence, which is always possible, has been written in explicitly. Atmospheric pressure has been set to zero. We know from our earlier scale analysis that the numerical values of η are likely to be only tens of centimeters. So setting the surface density to ρ_s, eq (27) is very nearly

$$p(x,y,z,t) = \rho_s g \eta(x,y,z,t) + \int_z^0 g\rho(x,y,z,t)dz \tag{28}$$

Thus the pressure field right at the seasurface (or at z=0 which is just below it) is $p_s = g\rho_s\eta(x, y, 0, t)$. But equations (15, 16) tell us how to compute the surface velocity given the surface pressure, so

$$-fv_s = -g\frac{\partial \eta}{\partial x} \tag{29}$$

$$fu_s = -g\frac{\partial \eta}{\partial y} \tag{30}$$

On the other hand, the thermal wind equations tell us that

$$\rho_s u(x,y,z = \eta \cong 0,t) = \frac{g}{f}\int_{z_0}^0 \frac{\partial \rho}{\partial y}dz + c(x,y,t,z_0) \tag{31}$$

$$\rho_s v(x,y,0,t) = -\frac{g}{f}\int_{z_0}^z \frac{\partial \rho}{\partial x}dz + b(x,y,t,z_0) \tag{32}$$

Thus if we knew the velocity at the reference level, the thermal wind equations permit us to compute the surface velocity and hence the surface elevation, relative to the geoid. Alternatively, if we knew the seasurface elevation (from here on I will omit the words "relative to the geoid"), we could compute the surface velocity, and use (31, 32) to find the integration constants b, c, and thus the absolute flow at all depths. This possibility, difficult as it is to contemplate, is what leads physical oceanographers to altimetry, and geodesists to worry about the ocean circulation.

The equations (29-32) show how measurements of the surface elevation connect directly, through the equations of motion, to oceanic flows at great depths. This property of altimetry is, to my knowledge, unique among quantities measurable from space, and thus it is quite wrong to say about altimetry that it "only measures the seasurface". It actually measures a quantity which reflects most of the water column. Equations (23-26) represent, as we have discussed, a reduction in the dynamics describing the ocean. If a slightly more complete set of equations is employed (e.g. Pedlosky, 1987; Gill, 1982), it is possible to estimate the depth to which the seasurface reflects the flow. Let $\rho_0(x, y, z)$ be the time average density field; then the quantity $N^2 = -g/\rho_0 \partial \rho_0/\partial z$ is a measure of the strength of the stratification (N has the dimensions of a frequency). Then geostrophically balanced flows with horizontal length scales, L, under wide conditions, have vertical length scales H such that $NH/fL \cong 1$, where $f = 2\Omega \sin\phi$. That is, if they manifest themselves at the surface as spatial variations over a distance L, then $H \cong fL/N$. If $f \sim 7.3 \times 10^{-5}/s$, $N \sim .01/s$, $L \sim 100km$, then, $H \cong 1km$. These motions are usually called "baroclinic" (because they depend intimately upon the stratification), and the quantity $R_D \equiv NH/f$ is often called the "baroclinic Rossby radius of deformation". It is possible simultaneously to have motions which are independent of the stratification, even when it is present. Such motions are sometimes called "barotropic". To a first approximation the barotropic oceanic response is depth *independent* (recall equations 21). One can simultaneously have barotropic and baroclinic motions of the same horizontal length scale L; but in both cases, the seasurface reflects motions which penetrate far into the water column.

We will return to all this at the end.

Scaling

Readers unfamiliar with fluid dynamics may be concerned that a huge leap was taken in going from the nearly complete set of equations (5-9) to the much reduced geostrophic set. A systematic method, commonly called "scale analysis", is widely employed to produce approximate sets of equations intended to be useful for particular fluid flow situations. This methodology is described in almost all textbooks on fluid dynamics; Pedlosky (1987) produces a large variety of examples in the context of "geophysical fluid dynamics".

Because the methodology is so important, and because it appears to be unfamiliar to most geodesists, I will sketch its application to equations (5-9), referring to Pedlosky (1987) for the carrying through of the complete analysis. At the outset, it must be emphasized that scale analyses are not an "automatic" procedure, rather they rely upon the scientist's insight into the physics he or she is attempting to describe. It is possible to go astray with scale analysis - but the reader may be assured that one is always capable of determining if that has happened.

Let us suppose that based upon observation or insight, that the motions we think are important to understand vary horizontally on a length scale L (assuming merely for simplicity of the example, that latitudinal and longitudinal scales are similar). That is on the sphere, we will suppose that the fields change significantly on a spatial scale $L \cong a\Delta\phi \cong a\Delta\lambda$. Suppose further that the vertical scale of variation is D (noting that D normally cannot exceed the ocean depth, where L could span the entire earth's circumference.) We suppose that the time scale over which the motions evolve is T, that the horizontal velocities u,v are about U in size, and that w is about W in magnitude. Again, a decision to treat w differently from u, v could be based upon observing certain motions which we seek to understand, and noticing that perhaps w, u are quite different in size. Let the pressure field have numerical magnitude P. With these labels, we can write

$$(u,v) = U(u',v'), \ w = Ww', \ p = Pp', \ \rho = \rho_0\rho'$$

and

$$\frac{\partial}{\partial t} = \frac{1}{T}\frac{\partial}{\partial t'}, \quad \frac{u}{a\cos\phi}\frac{\partial}{\partial\lambda} = \frac{U}{L}u'\frac{\partial}{\partial\lambda'}, \text{ etc.}$$

where all the primed variables are assumed to be dimensionless, and such that there largest magnitude in the particular fluid flow is not more than O(1) (it could be much less, even zero), and substituting into eq (1) produces

$$\frac{U}{T}\frac{\partial u'}{\partial t'} + \frac{U^2}{L}\frac{\partial u'}{\cos\phi\partial\lambda'} + \dots + \frac{WU}{D}\frac{\partial u'}{\partial z'} - \dots - 2\Omega\sin\phi Uv' = -\frac{P}{L\rho_0}\frac{\partial p'}{\rho'\partial\lambda'}$$

Now if we are centrally interested in motions believed to be geostrophic, that means we believe that the Coriolis term and the pressure term should be the dominant forces, and should nearly balance. We are free to render the coefficient of the Coriolis term of order unity (assuming that $\sin\phi$ is not too small) by dividing this last equation through by $2\Omega U$, thus producing

$$\frac{1}{2\Omega T}\frac{\partial u'}{\partial t'} + \frac{U}{2\Omega L}\frac{u'}{\cos}\phi\frac{\partial u'}{\partial\lambda} + \dots + \frac{W}{U^2 2\Omega L D}w'\frac{\partial u'}{\partial z'} - \sin\phi v' = -\frac{P}{L\rho_0 2\Omega U}\frac{\partial p'}{\rho'\partial\lambda'}$$

If we choose $P = 1/L\rho_0 2\Omega U$, then the Coriolis and pressure term are of roughly the same size. What can we say about the other terms? If the scaling is correct, the magnitudes of each of the other terms is given by the combinations of dimensional parameters multiplying each one. Notice that each of the combinations appearing is itself dimensionless. Consider the coefficient of the time derivative term: $1/(T2\Omega)$. If it is noticed that the time rate of change of the motions of interest is much longer than $1/2\Omega$, then $1/(T2\Omega \ll 1)$ and we might consider neglecting this term. Consider the coefficient of some of the non-linear terms, $U/2\Omega L$ which is clearly non-dimensional and appears so often in various guises that it has a name, the "Rossby number". At mid-latitudes, we notice that much of the most energetic motions occur on space scales of $L \cong 100km$ with velocities observed to be about 10 cm/s. Plugging these numbers in produces $U/2\Omega L \cong .01 \ll 1$ and we might consider dropping these terms relative to the geostrophic ones.

By considering the other equations, we can make inferences about the relative magnitudes of W, U (concluding $W \ll U$) and produce a set of equations much simpler than the full set, including for example, the geostrophic set in cartesian coordinates we use. People encountering such scaling arguments for the first time worry that something can go badly wrong if one makes a poor assumption. But there are some very powerful checks on the procedure. The simplified equations may well fail to describe the phenomenon one sees. Or, substitution of the solution to the simplified equation set back into the original set, may show that terms which were believed to be negligible are actually large, even dominant. In practice, scale analysis is an essential, and robust technique central to progress in almost all fluid dynamical problems.

Let us return now to the geostrophic equations, under the assumption that an appropriate scale analysis justifying them has been done. The transport of properties by a geostrophic flow is readily computed. Let the concentration of a property be given by $C(x, y, z)$ per unit mass. Then if the geostrophic flow is known, the flux of C between $x = x_1$ and $x = x_2$ lying between two depths $z_1(x)$ and $z_2(x)$ in the meridional direction is

$$V_c(y) \equiv \int_{x_1}^{x_2} \int_{z_1(x)}^{z_2(x)} \rho v C \, dz \, dx = \int_{x_1}^{x_2} \int_{z_1(x)}^{z_2(x)} \rho(v_R(x, y, z) + b(x, y))C(x, y, z) \, dz \, dx \quad (33)$$

which is made up of a part owing to the thermal wind, and a part owing to the reference level velocity. There is a corresponding expression for the flux in the zonal direction. If the property is mass itself, $C = 1$. For a given flow field, one can then compute the fluxes of heat, salt, carbon, etc. and their divergences to and from the atmosphere - with consequent great importance for understanding climate.

Water Masses

Oceanic water is "labelled" in a complex way by a number of measurable scalar properties, e.g. its temperature, salinity, oxygen and silicate concentration, its depth, latitude and longitude, etc. If there are N such properties, the water at any place in the ocean can be thought of as defining a point in an N-dimensional cartesian vector space. Because paper limits one to the representations of two-dimensions, much use has been made of depictions of oceanic water along selected two-dimensional subspaces or two-dimensional projections of three-dimensional subspaces of the N-space. Temperature and salinity are the water properties most easily and commonly measured, and so-called temperature-salinity diagrams have been much exploited. Fig 8 is Worthington's (1981) estimate, drawn as a histogram, for the temperature and salinity properties of the world ocean. Figure 2a,b are of course longitude-depth property diagrams, and one often encounters other choices of property, e.g. 6g is a silicate-salinity diagram, and 6h is one for oxygen-salinity.

It was noticed long ago that there is a tendency for water to lie in restricted regions of particular property diagrams. For example, in fig. 8 the single large peak, centered near 1.15°C and 34.685 psu, represents a large fraction of the total water in the entire ocean. Much of the remaining water lies on or near restricted curves in the temperature/salinity domain. This behavior of oceanic water led early in the history of the subject to the idea of "water masses". The concept has become somewhat blurred through varying usage. For our purposes, a water mass can be defined as any n-dimensional vector of measurable scalar property, e.g. $(\theta, S, Si, O, \ldots)$ where $N \geq 2$, Si, O etc. are the oxygen, silicate, etc. concentrations and where, by convention the geographical properties of depth, latitude, etc. are omitted. *Important* water masses are typically those for which a large quantity of water exists within a small distance of some central value $(\theta_0, S_0, \ldots)$ in the n-space; or, for which the n-tuple lies at the extreme end point of the existing curves, e.g. at the tips of the "y-shape" seen in fig. 8. The reasons for the particular water masses and abundances actually seen in the ocean remain a comparatively neglected part of theoretical understanding (e.g. why is there such a large peak, at the particular value observed in fig. 8?). (Recent work has emphasized the use of principal component analysis and related tools for describing oceanic properties (e.g. Mackas et al. 1987, Fukumori and Wunsch, 1991; Hamann and Swift, 1991)).

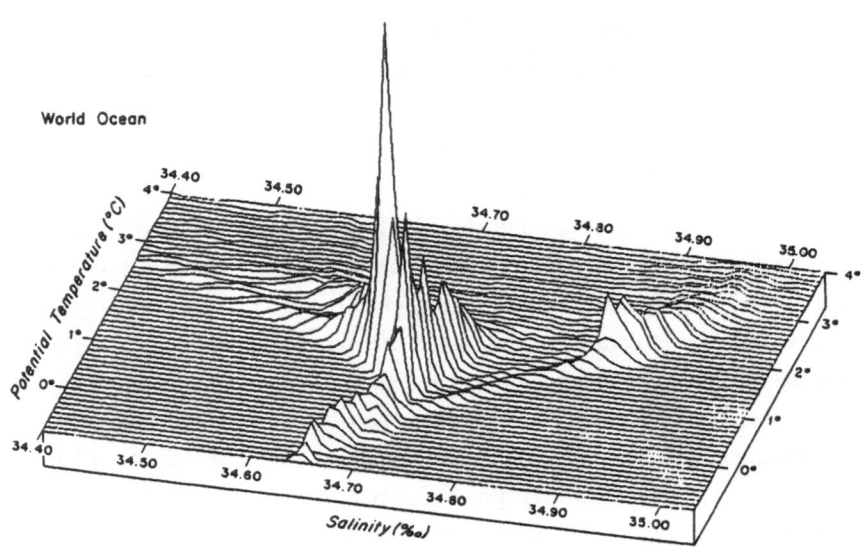

8. Histogram (Worthington, 1981) of the temperature-salinity characteristics of the world ocean, showing volumes of fluid in various intervals. There is no good explanation of the large peak.

Steady Circulation Pictures

Finding the integration constants in the thermal wind equations proved intractable for decades. To proceed, oceanographers made the plausible assumption that at great depth, the flow was sufficiently sluggish, that it could be taken as zero. This depth is the notorious "level-of-no-motion". Its plausibility depends in part upon the assertion that the oceanic circulation is dominated by wind-forced motions, which are expected to be strongest near the surface. It is reinforced by computation of the flow as a function of depth. Consider figure 9, where the level-of-no-motion was placed at about 2000 m depth. Shifting this depth several hundred meters in the vertical plainly produces little change in the near-surface velocities. To the extent that near-surface flows are of primary interest, one can argue that it really doesn't matter which depth is chosen, as long as it is deep enough.

Some serious attempts were made to find a rational choice of a level-of-no-motion. Property sections through the ocean (see Figures 6) strongly suggest to the eye that these "dyes", as one can think of them, delineate movements of water from a region where the properties are most intense, into the interior ocean. An appealing picture is that these property tongues indicate water mass origins - e.g. in the Circumpolar region, or the high northern Atlantic, or the mouth of the Mediterranean. One then argues that the flow must be from the source regions along the axis of the tongue into the oceanic interior. With some oversimplification, this is Wüst's "core-layer" idea, the tongue axes denoting the core-layers. Wüst then drew arrows along these tongues (see Figure 6a,b) and discussed the "spreading" ("ausbreitung" in German) of the water. His arrows are all of the same length, and he recognized that such ideas did not permit computation of flow magnitudes or of the distinction between flow and "mixing" (we return to mixing later).

But Wüst's ideas had a powerful impact on those who sought to employ the geostrophic balance and levels-of-no-motion. His picture supported the idea that the oceanic circulation was essentially "layer-like"; if in figures like 6, one identifies one of the water masses (e.g. the North Atlantic Deep Water) as moving from the Arctic regions, and one just overlying it, the Circumpolar Intermediate Water as coming from the Antarctic, then it is eminently reasonable to think that there must be a line of zero velocity between them. Such arguments have been widely employed; they finally broke down in a number of ways, including the recognition that the implied motions *below* the supposed level of no motion often made no sense, and with the availability of time series measurements, it became obvious that hydrographic sections could not be assumed to be repre-

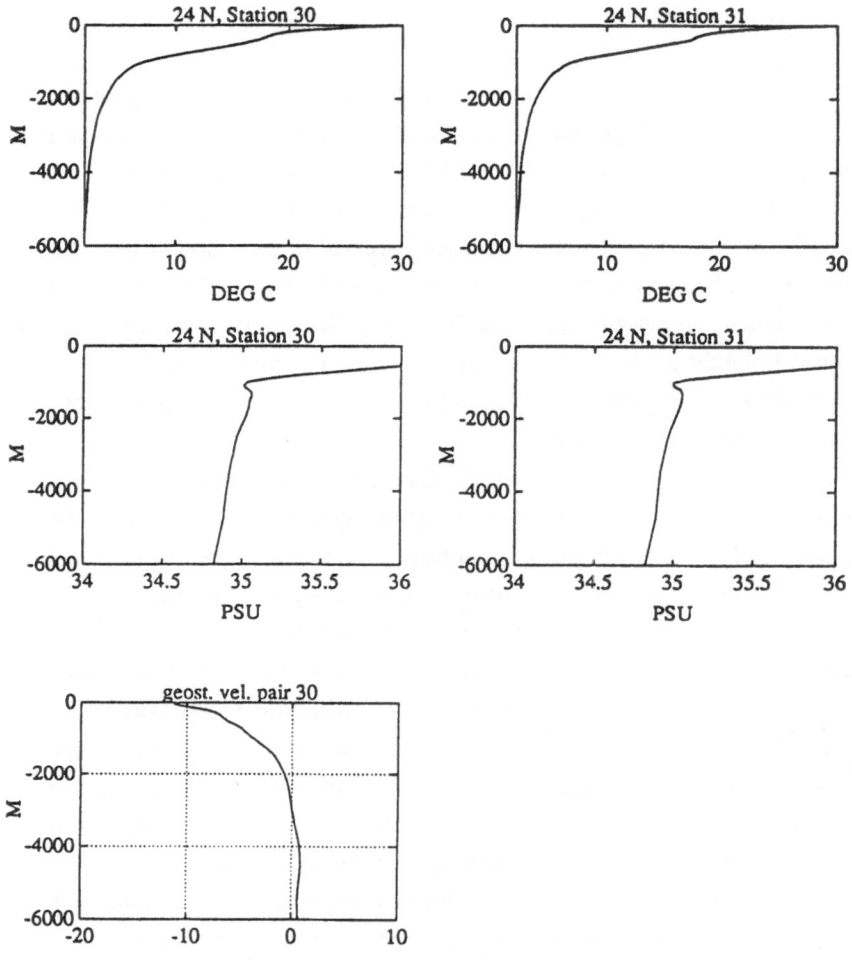

9. Temperature and salinity observed at two stations in the North Atlantic (part of the section depicted in fig. 6d). Upper two panels are temperature, middle two are salinity, and lowest panel is the meridional geostrophic flow computed from the thermal wind equations. A reference level near 3000 m has been used arbitrarily. Note weak vertical gradient in the deep water.

sentative of the time mean flows. Thus for example, figs. like 6b suggest the massive salinity tongue emanating from the Mediterranean is the result of a fluid flow from the Mediterranean, as depicted by Wüst's arrows. But the picture by itself provides no information as to whether the required flow did not actually take place long ago (millennia?), with either no flow having taken place since, or whether the flow continues to this day. (Indeed, present-day flows appear to be mainly *around* the tongue, rather than along its axis.) The problem is to distinguish between producing the "standing crop", i.e. the concentrations observed in the property charts, and the property fluxes required to maintain what is already there.

Although there were many other attempts at finding rational levels-of-no-motion. I will not take the trouble to discuss them further here. They led to various pictures of the circulation of the upper ocean (especially in the North Atlantic). A few of them are shown in figs 10. These results are difficult to interpret, mainly because they are purely qualitative. Error estimates are almost never provided, and despite the fact that the flow pictures often differ radically from each other, the reader has no way of knowing whether they differ because different data or methods were used, whether they refer to the ocean at different times, or are based upon different physics. Two pictures can disagree but be wholly consistent within error bars. One's problem is then very different than if they are not so consistent. The absence of error estimates has tended to mean that discussions of the ocean circulation do not converge - we get mainly a succession of different interpretations which may or may not be consistent with earlier ones. Thus the pictures do not improve. At least as important is the impossibility of making quantitative estimates of the all-important property fluxes (by "quantitative", I mean with uncertainty or error estimates).

2 The Steady Ocean Circulation

2.1 Making a Quantitative, Consistent Picture

In the last 15 or so years, a more quantitative, if not yet widely used, methodology has emerged, one which lends itself not only to the use of the historical hydrographic data but also to the ready application of altimetric data. I wish now to briefly discuss these methods.

First a piece of pure theory. The dynamic method does not use all the equations (15-19) which we have agreed apply to the circulation where it is nearly geostrophic. In particular, no specific use is made of equation (19a), the conservation of density (or any other scalar would work too). With hindsight, it is rather surprising how long it took to recognize that more information

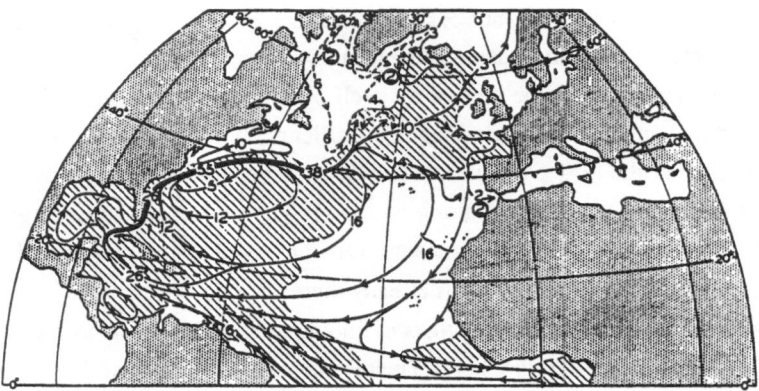

10a. Estimate of the near-surface circulation of the North Atlantic (Sverdrup et al., 1942) based upon the thermal wind and assumptions about the deep reference level.

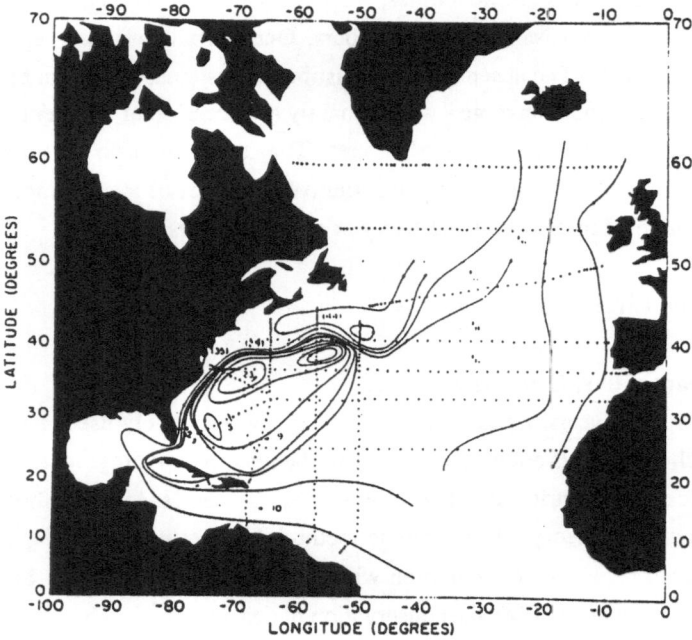

10b. Estimate of the pressure field in the thermocline (about 500 m depth) based upon application of kinematic conservation constraints to the thermal wind flow to determine the reference level velocity (Wunsch and Grant, 1982). Geostrophic flow would follow the contours.

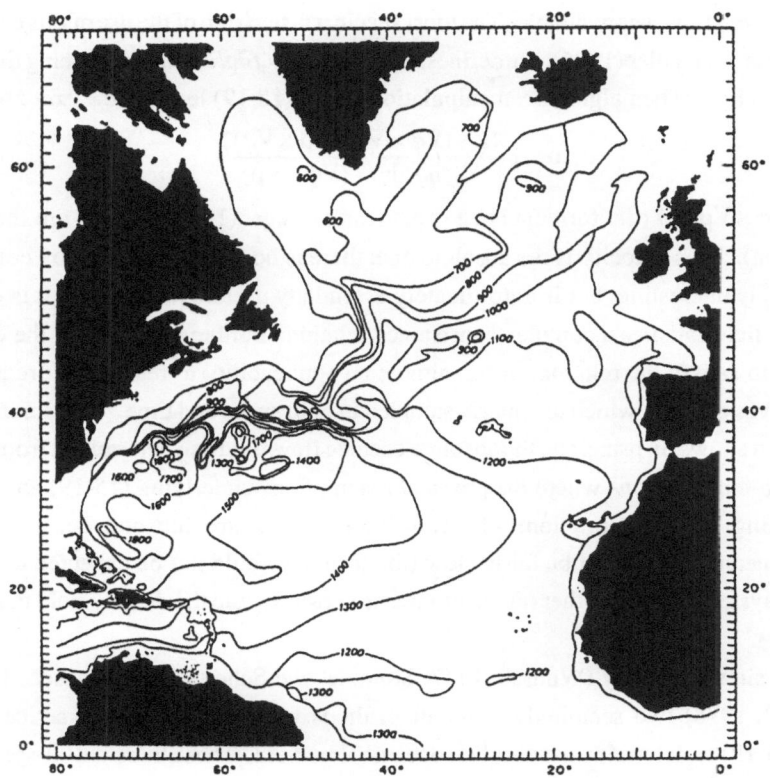

10c. Estimate of the near-surface pressure field (Stommel et al., 1978) based upon the thermal wind and the assumption of vanishing flow near 1500 m depth. Geostrophic flow would be along the contour lines.

about the required flow was available. The most succinct discussion of the use of this extra equation seems to be that of Needler (1985). He defines, $\mathbf{u} = (u, v)$, $q = f\partial\rho/\partial z$, the latter being the "planetary potential vorticity". Then algebraic manipulation of eqs (15-19) leads to the *exact* formula,

$$\mathbf{u} = g \frac{\{\hat{\mathbf{k}} \cdot (\nabla q \times \nabla\rho)(\nabla\rho \times \nabla q)\}}{\nabla(f\partial q/\partial z) \cdot (\nabla q \times \nabla\rho)} \tag{34}$$

as long as $\nabla q \times \nabla\rho \neq 0$ (the formula is an exact result of eqs. (15-19), which are themselves an approximation). This velocity is the absolute one; there is no missing integration constant.

Eq (34) is interesting, but is unfortunately essentially useless in practice. It is only correct if the density field satisfies a complex consistency relation. Furthermore even if the density field were known to satisfy the relation, it has almost no applicability to real data: it requires taking third derivatives of fields which are poorly sampled in both space and time. The chief importance of (34) is that it shows, in principle, the absolute oceanic flow field can determined from the density field alone, in those regions where the physical assumptions underlying (15-19) are valid. (It is worth repeating those assumptions: the flow must be geostrophic and density conserving. Geostrophy means that it must be fairly slow (the formula would not be expected to work in the Gulf Stream system or in any other region of intense currents), and the density field must represent a steady flow).

Nearly simultaneously (Wunsch, 1977, Stommel and Schott, 1977, Wunsch, 1978, Schott and Stommel, 1978) two seemingly different methods were proposed for practical use of the equations (15-19) to determine the absolute oceanic flow field. These were the "β-spiral", and the "inverse method". Shortly thereafter, it was recognized (e.g. Davis, 1978, Wunsch, 1989) that the two methods were essentially the same, differing in certain auxiliary statistical assumptions, and in convenience, the latter dependent mainly upon the spatial distribution of the available data.

Consider first the β-spiral. It is obtained by re-arrangement of the equations. Take (19a) and write it as

$$w = -u\frac{\partial\rho}{\partial x}/\frac{\partial\rho}{\partial z} - v\frac{\partial\rho}{\partial y}/\frac{\partial\rho}{\partial z} = u\frac{\partial h}{\partial x} + v\frac{\partial h}{\partial y} \tag{35a}$$

where

$$\frac{\partial h}{\partial x} = -\frac{\partial\rho}{\partial x}/\frac{\partial\rho}{\partial z}, \quad \frac{\partial h}{\partial y} = -\frac{\partial\rho}{\partial y}/\frac{\partial\rho}{\partial z} \tag{35b}$$

$z = h(x, y, \rho)$ defines the depth of a surface of constant density (an "isopycnal"). With this coordinate change, the thermal wind equations become

$$\frac{\partial u}{\partial z} = \frac{g}{f} \frac{\partial \rho}{\partial y} = -\frac{g}{f} \frac{\partial \rho}{\partial z} \frac{\partial h}{\partial y} \tag{36a}$$

$$\frac{\partial v}{\partial z} = -\frac{g}{f} \frac{\partial \rho}{\partial x} = -\frac{g}{f} \frac{\partial \rho}{\partial z} \frac{\partial h}{\partial x} \tag{36b}$$

where derivatives of ρ not multiplied by g have been omitted (the Boussinesq approximation). Differentiating (35a) and substituting the thermal wind values produces

$$\frac{\partial w}{\partial z} = u \frac{\partial^2 h}{\partial x \partial z} + v \frac{\partial^2 h}{\partial y \partial z} \tag{37}$$

But using the linear vorticity equation, (22), w can be eliminated, producing

$$u \frac{\partial^2 h}{\partial x \partial z} + v \left(\frac{\partial^2 h}{\partial y \partial z} - \frac{\beta}{f} \right) = 0 \tag{38}$$

called the β-spiral equation.

To see that a spiral is implied, we follow Bryden (1977). From the original thermal wind equations, eq (19a) is

$$w \frac{\partial \rho}{\partial z} = -\frac{uf}{g} \frac{\partial v}{\partial z} + \frac{vf}{g} \frac{\partial u}{\partial z} \tag{39}$$

Write $(u, v) = V(z) (\cos \alpha(z), \sin \alpha(z))$ and substituting into (39) we have

$$\frac{d\alpha(z)}{dz} = -\frac{g}{f} \frac{w \partial \rho / \partial z}{V^2} \tag{40}$$

The linear vorticity equation shows that if $v \neq 0$, i.e. if the flow is not purely zonal, then there must be a non-vanishing w. With $\partial \rho / \partial z < 0$, and $w > 0$, α would increase upward (spiral counterclockwise upward). If β vanishes, there is no spiral, and the system reduces to a statement of "parallel solenoids" (see Defant, 1961, pp. 476-482). (Stommel and Schott (1977) briefly noted that the sign of the spirals could be changed with a time dependent density term - e.g. if the ocean were cooling.) u, v in eq (38) refer to the total (absolute) velocity. But writing as usual,

$$u = u_r + c, \quad v = v_r + b \tag{41}$$

(38) is

$$c\frac{\partial^2 h}{\partial x \partial z} + b\left(\frac{\partial^2 h}{\partial y \partial z} - \frac{\beta}{f}\right) = -u_r\frac{\partial^2 h}{\partial x \partial z} - v_r\left(\frac{\partial^2 h}{\partial y \partial z} - \frac{\beta}{f}\right) \tag{42}$$

If we can estimate the partial derivatives of the density field from observations, then both the derivatives of h and the relative velocities are known. For each depth where such estimates are possible, we obtain an equation for (b, c). Suppose there are M such depths. Then combining all equations at a fixed location (x, y) produces a set of simultaneous equations

$$\mathbf{Ex + n = y}$$

of standard form, M equations in 2 formal unknowns which can be solved by many methods.

The Wunsch (1977, 1978) method starts with the same equations, but was motivated in the first instance by the fact that most of the high quality modern hydrography available was obtained on "long-lines" spanning the ocean as in figure 2 or 6d. Basically, all the method says is that if we consider a "box" surrounded by hydrographic stations, (fig. 11), then if C is any quantity for which we are prepared to write a conservation statement, then the transport of C into the box must equal what comes out (or we can specify sources/sinks if known). Quantities such as mass, or salt, will do. The statement that mass is conserved in volume is

$$\sum_{j}^{N}\sum_{q}^{Q}\rho_j(q)\,(v_{Rj}(q) + b_j)\delta_j\Delta a_j(q) \approx 0 \tag{43}$$

where we have denoted by $v_{Rj}(q)$ the thermal wind (relative velocity) in station pair j at depth interval q, $\Delta a_j(q)$ is the differential area in depth interval q in station pair j, and δ_j is the unit normal (± 1) for pair j for the volume under consideration. Eq (43) is a discrete approximation to the areal integrals over the boundary section and can be carried out in a variety of different approximations. Everything is known in (43) except for the reference level velocities. Eq (43) has been written as approximately equal to zero, rather than precisely so, in anticipation of the need to grapple with errors in the various terms of the sum.

Eq (43) is one equation in N unknowns b_j, and the addition of some further constraints, i.e. equations, would be helpful. One possibility is to require that the salt content of the volume should also be conserved. Let S be the salinity (salt/unit mass). Then we might require

$$\sum_{j}\sum_{q}\rho_j(q)S_j(q)\,(v_{Rj}(q) + b_j)\delta_j\Delta a_j(q) \approx 0 \tag{44}$$

Suppose we define the depth of fluid of density ρ_i as $z(\rho_i, x, y) \equiv h_i(x, y, \rho_i) = h_i(x, y)$. Then conservation of mass between density intervals $\rho_i \leq \rho \leq \rho_{i+1}$ is

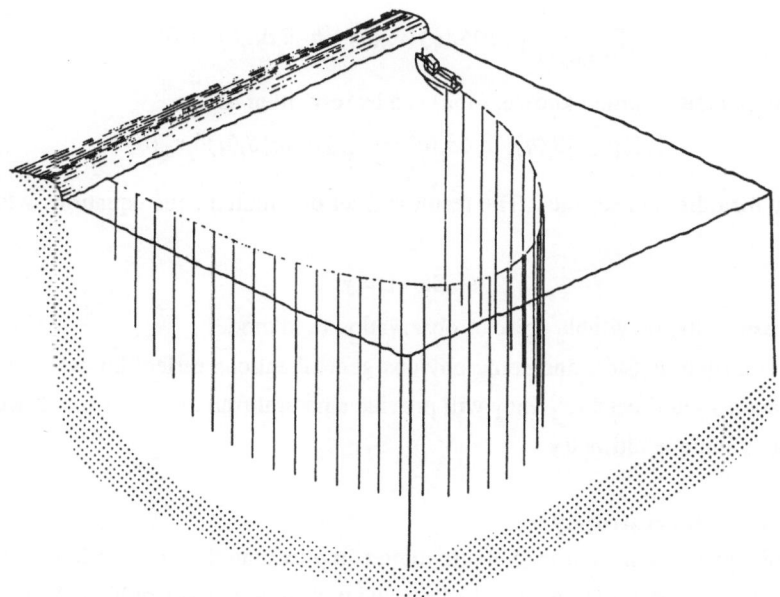

11. A sequence of hydrographic stations enclosing a volume of ocean permits the use of kinematic conservation ccnstraints to remove the ambiguity of the circulation owing to reference level assumptions.

$$\sum_j \sum_{h_i(x_j,y_j)}^{h_{i+1}(x_j,y_j)} \rho_j(q) S_j(q) (v_{Rj}(q) + b_j) \delta_j \Delta a_j(q) \approx 0 \qquad (45a)$$

Since $v_{Rj}(q)$ are assumed known, (45a) can be re-written as

$$\sum_j \sum \rho_j(q) S_j(q) b_j \delta_j \Delta a_j(q) \approx - \sum_j \sum \rho_j(q) S_j(q) \delta_j v_{Rj}(q) \qquad (45b)$$

and similarly for other constraints. The result is a set of simultaneous equations which we will write as

$$\mathbf{Ex + n = y} \qquad (46)$$

where n represents the inevitable noise of observation (and model).

Equations such as (46), and some obvious generalizations reflect knowledge of how the ocean circulation should behave. They will provide a natural framework in which we will eventually be able to include altimetry.

2.2 Deviations from geostrophy

The full equations of motion (5-9) are more complex than the ones we have just been considering, and it is essential to understand where and how simple geostrophic balance is expected to breakdown. Several general classes of breakdown are possible, singly, or in combination. Among them, we would list (1) time dependence, which is taken up in the next section, (2) nonlinearity owing to intense flows, (3) strongly dissipative flows. Any combination of these effects may occur simultaneously.

The reader should first recognize that deviations from pure geostrophy are absolutely essential in any real ocean circulation. The reduction in the system physics represented by the geostrophic approximation produces a flow field without any sources or sinks of energy or momentum, or the possibility of satisfying boundary conditions on arbitrarily specified solid boundaries.

The handling of the forcing owing to the windfield goes back to Ekman (1905). First, one notes that molecular friction can be totally ignored for this purpose. Then, to mimic the terms F_u, F_v of (3.3.1-2), Ekman (1905) appended eddy coefficient terms to eqs (15-17) producing a new system

$$-\rho f v = -\frac{\partial p}{\partial x} + A_v \frac{\partial^2 u}{\partial x^2} \tag{47}$$

$$\rho f u = -\frac{\partial p}{\partial y} + A_v \frac{\partial^2 v}{\partial y^2} \tag{48}$$

$$0 = -\frac{\partial p}{\partial z} - g\rho \tag{49}$$

The use of an eddy coefficient to represent the effects of scales smaller than the ones the investigator is interested in, is a matter of mathematical convenience, and tradition, and is perhaps "a triumph of hope over experience". That it can fail badly in general was documented at length by Starr (1968). The wind at the surface being supposed to exert a vector stress $\underline{\tau} = (\tau_x(x, y), \tau_y(x, y))$ in analogy to the stress boundary conditions for molecular viscosity, the boundary conditions on the solutions to equations (47-49) are

$$A_v \left(\frac{\partial u}{\partial z}, \frac{\partial v}{\partial z} \right)_{z=0} = (\tau_x, \tau_y), \quad w(0) = 0 \tag{50}$$

The "boundary layer" solution is Ekman's famous spiral of form

$$u_{BL} = \frac{\exp^{\sqrt{f/2A_v}\, z}}{\sqrt{2A_v f}} \{ (\tau_x + \tau_y) \cos \sqrt{f/2A_v}\, z + (\tau_x - \tau_y) \sin \sqrt{f/2A_v}\, z \} \tag{51}$$

$$v_{BL} = \frac{\exp^{\sqrt{f/2A_v}\, z}}{\sqrt{2A_v f}} \{ (\tau_y - \tau_x) \cos \sqrt{f/2A_v}\, z + (\tau_x + \tau_y) \sin \sqrt{f/2A_v}\, z \} \tag{52}$$

so that $(u_{BL}, v_{BL}) \to 0$ as $z \to -\infty$, , where the velocities have been labeled with the subscript BL as a reminder that they apply only in a near surface "boundary layer". Note a few salient features of this solution:

• The direct effects of the wind are confined to the Ekman depth $d_E = \sqrt{2A_v/f}$. Since the use of eddy coefficients is dubious, and the numerical values one is advised to use for A_v vary widely, it is fortunate that the square root reduces the sensitivity of the Ekman depth to the numerical choice. Extreme values of A_v lead to estimates of d_E ranging over about 25-200 m.

• The transport of water in this boundary layer is

$$U_{BL} = \int_{-\infty}^{0} \rho u \, dz = \tau_y / f \qquad (57a)$$

$$V_{BL} = \int_{-\infty}^{0} \rho v \, dz = -\tau_x / f \qquad (57b)$$

to the right of the wind, and is independent of the value of A_v - a result of greatest importance, again because of the artificial nature of an eddy coefficient. If we take plausible numerical estimates for the wind stress as 0.1 Pascals (about 1 dyne / cm^2), and a latitude of 30°, the amount of fluid predicted to move within the Ekman layer is approximately $1.4 \times 10^3 kg/m \, s$. Integrating over 1000 km, the flux is about 10^9 kg/s (about $10^6 m^3/s$ - 1 "Sverdrup").

• To the order of the approximations leading to the balance (51,52), there is no pressure signature of the Ekman layer (see Pedlosky, 1987). Pressure fields do not ordinarily form boundary layers; thus we do not expect to see any pressure field associated with the Ekman layer itself.

• Away from the surface boundary layer, the stress terms become negligible and the flow field reduces to a geostrophic "interior", satisfying (15-19) and whose flows we will label u_I, v_I, p_I.

• Equations (51,52) are linear, and we can write the total solution, Ekman + plus interior, as

$$u = u_{BL} + u_I, \quad v = v_{BL} + v_I, \quad p = 0 + p_I \qquad (54)$$

That is, the flow within the boundary layer can usefully be viewed as being the *sum of the Ekman flow plus the geostrophic interior*. The pressure field seen near the seasurface is that owing to the interior flow alone, a result of great importance in the use of satellite altimetry.

• Because the lateral transport of fluid within the Ekman layer is a function of position, owing to variations in the value of f, and the strength of the wind, there must be an exchange of fluid with the region below if the flow (and the seasurface) are to remain steady in time. The vertical velocity associated with this exchange is easily found to be

$$w_E = \frac{\partial}{\partial x}\left(\frac{\tau_y}{f}\right) - \frac{\partial}{\partial y}\left(\frac{\tau_x}{f}\right) \qquad (55)$$

and this flux must be absorbed, or provided by, the underlying fluid, which is presumed to be in geostrophic balance, and which sets this deeper fluid into motion.

The wind-driven general circulation of the ocean, away from the immediate vicinity of the equator, is driven by this mechanism, i.e. not by the wind, or the Ekman layer, but by the regional variations of the amount of fluid moving within the Ekman layer which sets the interior geostrophic fluid into motion. The underlying physics is probably best appreciated from the point of view taken by Stommel (1957, 1984), who avoided the Ekman layer entirely, and discussed how geostrophic flows could be driven by patterns of rainfall and evaporation (figure 12). Rain acts analogously to the way Ekman pumping does, with evaporation equivalent to Ekman suction. The Ekman layer itself is of secondary interest in the general circulation - although its transport of mass and other properties can be important, particularly at low latitudes. But its properties as a pump are the crucial element in understanding the general circulation.

From the point of view of a geodesist, interested in measurements of the seasurface topography, perhaps the most important message is that the surface intensified flow, represented by the Ekman layer, does not produce a noticeable elevation change - the directly wind-driven flow would be, to a first-order, invisible to an altimetric mission. Notice however, that in our linear ocean, we can *add* the flow owing to the geostrophic field to that of the Ekman layer - and thus an altimeter would see the pressure field owing to the geostrophic flow at the surface - *despite the presence of the Ekman layer*. Far from being a problem (one might think an invisible surface flow would be troublesome), it is a huge asset, because the surface geostrophic flow reflects the flow at great depth, while the Ekman flow at the surface only directly reflects that in a layer of order 100 m or less.

In 1948, Stommel first pointed out that the Gulf Stream was a physical phenomenon which needed explanation through the equations of fluid dynamics. More generally, he attempted to explain why all oceans seemed to have intensified flows on the western side. (Notice that pure geostrophy offers no explanation.) Instead of examining Stommel's (1948) solution let us instead explore the slightly more complicated model produced by Munk (1950). Munk used the same "Reynolds' analogy" employed by Ekman, but applied it to the horizontal mixing, so that the full equations are

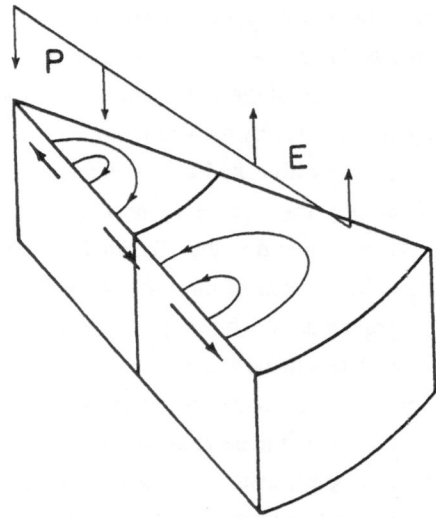

12. The driving of the ocean circulation by the wind field can be understood by recognizing that the Ekman layer acts as a "pump" for the geostrophic flow beneath it. This is Stommel's (1984) estimate of the circulation driven by a pattern of high latitude downward pumping (like a precipitation pattern, P), and lower latitude upward suction (like an evaporation pattern E).

$$-fv = -\frac{\partial p}{\partial x} + A_h \nabla_h^2 u + A_v \frac{\partial^2 u}{\partial z^2}$$

$$fu = -\frac{\partial p}{\partial y} + A_h \nabla_h^2 v + A_v \frac{\partial^2 v}{\partial z^2}$$

$$\frac{\partial u}{\partial x} + \frac{\partial v}{\partial y} + \frac{\partial w}{\partial z} = 0 \qquad (56a,b,c)$$

If these are *integrated vertically*, they produce

$$-fV = -\frac{\partial P}{\partial x} + \tau_x + A_h \nabla_h^2 U$$

$$fU = -\frac{\partial P}{\partial y} + \tau_y + A_h \nabla_h^2 V$$

$$\frac{\partial U}{\partial x} + \frac{\partial V}{\partial y} = 0 \qquad (57a-c)$$

The integrated continuity eq. (57c) can be satisfied identically with a transport stream function, defined as

$$U = -\frac{\partial \Psi}{\partial y}, \; V = \frac{\partial \Psi}{\partial x}$$

and which if substituted into (57a,b), and eliminating the pressure by cross-differentiation produces

$$A \nabla_h^4 \Psi + \beta \frac{\partial \Psi}{\partial x} = \hat{\mathbf{k}} \cdot \nabla_h \times \underline{\tau} \qquad (57d)$$

The solution to this partial differential equation is discussed in detail in all oceanographic textbooks (e.g. Pond and Pickard, 1983) and I will not reproduce the mathematical technology here (it is a classical example of a partial differential equation which succumbs almost trivially to the methods of singular perturbation theory - see Munk and Carrier, 1950). Suppose the oceanic dimensions are $0 \le x \le L$; then an approximate solution is

$$\Psi = \frac{L}{\beta} \frac{\partial \tau_x}{\partial y} \left\{ -\left(\frac{2}{\sqrt{3}} - \frac{\sqrt{3}}{kL} \right) e^{-1/2kx} \cos\left(\frac{\sqrt{3}}{2} kx + \frac{\sqrt{3}}{2kL} - \frac{\pi}{6} \right) + 1 - \frac{1}{kL}(kx - e^{-k(L-x)} - 1) \right\}, \quad k = \left(\frac{\beta}{A} \right)^{1/3}$$

$$(58)$$

which Munk (1950) plotted as shown in fig.13 (he plotted the exact solution using estimates of the real wind in a Pacific sized basin). The westward intensification is obvious and results from the first term in (58); the second term in (58) is what is called the "Sverdrup relation" - it relates the oceanic flow away from the western boundary directly to the local curl of the wind-stress. We are thus led to describe this solution as a "western boundary current", or "western boundary layer", with a "Sverdrup interior". Stommel's (1948) solution, and the one by Munk (1950), led in rapid succession to a host of "wind-driven" ocean current theories, which flowered as part of geophysical fluid dynamics over the next 10-20 years (e.g. Stommel, 1965). This development was tremendously exciting and greatly advanced the study of the ocean circulation, putting it for the first time on a dynamical footing, and making it a branch of physics, rather than an element of geography.

The huge attractiveness of these mathematical constructs means that one must constantly be aware of exactly what has been done. The Stommel/Munk/... models represent an ocean which is (a) steady, (b) without bottom topography, (c) linearized, (d) driven only by the wind, (e) dissipates momentum and vorticity through oversimplified parametric means. These models and their successors have been extremely important in providing insights into the elements that must make up the ocean circulation, but one should not confuse them with the real ocean.

3 Time Dependence. The Mesoscale

For many decades, the problem of observing the ocean was so difficult, that the only way one could attempt to describe it, was to lump together hydrographic observations from all expeditions ever made, often spanning years and decades. The earliest practitioners of oceanography recognized that the ocean was in fact time-dependent (e.g. Maury, 1855; Helland-Hansen and Nansen, 1920; see review in Wunsch, 1981), but any effort to distinguish the steady circulation from the time dependent one was defeated by the observational obstacles. A certain amount of theory grew up however, in the period from about 1948 onwards, to some extent reflecting meteorological experience. Some of the flavor of this work can be understood by considering the very simple situation of an ocean with a flat bottom, constant density, $\rho = \rho_0 \equiv 1$, and a "rigid lid" - i.e. the seasurface does not move. (This might be regarded as a particularly poor assumption in an altimetric context, but it is almost irrelevant. In a rigid lid ocean, the flow exerts pressures against the lid; to an excellent first approximation, they are equivalent to elevation changes,

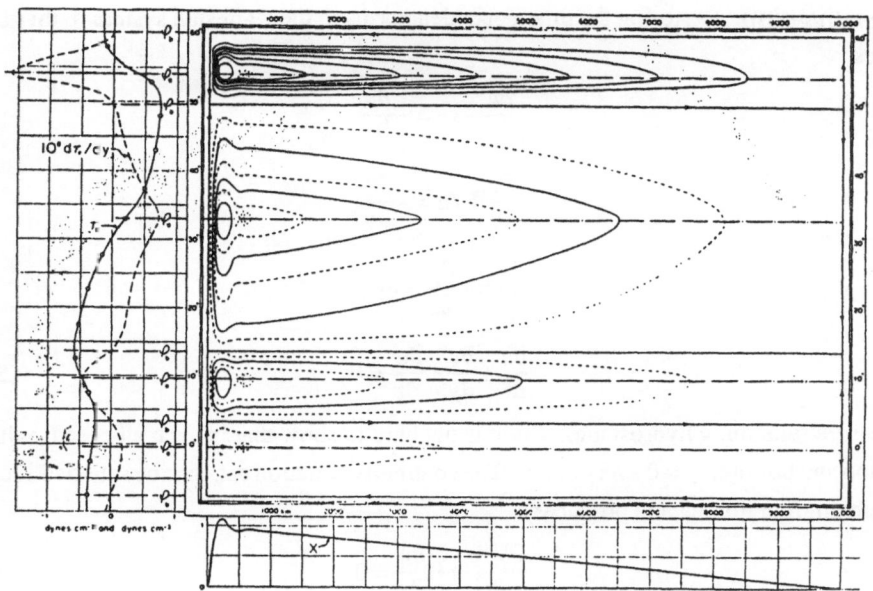

13. Munk's (1950) calculation of the wind-driven circulation in a rectangular basin given a realistic mean wind field. This is an important example of what became the field of theories of the ocean circulation based upon the equations of motion.

$\eta = p(z = 0)/g\rho(0))$. Restoring the time-dependent terms to the Cartesian system in this situation produces

$$\frac{\partial u}{\partial t} - fv = -\frac{\partial p}{\partial x}$$

$$\frac{\partial v}{\partial t} + fu = -\frac{\partial p}{\partial y}$$

$$0 = -\frac{\partial p}{\partial z} - g\rho_0$$

$$\frac{\partial u}{\partial x} + \frac{\partial v}{\partial y} + \frac{\partial w}{\partial z} = 0 \qquad (59a - d)$$

(the pressure remaining hydrostatic). But it is not hard to show that in such an ocean, with $w = 0$ at both top and bottom, $w = 0$ everywhere. The continuity equation can thus be satisfied identically with a stream function, ψ and produces

$$\frac{\partial \nabla^2 \psi}{\partial t} + \beta \frac{\partial \psi}{\partial x} = 0 \qquad (60)$$

a linear equation with constant coefficients, and which has solutions

$$\psi = A\, e^{ikx + ily - i\omega t}$$

which when substituted into (60) produces

$$\omega = -\frac{\beta k}{k^2 + l^2} \qquad (61)$$

which is the dispersion relation, plotted in figure 14. These waves, known variously as "Rossby waves", "planetary waves", or "oscillations of the second kind", have some strange properties. Notice that they propagate only westwards ($\omega/k < 0$), and the shortest waves (high k, l) have the longest period. (It is the phase velocity which is westward; the reader should confirm that the group velocity can be either eastward or westward.) With their strange properties, the waves are fundamental to understanding largescale geophysical fluids. If we take a bounded ocean, $0 \leq x \leq L$, $0 \leq y \leq M$, they can be superposed to produce $\psi = 0$ on the boundary (no flow into the boundary), as shown by Longuet-Higgins (1964), and depicted in fig.15. The solution is

$$\psi(x, y, t) = e^{-i\omega_{nm}t - i\beta k/2\omega_{nm}x} \sin\left(\frac{n\pi}{L}x\right) \sin\left(\frac{m\pi}{M}y\right) \qquad (62)$$

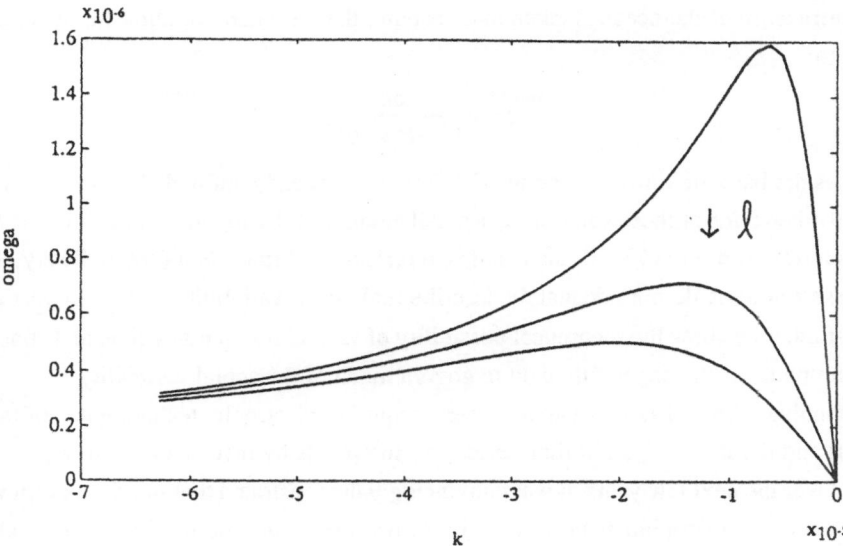

14. Dispersion relation for Rossby waves. Phase velocities are westward only, with the frequency generally decreasing with wavenumber (except near the origin). The group velocity necessarily can be either westward or eastward.

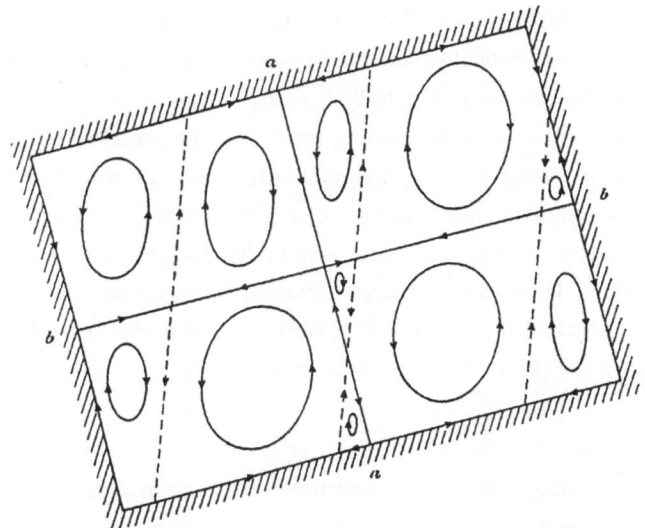

15. Longuet-Higgins (1965) calculation of the Rossby wave pattern in a closed basin. Phase lines shift westward.

If the stratification of the ocean is taken into account, the dispersion relationship is only slightly modified to

$$\omega = -\frac{\beta k}{k^2 + l^2 + 1/R_D^2} \tag{63}$$

where R_D is the baroclinic Rossby radius of deformation already defined. The reader is urged to examine the behavior of these solutions is a small amount of dissipation, in the form of Rayleigh friction is introduced into (59a-d); the result is surprising and important (see Pedlosky, 1987).

These solutions do not adequately describe real ocean variability. Their importance is at least two-fold: they show the theoretical possibility of variability on many time and space scales, and capture much of the physics thought to govern much of the actual variability.

Beginning about 1970, oceanographers acquired electronic technologies of sufficient robustness and duration to permit time series measurements by instruments left at sea untended by ships. Over the next few years, it was convincingly demonstrated that the circulation was fully turbulent on the geostrophic scales. Fig. 16 shows the results of the Mid-Ocean Dynamics Experiment (MODE Group, 1976) - the first convincing measurement of what became known as the "mesoscale eddy field" (an unfortunate nomenclature - because it corresponds to what is known in meteorology as the "synoptic scale", and where "mesoscale" means something else). Notice that the "eddy" propagates westward, just as the simple Rossby wave dispersion relationship suggests, albeit (e.g. McWilliams and Flierl, 1976) the motion is too rapid to be fully consistent with that simple model. The vertical structure is however, consistent with the horizontal structure suggested by the scale arguments leading to the Rossby radius.

From the point of view of understanding the ocean, the presence of this turbulence is an enormously complicating factor: the sampling requirements are formidable; and there is a distinct possibility, as in most turbulent fluids, that these smaller scales can drive, or control, the larger scales. One of the first applications of altimetry was to develop global mesoscale variability maps (e.g. Shum et al., 1990). There is now a large literature on the ocean mesoscale; for a couple of aging reviews, see Wunsch (1981); Robinson (1983). A more recent compilation is Schmitz and Luyten (1991). Holloway (1986) discusses some of the theory.

4 Puzzles of Climate and the Ocean

Because understanding of the ocean is so primitive, the ocean circulation has been somewhat under-emphasized until very recently in discussions of the climate system. Thus for example,

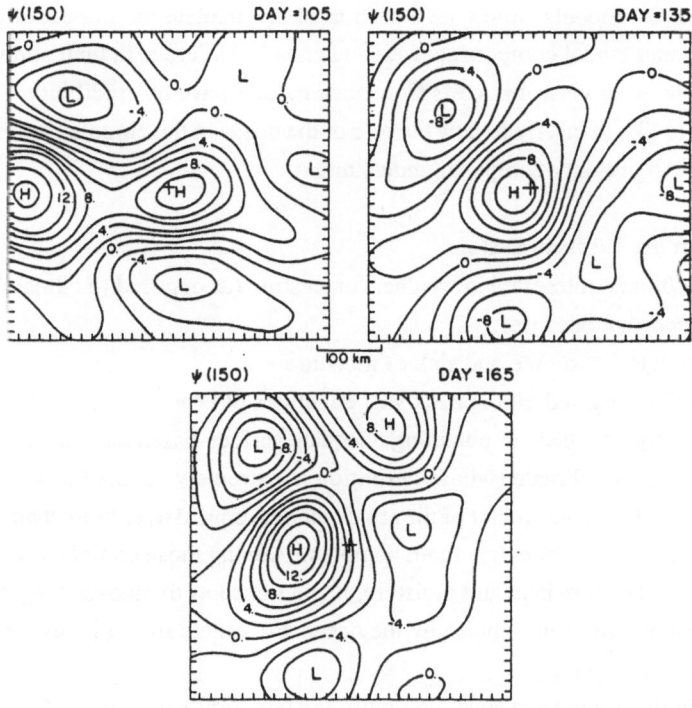

16. Observed pattern (McWilliams, 1976) of the mesoscale eddy field in a region southwest of Bermuda. Three different times are shown.

some climate forecast models which have been used to calculate the impact of a CO_2 doubling have contained ocean model components which are nearly ludicrous in their simplicity. From an oceanographer's point of view, forecasts from such models have no credibility.

Here I would like to briefly review the role of the ocean in the climate system, as it is a main driving force behind present attempts to understand it.

4.1 The Ocean in the Climate System

To a meteorologist interested in weather forecasting (as opposed to climate), the existence of the ocean is important because:

(1) It represents a massive source and sink of moisture

(2) It is capable of storing and giving back vast quantities of heat

(3) It represents a huge obstacle to obtaining adequate surface-based atmospheric observations.

When one turns to climate (whose definition is extremely vague, but which I will define here as any element of the state vector of the atmosphere averaged over more than about one year), the ocean's role becomes a good deal more complex. Among those complexities are

(1) That it not only can store heat and moisture, it can transport them over long distances so that the regions of heat and moisture uptake by the ocean are remote from the regions where it gives them back to the atmosphere.

(2) The ocean is both a source and sink, regionally and temporally varying, of "greenhouse gases".

(3) Whatever the ocean does today, or did yesterday in terms of the transports and storage of gases, heat, moisture, etc., the system is capable of *change* and there is no grounds whatsoever to assume that the system is in a steady state and/or will remain so.

(4) The extent to which the present ocean is in equilibrium with today's climate is unclear - is it still responding to climatic anomalies (e.g. Little Ice Age) of many years ago?

The ocean is clearly storing heat. The major features visible in figs. 2-4 appear to be permanent - the thermocline, etc. is visible in both the Challenger section, and the one made 80 years later. Thus heat is evidently not being exchanged with the atmosphere at a rate sufficient to modify the apparently stable gross stratification. But the near-surface parts of the ocean do exchange heat with the atmosphere, on time scales of seconds to days to seasons to interannually. Figure 17 shows (Bauer and Woods, 1984) the estimated annual march of heat stored in the upper ocean over the seasons. What is its significance to climate? First we note that the heat capacity of the entire atmosphere is contained in only the upper few meters of the ocean.

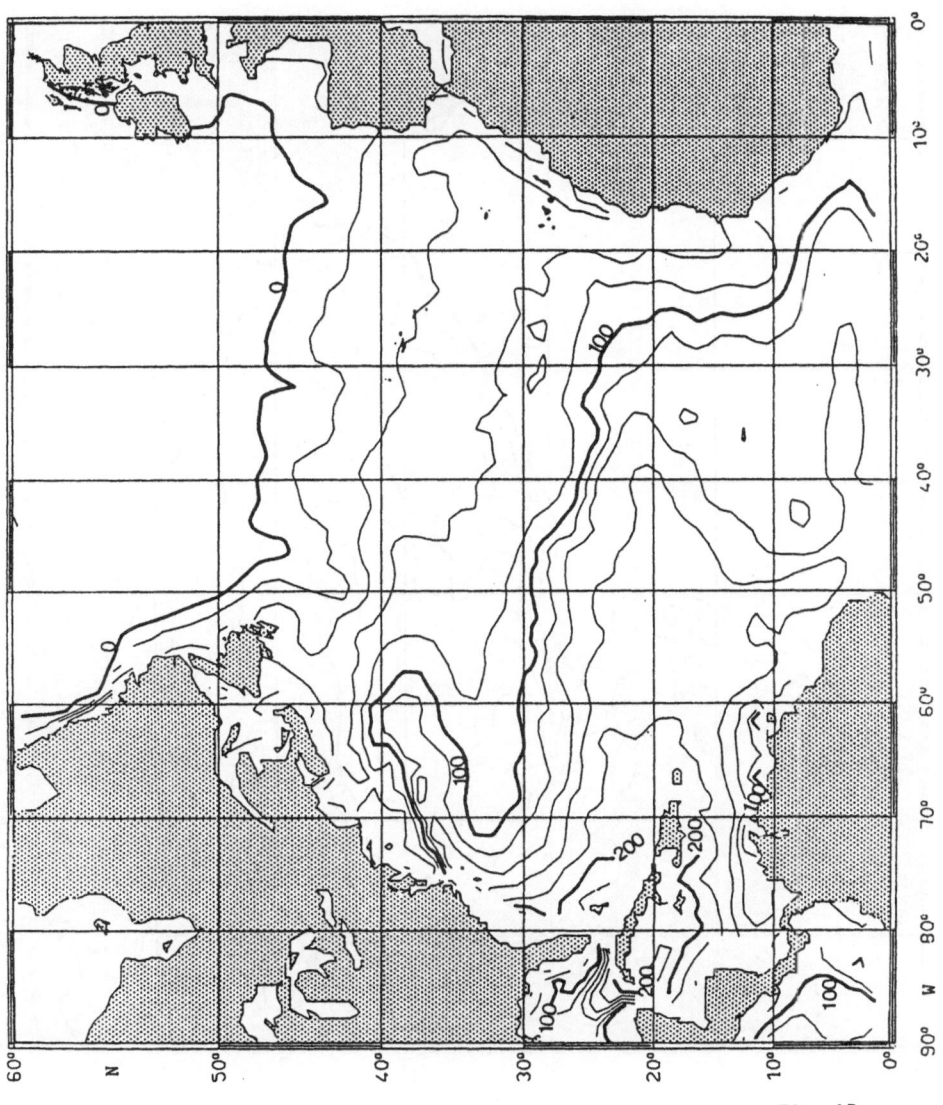

Fig. 17a

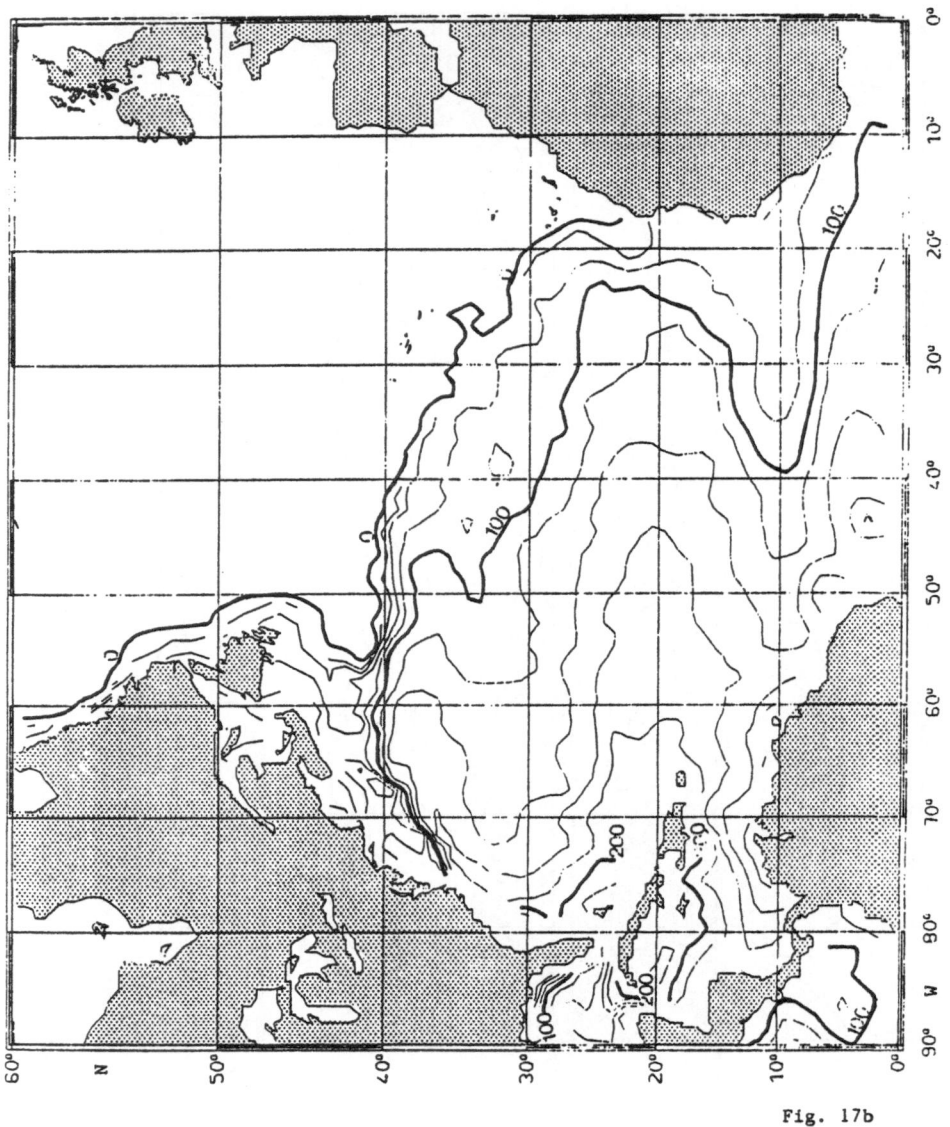

Fig. 17b

17a,b. Depth of a surface of constant density in the North Atlantic in July and December (Bauer and Woods, 1984) . Note the strong seasonal migration of the field, especially the zero line (the "outcrop" at the surface).

Where and how is this heat being exchanged between the two systems? Fig. 18 shows the Isemer et al. (1989) estimate of the annual average heat exchange in the North Atlantic. Taking this figure at face value, one sees that there is a slight heat gain in the tropics, and a heat loss, which reaches a maximum over the Gulf Stream system. A comparatively small amount of heat is lost by the ocean at high latitudes. Evidently, there is an important link between the oceanic general circulation, i.e. the Gulf Stream and heat exchange.

The ocean exchanges a variety of other substances with the ocean. Figure 19 displays the measured carbon dioxide content of the ocean along a line in the western Atlantic (Geosecs). Carbon dioxide is of course, a greenhouse gas and so a focus of intense interest. But the most important greenhouse gas of all is water vapor, and this too is exchanged with the ocean. Figure 20 shows an estimate of the net evaporation minus precipitation over the world ocean. The reader is warned not to take such figures too seriously.

4.2 Long-Term

Let us look at some "proxies" for the climate record, because the modern instrumental record is substantially less than 100 years long. Some of the most interesting evidence comes from ice cores. Falling precipitation traps air bubbles which are preserved when frozen at the surface. By drilling into the Greenland and Antarctic ice caps and through a variety of interesting and complex dating methods, (including simply counting annual 'growth rings') which I do not have the time to explore here (see, for example, Berger and Labeyrie, 1987) one can date the air and analyze its properties. The oxygen isotope ratio $^{18}O/^{16}O$ is controlled by the relative amount of water tied up as ice, and to a lesser degree by the mean temperature of the earth. The two effects can be separated to infer the ice volume, thus producing a global thermometer. Consider figure 21, showing a time history from the Antarctic ice cap of temperature, carbon dioxide, and methane content. This is an extraordinarily interesting figure: it shows us explicitly the last ice age, at which time the air temperature was about 9°C cooler than it is today. But it also shows us that the carbon dioxide content of the air was much lower than it is today. Thus we see that the last ice age was a period of an "anti-greenhouse". Unfortunately the data are not adequate yet to tell us whether the lowered CO_2 was a cause of the ice age, or a consequence - and a positive feedback on it. Nonetheless, there appears to be a very close coupling and correlation between global temperature change and atmospheric CO_2 concentration - one of the reasons of course, for the intense interest being stimulated by figure 21. Furthermore, figure 21 raises a series of problems. In particular, where

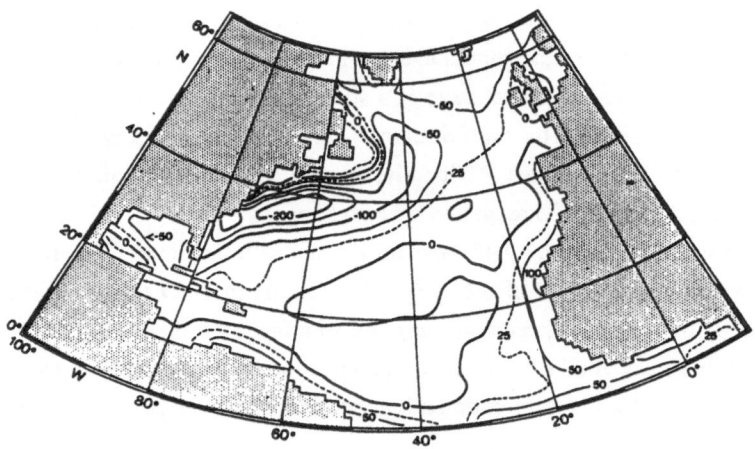

18. Estimate of the annual mean air-sea heat exchange (Isemer et al. 1989). Notice strong exchange over the Gulf Stream system. Such estimates are subject to large errors (30-50 W/m^2).

DEPTH IN METERS

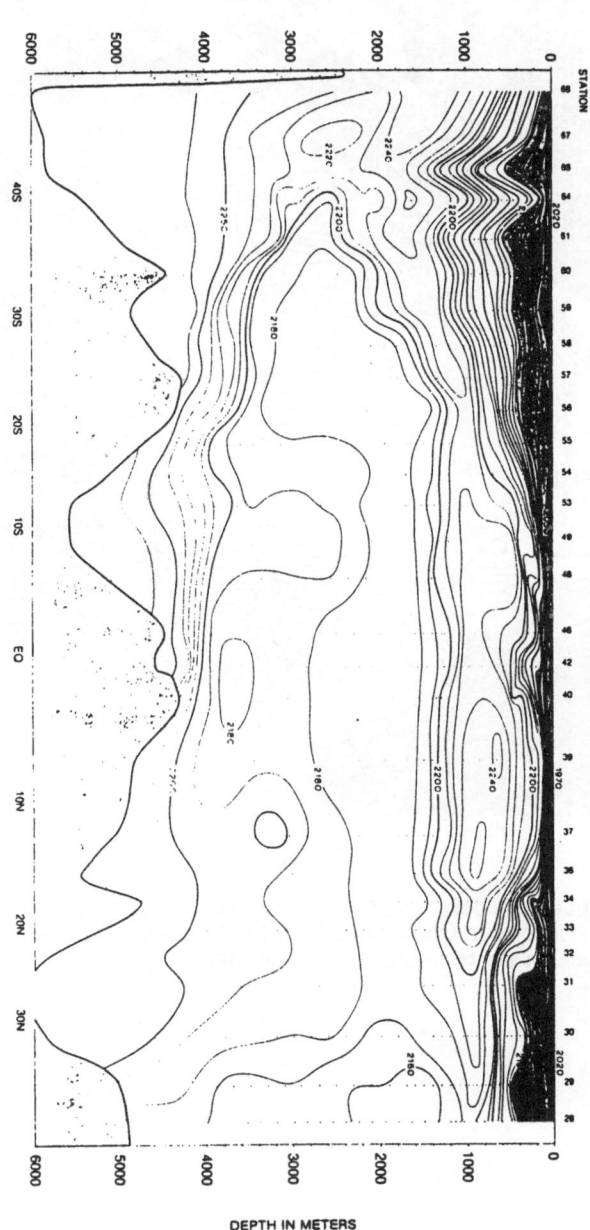

DEPTH IN METERS

19. Section down the western Atlantic Ocean (from the GEOSECS Expedition; Bainbridge et al., 19$) showing the total carbon dioxide content of the seawater. The natural values are so large that the increase from human contributions are not detectable.

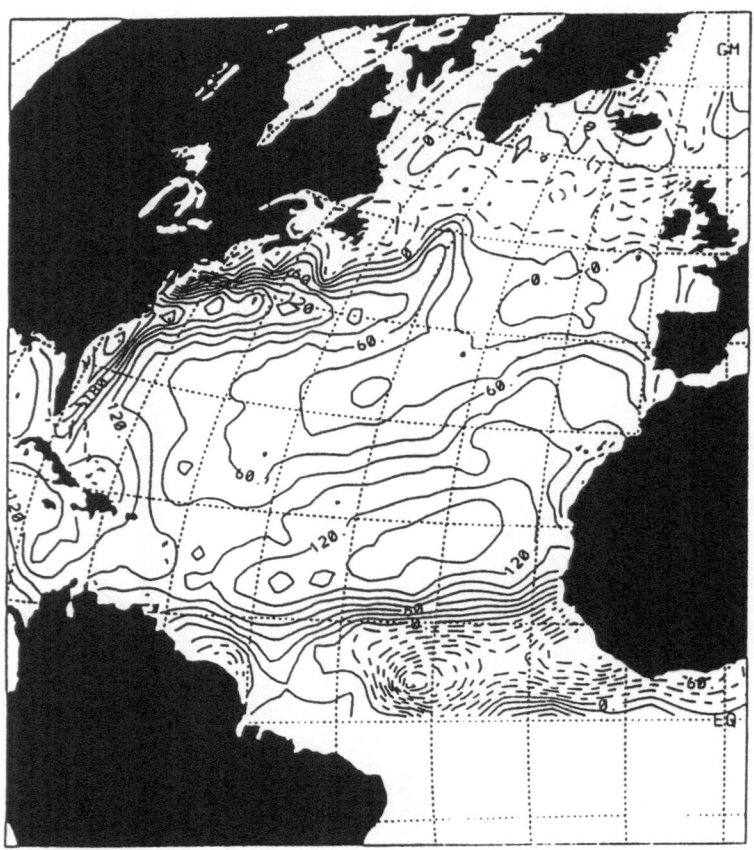

20. Estimate of net annual evaporation minus precipitation over the North Atlantic Ocean (after Schmitt et al., 1989). The North Atlantic is believed to be a region of net evaporation, which has important consequences for the resulting oceanic circulation.

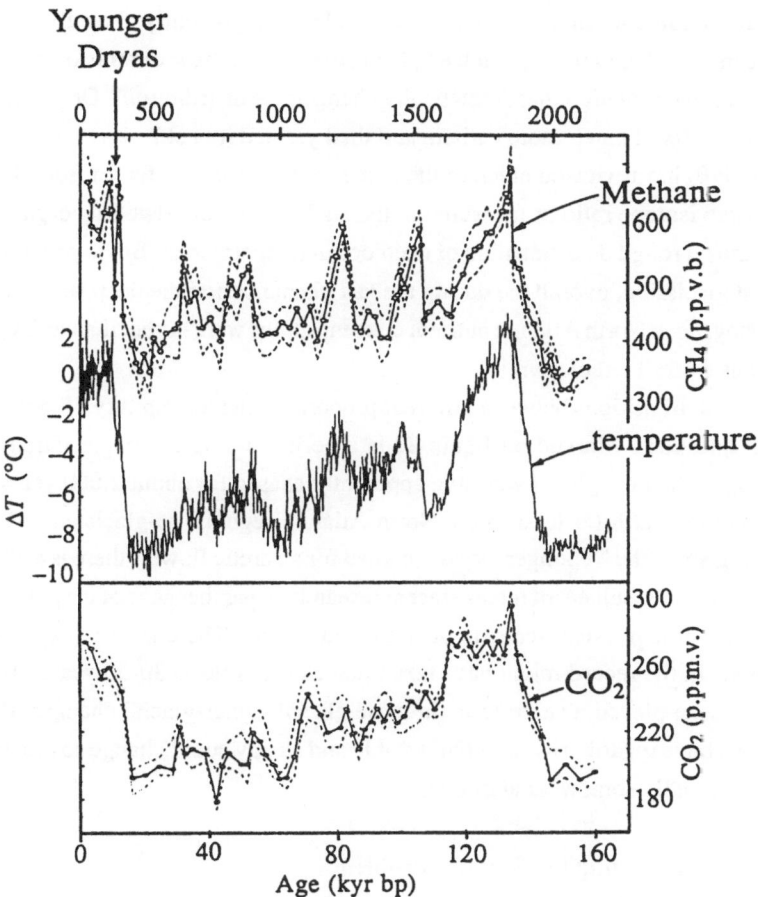

21. Estimates from Antarctic ice core (Vostok; Chappellaz, et al.,1990), of methane, temperature, and carbon dioxide content of the atmosphere over the pas 160,000 years. Note strong correlations of temperature with both CO_2 and methane, which are greenhouse gases. The massive amounts of "missing" carbon during the last glacial period about 18,000 years ago, are difficult to account for unless it was in the ocean. The Younger Dryas, which shows up most clearly in the methane curve, is indicated; this was a period in which the deglaciation ceased for about 2000 years, and the northern hemisphere (at least) cooled. The rapidity of this change is an important element in the climate discussion.

did all that carbon reside when it wasn't in the atmosphere? Apparently, the only place this carbon could have been stored, and then given back, is the ocean. But how did this occur? Was it some change in the ocean circulation that dramatically changed the distribution? Or was it some change in the biology that fixed much more carbon, and then yielded it back?

Boyle (1990) has reviewed much of the evidence that can be inferred from deep-sea cores. The same oxygen isotope ratio as measured in the shells of bottom dwelling organisms found in the cores, permits a rough determination of deep ocean temperatures. Boyle concludes that at the time of the last glaciation, overall the during the last glacial period the deep water was about $2°C$ cooler than today, deep North Atlantic nutrient concentrations were higher than today, and suggests a more vigorous overall circulation.

But there are transitions between different properties, and the rapidity of these transitions is an interesting question. Marked on figure 21 is a period during the deglaciation phase, about 12,000 years ago when the global warming apparently ceased for about 2,000 years, the glaciers re-advanced and the earth (at least in the North Atlantic regions) re-glaciated. This interval is known to geologists as the "Younger Dryas" (named for an arctic flower; there is an "Older Dryas" as well) and has been the subject of intense recent research, in part because of the publicity Broecker has given to a possible physical scenario of cause and effect. There is some evidence that many of the transitions in the record might have taken place in as little as 30-50 years. If we take as a given that the ocean played a central role in these glacial - inter-glacial changes, the possibility that a system we have regarded as essentially stable and steady could change so drastically in such a short time, is actually somewhat alarming.

4.3 Shorter Term Including the Sealevel Problem

The oceanic record really begins around 1870 (the longest extant instrument records are measurements of sealevel, beginning with the Dutch in the 1600's (e.g. Emery and Aubrey, 1991)). But given the Dutch propensity for modifying their shoreline, these records are not usable for studying trends in sealevel. A significant number of tide gauge records is available beginning about 1900 and a number of authors have attempted to determine whether there are trends in sealevel (Douglas, 1991; Barnett, 1990, Tushingham and Peltier, 1991). Although a simple problem to describe, because the instruments are basically very simple, the problem is a very difficult one, primarily because the continents do not represent stable platforms. Furthermore, the distribution of tide gauges is so irregular, that sampling considerations should make one suspicious of assertions that *global* trends could be computed.

Nonetheless, fig. 22 is Barnett's estimate of the trends in sealevel over the past 90 years. Assuming the trends are real, one confronts the question of what it means. Assuming that the tectonic contribution has been removed, sealevel reflects several physical properties:

(1) The total mass of the water in the ocean. A rise can occur owing to melting of glaciers, both continental and alpine.

(2) The mean temperature of the water. Water generally expands when warmed; seawater, if warmed at high pressures, expands more than it does at low pressures.

(3) The ocean circulation. Sealevel measured relative to one of the earth's gravitational equipotentials corresponds to the pressure field of the large scale ocean circulation. Fluctuations in the ocean circulation change the surface geostrophic currents, and hence sealevel.

This last contribution cannot change *global average* sea level; the second one could and the first one definitely does. The problem of interpreting global sealevel is compounded by the difficulties of separating these different contributions. From a societal point of view, there is a direct interest in future sealevel change, irrespective of its reflection of changing climate, and one would therefore like to better understand what has been going on for the past several hundred years. One might hope that knowledge of glacier movements would help us understand what fraction of the inferred rise can be attributed to melting. According to Meir (1984), the Greenland ice cap appears to be in balance, and the Antarctic cap may actually be growing; he attributes 1/3 to 1/2 of the observed sealevel rise to the incontrovertible melting of small mountain glaciers. This subject is fraught with difficulties.

What about other short-term climate changes? On a very short scale (less than a decade) there is one phenomenon, El Niño, which has been the focus of intense efforts for about 20 years (see Philander, 1990). Connections between El Niño and major meteorological shifts (droughts in normally wet areas, torrential rains in deserts, etc.) have been widely documented in the recent literature. From a climatological point of view, what is unsettling is that El Niño and related phenomena (particularly the so-called Southern Oscillation, SO) are not subtle. The Spanish Conquistadors evidently knew of El Niño from the middle of the 16th century; but it took 400 years to recognize that this obvious climatic phenomenon of the west coast of South America was but one element of an interesting, and major shift, in the entire global climate, one that directly connects oceanic physics to climate. The fact that it took so long to recognize what was going on,

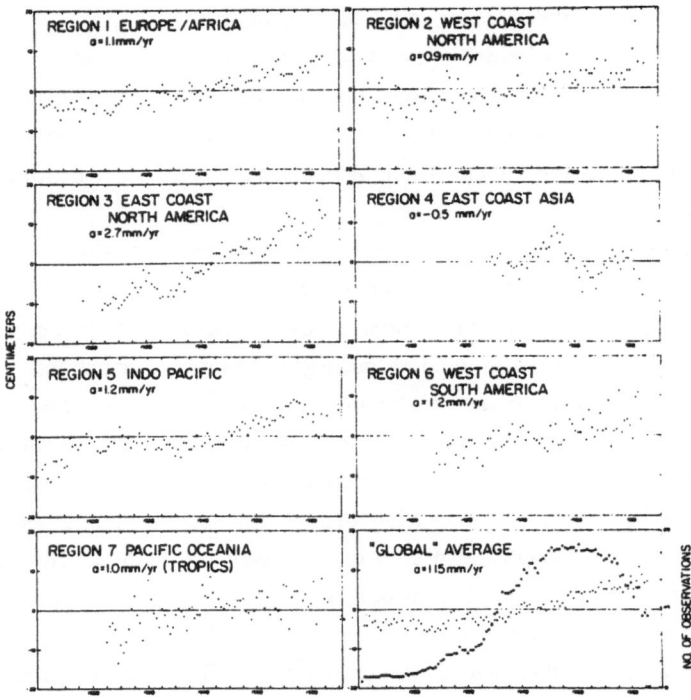

22. Estimated trends in regional sea level (Barnett, 1990). The results remain uncertain because of difficulties in computing the tectonic shifts of the gauges. Global trends are even more uncertain owing to the gross maldistribution of the tide gauge network.

leads one to ask: why? and what else is going on that may be of equal importance, but which remains unrecognized? We know very little, owing primarily to the very great difficulty of observations.

4.4 Heat and Moisture Fluxes

There are basically three methods for determining where, when, and at what rate heat, moisture, and various gases (e.g. CO_2) are exchanged between ocean and atmosphere. The first method, characterized by the work of Oort and Von der Haar (1976), is to compute the ocean storage, transport, and divergence as a residual of what purely atmospheric analyses produce. The Oort and von der Haar work was very important in calling attention to the apparently very large role of the ocean in the meridional flux of heat on a global basis. But residual methods are inherently very noisy; although they do not emphasize the point (relegating it to a footnote), the Oort and vonderHaar estimates for the ocean carry error bars overlapping zero meridional flux of heat.

One can try to estimate directly the transfer of heat between ocean and atmosphere on a point-wise or regional basis. There are many difficulties - the sign of the flux varies from day to day, and season to season and is made up of several different contributions, including long and short-wave radiation, conduction, evaporative fluxes, and is influenced by the local wind speed and cloud cover. Woods (1984), Isemer and Hasse (1985), and many others review the physics and methodologies. In regions like the subtropical Atlantic there are regions of net heating over the annual cycle, where current estimates suggest that the annual average is $O(150-200W/m^2)$. These estimates are made by calculating the different elements of the heating/cooling cycle (using so-called bulk formulas). The grave difficulty is that analysis suggests there are *systematic* errors approaching $50-80W/m^2$. Because the expected heating owing to the greenhouse warming is in the range of $4-10W/m^2$ there is a fundamental accuracy issue. Unfortunately, there is little prospect for any immediate improvement in these estimates.

One is left finally, and fundamentally (I believe) with the estimate of the air/sea fluxes from what the known ocean circulation *requires*. Consider the use of dynamic method calculations as discussed above. Given an estimate of the profile of velocity $\rho v_i(z_j)$ in station pair i, we can compute the meridional component of the mass flux owing to geostrophy as

$$M_g = \sum_{i,j} \rho v_i(z_j) a_{ij} \qquad (64)$$

where the summation is taken across the entire ocean basin and the a_{ij} are area elements (see eq. 45). Suppose to be specific, that we use the North Atlantic as an example, with the section in fig. 6d used, and that to a first approximation the ocean is regarded as closed to the north. Then the only other major contribution to the net meridional mass flux comes from the Ekman layer. Call the Ekman mass flux M_E. Then because we have assumed that the ocean is closed, it must be true that

$$M \equiv M_g + M_E = 0 \qquad (65)$$

How much heat does this mass flux carry? Since we know the water temperature, θ and the specific heat c_p we can write

$$H = \sum_{i,j} (\rho \theta_i(z_j) v_i(z_j) a_{ij} + \theta_E M_E \qquad (66)$$

where θ_E is the mean temperature of the fluid driven in the Ekman flux. A number of estimates (Hall and Bryden, (1982), Wunsch (1980), Roemmich (1980), Roemmich and Wunsch (1985)) have been made of the heat flux across 25°N in the Atlantic. They all agree that the result is about 1.1 PW poleward with a formal error of about ±0.1PW. If we divide these values by the area of the North Atlantic poleward of 25°N we find that *on the spatial average* the heat flux to the atmosphere must be about $50 \pm 9 W/m^2$ and must be transferred from the ocean to the atmosphere. This formal error is much smaller than that occurring in any of the residual or bulk formula estimates. Similar calculations have been made for other zonal sections (Wunsch, et al., 1983 in the South Pacific; Bryden et al., 1991 in the North Pacific; Fu, 1986 in the Indian Ocean; Rintoul, 1990 in the South Atlantic), where the formal errors are much larger, but nonetheless still much smaller than by any other means.

The point of all this is that a knowledge of the ocean circulation is probably going to be the only way in which we will determine the crucial air/sea fluxes which to a large extent govern the climate system. If we replace θ in eq (66) by any other scalar C we can similarly compute the flux of any other measured property, including such quantities as salt, nutrients and oxygen, and carbon. It should be plain however, that making such computations means we must first determine the oceanic general circulation, and that in turn requires vastly improved observations.

4.5 -Observations - The Future

I stipulate a few axioms:

(1) Shipboard measurements cannot solve the ocean-observing problem.

(2) Top-to-bottom observations are required.

(3) Any realistic ocean observing system will be built centrally around acoustical and spaceborne methods.

(4) Understanding the global ocean requires global synoptic observations.

A few comments are in order. Axiom (1) is probably self-evident. Shipboard measurements are too expensive, too manpower intensive and too slow to be the central element of a system purporting to adequately observe a global fluid. (This comment is *not* intended to imply that we will not continue to need ships.) I have already discussed the problem of observing the ocean in the absence of electromagnetic signalling capability. Any system which requires in situ observations to be obtained by recovering instruments which have stored information will be inadequate. First, one has no idea of what is going on until the instruments are recovered - meaning that one has neither information about whether the instruments are even working, nor any indication of what the climate system is doing. Second, instruments that need to be recovered tend to spend much of their working life ashore, awaiting re-deployment. The economics are against a vastly increased ship-borne system.

Axiom (2) is perhaps self-evident to oceanographers, but it is clearly not at-all-obvious to many others, including meteorologists. It is based upon the somewhat trivial observations that most of the climatologically important oceanic property fluxes (water, heat, nutrients, etc.) can only be computed as top-to-bottom integrals. Furthermore, the history of fluid dynamics (and we are dealing with the fluid dynamics of the ocean) does not support the pious hope that one can understand any fluid motion by *assuming* that large parts of the fluid volume are simply irrelevant to understanding of the piece one is most immediately interested in.

The observation problem is a difficult one. Historically, oceanographers have been slow to exploit the transparency of the ocean to sound. One can guess at the reasons. Underwater acoustics has been the domain of the military, with all that implies - the technology development has been dominated by military rather than civilian needs, and scientists have shied away from ocean acoustic science because of a combination of stigma, classification, and a sense that it was mere "engineering". But if one looks at the technology required to examine a global, electromagnetically opaque fluid, it seems hard to escape the conclusion that we need to become masters of acoustics - both for data transmission and direct probing.

What could a global ocean observing system look like? Any such system needs to be technically and economically feasible, and yet scientifically useful. The answers, I believe, like in understanding and investing heavily in certain technologies which are already with us, but which have not been developed to the point of true utility.

5 Quantitative Estimation Methods -Including Altimetry

Some of the estimation methods used by oceanographers will be familiar to geodesists, but using a different jargon. For some reason, what most of the world calls Gauss-Markov, or minimum variance estimation, is known to the geodetic community as "collocation", a term more familiar to oceanographers in its use by numerical modelers.

The central theme of this lecture is that observational "noise" or "error" is such an essential part of the real world, that it should be elevated to a central position in any discussion of what we know about the ocean. We have seen in an earlier lecture, that to a first approximation, the mathematical statement about the relationship between known ocean physics (and chemistry), and observations is reducible to a set of simultaneous equations of the very general form

$$\mathbf{E}\mathbf{x} + \mathbf{n} = \mathbf{y} \tag{67}$$

I would insist that the explicit inclusion of the "noise" term $\mathbf{n}$ is an essential step. To see that, suppose by good fortune, or supreme cleverness, that the matrix $\mathbf{E}$ is square, and has an inverse, so that if the equations were written conventionally as

$$\mathbf{E}\mathbf{x} = \mathbf{y}$$

we would be greatly tempted to write

$$\mathbf{x} = \mathbf{E}^{-1}\mathbf{y} \tag{68}$$

But this solution is generally, *wrong*, because it says $\mathbf{n} = \mathbf{0}$, which is hardly likely to be correct with any known observation method leading to determination of $\mathbf{y}$ - that is to say, no known (to me) measurement system of any type leads to perfect measurements.

Consider then the situation which the reader may feel is much more comfortable, in which we have more observations than knowns, like the β-spiral, so that $\mathbf{E}$ is of dimension $M \times N$, M > N, the classical overdetermined least-squares problem. The conventional solution is

$$\tilde{\mathbf{x}} = (\mathbf{E}^T\mathbf{E})^{-1}\mathbf{E}^T\mathbf{y} \tag{69}$$

assuming the matrix inverse exists, which produces information about the N elements of $\mathbf{x}$. Substituting this solution back into (67) produces an estimate

$$\tilde{\mathbf{n}} = \mathbf{y} - \mathbf{E}\tilde{\mathbf{x}} \tag{70}$$

which has M elements. So we have now have $M + N$ pieces of information, which we gained from only M equations. How can the situation be regarded as overdetermined then?

The answer is that the system is never actually overdetermined, unless we know for sure that the noise has some very special properties. Consider for example, that if we knew something about one of the noise elements in (70), that we would have to change the solution, $\tilde{x}$. I believe the best way to understand oceanographic estimation problems is to write the equations explicitly as in (67), and indeed to re-write them as

$$\mathbf{Ex} + \mathbf{n} = \mathbf{y}$$

$$\mathbf{E_1 x_1} = \mathbf{y}$$

$$\mathbf{E_1} = \{\mathbf{E} \quad \mathbf{I}_M\}, \quad \mathbf{x_1} = \begin{bmatrix} \mathbf{x} \\ \mathbf{n} \end{bmatrix}$$

so that the noise is formally part of the solution, $\mathbf{E_1}$ is dimension $M \times (M + N)$ and the equations are to be solved exactly.

If we adopt this point of view, several methods for estimating the solution are possible. Let us apply a version of this method to estimating the ocean circulation, so as to combine what we already know with altimetric information. Depending upon one's point of view, there are different reasons for doing this: (1) The altimetry + geodesy may improve on what is already known about the circulation, but there is little point in using altimetry to re-derive what we already know; (2) what we know about the ocean circulation can be used to improve altimetric estimates of the geoid, (3) both simultaneously - and is undoubtedly the central focus of future altimetric use. What follows is based upon Martel and Wunsch (1992, 1993).

The model is built mainly upon the following data and analyses:

(1) the hydrography, oxygen and nutrients of Fukumori et al. (1991), mapped as described in Fukumori and Wunsch (1991), using their so-called form-2 modes. Gridding data for any model raises a series of important practical questions some of which are taken up below.

(2) The wind stress compilation of Trenberth et al. (1989), obtained from the European Centre (ECMWF) global analyses. These data were averaged over 1980-86 and the resulting Ekman flux in the ocean estimated in the same way Trenberth et al. (1989) do.

(3) The water vapor flux estimates of Schmitt et al. (1989).

(4) The estimate of the annual surface heat flux climatology of Isemer, et al. (1989). Their value pinned to 1.2 PW (1.2×10^{15} W) at 25°N was used, leading as shown, to an estimate that most of

the North Atlantic is cooling on the seasonal average.

(5) River runoff climatology as estimated in the tables of Baumgartner and Reichel (1973).

(6) A compilation by N. Hogg (1990, personal communication) of long duration moored current meter records.

(7) The neutrally buoyant float observations, reduced to equivalent Eulerian velocities, and as discussed by Owens (1991).

(8) The absolute seasurface topography at long wavelengths as determined by Nerem et al. (1990).

The governing equation was, generically, a tracer conservation statement in the form

$$u_i \frac{\partial \rho C}{\partial x_i} = \frac{\partial}{\partial x_i} K_{ij} \frac{\partial \rho C}{\partial x_j} \tag{71}$$

for any scalar C and where summation over repeated indices is implied. The linearized inverse problem assumes the derivatives of ρC are perfectly known, with u, K_{ij} to be determined. Table 1 lists the layer depths chosen in the present computation.

Layer No.	Depth Range (decibars)
1	0-100
2	100-250
3	250-500
4	500-800
5	800-1100
6	1100-1400
7	1400-2000
8	2000-3000
9	3000-4000
10	4000-4500
11	4500-bottom

Table 1. Depth range of levels in model.

The velocity field appearing in (67) was assumed to be geostrophic, obeying the thermal wind equations as estimated from the gridded data set. The exception to this geostrophic balance occurs in layer 1, where an Ekman transport was added to the geostrophic flow. Apart from this surface layer, the horizontal velocities are of the form

$$u(i,j,k) = -\frac{g}{\rho f}\frac{\partial p}{\partial y} + b(i,j) \equiv u_R(i,j,k) + b(i,j) \tag{72a}$$

$$v(i,j,k) = \frac{g}{\rho f}\frac{\partial p}{\partial x} + c(i,j) \equiv v_R(i,j,k) + c(i,j) \tag{72b}$$

subject to continuity, eq (18).

Substituting into (71) results in a set of simultaneous equations, the formal unknowns of which are the "reference level velocities" $b(i,j)$, $c(i,j)$, the vertical velocity, w, and the unknown elements K_{xx}, K_{xy}, etc. of the mixing tensor, under the symmetry assumption, $K_{xy} = K_{yx}$, etc. The u_R, v_R are treated as known, but noisy.

Equation (71) was discretized in three dimensions as

$$\frac{u(i+1,j,k+1/2)\rho C(i+1,j,k+1/2) - u(i,j,k+1/2)\rho C(i,j,k+1/2)}{\Delta x} + \ldots +$$

$$\frac{w(i+1/2,j+1/2,k+1)\rho C(i+1/2,j+1/2,k+1) - w(i+1/2,j+1/2,k)\rho C(i+1/2,j+1/2,k)}{\Delta z} + \ldots$$

$$-K_{xx}(i+1,j,k+1/2)\frac{(\rho C(i+1,j,k+1/2) - 2\rho C(i,j,k+1/2) + \rho C(i-1,j,k+1/2))}{(\Delta x)^2} + \ldots$$

$$= 0 \tag{73}$$

The mixing terms in (71) require numerical second derivatives of the mapped fields - and it is desirable to reduce the differentiation of data insofar as is practical. The finite difference form (69) was thus integrated over a grid cell to produce a locally integrated balance:

$$\sum_{i,j,k \in \partial D} u(i+1,j,k+1/2)\rho C(i+1,j,k+1/2)a^x(i+1,j,k+1/2) -$$

$$u(i,j,k+1/2)\rho C(i,j,k+1/2)a^x(i,j,k+1/2) + \ldots$$

$$-K_{xx}(i,j,k+1/2)\frac{(\rho C(i,j,k+1/2) - \rho C(i-1,j,k+1/2))}{\Delta x}a^x(i,j,k+1/2) + \ldots$$

$$+n(i,j,k) = 0 \tag{74}$$

which involves only first derivatives of the data. The notation, $i,j,k \in \partial D$ denotes the indices defining the box boundary, ∂D, and $a^x(i,j,k)$ denotes the area of the interface normal to the x component of velocity $u(i,j,k)$ etc. $n(i,j,k)$ has been introduced to represent the inevitable noise in these "conservation" equations. Such noise elements are always present, even when not explicitly written.

Because of the existence of the mesoscale eddy field and other variability, the ocean tends to be noisiest on the smallest scales, i.e. at about 1° in the present resolution. On the other hand, the very large-scale hydrography of the North Atlantic is known to be very stable through time, and to consist of features (e.g., the whole subtropical gyre) which are well-determined. The signal to noise ratio on the largest scales appears to be significantly greater than on the 1° scale. The grid was thus thought of as a nested-one: the horizontal integration over the grid cell was taken over the entire North Atlantic as depicted in figure 23, as though the whole ocean were a single cell. The vertical integration was however, taken over each layer separately. But the flow in and out of this cell is written at a horizontal resolution of 1° along the zonal boundaries. This largest cell was then subdivided into four subcells as depicted in the figure, and the integration taken over each of them (at this stage there would be 5 basic balances written, including one redundant balance for each layer). The subdivision then continues until one reaches the maximum resolution of 1°, the computer capacity is exceeded, or the data noise becomes intolerable. In the present case, it was this last factor which was limiting. Recall that the error in the gridded data tends to grow away from the actual observation locations (see the error map in Fukumori et al., 1991). An attempt was made to keep the major sub-division lines close to the original hydrographic observation lines so as to minimize the error in the model. The subdivision stopped at the limit shown in figure 23 when we concluded that the balances (74) would be dominated by error.

The final model is thus on the grid in fig. 23 in a domain extending from 80°W to 10°W, and from 12°N to 60°N. At the final subdivision, the smallest regions are about 9° of longitude by 6° of latitude. Because of the complexity of topography and coastline, a large number of special treatments had to be accorded to various boxes at various steps of the subdivision. The final resolution is perhaps best described as a mixed-one. The resolution for the horizontal flows is 1° along each of the bounding surfaces, but the balance equations (constraints) are applied only on the box scales shown.

Conservation[2] equations were written for mass, heat, salt, oxygen, nitrate, phosphate and silicate in each cell. In addition, equations representing the net (i.e. zonally integrated) meridional flux of each property for each layer were written for latitudes at 6° intervals beginning with $12°N$. The constraint at $60°N$ includes a segment at $10°W$ between $54°N$ and $60°N$ necessary to close the northern boundary of the ocean. Further constraints were written at 60°N for the sum of layers 1 to 4 (above 800 decibars), and for layers 6 to 11 (below 1100 decibars) to diagnose the surface inflow and deep outflow to and from the Norwegian Sea.

Additional requirements were added that the Florida Straits mass flux should be $30 \pm 1 \times 10^9$

---

2 More precisely, "near-conservation", or "balance equations"

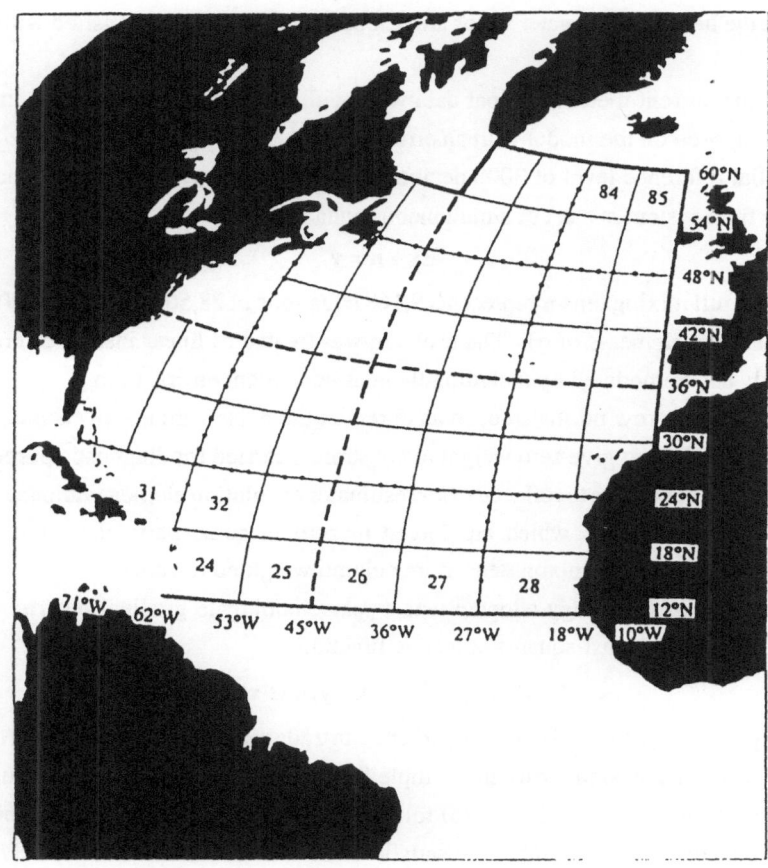

23. Model domain and bounding boxes used in inverse calculation of Martel and Wunsch (1993). Constraints based upon conventional oceanographic knowledge are written for the largest available box (entire North Atlantic) and then a series of nested sub-boxes is used. Basic resolution along box boundaries is 1°.

kg/s and that the heat, runoff, water vapor and Ekman fluxes should be satisfied within the stated error bars.

Where the current meter and float data were within a half-degree of a grid-line, the mean values were imposed on the model with an error estimate.

An initial reference level of 3000 decibars (or the bottom which ever was shallower) was chosen. The final system is a set of simultaneous equations

$$Ex + n = y \qquad (75)$$

which with the full mixing tensor represents 9,168 equations in 28,502 unknowns x (not counting the 9,168 unknown elements of n). The problem was treated as linear insofar as errors in E are only implicitly accommodated by including them as equivalent errors, n, in y.

The system was row normalized, so as to render the noise variance in each equation as far as possible identical, or to give zero weight to constraints carried for diagnostic purposes only. It was column normalized so as to reflect *a priori* estimates of solution element variance or to remove from the solution unknowns which are forced to zero in some particular solution (e.g., the off-diagonal elements of the mixing tensor are column weighted to zero).

The solution methodology adopted was a sparse conjugate gradient determination of the minimum of the tapered least-squares objective function

$$J = (Ex - y)^T W^{-1}(Ex - y) + \alpha^2 x^T S^{-1} x \qquad (76)$$

where $W = <nn^T>$, $S = <xx^T>$. By varying α^2 one can trade model residuals (the first term) against the solution norm (the second term), in a simple form of ridge regression (e.g., Wunsch, 1978), and the result is equivalent to solving (75) for differing a priori estimates of the noise variance. α^2 was chosen so that all constraints were satisfied within an accuracy of about 1 Sv.

The reference velocities are shown in fig. 24. We refer to Martel and Wunsch (1993) for a detailed description, but the most conspicuous features is the so-called deep western boundary current (DWBC) on the northern and western sides.

An estimate of the absolute surface topography, relative to the geoid is depicted in figure 25. Figure 26 is the surface elevation relative to the geoid in the North Atlantic extracted from the global estimate of Nerem et al. (1990) resulting from a joint solution for orbit, gravity and seasurface topography. Their estimate was computed as a spherical harmonic expansion to degree and order 10. No *a priori* oceanographic information was employed other than an estimate of the variance of the surface topography. Some elements of the subtropical and subpolar gyres are readily visible, and one of the questions we seek to answer is whether these features are consistent

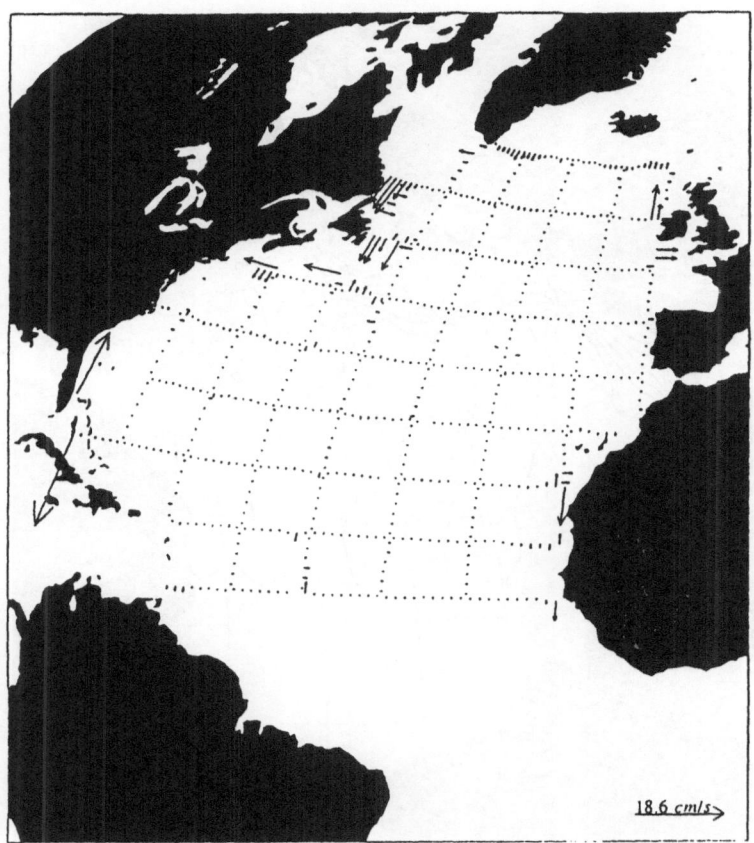

24. Reference level velocity (about 3000 m depth) computed from "inversion" of constraints, as described in Martel and Wunsch (1993). These velocities imply surface topography slopes.

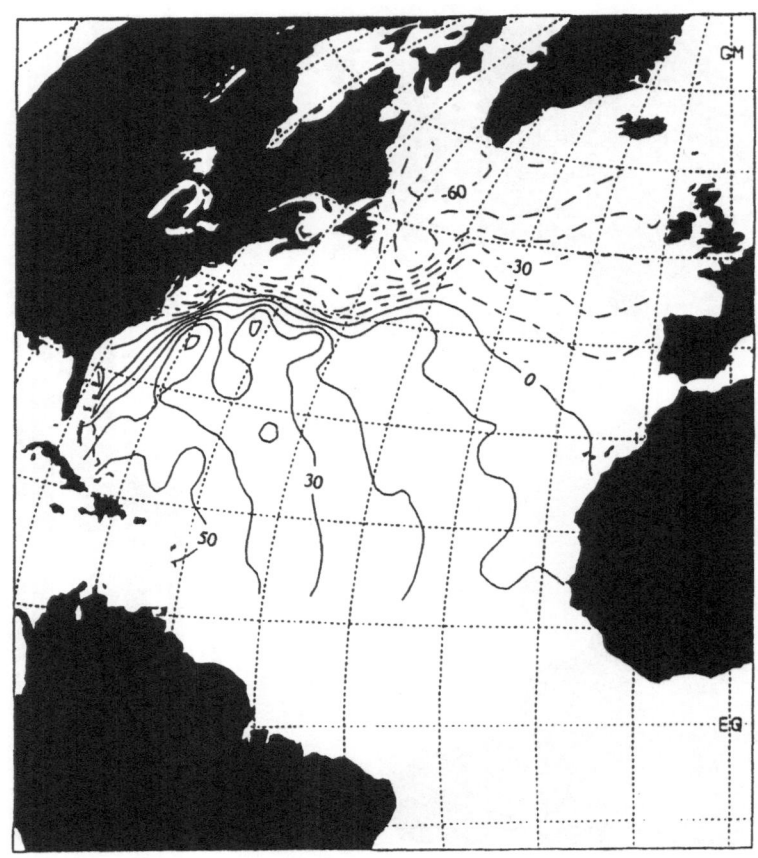

25. Absolute topography of the seasurface (up to an unknown constant) relative to the geoid, as computed from the velocity field in fig 24, and the thermal wind equations (Martel and Wunsch, 1992). Contour interval is 10 cm.

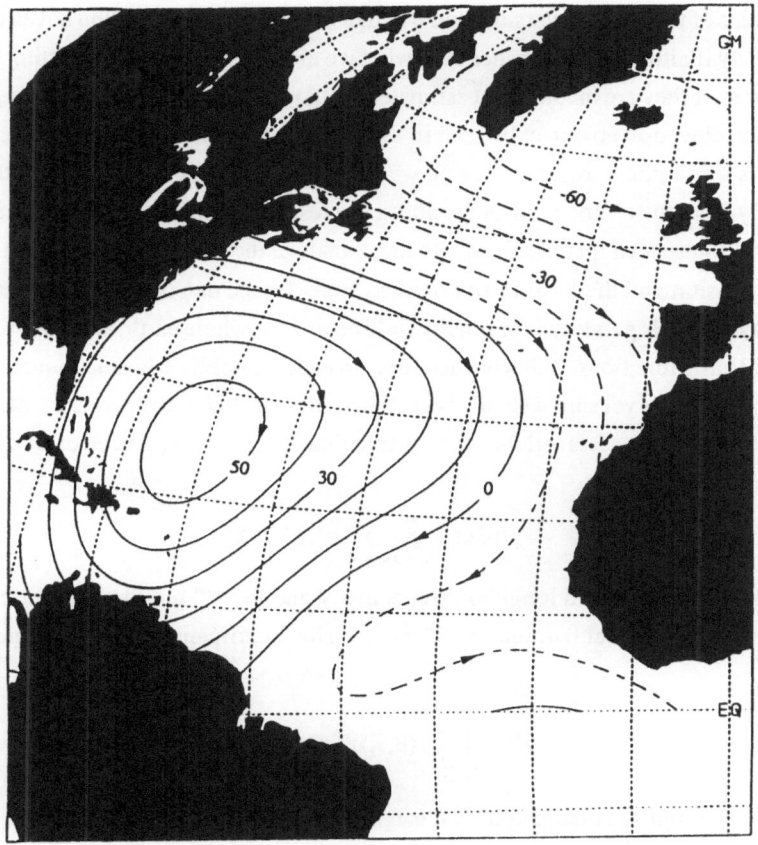

26. Estimate of absolute seasurface elevation relative to the geoid as estimated by Nerem et al. (1990) and which should be compared to fig. 25. Contour interval is again 10 cm.

with the physics of the *in situ* inversion, and whether they contain information not already included in it. Obviously if altimetry is no better than what we already know, we are wasting our time with it; and of course, if they are inconsistent estimates we have a real problem. The Nerem et al. (1990) estimate of the elevation error can be seen in their paper as Plate 3, with a value around 16 cm.

There is a qualitative resemblance between the *in situ* inversion surface and that computed independently from the altimetry. But the altimetric result is from a truncated spherical harmonic expansion and to proceed quantitatively we must analyze the relationship between the spherical harmonic expansion and the local *in situ* model. Here we see one of the zero-order problems of oceanography vis-à-vis geodesy - the gravity field covers the whole earth, and spherical harmonics are the natural basis functions. But what does one do with the ocean, which at best covers a fraction of the earth? We aren't yet sure what the best approach is, and what follows is a tentative solution.

The altimetric data set is a global one. A spherical harmonic expansion of an arbitrary global field, $f(\theta, \lambda)$, is

$$f(\theta, \lambda) = \sum_{n=0}^{\infty} \sum_{m=-n}^{n} \alpha_n^m Y_n^m(\theta, \lambda) \tag{77}$$

where θ, λ are co-latitude and longitude respectively and the Y_n^m are the conventional, complex (fully-normalized) spherical harmonics. The expansion coefficients α_n^m are obtained as a global integral,

$$\alpha_n^m = \int_0^{2\pi} \int_0^{\pi} f(\theta, \lambda) Y_n^m(\theta, \lambda)^* d\theta d\lambda \tag{78}$$

the * denoting complex conjugation. (We ignore the issue of completeness of the spherical harmonics of the first kind when used to expand a field defined only over the ocean.) If the sum in (77) is truncated to degree and order $m = n = 10$ we can write the result as

$$\bar{f}(\theta, \lambda) = \int_0^{2\pi} \int_0^{\pi} W(\theta, \lambda, \theta', \lambda') f(\theta', \lambda') d\theta' d\lambda' \tag{79}$$

where the averaging kernel

$$W(\theta, \lambda, \theta', \lambda') = \sum_{n=0}^{10} \sum_{m=-n}^{m=n} Y_n^m(\theta', \lambda') Y_n^m(\theta, \lambda)^* \tag{80}$$

This function is displayed in fig 27 for a location centered in the North Atlantic (there is a small location dependence). Apart from a weak antipodal secondary maximum, the truncated spherical harmonic expansion is a *local* spatial average. Because the altimetry contains no information about high wavenumbers, it is necessary to filter the model surface elevation to remove the short scales. The altimetry is global, and the model is local, and obtaining complete consistency in the filter pass-bands is difficult because of the boundary effects. An approximate filter averaging over 30° latitude by 30° of longitude was applied. The result is displayed in fig 28.

Both the altimetry and the spatially averaged model show the subtropical gyre structure, but it is much less intense in the model than as inferred from the altimeter data. The model shows a more developed subpolar gyre than does the altimetry, which more resembles an extension of the subtropical gyre into the area north of the North Atlantic Current. The maximum gradient between the centers of the gyres are about 0.55 m for the model and 1.1 m for the altimeter data. Note also the northwestward shift of the center of the subtropical gyre in the altimeter-derived pattern, associated with a strong southwestward surface flow in the middle and eastern part of the basin.

Given the two independent estimates of the seasurface topography/velocity, we can ask two central questions: (a) Are the altimetric and *in situ* observations mutually consistent? (b) If the answer to (a) is "yes", do the altimetric observations add any new information to the circulation as deduced from the conventional observations alone?

The easiest way to simultaneously answer both these questions is to re-solve the *in situ* model with the addition of the altimetric observations. Several options exist for combining quantitatively the altimetric and hydrographic constraints. In the present case, it proves most convenient to differentiate the altimetric elevation estimates, $\overline{h}$, to produce altimetric estimates of the surface geostrophic velocities as

$$\hat{\mathbf{v}}_{alt} = \frac{g}{f}\frac{\partial \overline{h}}{\partial s} \equiv \frac{g}{f}\left(-\frac{\partial \overline{h}}{a\,\partial \theta}, \frac{\partial \overline{h}}{a\,\sin\theta\,\partial\lambda} \right) \tag{81}$$

$f = 2\Omega\cos\theta = 2\Omega\sin\phi$ is the Coriolis parameter, and co-latitude, θ, rather than latitude is being used. The thermal wind velocities were then subtracted from $\hat{\mathbf{v}}_{alt}$ to produce a new set of constraints

$$\mathbf{E}_{alt}\mathbf{x} + \mathbf{n}_{alt} = \mathbf{y}_{alt} \tag{82}$$

on the reference level velocities - 33 constraints for the zonal ones and 33 for the meridional ones. The rows of $\mathbf{E}_{alt}$ represent the two-dimensional spatial averages of the unknown reference level velocities and are the same filtering operation as applied to the surface elevation, i.e. averaged over areas of 30° latitude by 30° longitude and computed every 10° of latitude and longitude. At

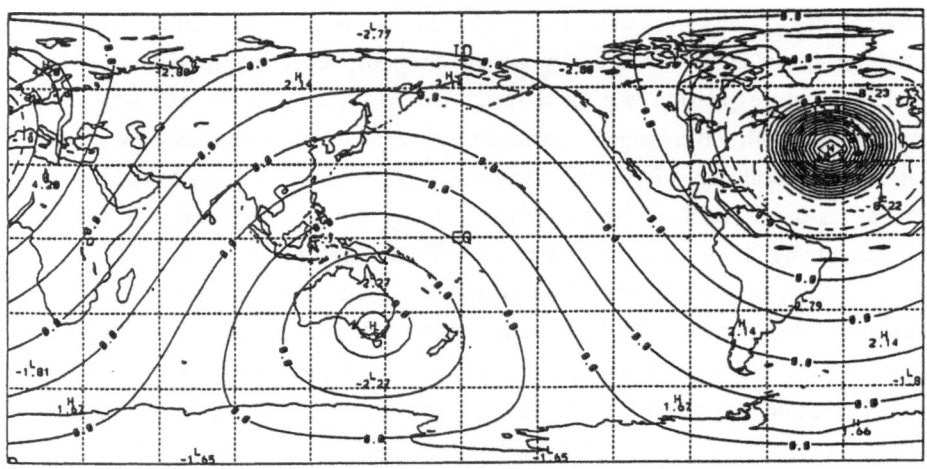

27. Spatial averaging function equivalent to truncating a spherical harmonic expansion at degree and order 10.

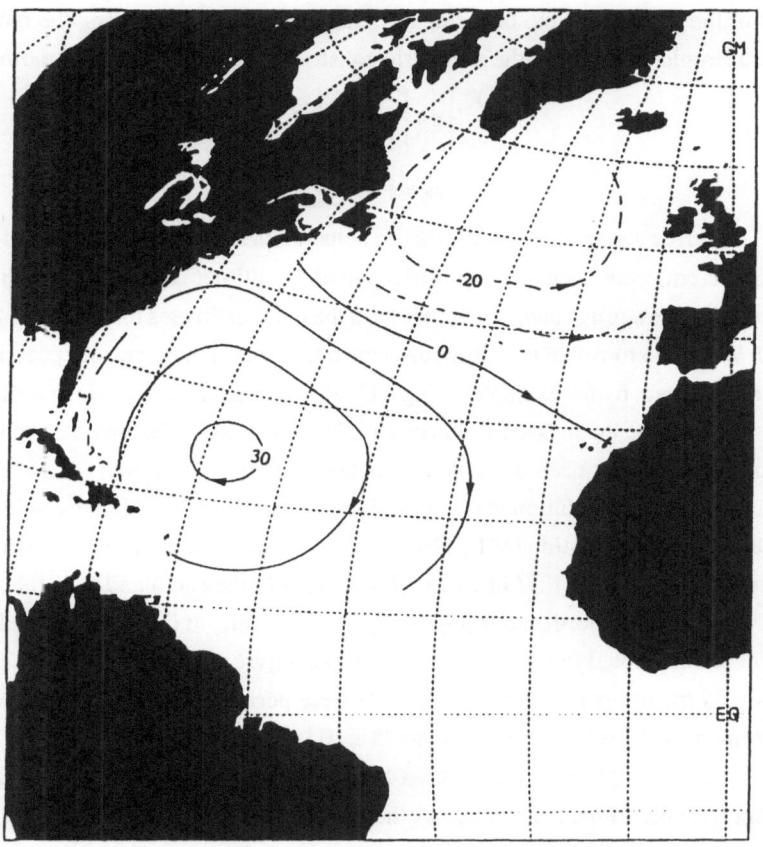

28. Seasurface topography in fig 25 approximately low-pass filtered spatially to be consistent with wavenumber content of fig. 26 .

least one half of the window had to be inside the smallest rectangle enclosing the model domain.

It proved simplest to append the altimetric constraints (82) to (75) in a combined system

$$\begin{Bmatrix} \mathbf{E} \\ \mathbf{E}_{alt} \end{Bmatrix} \mathbf{x} + \begin{bmatrix} \mathbf{n} \\ \mathbf{n}_{alt} \end{bmatrix} = \begin{bmatrix} \mathbf{y} \\ \mathbf{y}_{alt} \end{bmatrix}$$

$$\equiv \mathbf{E}_c \mathbf{x} + \mathbf{n}_c = \mathbf{y}_c \tag{83}$$

and re-solve. The error variance for the altimetric constraints used here, as computed from values provided by S. Nerem, were used directly. Strict use of the altimetric constraints requires that we properly account for the strong spatial correlations in the noise estimates for the altimetric velocity. One approach would be to rotate the altimetric constraints into a new vector space in which the errors were uncorrelated, by for example, using a Cholesky decomposition of the error covariance matrix. But because the error issue is fairly simple, as discussed below, this more elaborate computational recipe has not been deemed worthwhile at the present time.

(1) In this inversion, a solution was required that reproduced the altimetric velocities within the one standard error values estimated by Nerem et al. (1990). This solution forced a geostrophic mass flux across 30°N east of 71°W of $148 \pm 14 \times 10^9$ kg/s to the south and at 18°N east of 62°W of $197 \pm 15 \times 10^9$ kg/s to the north and is physically unacceptable. It is the direct result of the very strong gradient already noted in the altimetric subtropical gyre.

(2) When all the altimetric velocity residuals were permitted to rise to be twice as large as before, the strong mass fluxes were reduced, to $78 \pm 10 \times 10^9$ kg/s to the south across $30°N$ east of $71°W$ and $107 \pm 12 \times 10^9$ kg/s northward across $18°N$ east of $62°W$, which are still much too large to be consistent with the Ekman fluxes across these latitudes.

(3) When the altimetric errors were permitted to rise to five times those estimated by Nerem et al. (1990) the mass flux at $30°N$ east of $71°W$ was reduced to $49 \pm 9 \times 10^9$ kg/s to the south, which is acceptable and the mass flux across $18°N$ east of $62°W$ is $31 \pm 12 \times 10^9$ kg/s and possibly acceptable.

The reader should not however, infer that the errors in the altimetric estimates are five times larger than those estimated by Nerem et al. (1990). Strongly correlated errors, as are present in the altimetry, produce systematic errors in the transport constraints we are using to diagnose the ocean circulation. The central problem is that the altimetric elevation depicted in Figure 26 has a steep pressure head, about 85 cm. at 25°N along the zonal line from about 60°W to 10°W. In contrast, the *in situ* model produces a head of only about 25 cm over the same distance. The altimetry produces southward zonal average meridional surface velocity across this line of about 3.4 cm/s, whereas the model seems consistent with little more than about 1.2 cm/s. The formal error on the latter is somewhat smaller than 1 cm/s, as it is a zonal average across 50 degrees of longitude. Without using the full error covariance matrix for the altimetric velocities, we cannot rigorously estimate the zonal average error from the point errors of 2-3 cm/s stated above. But a standard error of 3-5 cm/s for the velocity average would clearly render the two results consistent. (After this work was completed, S. Nerem, private communication, September 1991, provided us with an improved altimetric estimate, fig. 29. In the sub-tropical gyre, the new estimate is quite similar to the model result, with the gradient considerably reduced. But in the subpolar gyre, the remaining gradient appears to be still somewhat too large to be consistent with the model ocean circulation.)

There are two simple conclusions. (1) The 2-4 cm/s rms error estimated by Nerem et al (1990) for their long-wavelength absolute altimetric velocity in this region appears optimistic by about a factor of 2. (2) The existing uncertainty of the ocean circulation is somewhat better than 1 cm/s rms on a roughly 1° scale, as inferred from the MW model. Thus at least on the scale of the North Atlantic Ocean, the altimetric accuracies must improve by about a factor of 2 in standard error before they can make a major difference in our ability to estimate the oceanic general circulation.

Although this last statement may sound rather pessimistic, one must recall that the GEOSAT mission data, upon which the present estimates were based, were generated by a system designed for a wholly different purpose without regard to general circulation studies. That we are already so close to the necessary accuracies, with what by present standards is a crude instrument system, is greatly encouraging. Had this study been carried out in the Pacific Ocean, it is probable that the altimetric information would have appeared much more useful; that ocean is sufficiently large that the greater accuracy of the geoid at long wavelengths, and the corresponding ignorance of the ocean circulation, would conspire to show utility of altimetry even in its present crude state.

In effect, the present study confirms the design parameters of the TOPEX/POSEIDON mission system - they are both realistic and necessary. That system includes not just the altimetric instrument but necessary improvements in the gravity field estimates and in a whole suite of physical corrections; see, e.g. Stewart et al. (1986).

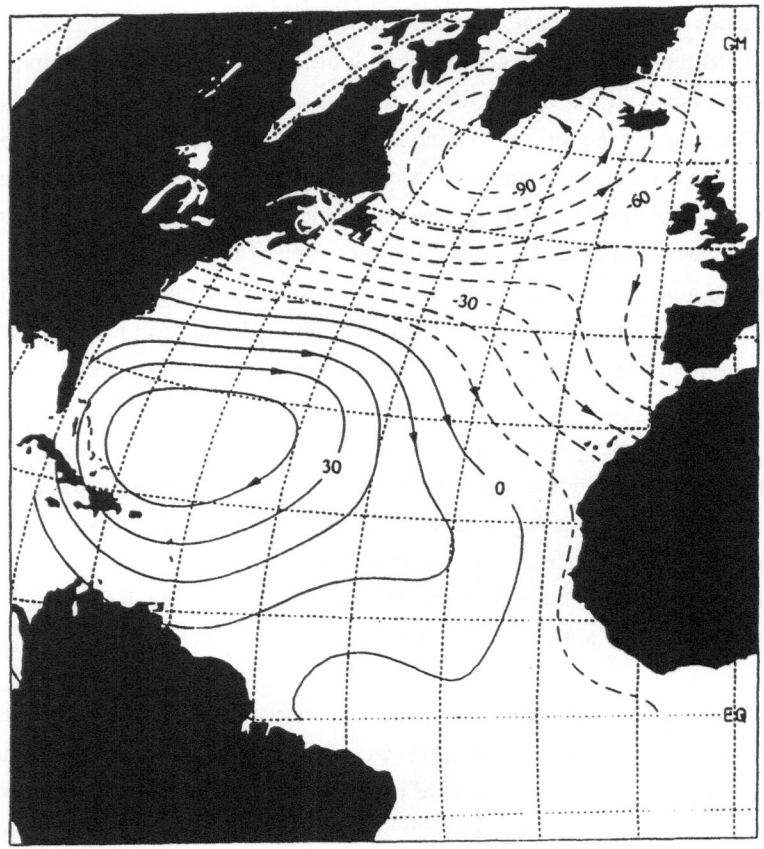

29. Revised version of fig. 26 (S. Nerem, private communication, 1991). Spatial gradients appear to be still too large to be consistent with the conventional oceanographic constraints.

A Final Word

To the oceanographer, the great attraction of altimetry, like all spaceborne measurements, is its global coverage - something we are unlikely ever to get from ships or even unmanned, in situ observations - the ocean is just too large, and the expense of measuring it too great. Like other space measurements, altimetry is confined to measurements of the seasurface. But because the surface elevation reflects oceanic motions to great depths, it carries information about the ocean unavailable from any other known physical phenomenon measurable from space. But the signals are small, less than 1 m, in the presence of geoidal variations of $O(100 \text{ m})$, and are masked by a host of geophysical noise sources described in other lectures. The extraction of the oceanographic signals from the geoidal and noise variations is one of the most formidable data handling problems known to me. On the other hand, altimetry has clearly worked a revolution in global determination of the earth's shape, especially over the sea with many of the remaining errors lying in oceanographic signals. Because small fluctuations in the geoid are believed to reflect vitally important physical processes deep within the earth, the removal of the oceanographic contamination becomes of great interest to the geophysical community. What it all means is that geodesists and oceanographers are fated to work closely together for a long time to come.

Acknowledgements. Supported in part by the National Aeronautics and Space Administration Grant NAGW-918, and the Jet Propulsion Laboratory, Contract 958125. Contribution from the World Ocean Circulation Experiment.

References

Bainbridge, A. E., 1980. GEOSECS Atlantic Expedition. Vol. 2, Sections and Profiles. National Science Foundation, U. S. Government Printing Office, Washington DC, 198 pp.

Baker, D. J, jr., 1981. Ocean instruments and experiment design. in Evolution of Physical Oceanography. Scientific Surveys in Honor of HenryStommel, B. A. Warren and C. Wunsch, The MIT Press, Cambridge,396-433.

E. I. Balazs and B. C. Douglas, 1979. Geodetic leveling and the sealevel slope along the California coast. J. Geophys. REs., 84, 6195-6206.

Barnett, T. P., 1990. Low frequency changes in sealevel and their possible causes. in, The Sea, Vol. 9, B, B. L. LeMehaute and D. M. Hanes, eds, Wiley-Interscience, New York, 841-867.

Bauer, J. and J. D. Woods, 1984. Isopycnic Atlas of the North Atlantic Ocean. Tech. Rept. No. 132, Institut for Meereskunde und der Christian-Albrechts Universitat, Kiel, 173 pp.

Baumgartner, A. and E. Reichel, 1975. The World Water Balance. Elsevier, Amsterdam, 179 pp.

Berger, W. H. and L. D. Labeyrie, eds., 1987. Abrupt Climate Change. Evidence and Implications. NATO ASI Series, Vol 216. Reidel, Dordrecht, 425 pp.

Born, G. H., B. D. Tapley, J. C. Ries and R. H. Stewart, 1986. Accurate measurement of mean sea level changes by altimetric satellites. J. Geophys. Res., 91, 11,775-11,782.

Boyle, E. A., 1990. Quaternary deepwater paleoceanography. Science, 249, 863-870.

Broecker, W. S. D. Peteet and D. Rind, 1985. Does the ocean-atmosphere system have more than one stable mode of operation? Nature, 315, 21-26.

Broecker, W. S., T. Takahashi and Y.-H. Li, 1976. Hydrography of the central Atlantic-I. The two-degree discontinuity. Deep-Sea Res., 23, 1083-1104.

Bryan, F. O. and W. R. Holland, 1989. A high resolution simulation of the wind- and thermohaline-driven circulation in the North Atlantic Ocean. in, Parameterization of Small Scale Processes, Proceedings 'Aha Huliko'a, Hawaiian Winter Workshop, U. of Hawaii at Manoa, January 17-20, 1989, Hawaii Inst. of Geophys. Spec. Pub., P. MÜller and D. Henderson, eds., 99-115.

Bryden, H. L., D. H. Roemmich, and J. A. Church, 1991. Oceanic heat transport across 24°N in the Pacific. Deep-Sea Res., 37, 297-324.

Bryden, H. L., 1977. Geostrophic comparisons from moored measurements of current and temperature during the Mid-Ocean Dynamics Experiment. Deep-Sea Res., 24, 667-681.

Bryden, H. L., 1980. Geostrophic vorticity balance in midocean. J. Geophys. Res., 85, 2825-2828.

Chappellaz, J. M., D. Raynaud, Y. S. Korotkevich, and C. Lorios, 1990. Ice-core record of atmospheric methane over the past 160,000 years. Nature, 345, 127-131.

Davis, R. E., 1978. Estimating velocity from hydrographic data. J. Geophys. Res. 83, 5507-5509.

Deacon, M., 1979. Scientists and the Sea. 1650-1900. A Study of Marine Science 445 pp., Academic, London

Defant, A., 1961. Physical Oceanography, Vol.1. Pergamon, N.Y., 598 pp.

Douglas, B. C., 1991. Global sea level rise. J. Geophys. Res., 96, 6981-6992.

Drake, C. L., J. Imbrie, J. A. Knauss, and K. K. Turekian, 1978. Oceanography. Holt, Rinehart and Winston, New York, 447 pp.

Ekman, V. W., 1905. On the influence of the earth's rotation on ocean-currents. Arkiv for Matematik, Astronomi och Fysik 2(11), 52 pp.

Emery, K. O. and D. G. Aubrey, 1991. Sealevels, land levels and tide gauges. Springer-Verlag, New York, 237 pp.

Fu, L., 1986. Mass, heat and freshwater fluxes in the South Indian Ocean. J. Phys. Oc., 16, 1683-1693.

Fukumori, I. and C. Wunsch, 1991. Efficient representation of the North Atlantic hydrographic and chemical distributions. Prog. in Oceanog., 27, 111-195.

Fukumori, I., F. Martel and C. Wunsch, 1991. The hydrography of the North Atlantic in the early 1980's. An atlas. Prog. in Oceanog., 27, 1-110.

Gill, A. E., 1982. Atmosphere-Ocean Dynamics. Academic Press, New York, 662 pp.

Hall, M. M. and H. L. Bryden, 1982. Direct estimates and mechanisms of ocean heat transport. Deep-Sea Res., 29, 339-359.

Hamann, I. M. and J. H. Swift, 1991. A consistent inventory of water mass factors in the intermediate and deep Pacific Ocean derived from conservative tracers. Deep-Sea Res., 38 (Supplement. J. L. Reid Volume), S129-S170.

Hastenrath, S., 1984. On meridional heat transports in the world ocean. J. Phys. Oc., 12, 922-927.

Heinmiller, R. H., 1983. Instruments and methods. in, Eddies in Marine Science, A. R. Robinson, ed., Springer-Verlag, Berlin, 542-567.

Helland-Hansen, B. and F. Nansen, 1920. Temperature variations in the North Atlantic Ocean and in the atmosphere. Smithsonian Misc. Collect. 70:4, 408pp.

Holloway, G., 1986. Eddies, waves, circulation and mixing: statistical geofluid mechanics. Ann. Revs. Fl. Mech., 18, 91-147.

Isemer, H.-J. and L. Hasse, 1985. The Bunker Climate Atlas of the North Atlantic Ocean. Vol 1. Observations. Vol 2 Air-Sea Interactions. Springer-Verlag, Berlin, 218pp and 252pp.

Isemer, H. J. Willebrand and L. Hasse, 1989. Fine adjustment of large scale air-sea energy flux parameterizations by direct estimates of ocean heat transport. J. Climate, 2, 1173-1184.

Kawase, M. and J. L. Sarmiento, 1986. Circulation and nutrients in middepth Atlantic waters, J. Geophys. Res., 91 9749-9770.

Knapp, G. P. and H. Stommel, 1985. Hydrographic Data from R/V Oceanus Cruse 129. WHOI 85-38, Woods Hole Oceanographic Institution Technical Report, Woods Hole, MA, 107.

Lambeck, K., 1988. Geophysical Geodesy. Oxford U. Press, New York, 718 pp.

Levitus, S., 1982. Climatological Atlas of the World Ocean. NOAA Professional Paper 13, 173 pp.

Longuet-Higgins, M. S., 1964. Planetary waves on a rotating sphere. Proc. Roy. Soc. A, 279, 446-473.

Mackas, D. L. K. L. Denman, and A. F. Bennett, 1987. Least squares multiple tracer analysis of water mass composition. J. Geophys. Res., 92, 2907-2918.

Martel, F. and C. Wunsch, 1992. Combined inversion of hydrography, current meter data and altimetric elevations for the North Atlantic circulation. unpublished document.

Martel, F. and C. Wunsch, 1993. The North Atlantic circulation in the early 1980's - an estimate from inversion of a finite difference model. J. Phys. Oc., in press.

Maury, M. F., 1855. The Physical Geography of the Sea and Its Meteorology. Harper and Bros., New York. (reprinted by Harvard University Press, J. Leighly ed., 1963), 432pp.

McWilliams, J. C. and G. R. Flierl, 1976. Optimal, quasi-geostrophic wave analysis of MODE array data. Deep-Sea Res., 23, 285-300.

Meir, M. F., 1984. Contribution of small glaciers to global sea level. Science, 226, 1418-1421.

Millero, F. J., C.-T. Tung, A. Bradshaw and K. Schleicher, 1980. A new high pressure equation of state for seawater. Deep-Sea Res., 27A, 255-264.

MODE Group, The 1978. The Mid-ocean dynamics experiment. Deep-Sea Res. 25, 859-910.

Munk, W., 1950. On the wind-driven ocean circulation. J. of Meteor., 7, 79-93.

Munk, W., 1981. Internal waves and small-scale processes. in Evolution of Physical Oceanography. Scientific Surveys in Honor of Henry Stommel, B. A. Warren and C. Wunsch, eds, The MIT Press, Cambridge, Ma, 264-291.

Munk, W. H. and G. F. Carrier, 1950. The wind-driven circulation in ocean basins of various shapes. Tellus, 2, 158-167.

Needler, G. T., 1978. The absolute velocity as a function of conserved measurable quantities. Prog. Oceanog., 14, 421-429.

Nerem, R. S., B. D. Tapley and C. K. Shum, 1990. Determination of the ocean circulation using Geosat altimetry. J. Geophys. Res., 95, 3163-3180.

Oort, A. H. and T. H. Vonder Haar, 1976. On the observed annual cycle in the ocean-atmosphere heat balance over the northern hemisphere. J. Phys. Oc., 6, 781-800.

Owens, B., 1991. A statistical description of the mean circulation and eddy variability in the Northwestern Atlantic using SOFAR floats. Prog. Oceanog., 28, 257-303.

Pedlosky, J., 1987. Geophysical Fluid Dynamics, Second edition. Springer-Verlag, 710 pp.

Philander, S. G., 1990. El Niño, La Niña, and the Southern Oscillation. Academic, San Diego, 289 pp.

Phillips, N. A., 1963. Geostrophic motion. Revs. Geophys., 1, 123-176.

Phillips, N. A., 1966. The equations of motion for a shallow rotating atmosphere and the "traditional approximation" J. Atm. Scis., 23, 626-628.

Pickard G. L. and W. Emery, 1982. Descriptive Physical Oceanography. An Introduction. 4th Edition. Pergamon, Oxford, 249 pp.

Pond, S. and G. L. Pickard, 1983. Introductory Dynamical Oceanography, second edition. Pergamon, Oxford, 329pp.

Reid, J. L. 1965. Intermediate Waters of the Pacific Ocean. Johns Hopkins Oceanographic Studies No. 2, Johns Hopkins U. Press, Baltimore, 85 pp.

Reid, J. L., 1989. On the geostrophic circulation of the South Atlantic Ocean: flow patterns, tracers, and transports. Progress in Oceanog.,23, . 149-244.

Rintoul, S., 1990. South Atlantic interbasin exchange. J. Geophys. Res., 96, 2675-2692.

Robinson, A. R., ed, 1983. Eddies in Marine Science. Springer-Verlag, Berlin, 609 pp.

Roemmich, D., 1980. Estimation of meridional heat flux in the North Atlantic by inverse methods. J. Phys. Oc., 10, 1972-1983.

Roemmich, D. and C. Wunsch, 1985. Two transatlantic sections: Meridional circulation and heat flux in the subtropical North Atlantic Ocean. Deep-Sea Res. 32, 619-664.

Schmitt, R., P. Bogden, and C. E. Dorman, 1989. Evaporation minus precipitation and density fluxes for the North Atlantic. J. Phys. Oc., 19, 1208-1221.

Schmitz, W. J. Jr. and J. R. Luyten, 1991. Spectral time scales for mid-latitude eddies. J. Mar. Res., 49, 75-107.

Schott, F. and H. Stommel, 1978. Beta spirals and absolute velocities in different oceans. Deep-Sea Res., 25, 961-1010.

Shum, C. K., R. A. Werner, D. T. Sandwell, B. H. Zhang, R. S. Nerem and B. D. Tapley, 1990. Variations of global mesoscale eddy energy observed from Geosat. J. Geophys. Res., 95, 17,865-17,876.

Starr, V. P., 1968. Physics of Negative Viscosity Phenomena. Mc-Graw Hill, New York, 256 pp.

Stommel, H., 1948. The westward intensification of wind-driven ocean currents . Trans. Am. Geophys. Un., 29, 202-206.

Stommel, H., 1957. A survey of ocean current theory. Deep-Sea Res., 4, 149-184.

Stommel, H., 1965. The Gulf Stream: A Physical and Dynamical Description, 2nd ed. U. Calif. Press, Berkeley, 248pp.

Stommel, H., 1984. The delicate interplay between wind-stress and buoyancy input in ocean circulation: the Goldsbrough variations. Tellus, 36A, 111-119.

Stommel, H. and F. Schott, 1977. The beta spiral and the determination of the absolute velocity field from hydrographic station data. Deep-Sea Res., 24, 325-329.

Stommel, H., P. Niiler and D. Anati, 1978. Dynamic topography and recirculation of the North Atlantic Ocean. J. Mar. Res., 36, 449-468.

Sturges, W., 1974. Sea level slope along continental boundaries. J. Geophys. Res., 79, 825-830.

Sverdrup, H. U., M. W. Johnson and R. H. Fleming, 1942. The Oceans. Prentice-Hall, Englewood Cliffs, N. J., 1087 pp.

Trenberth, K., J. G. Olson, and W. G. Large, 1989. A global ocean wind stress climatology based on ECMWF analysis. NCAR/TN-338TSTR, Aug. 1989, 93pp.

Tushingham, A. M. and W. R. Peltier, 1991. Ice-3G: A new global model of late Pleistocene deglaciation based upon geophysical prediction aof post-glacial relative sealevel change. JGR, 96B, 4497-4523.

Veronis, G., 1981. Dynamics of large-scale ocean circulation. in, Evolution of Physical Oceanography. Scientific Surveys in Honor of Henry Stommel, B. A. Warren and C. Wunsch, eds., The MIT Press, Cambridge, Ma, 140-183.

Weiss, R. F., J. L. Bullister, R. H. Gammon and M. J. Warner, 1985. Atmospheric chlorofluoromethanes in the deep equatorial Atlantic. Nature 314, 608-610.

Weyl, P. K., 1968. The role of the oceanis in climatic change: a theory of the ice ages. Meteor. Monographs, 8, 37-62.

Woods, J. D., 1984. The upper ocean and air-sea interaction in global climate. in, The Global Climate, J. T. Houghton, ed., Cambridge U. Press, 141-187.

Worthington, L. V., 1981. The water masses of the world ocean: some results of a fine-scale census. in, Evolution of Physical Oceanography. Scientific Surveys in Honor of Henry Stommel, B. A. Warren and C. Wunsch, eds., The MIT Press, Cambridge, 42-69.

Wunsch, C., 1977. Determining the general circulation of the oceans: A preliminary discussion. Science, 196, 871-875.

Wunsch, C., 1978. The North Atlantic general circulation west of 50°W determined by inverse methods. Revs. Geophys. and Space Phys., 16, 583-620.

Wunsch, C., 1980. Meridional heat flux of the North Atlantic Ocean. Proc. Nat. Acad. Scis., USA, 77, 5043-5047.

Wunsch, C., 1981. Low frequency variability of the sea. in Evolution of Physical Oceanography: Scientific Surveys in Honor of Henry Stommel, B. A. Warren and C. Wunsch, eds., The MIT Press, Cambridge, 342-374.

Wunsch, C., 1989. Tracer inverse problems. in, Oceanic Circulation Models: Combining Data and Dynamics, D. L. T. Anderson and J. Willebrand, eds., Kluwer, Dordrecht, 1-77.

Wunsch, C. and B. Grant, 1982. Towards the general circulation of the North Atlantic Ocean. Prog. in Oceanog., 11, 1-59.

Wunsch, C., D.-X. Hu, and B. Grant, 1983. Mass, heat, salt and nutrient fluxes in the South Pacific Ocean. J. Phys. Oc., 13, 725-753.

Wüst, G., 1935. Schichtung und Zirkulation des Atlantischen Ozeans. Die Stratosphare. . Wissenschaftliche Ergebnisse der Deutschen Atlantischen Expedition auf dem Forschungs-und Vermessungsschiff "Meteor" 1925-1927, 6: 1st Part, 2, 180 pp. (reprinted as The Stratosphere of the Atlantic Ocean, W. J. Emery, ed., 1978, Amerind, New Delhi, 112 pp)

Theory of Ocean Tides with Application to Altimetry

David E. Cartwright

3, Borough House, Borough Road,
Petersfield, Hampshire, England GU32, 3LF,
United Kingdom

1. HISTORICAL INTRODUCTION

The history of tidal studies is as long as the history of geodesy, much longer than the relatively recent subject of physical oceanography. In 1790, Laplace declared the tides to be " . . . ce problème, le plus épineux de toute la mécanique céleste. " Few people are aware that some of the problems identified by Laplace have remained unsolved until the use of modern computers, or that the subject has been radically advanced by space geodesy. A rushed survey of tidal history will serve to introduce the principal concepts, even without a detailed bibliography. Classical ideas are well summarised in Darwin (1910); more recent trends relevant to geodesy may be found in Lambeck (1980) and other modern geophysical texts. For a wider range of modern hydrodynamic topics, see Parker (1991).

The only observational material has traditionally been the rise and fall of the sea surface and its horizontal ebb and flood near the coasts. For centuries, seamen had associated these phenomena with the passage of the Moon and its phase, but nobody understood why there are two tides each day or could predict their amplitudes or timing with any accuracy.

After false hypotheses by several distinguished authors, certain Propositions of Isaac Newton's *Principia* (1687) gave the first correct theory for the **forces** which generate the tides, as the **difference** between the Moon's (or Sun's) gravitational attraction at a field point and at the Earth's centre, (Figure 1). These qualitatively explained the main observed features - the twice-daily periodicity, the daily inequality between successive tides, the association of 'spring' tides and 'neap' tides with lunar phases - but it did not address the crucial problem of how the ocean responds to the prescribed force-field.

Some 50 years later, Daniel Bernoulli elaborated a concept which he called the *Equilibrium Tide*, a prolate spheroid which would hold an ocean covering the globe in (almost) static eqilibrium with the tide-raising forces. In the notation of Figure 1, its surface differs from a sphere by a term proportional to $P_2(Z)$, with mean (lunar) amplitude about 0.36m. Although it bears no simple relationship to the natural ocean tide, the Equilibrium Tide survives as a convenient reference function for empirical data and predictions.

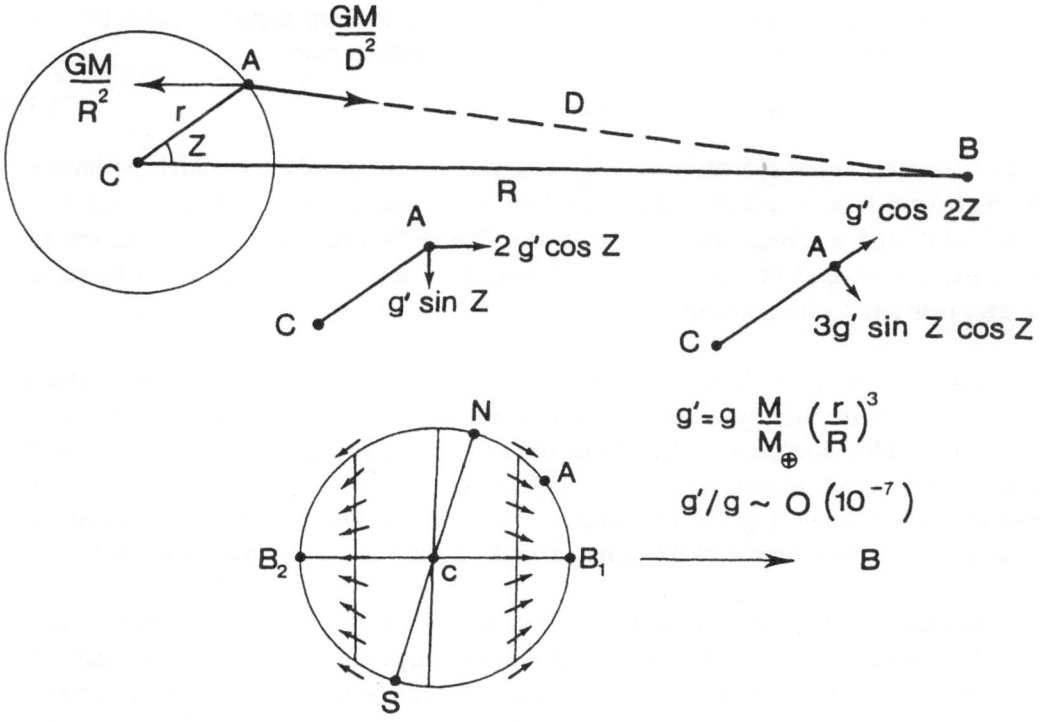

Figure 1: (Top) According to Newton, a particle of unit mass at A on the Earth's surface experiences a tidal force equal to the **difference** in attraction of a celestial body B of mass M at A and at C. (Middle) Components of tidal force at A to first order, parallel and normal to CB (left) and locally vertical and horizontal (right). (Bottom) The **horizontal** component which drives the ocean tide at all points of the surface, is directed towards B_1 and B_2 where it is zero. It has greatest magnitude along zones Z = 45°,135° as indicated. If B_1B_2 is not normal to the polar axis NS, A experiences a daily as well as a twice-daily cycle as the Earth rotates.

Pierre, Marquis de Laplace published long analytical essays on "le flux et le réflux de la mer" from 1776 onwards, culminating in the tidal chapters of his *Traité de Mécanique Celeste* (1799-1823). His work laid the foundations for modern tidal research. In briefest terms, Laplace's main contributions were : (1) the separation of tides into distinct *Species* of long period, daily and twice-daily (and higher) frequencies; (2) the (almost exact) dynamic equations linking the horizontal and vertical displacements of water particles with the horizontal components of the tide-raising force; (3) the hypothesis that, owing to the dominant linearity of these equations, the tide at any place will have the same spectral frequencies as those present in the generating force. Laplace derived solutions for the dynamic equations only for oceans and atmospheres covering a globe, but found them to be strongly dependent on the assumed depth of fluid. Realistic bathymetry and continental boundaries rendered their solution mathematically intractable.

William Thomson (1868), later named Lord Kelvin, took up Laplace's point (3) and proposed a formal empirical expression for the tide at any given place P :

$$\zeta_P(t) = \sum_n H_n \cos (\varepsilon_n + \sigma_n t - G_n) \qquad (1)$$

where ε_n, σ_n are a set of epochs and speeds determined from luni-solar celestial mechanics and H_n, G_n are local amplitudes and phase-lags to be determined empirically from past records at P. Such a scheme, as developed by Sir George Darwin (1883), became known as the *Harmonic Method* of tidal analysis and prediction, and was soon shown to be more accurate than any previous formulation.

The success of the Harmonic Method as a prediction tool removed a major practical incentive for research into tides. However, growing awareness of related geophysical problems provided another incentive, to define and understand the *spatial distribution* of tides in the world oceans, either by solution of Laplace's equations or by direct measurement. Dependent problems included the elastic loading of the earth's crust by ocean tides, and the influence of tidal friction on the Earth's rotation and the Moon's orbit.

Sydney Hough (1896-98) improved Laplace's solutions for spherical oceans, and identified a 'second class' of solutions, governed by vorticity rather than vertical elevation. In the early 20th century, George Goldsbrough, Joseph Proudman and Arthur Doodson devised analytical solutions for oceans bounded by meridians and parallels of latitude. Proudman, in particular, formulated expansion theorems for tides in terms of Lagrangian normal modes, valid in principle for oceans of general shape.

On the observational side, tidal data collection from ports of commerce steadily accumulated from 17th century beginnings, especially after the invention of automatic recorders around 1830. *Harmonic Constants* (H_n, G_n) for each port were accumulated from about 1880, and now number a few thousand sets. Early attempts to plot phase-lags or time-lags across oceans between ports were doomed to inadequacy, as the comparison of 19th and late 20th century maps in Figure 2 shows. Plotting amplitude was out of the question before modern computers. However, the existence of nodal points of zero amplitude, known as *amphidromes*, was first pointed out by William Whewell in 1833.

Realistic tidal maps for shallow, semi-enclosed seas were advanced in the period 1920-1950 by combinations of measurement and computation, treating the tides as *free waves* with empirical boundary conditions, an easier problem. Geoffrey Taylor's (1919) analysis of tidal energy dissipation in the Irish Sea deserves special mention because of its wider implications. It was open to question whether the tides could dissipate enough energy to account for the known deceleration of the Moon's longitude; if so, it most likely took place in shallow seas with strong turbulent currents. Taylor's calculated dissipation rate in the Irish Sea was numerically small, but when scaled in proportion to the total area of all the world's shallow seas, it was seen to be quite adequate. Worldwide estimates of tidal dissipation by Harold Jeffreys and Weikko Heiskanen soon followed, but precise quantification has been debated until the present day.

Application of electronic computers to solving Laplace's tidal equations in separate seas and oceans began in the 1950's with the pioneering work of Walter Hansen of Hamburg University, and a first tentative solution for the world ocean by Chaim Pekeris of the Weiszman Institute (Rehovot) was presented in 1960. (From this date, names are too frequent and too recent to quote.) A succession of computed tidal maps followed, unfortunately differing in detail from each other and from known measurements. These maps purported to give contours of amplitude $H_n(\theta,\phi)$ and phase lag $G_n(\theta,\phi)$ for one or more of the principal Harmonic Constituents defined in (1), as functions of colatitude θ and longitude ϕ. Results differed according to computational method and whether empirical data were used to constrain the solution. There was a period of re-appraisal of how friction and boundaries should be represented, and of the roles of crustal elasticity and self-gravitation. A second generation of computed tide maps around 1978-83 showed better agreement.

Assessing the relative merits of different computer models has always been hindered by the dearth of direct tide measurements in the open ocean. The invention of deep-ocean *bottom pressure recorders* in the mid 1960's was therefore an important technological advance, greeted for a time with international interest by oceanographers. However, the instruments proved expensive and required a lot of ship-time to reach sites of global interest. Between 1965 and 1985, about 200 tidal pressure records were taken, mostly in the northern hemisphere. Since that time, ocean pressure recording has been pursued mainly for non-tidal interests.

Another conceptual advance due to modern computers was the calculation of the ocean's *eigenfunctions* and their associated natural frequencies, of which there are a great (theoretically infinite) number. Solutions, appearing about the same time (early 1980's) in USA and Russia, confirmed a suspicion that the natural frequencies are clustered in the same part of the spectrum as the astronomically imposed tidal frequencies, implying near-resonance. The associated *Quality Factors* of such resonances are still a matter for speculation and interest.

Finally, the impact of *space geodesy* has opened previously unimagined ways of tackling the outstanding problems. Being essentially global in character, satellite methods apply particularly well to just those aspects of tides which are hard to ascertain from *in situ* measurements. First to come to light were the low-degree spherical harmonics of the tidal gravity field at satellite heights. The harmonics of degree 2 are precisely related to the global dissipation of the tides, which had been sought for decades by inefficient methods.

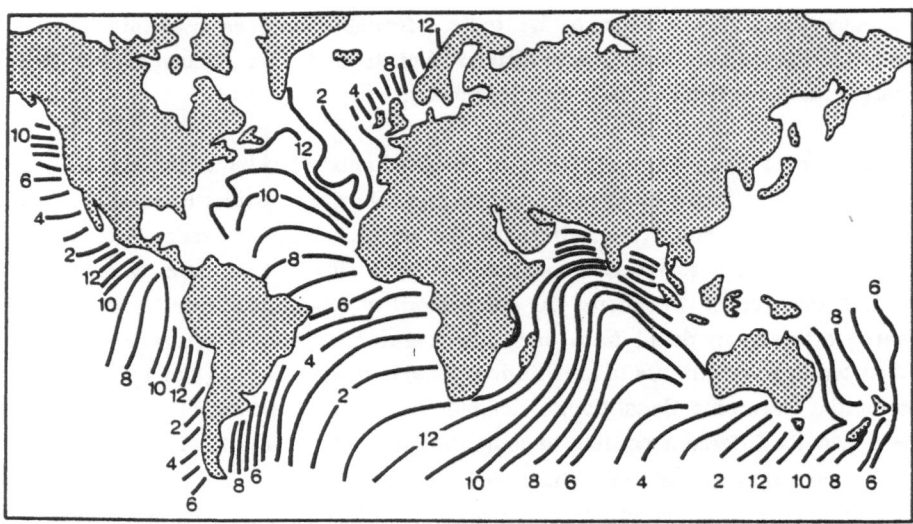

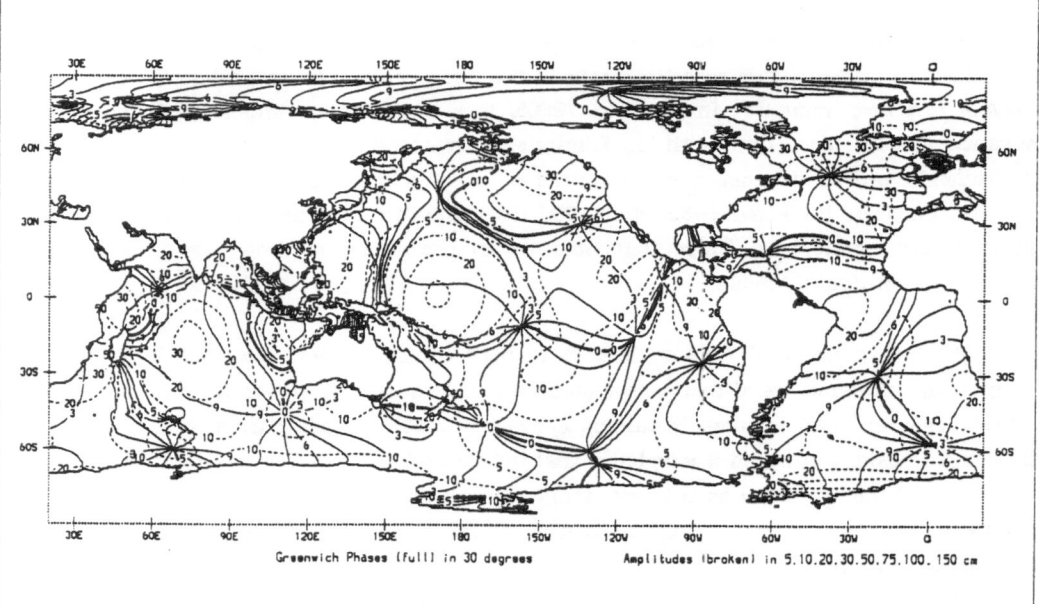

Greenwich Phases (full) in 30 degrees   Amplitudes (broken) in 5,10,20,30,50,75,100,150 cm

Figure 2: (Top) A 19th century "cotidal map", showing contours of Greenwich Time (hours) of High Tide on days of Full and Change of the Moon, as redrawn by Airy from Whewell's original essay,(adapted from G.B.Airy: *Tides and Waves*, Encyclopedia Metropolitana,1842.) (Bottom) A modern computed map, with data assimilation, of the principal solar tide S_2 showing contours of both Greenwich time-lag in hours (equivalent to $G_n/30$) and amplitude H_n in cm, as in equation (1) with $2\pi/\sigma_n = 12$ h. (D.Grawünder, Institut für Meereskunde, Hamburg,1991, by permission.)

The other principal space technique is *satellite altimetry*. With suitable analysis, altimetry can provide direct mapping of $H_n(\theta,\Phi)$, $G_n(\theta,\Phi)$ for leading values of n, over all sea areas within the latitude limits of the orbital inclination. The results may be used as independent data, or for assimilation into rigorous dynamic models. At the same time, the whole subject of altimetry has increased the appetites of oceanographers who analyse the non-tidal variations of sea surface topography, for better definition, and hence elimination, of the pervasive tide signal.

2. THE TIDE POTENTIAL AND ITS USES

2.1 EXPANSION OF THE PRIMARY POTENTIAL

We shall expand the scalar potential of the Newtonian tide-generating force as a function of geographical coordinates and time. From fundamental definitions - see Figure 1 - we have $T(A) = F(A) - F(C) = \text{grad } U(A)$, where

$$U(A) = GM [D(A)^{-1} - r(A).R\,R^{-3} - R^{-1}] \tag{2}$$

Expanding R/D in Legendre polynomials of cos Z then gives

$$U(A) = U_2(A) + U_3(A) + \ldots,$$
$$U_n(A) = (GM/R) (r/R)^n P_n(\cos Z) . \tag{3}$$

It is customary to normalise with respect to the equatorial radius a, so that $a/R = \Pi$, the 'sine equatorial parallax' of the Moon (Π' for Sun), and $GM_\oplus/a^2 = g_\oplus$, a proxy for equatorial gravity. We thus have

$$U_n(A) = g_\oplus K_n (r/a)^n (\Pi/\bar{\Pi})^{n+1} P_n(\cos Z),$$
$$K_n = a (M/M_\oplus) \bar{\Pi}^{n+1} \tag{4}$$

where $\bar{\Pi}$ is the *mean* sine equatorial parallax of the body B. Numerical values of K_n (Moon) and K_n' (Sun) are :

$$K_2 = 35.8373 \text{ cm}, \quad K_2' = 16.4570 \text{ cm}, \quad K_3 = 0.5946 \text{ cm}, \quad K_3' = 0.0007 \text{ cm}.$$

One sees that the solar tide factor K_2' has about 0.46 times the strength of the leading lunar factor K_2, and the Sun's 3rd degree factor K_3' is negligible. The Moon's 3rd degree factor K_3 is not negligible, but it is usually omitted from discussion of the major tidal terms.

We first need to express the zenith angle Z in terms of the colatitude θ and Greenwich longitude ϕ of the field point A, and the corresponding coordinates Θ, Φ of the body B. In the spherical triangle NAB (Figure 3a) we have

$$\cos Z = \cos \theta \cos \Theta + \sin \theta \sin \Theta \cos (\phi - \Phi). \tag{5}$$

Substituting (5) into (4) gives

$$\begin{aligned}
U_2(A) = g_\oplus K_2 (r/a)^2 \rho^3 [\ &(1/4)(3\cos^2\theta - 1)(3\cos^2\Theta - 1) \\
&+ (3/4) \sin 2\theta \sin 2\Theta \cos(\phi - \Phi) \\
&+ (3/4) \sin^2\theta \sin^2\Theta \cos 2(\phi - \Phi)],
\end{aligned} \tag{6}$$

where $\rho = \Pi/\bar{\Pi}$ for the Moon. $U_3(A)$ has a similar expansion up to $\cos 3(\phi - \Phi)$.

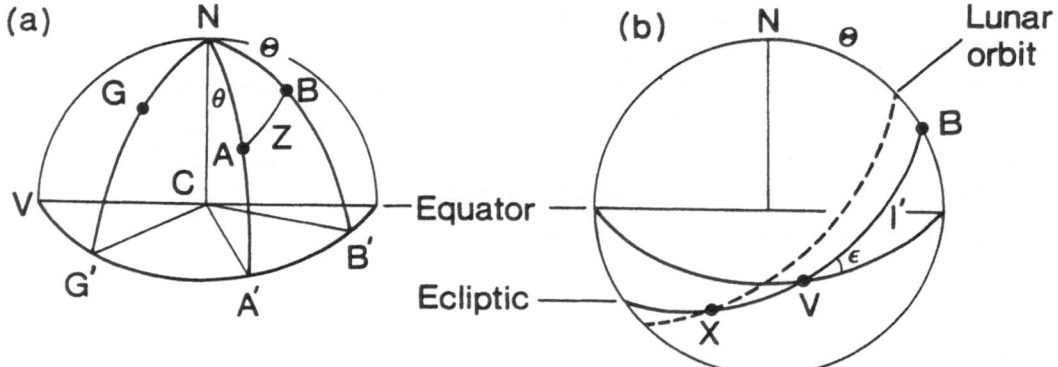

Figure 3: (a) Configuration of field point A colatitude θ, celestial body B codeclination Θ, Greenwich meridian NGG', and vernal equinox V. Angles G'A'=φ, G'B'=Φ, VB'=α, VG'=α_G.

(b) Configuration of equator, ecliptic and lunar orbit, with B drawn in a solar position, celestial longitudeVB=l', latitude d'=0. X is the ascending node of the Moon's orbit with angle VX(eastward) = N(T).

The three terms in [ ] in (6) divide the variation into Laplace's three 'species', or frequency bands, because the leading frequencies of ρ and Θ are monthly, while

$$\Phi = \alpha - \alpha_G ,$$

with α the Right Ascension of B and α_G the Right Ascension of the Greenwich meridian, has the much faster frequency of a cycle per lunar (solar) day. Thus, the first term of (6) has only low frequencies, (i.e. a few cycles per month(year)), the second term has frequencies around one cycle per day (cpd), and the third around 2 cpd. U_3 contributes small increments to the same species, and adds a fourth term around 3 cpd.

The general form of expansion (6) can be written

$$U_n(A) = g_\Theta (r/a)^n \, \mathrm{Re} \sum_{s=0}^{n} C_{ns}(t) \, f_{ns} P_{ns}(\cos \theta) \, \exp(is\phi),$$

where

$$C_{ns}(t) = K_n \, \varepsilon_s \, \frac{(n-s)!}{(f_{ns})^2 \, (n+s)!} \, \rho^{n+1} f_{ns} P_{ns}(\cos \Theta) \, \exp(-is\Phi) \qquad (7)$$

is a temporal function, specified in terms of the sidereal time angle α_G the lunar ephemerides ρ,Θ,α with K_n, (or the solar ephemerides ρ',Θ',α' with K_n'). ε_s is 1 when s=0, 2 when s>0, and f_{ns} is a normalising factor associated with the Legendre function

$$P_{ns}(x) = [(1-x^2)^{s/2}/2^n \, n! \,] \, (d/dx)^{n+s} \, [(x^2-1)^n]. \qquad (8)$$

Equation (6) uses $f_{ns}=1$, but three other normalising conventions occur in tidal literature, namely Doodson's: $f_{ns} = \pm(\max |P_{ns}|)^{-1}$ (minus for f_{20}, f_{30}, f_{31}) ,

meridional: $f_{ns} = (N_{ns})^{-1/2}$, and global: $f_{ns} = (2\pi N_{ns})^{-1/2}$,

where $N_{ns} = \dfrac{2}{2n+1} \dfrac{(n+s)!}{(n-s)!}$

Note : Cartwright & Tayler (1971) and related papers use global normalisation x $(-1)^s$.

Expansions in time and frequency

$U_n(A)$ or its individual species components s =0,1,2,3 may be accurately computed for any time t from equations (7) using official astronomical ephemerides for the Moon and Sun, and the results added together. However, for the practical analysis of physical tidal data it is more important to develop $C_{ns}(t)$ in harmonic series. Although no longer used in modern astronomy, harmonic series for ρ,Θ,α and for ρ',Θ',α' are known from the classical tables of E.W.Brown and S.Newcomb, respectively. Both tables contain many hundreds of harmonic terms necessary for precision of order 10^{-7} radians. Tidal work requires only 10^{-8} to 10^{-5} radians precision and therefore fewer terms, but their incorporation into equation (7) is still very cumbersome, especially for the Moon, so one has to have recourse to specially prepared tables such as Doodson (1921) - based on epoch 1900 - or Cartwright & Edden (1973). Here, I can only briefly sketch the principles involved.

Predictive formulae for the varying positions B of the centroid of the Moon (or Sun) are expressed as *celestial longitude* 1 (1'), *celestial latitude* d (d'), with respect to the ecliptic plane (Figure 3b), inclined to the equator at the ascending node V by the *obliquity* ε, where

$$\varepsilon = 23.4393 - 0.0130 \text{ T degrees.} \tag{9}$$

Here, T is a conventional time unit in *Julian Centuries* of 36525 days with origin at AD 2000, January 1.5, (i.e. noon of January 1). For harmonic expansion, longitude and parallax are expressed as

$$l = q + \Delta l, \qquad l' = q' + \Delta l',$$
$$\rho = \Pi/\overline{\Pi} = 1 + \Delta\rho, \qquad \rho' = \Pi'/\overline{\Pi'} = 1 + \Delta\rho', \tag{10}$$

where q,q' denote the *mean longitudes* which increase linearly in time with periods of a *tropical month* (27.3216 days) and a *tropical year* (365.2422 days), respectively.

In (10), $\Delta l, \Delta l', \Delta\rho, \Delta\rho'$ are small perturbations of zero mean value, depending mainly on the ellipticity of the orbit but also on the relative positions of Moon and Sun. All four are expressible by convergent harmonic series of general form :

$$\Delta x = \sum_j A_j (\sin, \cos) [\ k_j^1(q-p) + k_j^2(q'-p') + k_j^3(q-N) + k_j^4(q-q')\ ] \tag{11}$$

where the A_j are numerical coefficients derived from celestial mechanics, and the k_j^i are small ± integers or zero. Sines are generally used for longitudes and cosines for parallax, because their perturbations are in quadrature. The three new arguments in (11) are :

p : the mean longitude of lunar perigee,

p' : the mean longitude of the Sun at perihelion,

N : the mean longitude of the Moon's ascending node.

all measured, like q,q', eastward along the ecliptic from V. They advance linearly in time with cycles of 8.85y, 20,940y, and -18.61 y respectively. The Moon's node *regresses* westwards. Parameters defining q,q',p,p', and N in the form $(c_0 + c_1 T)$ degrees (neglecting very small terms in T^2) and their speeds in degrees per day are listed in Table 1.

The celestial latitude of the Sun, d', is effectively zero to adequate accuracy, but the Moon's latitude d oscillates between ±5.3°, its leading term in (11) being 5.1°sin(q−N).

TABLE 1 : Linear secular coefficients of the mean longitudes $(c_0 + c_1 T)°$, epoch 2000.

Object	Symbol		c_0	c_1	Speed (°/day)
Mean Moon	q	•	218.3164	481,267.8814	13.176 3965
Mean Sun	q'	•	280.4661	36,000.7698	0.985 6474
Lunar Perigee	p		83.3535	4,069.0139	0.111 4035
Lunar Node	N		125.0445	−1,934.1363	−0.052 9538
Solar Perigee	p'		282.9384	1.7195	0.000 0471

• The usual (Darwin's) symbols for q, q' are s, h respectively.

Finally, we convert from celestial longitude l and latitude d to *Right Ascension* α and *Codeclination* Θ by using the standard formulae :

$$\cos \Theta = \cos d \sin l \sin \varepsilon + \sin d \cos \varepsilon ,$$
$$\cos \alpha \sin \Theta = \cos d \cos l ,$$
$$\sin \alpha \sin \Theta = \cos d \sin l \cos \varepsilon - \sin d \sin \varepsilon , \tag{12}$$

while the Right Ascension of the reference meridian is accurately defined by

$$\alpha_G = 360° \times (\text{Universal Time in days}) + q'(T) - 180° \tag{13}$$

Application of (10-12) to equation (7), either algebraically (as in Doodson, 1921), or by numerical spectral analysis (Cartwright & Tayler, 1971), leads to a general expansion of the form :

$$C_{ns}(t) = g_\Theta \sum_r B_r \cos (\sin) [s\tau + i_r q + i_r' q' + j_r p + j_r' p' - k_r N] \tag{14}$$

where $\tau = (\alpha_G - q)$ is mean lunar time, a conventional proxy for UT, and i,i',j,j',k are sets of small ± integers or zero. The B_r are a set of harmonic amplitudes, practically constant on a time scale of decades, but in reality slowly varying with the obliquity ε (9). Algebra dictates that cosines apply to all expansions with (s+n) even, sines to (s+n) odd.

A full set of coefficients of (14), of accuracy sufficient for all physical applications, is tabulated in Cartwright & Edden (1973), with reference to Cartwright & Tayler (1971) for certain details. These tables give amplitudes B_r for three epochs, 1874, 1924 and 1960, from which values for later epochs may be deduced by linear extrapolation, if necessary for high precision - the changes per century are quite small. Table 2 below lists a reduced set of leading terms of degree n=2, adjusted to epoch 2000, sufficient for an overall accuracy of about 3 percent in total amplitude. In the three species s=0,1,2, only i,i',j are represented for brevity; two terms requiring the addition of j'=1 are noted by •. D is an additive constant in order to convert all terms to the form cos (sτ+iq+i'q'+jp+j'p'+D).

The terms of (14) involving the Moon's nodal longitude N are defined explicitly in the references cited above by use of the increment kN^1 $(N^1=-N)$. Their amplitudes are relatively small but they are not negligible, and they provide the wellknown 'nodal modulations' of all constituents, with period 18.61y. In Table 2 the nodal terms are expressed (unusually in tidal literature) as nondimensional modulatory factors δ_1^+, δ_1^-

associated with each constituent $B_r \cos(x)$ in the formalism :

$$B_r [\cos (x) + \delta_1^+ \cos (x+N) + \delta_1^- \cos (x-N)] \tag{15}$$

TABLE 2 : The principal harmonic constituents of $C_{2s}/g_\oplus$ and their nodal modulations. Amplitudes B_r (cm) are correct around Epoch 2000. Their 'names' are the traditional symbols assigned by Darwin (1883).

s	i	i'	j	D	f	Name	B_r	δ_1^+	δ_1^-
0	0	0	0	180	0.0	-	-31.46		-0.089
0	0	2	0	180	0.0055	Ssa	-3.10		
0	1	-2	1	180	0.0314	MSm	-0.67	-0.071	-0.065
0	1	0	-1	180	0.0363	Mm	-3.52	-0.066	-0.065
0	2	-2	0	180	0.0677	MSf	-0.58	0.071	-0.065
0	2	0	0	180	0.0732	Mf	-6.66		0.415
0	3	0	-1	180	0.1095	Mt	-1.28		0.415
1	-3	0	2	-90	0.8570	2Q	0.66	0.188	
1	-3	2	0	-90	0.8618	σ	0.80	0.188	
1	-2	0	1	-90	0.8932	Q	5.02	0.188	
1	-2	2	-1	-90	0.8981	ρ	0.95	0.188	
1	-1	0	0	-90	0.9295	O	26.22	0.188	
1	0	0	-1	-90	0.9658	-	-0.74	0.188	
1	0	0	1	-90	0.9664	M_1	-2.06		0.201
1	1	-3	0°	-90	0.9945	π	0.71		
1	1	-2	0	-90	0.9973	P	12.20		
1	1	0	0	-90	1.0027	K_1	-36.87		0.136
1	1	2	0	-90	1.0082	φ	-0.52		
1	2	-2	1	-90	1.0342	θ	-0.40		0.200
1	2	0	-1	-90	1.0390	J	-2.06		0.200
1	3	0	0	-90	1.0759	OO	-1.13		0.641
2	-3	2	1	0	1.8283	-	0.47		
2	-2	0	2	0	1.8597	2N	1.60		
2	-2	2	0	0	1.8645	μ	1.93		
2	-1	0	1	0	1.8960	N	12.10	-0.037	
2	-1	2	-1	0	1.9008	ν	2.30		
2	0	0	0	0	1.9323	M_2	63.20	-0.037	
2	1	-2	1	0	1.9637	λ	-0.47		
2	1	0	-1	0	1.9686	L	-1.79	-0.037	
2	2	-3	0°	0	1.9973	T	1.72		
2	2	-2	0	0	2.0000	S_2	29.40		
2	2	0	0	0	2.0055	K_2	7.99		0.298
2	3	0	-1	0	2.0418	-	0.45		

° : add p' to the argument

Modulating factors $\delta_2^\pm$ with argument 2N also exist, but they are generally within the accuracy assumed in Table 2.

Finally, the frequency (or 'speed') of each constituent is listed in column 5 of Table 2 from the formula :

$$f = 0.966137 \, s + 0.036601 \, i + 0.002738 \, i' + 0.000309 \, j \quad cpd$$

whose numerical coefficients are the solar-time equivalents of a cycle per lunar day, tropical month, year, and perigee cycle (8.85y), respectively.

2.2 RELATING OBSERVED TIDAL PHENOMENA TO THE POTENTIAL

Bypassing much hydrodynamic theory (to be considered in detail in §3), we may generalise the relation between a physical tide variation $\zeta(t)$ at some point P and the tide-generating potential U(A), equation (4) by a form :

$$\zeta(t) = \mathrm{Re} \left[\sum_n \sum_s \int_0^\infty C_{ns}(t - \tau) \, w_{ns}(\tau) \, d\tau \right] \tag{16}$$

where $w_{ns}(\tau)$ is a response function for the point P. It is necessary to discriminate for n and s because the ocean responds quite differently to different spherical harmonic generating forces. However, $w(\tau)$ is difficult to define as a continuous function, and it is easier to work with discrete values of its fourier transform, namely the *Admittance* :

$$Z_{ns}(f) = \int_0^\infty w_{ns}(\tau) \, e^{-2\pi i f \tau} d\tau = H_{ns}(f) / \, \Gamma_{ns}(f) \tag{17}$$

where $H(f)$, $\Gamma(f)$ are (complex) fourier spectra of $\zeta_T(t), C(t)$ respectively, $\zeta_T(t)$ being the tide-coherent, i.e. 'noise-free' part of the variation $\zeta(t)$.

Since, from § 2.1, $\Gamma_{ns}(f)$ is expressible as a discrete sequence $\Gamma_{ns,r}$ associated with a a set of precisely known frequencies f_r , so too is the spectrum of the physical tidal process $\zeta(t)$ expressible as a sequence

$$H_r = Z_{ns}(f_r) \, \Gamma_{ns,r} \tag{18}$$

In other words, with reference to equation (14),

$$\zeta(t) = \sum_r |H_r| \, (\cos,\sin) \, [s\tau + i_r q + i_r' q' + j_r p + j_r' p' - k_r N - G_r] \tag{19}$$

where $G_r = -\mathrm{Arg}[Z_{ns}(f_r)]$, a 'phase-lag' in accordance with the Harmonic Method of tidal analysis, equation (1). Once evaluated, $|H_r|$ and G_r are known as the *Harmonic Constants* of the process $\zeta_T(t)$ at field point P.

Data banks of sets of tidal Harmonic Constants for sea surface elevation at some thousands of coastal ports are held at the Canadian Environmental Data Service, Ottawa, on behalf of the International Hydrographic Bureau, Monaco. Similar constants for a few hundred open ocean sites are held at the Proudman Oceanographic Laboratory, Birkenhead, U.K., on behalf of the International Association of the Physical Sciences of the Ocean.

Note that, with $\tau = \alpha_G - q$ and α_G defined by (13), the phase-lag G_r is implicitly associated with the origin of time, assumed in (13,19) to be that of Universal Time, that is Greenwich (mean) midnight. Other conventions are often used in tidal literature, depending on source, notably

$$g = G - 15°t_L$$

where t_L is the 'Local Time Zone' of P in hours west of Greenwich. Another type of phase-lag is $\kappa = G + s\phi°$, (s=species, ϕ=east longitude of P) which refers effectively to transits of the mean Sun at the precise point P.

Mariners' tables usually give phase-lags in the 'g' convention, together with t_L. The 'κ' convention originated from Darwin (1883) and is common in early 20th century (especially American) literature. The κ phase-lag is directly relevant to calculation of the *action* of the tidal forces on the ocean (§ 4.1). However, Greenwich phase-lag G or 'time-lag' (G/15s) have become the conventional quantities to plot in global ocean tide maps. Clearly, care has to be taken in defining or interpreting tidal harmonic phase-lags. Mistakes over Time Zone are particularly common.

Not surprisingly, traditional methods of data analysis to establish Harmonic Constants have the character of spectral time-series analysis. Spectral methods are most elegant and exact when the data are sampled at high frequency and at equal time increments. For satellite data, this is unlikely to be the case, but with care for aliassing problems, similar results may be achieved by suitably designed least-squares or inverse methods.

Whatever scheme for spectral filtering is adopted, the *Rayleigh criterion* must be respected, whereby no information is sought at a frequency separation less than $\Delta f = 1/\Delta t$, where Δt is the total time span of the data. This means in effect that the admittance function $Z(f)$ must be treated as a sequence of constant steps of width Δf. Such an assumption is nearly always made with regard to the fine structure of the 'nodal modulations', in that the same parameters $\delta^{\pm}$ (equation 15) are assumed to apply to both the tide potential and the data. Less justifiably, if Δt is considerably less than 1y, one is obliged to assume that the relatively strong solar line-clusters $(P,S,K)_1$, $(T,S,K)_2$, whose frequencies differ by 1cpy, have uniform admittances, and hence uniform phase-lags. Rigorous analyses of long spans of tide-gauge data show that such assumptions are inaccurate, on account of non-gravitational (or 'radiational') effects peculiar to the solar tides. In general, at least one year of data has to be analysed to obtain accurate tidal constants or admittances.

Let $\Delta C_r(t)$ denote the sum of those harmonic terms of the potential function (14) whose frequencies f are within $(f_r \pm \Delta f/2)$, where f_r is the frequency of a major harmonic constituent and Δf^{-1} is the time span of available data of a variable $\zeta(t)$, whose Harmonic Constants are sought. The observational equations to be solved are the 'normal equations' of

$$\text{Re} \left[\sum_{r=1}^{N} Z(f_r) \, \Delta C_r(t) \right] - \zeta(t) = 0 \qquad (20)$$

whose matrix yields a sequence of N solutions $Z(f_r)$ for the admittance at the resolved groups of frequencies. The Harmonic Constants within each group r are then simply derived from the formulae

$$H_r = B_r \, |Z(f_r)|, \quad G_r = -\text{Arg}[Z(f_r)] \, ,$$

where B_r is the potential amplitude, taken for example from Table 2 or more detailed tables.

An alternative approach is to parametrise the admittance $Z_{ns}(f)$ as a smooth function of frequency across the bandwidth (roughly 0.25 cpd) of each species s. This is the basis of the *Response Method* of Munk & Cartwright (1966) - see also Cartwright (1968) for improvements to the initial scheme. Omitting the suffixes n,s for convenience, $Z(f)$ is represented as the sum of a few terms of a fourier series,

$$Z(f) = \sum_{m=-M}^{M} w_m \exp(-2i \, f \, m \, \pi \, \Delta t) \qquad (21)$$

where $\Delta t = 48$ hours is a carefully chosen time lag, w_m is a set of complex 'weights', to be evaluated, and $M = 1$ or 2, depending on the degree of variability (or 'wiggliness') of admittance permitted by the quantity of data and their noise level. In the time domain, (21) is equivalent to the simple relationship

$$\zeta(t) = \text{Re} \left[\sum_{m=-M}^{M} w_m C_{ns}^{*}(t - m\delta t) \right] \qquad (22)$$

The normal equations of (22), summed over the relevant tidal species s, (and in principle over degree n, but more usually for n=2 only), are solved for the weight parameters w_m. Hence $Z_{ns}(f)$ from (21) and Harmonic Constants for $f = f_r$.

The advantages of the Response Method are that (a) equations (21,22) express the tide at P with a smaller number of arbitrary terms than are necessary for a full harmonic expansion of $\zeta(t)$, and (b) resolution of the weights w_m is in principle independent of the Rayleigh criterion for minimum duration of data. A disadvantage is that the elements of $C(t-m\delta t)$ for different m are not orthogonal, so solutions for w_m (but not for $Z(f)$) are strongly dependent on choice of M. A set of orthogonal functions related to $C(t-m\delta t)$ were derived by Groves & Reynolds (1975) and have been found useful in tidal interpretation of satellite altimetry, (Cartwright & Ray,1990).

The **radiational** anomaly affecting solar tides, mentioned earlier, introduces a distortion of the smooth admittance functions assumed in (21,22). In order to accommodate such effects in the 'Response' formalism, Munk & Cartwright (1966) introduced a *Radiational Potential* (RP) to be added to (22) with separate weight parameters. The RP, which models the the spatial pattern of solar radiation falling on the Earth's surface, (zero at night), contributes to all tidal species at or very near to s cpd, and has been extensively applied to tide-gauge records of long duration.(Cartwright, 1968; Cartwright et al, 1988).

Whether the RP effect can be adequately resolved from altimetry remains to be seen. The principal anomaly occurs at the main solar constituent S_2, where it has been shown to be more than 10 percent of the gravitational tide ; there is also a smaller effect at S_1, whose purely gravitational tide is practically negligible. Both these anomalies are likely to be the result of the action of the predominantly radiational air tide on the entire ocean surface. The annual cycle in sea level, usually called 'Sa', can also be related to the RP, but Sa is in fact due to a complex mix of seasonal processes in both ocean and atmosphere. If the RP is ignored, the admittance $w_{22}(f)$ will reflect the combined response at S_2, at the expense of a small error in the apparent admittance at K_2.

I shall ignore **nonlinear** tidal effects here, because they are known from theory and observation to be generally weak except in coastal and shelf seas, especially river estuaries, to which satellites have only limited application. Where nonlinear effects are sensible, they appear as harmonic terms with compound frequencies of the linear harmonics, for example M_4 at twice the frequency of M_2 and MS_4, MSf at the sum- and difference- frequencies of M_2 and S_2. Friction tends to induce triple interactions, such as $(M+M-S)_2$ having the same frequency as a small linear term known as μ_2 (Table 2). A scheme for dealing with nonlinear Response analysis has been evolved (e.g. Cartwright,1968), but if nonlinear effects should prove relevant to altimetry, it would be safer to treat them by the classical Harmonic Method, which routinely includes leading nonlinear terms.

2.3 SECONDARY TIDE POTENTIALS

Having explained the formalism by which the ocean (or earth) tides are related to the primary tide-generating potential, we are in a position to define the additions to the potential induced by the tidal deformations themselves. These so-called *secondary* potentials are by no means negligible in magnitude, but being linearly related to the primary potential they may be included in the same analytical process.

The largest increment results from the elastic yielding of the Earth's crust to the tidal forces. Unlike the ocean, the Earth's eigenfrequencies, revealed by seismology, are an order of magnitude higher than the tidal frequencies, and so the deformation is direct and effectively instantaneous. Any phase lag is reckoned to be <0.25° (Zschau,1986),and may be ignored in the present context. From static computations of spherically symmetrical models of the Earth's interior it has been shown (see Melchior,1983) that a spherical harmonic primary tide potential U_n produces an 'equilibrium' deformation of the Earth's figure equal to $h_n U_n/g_o$, which in turn increases the tide potential at radius r by $k_n U_n$. Here, h_n, k_n are the *Love numbers*, which are roughly speaking independent of frequency. Owing to the displacement of a surface particle from r to $(r + h_nU_n/g)$, the potential *in situ* is diminished by h_nU_n, making the total effective potential at the physical surface to become γ_nU_n , where

$$\gamma_n = 1 + k_n - h_n.$$

Numerical estimates of k_n and h_n vary according to the Earth model used, but those due to Wahr(1981) are now commonly accepted. Clearly, the most important are :

$$k_2 = 0.302, \ h_2 = 0.609, \text{ giving } \gamma_2 = 0.693$$

valid at most tidal frequencies. However, owing to a sharply tuned resonance of the Earth's core near 1 cycle/sidereal day, the best values at the frequency of K_1 are 0.26, 0.52, 0.74 respectively. Thus, the effective potential for generating the ocean tides is some 31% (26% at K_1) less than the primary potential U_2.

The quantity h_2U_2 (+h_3U_3 , where $h_3 \approx 0.29$) is known as the *Body Tide* of the solid Earth. A more complicated secondary effect results from the elastic loading of the crust by the ocean tide itself. This is called the *Load Tide*. The Load Tide may be expressed as a Green's Function of angular distance from each incremental tidal load, effective up to 180°, but given global definition of the ocean tide it is more convenient to express it in terms of a sequence of load-Love numbers k_n', h_n' times the spherical harmonics of degree n of the ocean tide. If $\zeta_n(\theta, \lambda, t)$ denote any nth degree spherical harmonic of ζ, the secondary potential and bottom displacement due to elastic loading are

$$g(1+k_n')\alpha_n\zeta_n \text{ and } h_n'\alpha_n\zeta_n \text{ respectively, where}$$

$$\alpha_n = \frac{3}{(2n+1)} \times \frac{\text{mean ocean density}}{\text{mean Earth density}} = \frac{0.563}{(2n+1)} \qquad (23)$$

The essential difference here from the formulation of the Body Tide is that the spherical harmonic expansion of the ocean tide itself requires terms up to very high degree n, for adequate definition.

Farrell's (1972) calculations of the load Love numbers, based on the Gutenberg-Bullen Earth model, are frequently used. Table 3 lists a selection of both Farrell's numbers and those from a more advanced calculation by Pagiatakis (1990), based on the PREM model.

TABLE 3 : Factors α_n (equation 23), and loading Love numbers computed by Farrell (1972) and by Pagiatakis (1990).

		Farrell		Pagiatakis	
n	α_n	$-h_n'$	$-k_n'$	$-h_n'$	$-k_n'$
1	0.1876	0.290	0	0.295	0
2	.1126	1.001	0.308	1.007	0.309
3	.0804	1.052	.195	1.065	.199
4	.0625	1.053	.132	1.069	.136
5	.0512	1.088	.103	1.103	.103
6	.0433	1.147	.089	1.164	.093
8	.0331	1.291	.076	1.313	.079
10	.0268	1.433	.068	1.460	.074
18	.0152	1.893	.053	1.952	.057
30	.0092	2.320*	.040*	2.411	.043
50	.0056	2.700*	.028*	2.777	.030
100	.0028	3.058	.015	3.127	.016

* Interpolated from data at n = 32,56

As a crude 'rule-of-thumb', with the typical horizontal scales of the tides in the open ocean, the load potential and the bottom displacement approximate to 0.05gζ and -0.07ζ respectively - see, for example, the computed maps of Parke & Hendershott (1980). These approximations do not apply over shelf seas, where the horizontal scale is much shorter. In reality the loading effects extend over all continents, where ζ=0 by definition.

3. HYDRODYNAMIC MODELS

3.1 DYNAMIC EQUATIONS AND IDEALISED SOLUTIONS

Every hydrodynamic model of the tides is a compromise between physical accuracy and mathematical tractability. Proudman (1941), Miles (1974), and Schwiderski (1980a) discuss several minor terms in the equations of momentum which are customarily omitted for convenience. We shall here consider only oceans of depth greater than about 0.2 km, that is, most of the open ocean excepting continental-shelf seas. Shelf seas amplify the tidal currents, causing nonlinear friction and other second-order terms to take effect. Vertical acceleration may be ignored in comparison with gravity, and we ignore also internal motions due to vertical density structure, which have shorter wavelength and negligible surface elevation. See various papers in Parker (ed.1991) for modern reviews of most of these neglected topics.

With the above omissions, a familiar and well-tried representation of the relation between southward (u,θ) and eastward (v,ϕ) components of depth-mean velocity, surface elevation ζ, and the generating potential U over the ocean at large may be written :

$$\frac{\partial u}{\partial t} - 2\Omega \cos\theta.v = \frac{-\partial}{a\partial\theta}\left[g\zeta - \gamma U - \Sigma\gamma_n'\alpha_n\zeta_n\right] - F_\theta ,$$

$$\frac{\partial v}{\partial t} + 2\Omega \cos\theta.u = \frac{-\partial}{a \sin\theta\partial\phi}\left[g\zeta - \gamma U - \Sigma\gamma_n'\alpha_n\zeta_n\right] - F_\phi ,$$

$$\frac{\partial\zeta}{\partial t} + \frac{1}{a \sin\theta}\left[\frac{\partial(hu \sin\theta)}{\partial\theta} + \frac{\partial(hv)}{\partial\phi}\right] = 0 . \tag{24}$$

where Ω = 6.3004 rad/solar day is the Earth's sidereal rotation (i.e. K1) frequency, γU is short for the *total* tide-generating potential, and the Σ term is the load potential summed over the spherical harmonics of ζ by degree n. The F components represent friction, usually specified in terms of u,v in a variety of possible ways, to be considered later.

The first two equations express the balance of horizontal acceleration on a rotating sphere ; the third equation, involving the local ocean depth h, expresses mass continuity of the motion in a virtually incompressible fluid. Note that ζ here strictly represents (surface elevation - bottom elevation - h),

as measured *in situ* by tide gauges. Except for the inclusion of secondary potentials, equations (24) are essentially the same as *Laplace's Tidal Equations* .

In application to any model of the real ocean, solutions have to satisfy a boundary condition $u . n = 0$, where u is the horizontal velocity vector at a land boundary with unit normal vector n. In some circumstances, where it is necessary to specify a boundary at the open entrance to a shelf sea area, the no-flow condition is replaced at such a boundary by an expression like

$$<(u.n) \zeta> = E >0,$$

where < > denotes a time average over a tidal cycle and E denotes a known flux of energy, as discussed for example by Accad & Pekeris (1978). Otherwise, conditions at open ocean boundaries have to be specified in terms of measured data, if available.

Free waves in flat rotating seas

Before describing the mathematical approach to solving (24) in oceans of global scale, it is instructive to consider some canonical forms of free-wave solutions which may exist in the simplest of idealisations, namely a flat sea of uniform depth with straight boundaries. Replacing the spherical distance elements ($a\partial\theta$, $a \sin\theta\partial\phi$) by local Cartesian elements ($\partial x, \partial y$) - rotatable to any horizontal orientation - and the Coriolis parameter $2\Omega \cos\theta$ by its usual symbol f, the equations without driving potentials or friction reduce to :

$$\frac{\partial u}{\partial t} - fv = - g\frac{\partial \zeta}{\partial x}, \qquad \frac{\partial v}{\partial t} + fu = - g\frac{\partial \zeta}{\partial y}, \qquad \frac{\partial(hu)}{\partial x} + \frac{\partial(hv)}{\partial y} = - \frac{\partial \zeta}{\partial t} \qquad (25)$$

Seeking wave-like solutions in which (ζ, u, v) are all proportional to

$$\exp [i(\kappa x + \mu y - \sigma t)], \qquad (26)$$

equations (25) with constant h easily yield a dispersion relation :

$$\kappa^2 + \mu^2 = (\sigma^2 - f^2)/gh , \qquad (27)$$

corresponding to plane waves of fixed wavenumber (given σ, f, h), propagating in an arbitrary direction. Note that, if κ and μ are to be real, then no such solutions exist for $\sigma < f$. Thus, there are no free-wave solutions of uniform amplitude in uniform depth for the long-period tides, or for the diurnal tides poleward of about 30° latitude since $2\Omega \cos\theta > 2\pi/\text{sidereal day}$, the frequency of K_1. However, tides exist at all latitudes.

Solutions of type (26) are most meaningful if combined to satisfy a valid boundary condition, say v=0 along the x-axis. Solutions with $\pm\mu$ may be so combined to make $v \propto \sin \mu y$ and $\qquad \zeta = \zeta_0 (\sigma\mu \cos \mu y - f\kappa \sin \mu y) \exp [i(\kappa x - \sigma t)] \qquad (28)$ These are the *Poincaré Waves* (Poincaré, 1911) with sinusoidal crests and a periodic array of nodal lines in elevation, parallel with the boundary. The possible condition $\mu = 0$ gives a special case of plane waves, known as *Sverdrup Waves* (see Marchuk & Kagan, 1989).

A more commonly identified type of solution is the *Kelvin Wave* , (Thomson, 1879), which by assigning a purely *imaginary* value to μ, namely

$$\mu = if/\sqrt{gh} ,\qquad(29)$$

makes v = 0 everywhere, as can be verified from (25). The relation (27) becomes nondispersive : $\kappa = \sigma/\sqrt{gh}$; waves with sloping crests,

$$\zeta = \zeta_0 \exp(-|\mu| y) \cos(\kappa x - \sigma t)$$

progress at speed $\sqrt{gh}$ with the boundary y=0 to their right (left) in the N (S) hemispheres. Despite the apparently restrictive assumptions, Kelvin Waves retain their character along irregular and curved coastal boundaries and variable depth profiles, and they account at least qualitatively for many observed tidal features, including amphidromes. Munk, Snodgrass & Wimbush (1970) successfully synthesized diurnal and semidiurnal tides west of California in terms of an arbitrary mix of Kelvin and Poincaré Waves.

(N.B. Modern oceanographers often refer to 'Kelvin Waves', meaning Equatorial- or Double-Kelvin Waves, which can propagate eastwards along the Equator (x'=x-aπ/2=0) with lowest-mode profile exp(-rx'^{2}). These depend on vertical density structure, and have no simple application to tides.)

Kelvin Waves (in the original sense) and Poincaré Waves are essentially *gravity waves* , that is waves controlled by the vertical displacement of water. Their energy is more or less equally divided between potential energy $\rho g \zeta^2/2$ and horizontal kinetic energy $\rho h(u^2+v^2)/2$. At low frequencies, there also exist a second class of solutions to (25), controlled by conservation of **potential vorticity**,

$$[f + \frac{\partial v}{\partial x} - \frac{\partial u}{\partial y}]/h$$

rather than by vertical displacement, and in which kinetic energy predominates. Such waves manifest themselves mainly as waves of current, but they do possess detectable profiles of surface elevation. Allowing for the variation of f with colatitude (x) and solving (25) with $\zeta=0$ in the third equation gives rise to *Planetary* or *Rossby Waves* (see Longuet-Higgins, 1964). Such waves have strictly westward velocity (κ<0 at any latitude), governed by the dispersion relation : $\kappa^2 + \mu^2 = -\beta\kappa/\sigma$, where $\beta = -\partial f/\partial x > 0$.

Another type of 'Class 2' wave, with shorter wavelength, is the *Topographic Rossby Wave*, associated with steep variation in depth h. Trapped along shelf-edges, these are known as *Continental Shelf Waves* (Mysak, 1967). Along a straight coastal boundary y=0 with uniform shelf profile h(y) and longshore wavenumber κ, their governing equation deducible from (25), is :

$$(h\zeta_y)_y - \zeta(\kappa^2 h - \kappa f h_y/\sigma) + (\sigma^2 - f^2)\zeta/g = 0 \qquad(30)$$

where suffixes y denote differentiation. In contrast to Kelvin- and Poincaré Waves, Continental Shelf Waves have short horizontal scales (typically a few 100km) and can exist only for σ < f, therefore not a feature of semidiurnal tides. In general, Topographic Rossby

Waves have been shown to be important features of long-period tides (Wunsch, 1967) and of diurnal tides along shelf-edges at high latitudes (Cartwright et al, 1980).

Solution for tides in a hemispherical sea

We now return to the more realistic spherical geometry of equations (24), and consider the simplest case of a large bounded sea, namely, a hemispherical ocean of uniform depth, bounded by two complete meridians, 180° apart. Solutions for a fixed frequency without friction were first given by Doodson (1938); the complete set of eigenfrequencies was first computed by Longuet-Higgins & Pond (1970). We here follow the analysis of Webb (1980), who first solved for *forced* tides with continuously variable frequency and linear friction: $F_\theta = ku/h$, $F_\phi = kv/h$, where k>0 is an arbitrary constant. It is convenient to define a vector $\mathbf{X}$ to represent horizontal displacement, so that $\partial \mathbf{X}/\partial t = \mathbf{u} = (u,v)$.

Following a standard procedure, $\mathbf{X}$ can be expressed as the sum of the gradient of a scalar potential Φ and the curl of a vertical stream function vector $\mathbf{\Psi}$:

$$\mathbf{X} = \nabla\Phi + \nabla_\wedge\mathbf{\Psi} \tag{31}$$

Then, because the final term of (31) has no divergence, the mass conserving equation of (24) gives $\zeta = -h\nabla^2\Phi$ as the 'Class 1' or 'gravity wave' solution, while the vertical component of $\nabla_\wedge\nabla_\wedge\mathbf{\Psi}$ defines the the 'Class 2' or 'vorticity wave' solution. Along a boundary, $\mathbf{n}.\nabla\Phi = 0$, and $|\Psi|$ = constant.

The velocity potential Φ is further expanded as the sum of the eigenfunctions of the equation: $\nabla^2\phi_r + \alpha_r\phi_r = 0$, subject to $\nabla\phi.\mathbf{n} = 0$ at the boundary, and the stream function Ψ (vertical of $\mathbf{\Psi}$) as the sum of a similar set of eigenfunctions of:
$$\nabla^2\psi + \beta_r\psi_r = 0, \text{ subject to } \psi = 0 \text{ at the boundary.}$$

Thus, $\Phi(\theta,\phi,t) = \sum_r p_r(t)\,\phi_r(\theta,\phi)$, $\Psi(\theta,\phi,t) = \sum_r p_{-r}(t)\,\psi_r(\theta,\phi)$,

where the time factors p_r, p_{-r} are defined below. ϕ_r and ψ_r may be termed the normal modes of the basin *without rotation*. They are sometimes called *Proudman Functions*, after Proudman (1917), who first showed their general applicability as basis functions for all classes of tidal motion in seas of arbitrary shape. In the present simple geometry they are merely spherical harmonics with explicitly defined eigenfrequencies, α_r,β_r, (Webb, 1980, Appendix 1).

To determine the p_r factors, one multiplies the accelerative equations of (24) by $\nabla\phi_r$ and $\nabla\psi_r$, assume time dependence $e^{i\sigma t}$ and integrate over the basin, giving finally:

$$-(\sigma^2 + i\,k\sigma/h)\,p_r - 2i\sigma\Omega\alpha_r^{-1}\sum_{s=-\infty}^{\infty}\gamma_{r,s}p_s + gh\alpha_r p_r = g\zeta_r',$$

$$-(\sigma^2 + ik\sigma/h)\,p_{-r} - 2i\sigma\Omega\beta_r^{-1}\sum_{s=-\infty}^{\infty}\gamma_{-r,s}p_s = 0 \tag{32}$$

where the $\gamma_{r,s}$ are coefficients defined as areal integrals of products such as $(\nabla\phi_r.\nabla\psi_s)$, and ζ_r' is the areal integral of $(\gamma U\nabla\phi_r)$. Evidently, the presence of rotation Ω greatly complicates the mechanical system and its solution. Equations (32) are a linear infinite set from which p_r may be solved. In practice, they are of course truncated to a finite set which defines the longer waves.

Figure 4(a) shows the rms amplitude of the tides forced by the P_{22} spherical harmonic of U/g in Webb's hemispherical ocean of depth h=4.4 km, as function of frequency σ>0, corresponding to the Moon travelling from east to west relative to the Earth ; (computations were also carried out for the hypothetical case σ<0). Results for three friction parameters ('decay times'), h/k = 50,30,12 hours, are shown. For such a simple geometry there is a surprising number of resonant peaks, corresponding to the eigenfrequencies of φ and ψ. Close peaks merge together as the decay time increases, that is for stronger friction.

Figure 4(b) shows the spatial form of the solution for the 'realistic' case of a nearly resonant case 2 cpd (that is S_2), with h/k =30 hours. If k were zero (no friction),the solution would be symmetrical about the central meridian ; friction skews the solution towards the west. The concentration of large amplitudes on the equator is not dissimilar to the tides of the Pacific Ocean. The corresponding solution for the diurnal potential P_{21} is anti-symmetric about the equator.

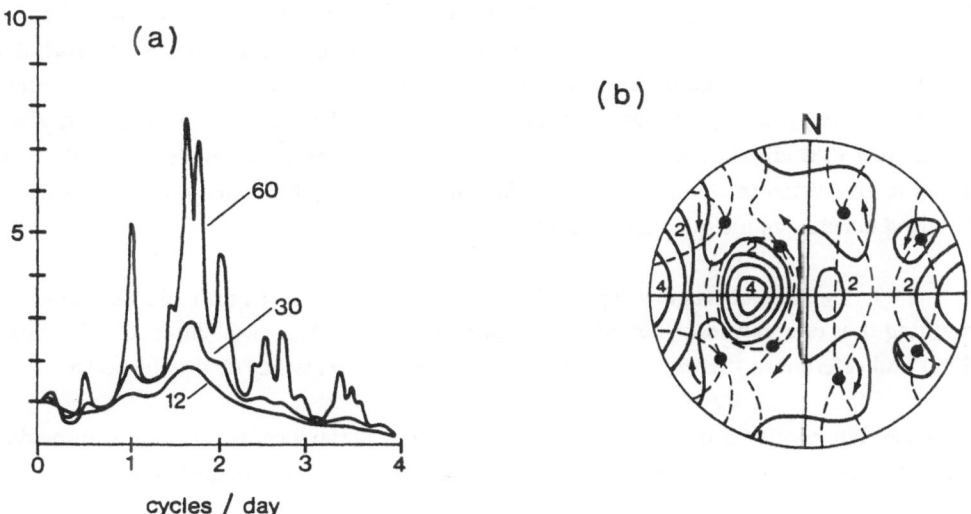

Figure 4: (a) Rms amplitude in a hemispherical ocean of uniform depth 4.4km as function of frequency σ/2π, for three 'frictional decay times' h/k (hours).   (b) Tidal map for 2 cpd and 30 hours. Full lines are contours of amplitude relative to 'equilibrium' amplitude (zero frequency) ; dashed lines are contours of phase lag 0,90,180,270°, with progression indicated by arrows. Heavy points are 'amphidromes' (zero amplitude). Adapted from Webb (1980) with permission.

3.2 REALISTIC GLOBAL OCEAN MODELS

From the preceding account it will be clear that any tide model involving the real shape and bathymetry of the ocean can only be tackled by quite heavy numerical computation. Apart from the coastal configuration, the depth of the ocean varies irregularly from <1 km to >6 km and is bounded by significant areas of continental shelf seas of depth 0–0.2 km. On these shelves, the reduced wave velocity $\sqrt{gh}$ produces a seven-fold reduction in the horizontal scale of tidal features requiring high resolution, and a ten-fold increase in vertical amplitude and current associated with most of the ocean's dissipation of energy through friction.

There have been many global tide models since about 1970. Time and space does not permit a comprehensive account of every one of them. Instead, we shall concentrate on certain broad categories which distinguish the different approaches.

Friction

Physical studies of flow in rivers and channels show that the friction terms in (24) should strictly take the form :

$$F_\theta = (c/h)u\sqrt{u^2+v^2}, \qquad F_\phi = (c/h)v\sqrt{u^2+v^2} \qquad (33)$$

where c is a universal constant in the region of 0.0025±0.0005. The factor h^{-1} appears because the effective body force per unit mass is a bottom stress distributed by turbulence over the whole water column, emphasizing still more the importance of shallow seas in dissipating energy. Most shallow sea models and some which cover both deep and shallow regions use (33) directly, but its nonlinear form precludes spectral decomposition into harmonic constituents. Solution has therefore to proceed by time-stepping, with a number of leading harmonics included simultaneously.

Le Provost has developed a linearised form of (33) by harmonic decomposition, assuming that one harmonic constituent dominates, usually M_2 - see Vincent & Le Provost (1988) and references therein. For a chosen harmonic of frequency σ', this leads to

$$F_\theta = r_1u' + r_2v', \qquad F_\phi = r_3u' + r_4v',$$

where the four coefficients r_i are specific to σ' and the spatial locality and depth considered. This complexity at least enables spectral solutions to proceed on a linear basis.

At the other extreme, Accad & Pekeris (1978) and Parke & Hendershott (1980) treat the deep-ocean field equations as frictionless ($F_\theta=0=F_\phi$), but allow energy to radiate outwards across certain open boundaries with shallow seas. This is done by relaxing the boundary constraint u.n=0, and either specifying a phase relation between u.n and ζ, or allowing u to be a free variable at such boundaries.

A few authors (e.g. Zahel,1977; Schwiderski,1980a) have felt justified in including, besides bottom stress, a term representing horizontal eddy viscosity, essentially to dampen excessive values of $|\nabla u|$. The precise formalism is fairly complicated, and involves an arbitrary coefficient whose adopted values vary widely, but at least such terms provide a form of numerical stability to the solution, which seems necessary to some schemes.

Loading potentials

At face value, the terms $\Sigma\gamma_n'\alpha_n\zeta_n$ in (24) greatly complicate the solution, because ζ_n implies global integration of the entire solution $\zeta(\theta.\phi)$, in effect producing an integro-differential equation. Some early attempts at a full solution by iteration, starting with neglect of the secondary potentials, (Hendershott,1972), failed to converge, but Parke & Hendershott (1980) found a compromise solution by combining basis functions.

Accad & Pekeris (1978) found a slowly convergent scheme, but they speeded up the rate of convergence by making use of the similarity of the load potential $\Sigma\gamma_n'\alpha_n\zeta_n$ to $Kg\zeta$, with K an arbitrary constant about 0.085 with a small imaginary part, as was mentioned in §2.3. Schwiderski (1980a) also simplified his equations by adopting K=0.10. However, such relations are not exact, and their use has been criticised. Other modellers (e.g. Vincent & Le Provost (1988)) have used Schwiderski's global solution for ζ to calculate the load potential as an explicit addition to the primary potential.

Normal modes and Proudman Functions

Instead of aiming, at a given frequency, for a direct solution of the dynamic equations and boundary conditions, a few authors have solved first for the *normal modes* of the ocean and their resonant frequencies, as shown in principle in §3.1. Normal modes of the global ocean have been computed by Gotlib & Kagan (1980) on a 5° grid, and by Platzman et al.(1981) in a finite-element model of similar average mesh size. Sanchez (1991) computed the more primitive *Proudman Functions* for a 4°-grid global ocean, analogous to the eigenfunctions ϕ_r, ψ_r defined for a hemisphere in §3.1. As in §3.1, but with the full complexity of a realistic ocean, equations (24) are solved with the generating potential set equal to zero and non-radiative boundary conditions. Friction has also been excluded, but the eigenfrequencies, (though not the normal modes themselves), appear to be little affected by moderate friction.

Figure 5 gives an idea of the most important normal-mode frequencies affecting the tides of species 1 and 2, from the results of Platzman et al (1981). The quantity plotted as 'spectral density' is actually the 'input coefficient' c_k for mode number k (related to ζ' in equation 32) :

$$c_k = A^{-1} \int \zeta_k{}^* C_{2s} dA,$$

where A is the ocean area. For a given spherical harmonic potential C_{2s} , c_k is the factor which expresses whether the modal wave form is spatially matched to the potential. The 'output' also depends on the difference between the natural frequency σ_k and the tidal frequency σ, making the total tide :

$$\zeta(\theta,\phi,t) = \sum_k [1 - \sigma/\sigma_k]^{-1} c_k \, \zeta_k(\theta,\phi) \cdot \exp(i\sigma t) \qquad (35)$$

See Platzman (1984a) for a formal mathematical analysis.

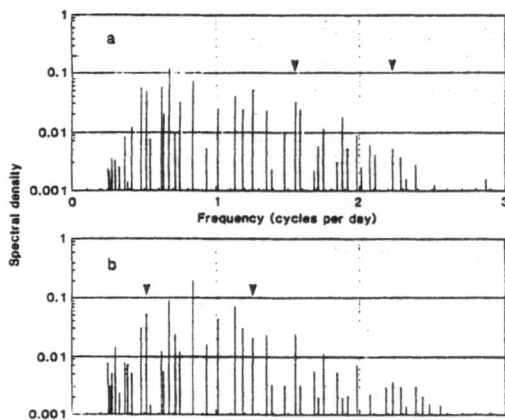

Figure 5 : The spectral density of 'input coefficients' (see text) for the spherical harmonic potentials of spatial form C_{22}(upper panel) and C_{21}(lower panel), as determined from the normal modes of the world ocean (from Platzman, 1983 - by permission). The small vertical arrows denote the effective frequency limits within which the modes contribute to the tides according to the filter (35).

Frictionless models give real frequency σ_k, but (35) may be extended to more realistic physical conditions by adding a complex component

$$\sigma_k(1 + i/2Q_k) \qquad (36)$$

where Q_k is the overall 'Quality Factor' of the mechanical system of mode k. Q_k accords with the usual definition :

$$Q_k = \sigma_k \, E/(-dE/dt)$$

(37) where E is the mean energy of the motion and (-dE/dt) its mean rate of dissipation. Platzman (1984a) deduced appropriate values of Q to be about 15 for M_2, 5 for O_1, in keeping with estimates deduced by Parke (1982) from a quite different calculation.

Sanchez, Rao & Woolfson (1985) proposed a tidal expansion in terms of the Proudman Functions of the basin, originally worked out for very large lakes but later successively extended to ocean basins of global scale. Since Proudman Functions are merely the eigenfunctions of the mass-conservation equation with no flow across boundaries, they are more primitive than the normal modes, and in a sense less restrictive to imposed physical assumptions. Sanchez & Cartwright (1988) fitted altimetric data from Seasat over the Pacific Ocean with a set of 50 Proudman Functions ϕ_n in the form :

$$\sum_{n=0}^{49} [\, a_n \, c(t) + b_n \, s(t) \,] \, \phi_n(\theta,\phi)$$

where c and s are specially chosen, nearly periodic, tide-like functions and a_n, b_n are arbitrary numerical coefficients, fitted to the data by least-squares. The results for M_2 and O_1 were quite creditable for the limited amount of data, and were shown to be slightly better than a similar fit by spherical harmonics. The set of Proudman Functions for the world ocean computed by Sanchez (1991) has yet to be exploited in this way.

<u>Data assimilation</u>

From a scientific point of view one would like a synthesis for the tides to be based entirely on dynamic principles, without constraints imposed to fit empirical data. Unfortunately, the best available 'unconstrained' models (e.g. Zahel (1978); Accad & Pekeris (1978)) show only qualitative agreement with the patterns of coastal measurements. For general accuracy of ζ better than about 0.1m, one must use a model constrained to agree with at least some of the measured data. The models of Parke & Hendershott (1978) and Parke (1982) are so constrained at continental coasts, but not at islands, thus leaving the island data free as tests of model accuracy. Schwiderski(1980b) applied constraints to force agreement with practically all known data, making it harder to assess accuracy except by means of new sources of data from the open ocean. Nevertheless, Schwiderski's maps of 11 harmonic constituents are at present considered to be the best available for general accuracy, (with some reservations).

Constraining solutions to agree with measured data has disadvantages : (a) as discussed by Schwiderski (1980b), one has to compromise with the condition of no-flow across the land boundary between adjacent data and hence on global mass conservation; (b) some data are distorted by fine coastal irregularities such as bays and capes, not resolved on the model grid; (c) some data are simply inaccurate, especially for minor constituents. Comparison of Schwiderski's maps with widespread measurements from the altimetry of Geosat (Cartwright & Ray,1990,1991) showed large areas of apparent discrepancy, possibly attributable to errors of type b or c, or to lack of coastal data in remote areas. At the other extreme, accepting an entire map of empirical data from altimetry allows measurement noise to contaminate the assumed tidal field. Noise errors tend to be most serious when the elevation field is differentiated to obtain currents or energy flux.

Such a paradox has led Zahel (1991) and others to study the assimilation of tidal data into a mathematical model, as an inverse problem - see Bennett & McIntosh (1980). Given elevation data $\zeta_j (j=1,m)$, Zahel's scheme is to minimise by variational methods a functional J of $u = (u,v)$, namely

$$J(u) = \langle [\, \iint |L(u)|^2 dA + \sum_{j=1}^{m} w_j^2 |\partial \zeta_j/\partial t + \nabla.(hu)|^2 \,] \rangle \tag{38}$$

where $L(u)$ is the residual of the equation in u formed by eliminating ζ from equations (24), including the load integrals. The last squared bracket is the residual of the divergence equation (3rd equation of (24)) for each elevation data value, and the w_j are suitable weighting parameters. The solutions incorporate a strict condition of global mass conservation, absent from all other known models. By such means it is hoped to achieve an ideal compromise between exact solutions of an assumed dynamic formulation and exact agreement with measured elevations.

Jourdin et al (1991) have also pursued an inverse method to the assimilation of a wider variety of tidal data, including not only tide-gauge records but also tidal variations of gravity at land stations and satellite altimetry. Their model is, however, purely empirical, without embodying any dynamic constraints; the authors hope eventually to test its efficacy by using its solutions as boundary input to a hydrodynamic model.

4. TIDAL ENERGETICS AND ORBITAL EVOLUTION

4.1 ENERGY DISSIPATION IN THE GEOSPHERE

The determination of the total rate of dissipation of the Earth's rotational energy in the tides is, to some geophysicists, a more important goal than accuracy of detail in the tidal field itself. The subject has gained interest in recent decades owing to the great precision with which the rotation is now monitored, the greater awareness of the many physical causes of irregularity in the length of day, and the growth of space geodesy in general. In this section, we shall broaden our view of the physical process of tidal dissipation and momentum exchange at the Earth's surface. But first, it is necessary to review the overall problem in a setting of celestial mechanics.

Figure 6 is a version of the classic picture of tides on the Earth's surface retarded by friction and hence experiencing a (second-order) torque in the opposite sense to the rotation, caused by the tidal forces on the tidal mass itself. The torque decelerates the Earth's spin Ω, and in order to conserve angular momentum increases the orbital acceleration of the Moon in accordance with Kepler's law :

$$n_{\mathfrak{c}}^2 a_{\mathfrak{c}}^3 = \text{constant}, \ (n_{\mathfrak{c}} = dq/dt) \tag{39}$$

(39) is satisfied by a deceleration of the Moon's mean longitude q at about -25"/cy^2, (cy = century), and a positive increase of the Moon's semi-major orbital axis $a_{\mathfrak{c}}$ by about 3.7m/cy - the numerical estimates have only been agreed recently and may yet be modified. Interest in $d\Omega/dt$ stems from the need to remove the longterm trend due to tides from observed variations. Interest in $da_{\mathfrak{c}}/dt$ stems from interest in the history of the Moon's orbit.

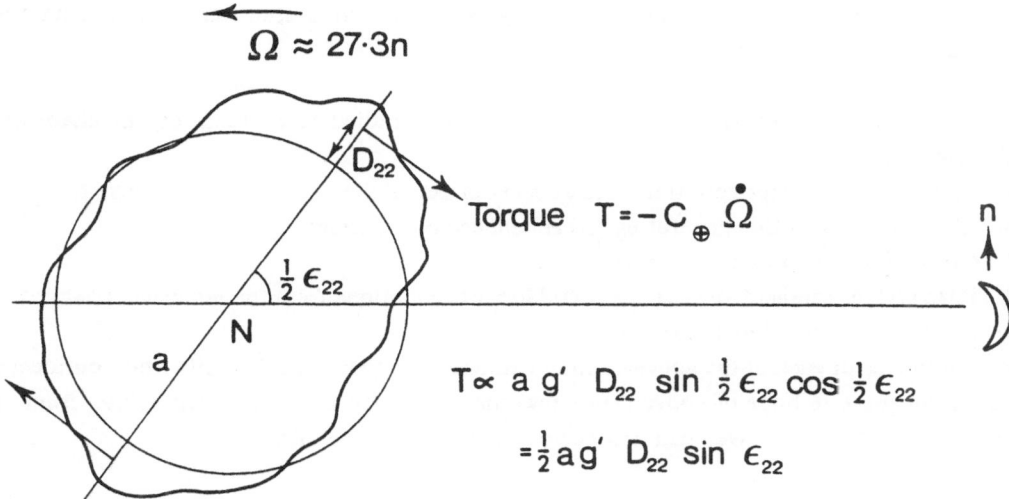

$$\Omega \approx 27 \cdot 3n$$

$$\text{Torque} \quad T = -C_\oplus \dot{\Omega}$$

$$T \propto a\, g'\, D_{22}\, \sin \tfrac{1}{2}\epsilon_{22} \cos \tfrac{1}{2}\epsilon_{22}$$

$$= \tfrac{1}{2} a\, g'\, D_{22}\, \sin \epsilon_{22}$$

Figure 6: Friction in the ocean tide (wavy line - exaggerated scale) induces a phase-lag ϵ_{22} relative to (2x Earth's longitude) - hence $\epsilon_{22}/2$ in geometric plan. Tidal forces on mean bulge, amplitude D_{22} apply torque T in opposite sense to rotation Ω, proportional to $-d\Omega/dt$. Notation of formulae as in Figure 1.

The usual diagram, which goes back to Lord Kelvin, shows the tide as a smooth tilted oval suggesting a lagged Body Tide. That is misleading because the Body Tide is now known to have negligible phase-lag; practically all the lag resides in the ocean tide, which has a very irregular profile. The axis shown in Figure 6 is that of the P_{22} harmonic of the ocean tide, whose M_2 amplitude is only about 2 cm. Figure 6 does not apply to the *zonal* tide potential P_{20} which being radial cannot brake the Earth's rotation; the zonal constituent Mf does however have a small *positive* accelerative effect on the Moon's longitude.

The solar oceanic tide similarly induces a braking torque, of smaller magnitude, but the corresponding deceleration of the Sun's longitude may be shown to be undetectably small. The solar tide in the **atmosphere** is principally induced by a *thermal* mechanism, causing it to *lead* the Sun. ($\epsilon_{22(air)} = -119°$), and thereby accelerate the Earth by a small but detectable amount - the braking effect of the gravitational oceanic tides of course predominates.

A question which has been curiously neglected is : how is the torque on the ocean tide transmitted to the solid Earth ? Bottom friction on steady tide-induced westward currents along the equator, with return flow at higher latitudes, are conceivable, but the more likely mechanism is a small differential pressure on west-facing relative to east-facing continental slopes. The details have never been thoroughly examined. Of possible observation interest is the fact that, since the torque and its associated pressures vary with (tide amplitude)[2], they will exhibit 14.77days periodicity with the spring-neap cycle (MSf as distinct from Mf), which may be more easily detectable on a short-term basis than the constant effect.

The following relevant features may be studied to evaluate the mechanisms discussed above:

A. Physics of tide dissipation in the ocean, and horizontal divergence of power-flux;

B. Rate of working of tide-generating forces and sea-bed motion;

C. Perturbations of orbits of artificial satellites;

D. Times and locations of solar eclipses and lunar occultations from the historical past;

E. Long-term series of lunar laser-ranging.

Measurements of each of the above distinct features have their own diversity and confidence limits, but there is hope of convergence towards definitive and consistent values from at least the last four items. We shall now review their essential details.

A. Dissipation in the ocean

Applying the general drag formula (33) to an area dA of sea with near-bottom tidal velocity u, one gets for the local rate of energy dissipation

$$D = <c\rho |u|^3 dA> \tag{40}$$

where $< >$ as usual denotes the average over a tidal cycle. For a single harmonic constituent of amplitude u_0, considered in isolation, (40) becomes

$$D = (4/3\pi)c\rho u_0^3 dA, \tag{41}$$

but addition of other constituents requires more elaborate calculation of cross-terms with averaging over a synodic period, especially if the constituent current ellipses have different orientations. If (41) is used with a mean spring-tide amplitude (M_2+S_2) for u_0 (as commonly supplied in mariners' tables), the result is roughly twice the value of D for M_2 alone.

Rough integration of (41) over all the world's shallow seas, carried out around 1920, was the first method to give credible results for global dissipation, but the patchy information about u together with uncertainty of the coefficient c confine this method to the category of 'historic interest'.

A more reliable approach is to calculate the average flux of tidal power across a linear element ds of arbitrary boundaries separating dissipative sea areas from the deep ocean. Denoting distance along such a boundary by s, depth by $h(s)$, and n the inward unit normal to s (towards the dissipative area), the dissipation within is estimated as

$$D = \rho g \int h(s) < n.u(s,t) \ \zeta(s,t)>ds \tag{42}$$

For a harmonic constituent $\zeta = \zeta_0 \cos(\omega t-G)$, $u.n = u_0 \cos(\omega t-G^*)$, the time average in (42) becomes $0.5 \ u_0\zeta_0 \cos (G-G^*)$. Increments from other constituents are simply additive.

The only attempt to apply (42) globally, over all reasonably documented shallow seas, was by G.W.Miller (1966). Miller estimated the total rate of dissipation by M_2 to be 1.4 Terawatts (1 TW $=$ 10^{12} Watts), possibly stretched to 1.7 TW, but no further in his estimation. Cartwright et al (1980) and Cartwright & Ray (1989) showed detailed applications to certain seas in the Atlantic Ocean bounded by more reliable direct measurements than were available to Miller, obtaining somewhat different results but not seriously affecting the total. Sündermann (1977) showed by means of a hydrodynamic model that the Bering Sea, previously considered by all investigators to be a major sink of tidal power, is in fact a rather trivial dissipator. Subtraction of the Bering Sea brings Miller's mean total estimate down to close to his lower figure of 1.4TW. It will be seen below that such a figure is significantly less than the presently well-agreed M_2 dissipation rate of about 2.50±0.05TW as deduced from geodetic and astronomical methods. The difference presents a certain oceanographic challenge : What is the physical nature and distribution of the unidentified oceanic sink?

B. Work done by the external tide-generating forces

The rate of working by the external forces per volume element of ocean may be expressed as $\rho u.\nabla_3 U'$, where ∇_3 denotes the 3-dimensional gradient vector, and U' the *total* tide potential. Since $\nabla_3.u=0$, this is the same as $\rho\nabla_3.(uU')$. Integrating over a thin water column of depth (h+ζ) and adding the work done by the bottom elevation ζ_b gives a mean rate of working of all external forces per surface area dA :

$$W = \rho <\left[U'd\zeta/dt + g(h+\zeta)d\zeta_b/dt + \nabla.\{u(h+\zeta)U'\}\right]>dA \qquad (44)$$

where $\nabla.$ is the horizontal divergence operator.

For simplicity, neglect the 3rd degree potential U_3', so that

$$U' = (1+k_2)U_2 + U_L; \quad \zeta_b = h_2U_2/g + \zeta_L ,$$

where U_L is the load potential and ζ_L is the load tide (equations 23). Since for any pair of variables x,y we have $<xdy/dt> = - <ydx/dt>$, equation (44) can then be written :

$$W = \rho <\left[(\gamma_2U_2+U_L-g\zeta_L)d\zeta/dt + \nabla.(uhU')\right]>dA \qquad (45)$$

where, in the last product, ζ has been neglected in comparison with the depth h. Note that here ζ strictly represents the surface tide relative to the bottom tide.

Considering the first term of (45), the substantial transfer of power through the ocean floor implied by the partial product ($-h_2U_2d\zeta/dt$) does not imply an energy sink in the solid Earth; it is merely part of the mechanism whereby the tidal forces act on the Earth and ocean together. Insofar as U_L and ζ_L are nearly proportional to ζ without phase-lag, these terms contribute a negligible amount to the time average. Invoking *complex* loading Love numbers would indeed imply a possible energy sink in the solid Earth due to inelasticity. Such sinks have been proved to be small (Zschau,1980), in fact negligible compared with the oceanic power loss at present levels of uncertainty.

When (45) is integrated over a large ocean area, the divergence term becomes a line integral of the normal flux of power (equation 42) across the bounding periphery. Over the entire oceans and seas this integrated flux vanishes because $u.n = 0$ at every land boundary. With neglect of the products of U_L and ζ_L, (45) then reduces to

$$W_{total} = \rho \int <(\gamma_2 U_2 d\zeta/dt)> dA \tag{46}$$

W_{total} represents the total rate of working of the tidal forces and hence the total power loss in the ocean through all physical mechanisms. The various harmonic constituents contribute to it by linear addition of their respective integrals.

On the other hand, if the integral (46) is confined to an area wholly or partially open to other seas and oceans, then the power flux across the open boundary has to be added from (42,43), with the divergence term in (45) added to replace ζ in (42) by $(\zeta - U/g)$. The power loss in the designated area is then the sum of (46) and the net inward power flux across its boundaries.

(46) is certainly the most robust method of calculating the dissipation rate by direct inference from observable ocean dynamic features, as distinct from indirect astronomical inference. Different estimates are due to differences in the models used to formulate ζ, and to the degree of empiricism used to constrain solutions. When applied directly to a given tidal model, the integrand is found to vary spatially between + and - values on a relatively small scale. The overall positive work rate is due to a small systematic tendency to positive values, in the case of semidiurnal tides concentrated near the equator - see Cartwright & Ray, 1991, Figure 9. There is no relation between the distribution of larger positive values of the integrand and the location of shallow seas where the actual sinks of energy supposedly lie. The tides advect energy horizontally in a complicated system.

Since (46) consists of a convolution of $d\zeta/dt$ with U_2 only the degree-2 spherical harmonics of ζ are relevant to the dissipation integral - the contribution from U_3 is quite negligible here - hence the relevance of Figure 6. More specifically, if the P_{22} harmonic of ζ is written :

$$\zeta_{22} = D_{22} P_{22}(\sin \theta) \cos(\sigma t + 2\phi - \varepsilon_{22}),$$

corresponding to a semidiurnal constituent of U_2 of frequency σ :

$$U_{22} = \sqrt{15/32\pi}\ g_e H' \sin 2\theta \cos(\sigma t + 2\phi)$$

with "globally normalised" amplitude H' (0.632m for M_2), then application of (46) gives :

$$W_{total} = \sqrt{24\pi/5}\ \rho\sigma\gamma_2 GM\ H'\ D_{22} \sin \varepsilon_{22}\ (Terawatts),$$

where GM is the fundamental gravitational constant $398.6 \times 10^{12}\ m^3 s^{-2}$.
For M_2, with D_{22} in cm, this becomes numerically

$$W_{total} = 0.986\ D_{22} \sin \varepsilon_{22}\ \ (TW)$$

(The above notation for degree-2 amplitude D and phase lag ε is that of Lambeck (1980). Most satellite geodesists give results for the same quantity in terms of a 'phase' equal to $(450° - \varepsilon_{22})$.)

There have been many estimates of W_{total} for M_2 , and increasingly for other constituents, from various models of $\zeta(\theta,\phi)$. Marchuk & Kagan (1989), in their Tables 8.4, 8.5, quote many pre-1980 results, mostly between 2 and 4 TW, but with a few wild values up to 7 TW. Lambeck (1977) favored values greater than 3 TW, which happened to be close to then high current estimates from satellites. Platzman (1984b) pointed out that post-1980 models for M_2 have tended towards lower values in the range 1.8-2.8 TW. Table 4 (below) lists a selection of the most recent estimates of W_{total} for M_2, together with values of $(D,\varepsilon)_{22}$ where known, from four dynamic/empirical models of $\zeta(\theta,\phi)$ with integration (46), and from three independent analyses of satellite orbits, (§4.2), for comparison. The four models cited vary from the purely dynamic calculation by Accad & Pekeris (1978), (whose agreement with the best satellite-derived values for W may well be fortuitous), through the data-constrained models of Parke (1982) and Schwiderski (1983), to the purely empirical model of Cartwright & Ray (1991) derived from Geosat altimetry. Except for the low value of W from the otherwise very good model of Schwiderski (1983), there are strong signs of convergence of independent estimates towards W_{total} $(M_2)=2.5\pm0.1$TW. By adding similar results from the other major harmonic constituents including diurnals, a good modern estimate for the *overall tidal dissipation rate* is 3.5 ± 0.2 TW.

TABLE 4 : Parameters of P_{22} harmonic of the tide constituent M_2 with rate of working W

Reference	Method	D_{22}	ε_{22}	TW
Accad/Pekeris 1978	dynamic			2.55
Parke/Hendershott 1982	semi-empirical			2.22
Schwiderski 1983	semi-empirical	2.98	138°	1.94
Cartwright/Ray 1991	altimetry	3.51	133°	2.54
	Satellites			
Gendt/Dietrich 1988	LAGEOS	3.28	131°	2.44
Cheng et al 1990	Starlette	3.22	131°	2.41
Marsh et al 1990	GEM-T2	3.32	129°	2.55
(rms uncertainty)	GEM-T2	0.05	0.8°	0.05

4.2 SATELLITE ORBITOGRAPHY AND ASTRONOMICAL METHODS

C. Perturbation of satellite orbits

In § 4.1 we approached the problem of tidal braking of the Earth's rotation from its fundamental origin in oceanic dissipation. We now consider supporting ways of estimating the same or closely related quantities, which do not involve direct observation of surface elevation or analysis of ocean dynamics.

The (practically) complete equation for the exchange of momentum between the Earth's spin and the Moon's orbit is derived as follows : The total angular momentum parallel to the Earth's axis is

$$A = C_\oplus \Omega + \mu^{-1} m_\ell a_\ell^2 \sqrt{1-e^2} \, n \cos i, \quad (\mu = m_\ell/m_\oplus = 1.0123), \tag{47}$$

where m_ℓ, a_ℓ, e, n, i are the mass, major axis, eccentricity, mean angular velocity, and mean inclination (= obliquity ε) of the Moon's orbit, and $m_\oplus$, $C_\oplus$, Ω are the mass, principal moment of inertia and spin rate of the Earth. The relatively small angular momentum of the Moon's monthly spin has been neglected. Constancy of A with Kepler's Law (39) then gives :

$$C_\oplus (d\Omega/dt)_{LT} = \mu^{-1} m_\ell a_\ell^2 [(1/3)\dot{n} \cos i + n e \, \dot{e} \cos i + n \sin i \, (di/dt)] \tag{48}$$

The last two terms of (48) are much smaller than the others in magnitude, so the equation is approximately a relation between $\dot{n}$ and $(d\Omega/dt)_{LT}$, where the suffix LT denotes the part due only to the lunar tides.

The rate of increase of rotational energy gives another fundamental relation :

$$(dE/dt)_{LT} = C_\oplus \Omega (d\Omega/dt)_{LT} - (1/3) m_\ell a_\ell^2 n \dot{n} \tag{49}$$

This is generally negative; its magnitude has to be equated to the oceanic dissipation rate from lunar tides, discussed in §4.1. There is a similar increment (without the last term) from the solar tides, including the solar atmospheric tides. Observational data for changes in Ω and in the 'length of day' (lod) also include the effects of momentum exchanges between atmosphere and lithosphere and between crust and core, on a wide spectrum of time scales.

We shall later consider the direct evaluation of $\dot{n}$, but all the terms on the right hand side of (48) may be evaluated from analysis of the orbits of artificial satellites, in effect, the tidal parameters D_{2m}, ε_{2m} (m=0,1,2) defined in §4.1. The reasoning is as follows, leading to equations (53-55).

With a constituent of the ocean tide, frequency σ, expanded in general terms as

$$\zeta_\sigma(\theta,\phi,t) = \sum_{l=1}^{\infty} \sum_{m=0}^{l} \left[D_{lm}^{+} \cos(\sigma t + m\phi - \varepsilon_{lm}^{+}) + D_{lm}^{-} \cos(\sigma t - m\phi - \varepsilon_{lm}^{-}) \right] P_{lm}(\cos\theta), \tag{50}$$

the associated disturbance to the gravitational potential δU to be added to the Earth's own potential at geometric distance r may be written :

$$\delta U = 4\pi\rho G a \sum_{l} \sum_{m} \frac{1+k_l'}{2l+1} \left(\frac{a}{r}\right)^{l+1} \left[D^{+} + D^{-}\right] P_{lm}(\cos\theta) \tag{51}$$

where a is Earth's equatorial radius, and D^{+}, D^{-} are short for the terms in [ ] in equation 50. The load factor $1+k_l'$ happens to be identical to $\gamma_l = 1 + k_l - h_l$, (Molodenskii, 1977).

In order to calculate the perturbations to to the orbit of a satellite (the Moon, or an artificial one) from the increment of potential δU, the argument $m\phi$ is transformed into the Keplerian elements of the orbit according to Kaula's theory (Kaula, 1966). In this theory, r in (51) is replaced by the satellite's major semi-axis a_s, and each spherical harmonic is expanded in a further double summation of the form :

$$\sum_{p=0}^{1} \sum_{q=-\infty}^{\infty} F_{lmp}(i)G_{lpq}(e) \ (\cos,\sin)(\sigma t - \varepsilon_{lm}^{\pm} \pm \Psi_{lmpq})$$

with $\Psi_{lmpq} = (l-2p)\omega_{s} + (l-2p+q)M_{s} + m(N_{s}-\phi_{s})$, $\qquad\qquad$ (52)

where i,e,ω,M,N, ϕ are the satellite's inclination, eccentricity, perigee, mean 'anomaly', ascending node and sidereal angle, respectively. The (cosine,sine) arguments apply to (even,odd) values of (l+m). F(i),G(e) are simple polynomials in cos i, sin i, and e respectively, quoted by many authors without definition, but tabulated in Kaula (1966) and Lambeck (1980, p.114).

The whole construction is simplified enormously by the fact that the only perturbations producing a significant response in the orbit are those of very low or zero frequency. These have the $^{+}$ superfix and a certain combination lmpq for each tide constituent; as leading examples, 2200(M_2), 2201(N_2), 2210(K_2), 2100(O_1), 2110(K_1). Further details may be found in Lambeck (1980). I cite here only the perturbations in semi-major axis a_s and inclination i_s :

$$(\dot{a}_s)_{lmpq} = 2 K_{lmpq} \ (l-2p+q) \ (\sin,\cos) \ \varepsilon_{lm}^{+} \qquad\qquad (53)$$

$$(di_s/dt)_{lmpq} = K_{lmpq} \frac{(l-2p)\cos i - m}{a_s \sin i \sqrt{1-e^2}} \ (\sin,\cos) \ \varepsilon_{lm}^{+} \qquad\qquad (54)$$

with $K_{lmpq} = 3\sqrt{Gm_\oplus/\mu a_s^3} \dfrac{\gamma_l}{2l+1} \dfrac{\rho}{\rho_\oplus} (a/a_s)^{l-1} F_{lmp}(i) \ G_{lpq}(e) \ D_{lm}^{+} \qquad\qquad (55)$

As useful examples of F,G; $F_{220} = (3/4)(1+\cos i)^2$, $F_{210} = (3/4)\sin i \ (1+\cos i)$, $G_{200} = 1-(5/2)e^2$, $G_{210} = 7e/2$. The density ratio $\rho/\rho_\oplus$ is 0.1876.

Applying (53) to the Moon itself, we get secular (zero frequency) changes in $a_{\langle}$ which translate simply to $\dot{n}$ through Kepler's Law (39). The relation of $\dot{n}$ to $D_{22} \sin \varepsilon_{22}$ is the precise formulation of that suggested heuristically in Figure 6. Equation (54) gives $(di/dt)_{\langle}$, and a similar equation gives $\dot{e}_{\langle}$ .Thus we can express all the elements on the right of equation (48) in terms of the tide harmonics $(D,\varepsilon)_{2m}$ and so determine $(d\Omega/dt)_{LT}$.

Applying (54) to an artificial satellite, relatively close to Earth, instead of secular terms the circular functions (52) have finite low frequency arguments depending on the orbit configuration and the parameters σ,l,m,p,q of the tide harmonic. As example, the satellite 'Starlette' shows perturbations in i_s of periods 10.5, 36, 11.8 and 90 days, produced by the tide constituents M_2, S_2, O_1, K_1 respectively. These and similar variations in N_s have amplitudes of order 0.1" - 1.5", which may be extracted from spectral analysis of long series of observations, finally giving the amplitudes and phases of the respective tide harmonics which in turn give the secular terms in the lunar orbit, as described above. Unlike the lunar case, the ratio (a/a_s) in (55) is not very much less than unity, so terms of degree l>2 are not negligible and add to the spectral perturbations. Ambiguities can, in principle, be resolved by analysing the orbits of several satellites.

The orbits of several geodetic satellites have been extensively analysed in the above manner by Cazenave & Daillet (1981, and earlier references therein). However, in recent years, more precise results for the tide harmonics and secular accelerations have been obtained by teams in the USA, using non-spectral methods. Millions of tracking data from several satellites are inverted to solve large matrices of spherical harmonic coefficients of the Earth's gravity field, including tidal perturbations (51) along with the fixed geographic expansion. As an impressive example, the 'Goddard Earth Model' GEM-T2 (Marsh et al, 1990,1991) used over two million tracking observations from 1130 orbital arcs of 31 satellites to produce a 90-term development of 12 major tide harmonics along with a high precision fixed gravity field to degree and order 36. A few results are included in Tables 4 and 5. Table 4 also shows similar results from single satellite orbits by other groups of authors, with evident close agreement in the leading tidally induced terms.

(D.E) Direct observations of lunar acceleration

An acceleration of the Moon, (or strictly, a deceleration of its mean longitude), has been known since the late 17th century when Edmond Halley observed that the times recorded for ancient eclipses did not tally with current Moon positions and times. But only in the early 20th century was it realised that the Earth's rotation has been slowing down too, as well as fluctuating. The increase in lod required a radical revision of early lunar timings and positions, because the time itself had been assumed to be uniformly geared to the Earth's rotation. The two decelerations (Moon and Earth) had to be disentangled before any physical sense could be made of either. The trick, in brief, is to analyse the 'Weighted Discrepancy Difference' (Munk & MacDonald, 1960) :

$$WDD(T) = \Delta\lambda_\text{C} (T) - (n_\text{C}/n_O)\Delta\lambda_O(T) \tag{56}$$

where $\Delta\lambda_X$ is the discrepancy between a measured longitude of celestial object X (O for Sun) and its calculated ephemeris value, as a function of time T in centuries. The longitudes of Mercury and Venus have also been used in place of Sun's longitude.

The WDD is free from the effects of secular changes in Earth rotation; we may assume that WDD $\approx$ (1/2) $\dot{n}_\text{C} T^2$ + measurement errors. Hence, $\dot{n}_\text{C}$ (assumed constant on a long time scale), may be evaluated by a least-squares parabolic fit to WDD data extending over a few centuries. The resulting value may then be used in turn to remove the lunar acceleration from $\Delta\lambda_\text{C}(T)$ or eclipse timings, and hence obtain unbiassed estimates of $(\delta\Omega/\delta t)$ and its variations at various epochs.

The available observations of sufficient precision consist of timed occultations of stars by the Moon and transits of the Sun and Mercury since the invention of the telescope about 1620, and timings of total solar eclipses from the millennium around 0 BC, selected from variably reliable Chinese and Babylonian sources, (Stephenson & Morrison, 1984). However, the selection of 'reliable' data is somewhat subjective, so the results of different authors have varied. For many years, the results of Spencer-Jones were accepted, namely $\dot{n}$ $=-22.4''/cy^2$ (arcseconds per century^2) from telescopic data and $-38''/cy^2$ from certain Babylonian eclipse records - see Munk & MacDonald (1960). Lambeck (1977) quotes 17 other estimates from the 1970's, varying from -27 to $-79''/cy^2$. Most aficionados now seem to prefer Morrison & Ward's(1975) value of -26, which Stephenson & Morrison use to deduce a longterm average of ($\delta\Omega/\delta t$) of -6.4×10^{-22}rad/sec^2, equivalent to a *tidal* increase of lod at 2.4msec/cy, consistent with both telescopic and ancient data. (The quoted units are 'traditional'; 10^{-22} rad/s^2 = $205''/cy^2$.) The well observed rate of increase of lod in the 20th century averages only 1.4ms/cy, showing that non-tidal influences are now causing the Earth to *accelerate* on a timescale of several decades, supposedly reflecting momentum exchange between crust and core.

Lastly, another independent technique for estimating the tidally induced lunar acceleration is *LunarLaserRanging* to the reflectors deposited on the Moon by the 'Apollo' astronauts in 1969. The measure sought, relevant to us, is $\dot{a}_{\mathfrak{l}}$. Since this is only a few cm/y, extraordinary precision is required in orbital and rotational corrections, to be maintained over many years. Having now completed a nutational cycle of 19 years, the latest results from the JPL team give $\dot{a}_{\mathfrak{l}}$ =3.8±0.2 cm/y, equivalent to $\dot{n}$ = $-25.6\pm1''/cy^2$ according to Kepler's Law. The figure $-24.9''/cy^2$ quoted in Table 5 from Newhall et al (1988) were presented at a Workshop in 1988; the later estimate, $-26.1\pm1''/cy^2$, is from a 1991 Abstract.

TABLE 5 : Recent estimates of total tidal accelerations of the Moon and Earth

Reference	Method	$\dot{n}$ $''/cy^2$	$(\delta\Omega/\delta t)$ $10^{-22}/s$	Δ(lod)$_T$ ms/cy
Cartwright & Ray, 1991	Altimetry	-24.8	-6.3	2.4
Cazenave, 1982	Orbit spectroscopy	-26.1	-	-
Marsh et al, 1991	Satellite ranging	-24.9	-5.8	2.2
Morrison & Ward, 1975	Occultations	-26	-	-
Stephenson & Morrison, 1984	Historical data	-	-6.4	2.4
Newhall, Williams, Dickey, 1988	L.L.R.	-24.9	-	-
Dickey, Williams, Newhall, 1991	L.L.R.	-26.1	-	-

Table 5 summarises the most recent estimates known to the writer from all the techniques discussed above. Note that they strictly refer to the *total tidal* effects, not the partial effects of single harmonic constituents. Those values derived by spectral methods have been summed over relevant constituents, namely M_2, N_2, O_1, Q_1, (Mf if available) for $\dot{n}$, and both lunar and solar constituents for ($\delta\Omega/\delta t$) and Δ(lod), including a small increment from the atmospheric S_2 tide. The values quoted from altimetric analysis of the ocean tides (line 1) are not published directly in Cartwright & Ray (1991) but have been calculated from the published values of $(D,\varepsilon)_{2m}$ and the energy equation (49). The value for $\dot{n}$ from Cazenave (1982) epitomises several authors' work on orbital spectroscopy, agreeing closely in the value $-21.9\pm1.5''/cy^2$ for M_2 alone. There is an encouraging degree of convergence in all three columns.

5. DIRECTIONS FOR FUTURE RESEARCH

5.1 AREAS WHERE IMPROVED KNOWLEDGE IS NEEDED

The most promising directions are to exploit the advances in knowledge of tides recently acquired through satellite geodesy. These may be divided into two categories : (a) large scale global tidal features derived from high precision orbitography, and (b) small scale features which may be directly observed by altimetry. Ideally, (a) and (b) should merge into a unified definition of the ocean tide field, but in practice each is more accurately defined at its particular end of the wavelength scale.

Large scale features are best described by expansion in spherical harmonics from degree 1 up to a practical limit; there should be no (0,0) harmonic because of mass conservation. The GEM-T2 gravity field, derived from the orbits of 31 satellites, expands the ocean tide elevation/phase field up to degree 6 for 12 constituent frequencies. Higher expansion from pure orbitography is limited by the relatively small amplitude of the tidal perturbations at orbit height, on account of the factor $(a/r)^l$ in the potential at geocentric distance r. High order expansions from combined analysis of tracking and altimetry are being explored.

The principal need for accurate low-degree tide expansion is in precision orbit determination, which of course has many geodetic applications. Terms of degree 2 require additional precision on account of their relevance to Earth braking and secular changes in the Moon's orbit. We saw in §4 that the leading results from altimetry and other methods are converging, but further research is needed to achieve 1 percent accuracy in all major constituents, especially those affected directly or indirectly by solar radiation.

Satellite altimetry is obviously the best tool for determining medium and short wavelength detail. Although this too depends on accurate removal of tide perturbations in the orbital radius, modern p.o.d. packages seem to be accurate enough (<1cm in tidal detail) for the orbital influence to be neglected. Cartwright & Ray (1991) showed the kind of detail which can be achieved from one satellite mission, but a lower noise level is clearly desirable. Further determinations are required from other altimetric satellites and new experiments in combining data assimilation with dynamic models.

Some outstanding goals are :

(a) The definition of major harmonic constituents of the tide elevation field with sufficient accuracy for determining horizontal fluxes, by inversion of the momentum equations (24). The tidal mass flux or the depth-averaged current are required for acoustic tomography, and also for determining the field of power flux and its divergence, in order to identify the energy sinks. We need a **physical** explanation of where and how the ocean dissipates 3.5 TW over the tide spectrum.

(b) Fine detail in high amplitude coastal and shelf seas. Because of very short wavelengths and altimeter locking problems near land, this may require a combination of shelf sea dynamic models with oceanic boundary conditions supplied by altimetry.

(c) To complete global definition by a linked dynamic model of the Arctic Ocean (never completely covered by altimetry), and ice-covered parts of the Southern Ocean. The Mediterranean Sea also requires separate modelling.

Besides their intrinsic interest, these goals also have a practical motive, to provide the best possible synthesis of the tide elevation field to subtract from altimetry for general oceanographic interpretation.

5.2 SOME ASPECTS OF THE TIDE SIGNAL IN ALTIMETRY

We conclude with some points of difficulty which must be considered when examining a long record of altimetric data for the purpose of extracting and specifying its tidal signal.

Components of the signal

After applying several more-or-less known corrections for ionospheric, atmospheric and surface conditions to the pulse travel-time, and having subtracted the resulting altimetric height from the computed (p.o.d.) orbital radius, the geocentric (ellipsoidal) sea surface elevation SSE can be expressed :

$$SSE(\theta, \phi, t) = z_{geo} + z_{orb} + z_{tid} + z_{dyn} + \text{instr. noise} \tag{57}$$

where the four z variables represent respectively geoid height (30), radial orbit error (0.3), geocentric tide (0.5), and dynamic ocean topography (0.1-1.0), bracketed numbers being typical spatial or temporal rms variability in meters. The last term, instrumental noise (0.1) is supposed to include errors in transmission corrections; it will not be discussed here.

Removal of z_{geo} is clearly a prime necessity. There are various schemes involving subtraction of an 'optimal time-averaged surface', but analysis of

$$S'(\theta,\phi,t_1,t_2) = SSE(\theta,\phi,t_1) - SSE(\theta,\phi,t_2) \qquad (58)$$

has many advantages for tidal analysis, especially if t_1-t_2 = constant. This condition applies everwhere only in 'repeat' orbits; any harmonic constituent is thereby transformed in amplitude and phase by an elementary and precise rule. With care and precision, S' is free of both geoidal undulations and geographically correlated errors in z_{orb}.

The figure 0.3m quoted above for a typical rms of z_{orb} (0.42 in S') is optimistic for 1992 state-of-art orbit definition; errors >1m were common in earlier practice. Further reduction of the dominant 'once-per-revolution' error in z_{orb} is however desirable, in order that ascending and descending passes, differing by half a revolution near the equator, may be lumped together in the same analysis. Unified treatment of ascending and descending passes is essential for separation of the diurnal and semidiurnal species, which otherwise have very similar aliasses. I shall leave discussion of the general problem of reducing orbit error to other experts at this Summerschool, with a warning that any method which relies on the altimetry itself must be free of tidal influence from all parts of the ocean.

Having beaten down the two most serious noise sources, we are left with the tide signal superposed on the oceanographer's 'signal' z_{dyn}. In the context of this Course, z_{dyn} has to regarded as 'noise'. This is much the same situation as in analysing the records from a conventional tide gauge, except that the altimetric sampling regime involves severe aliassing of the tidal signal into the most energetic part of the noise spectrum. On the other hand, there are some advantages to be gained from the **spatial** coherence of the tide field. The main strength of tidal analysis, of course, lies in the strong **temporal** coherence with the generating potential.

Aliassing

Tidal analysis may be organised in various ways; it is convenient to think in terms of spectral resolution of the major harmonic constituents, especially $(Q,O,P,K)_1$, $(N,M,S,K)_2$. It is fundamental that in order to obtain direct information about these harmonics one must sample each of them at a fairly even distribution of phase arguments between O and 2π. Consider a satellite of period T which repeats its ground track every N revolutions in time NT = M nodal days, where N,M are integers. A nodal day is $(1-\delta)$ sidereal days, δ being small in magnitude and >0 for prograde ($i<90°$), <0 for retrograde orbits ($i>90°$), according to a wellknown formula relating regression rate to inclination and orbital radius. Between successive passes over a given field point P a constituent of frequency σ will advance in phase by

$$\Delta\Phi_0 = (180NT\sigma/\pi) \text{ [mod } 360°]; \quad (-180<\Delta\Phi_0<180) \qquad (59)$$

In order to sample this constituent adequately in, say a year, $|\Delta\Phi_0|$ should be >NT in days.

If this is not the case, one may be able to use the phase change from P to P^1, the next *adjacent* pass at the same latitude. The smallest longitude separation is 360°/N and the time interval is

$$N^1T = (kN \pm 1)T/M,$$

where k is the smallest integer making N^1 integral,(± for West,East). The phase from P to P^1 is $\Delta\Phi_1$ defined similarly to $\Delta\Phi_0$ with N^1 in place of N. $|\Delta\Phi_1|$ should again be not too small, but its least permissible value depends on the zonal gradient of the tidal phase lag, and on how the spatial tide field is modeled.

By way of illustration, Table 6 lists some numerical values of $\Delta\Phi_0$ and $\Delta\Phi_1$ for (i) the planned TOPEX/POSEIDON orbit (T=0.07804d, 1 nodal day = 0.9912d, N,M,k,N^1 = 127,10,3,38), and (ii) a hypothetical sun-synchronous orbit with the same integral parameters, but T=10/127 = 0.07874d, 1 nodal day = 1 solar day.

TABLE 6 : Phase increments (degrees) for some tide constituents, on two repeat orbits.

Constituent symbol		O_1	K_1	M_2	S_2
180σ/π (°/d)		334.633	360.986	695.618	720.000
TOPEX/POSEIDON :	$\Delta\Phi_0$	77	-22	55	-63
TOPEX/POSEIDON :	$\Delta\Phi_1$	-87	-9	-97	-24
Sun - synchronous :	$\Delta\Phi_0$	106	10	116	0
Sun - synchronous :	$\Delta\Phi_1$	-79	0	-79	-6

All phase increments for the T/P orbit are respectably large, since that orbit was specifically planned to allow efficient tide elimination and analysis. In the Sun-synchronous (SS) orbit, phase increments for the lunar constituents O_1, M_2 are good, but S_2 is 'frozen' in sampled time and space, and K_1 is only just resolvable in a year, by aliassing into the seasonal cycle frequency. This is characteristic of all SS orbits, as are the small values for solar $\Delta\Phi_1$.

When a minor tide constituent is 'frozen' by the orbital sampling scheme, i.e. $\Delta\Phi_0 \approx 0$, it may be recoverable by invoking the 'smooth admittance' hypothesis to interpolate between other well-determined major constituents. As example, in the analysis of Geosat altimetry by Cartwright & Ray (1990,1991) the 'frozen' constituent P_1 was well defined by a smooth admittance between O_1 and K_1. If, however, a major constituent is frozen, as in SS orbits, any form of inference from minor terms with low signal/noise ratio is unlikely (in my view) to be physically reliable. Nevertheless, Mazzega (1989) has proposed a scheme for recovery of solar tide information from SS orbits by means of an elaborate analytical scheme involving spatial covariances between individual constituents and the use of *a priori* data from a dynamic tide model. See Mazzega & Jourdin (1991) for a more general account of covariance inversion of tide data.

Spatial resolution

Resolution of spatial detail from an altimetric tide analysis is obviously limited by the spacing of adjacent passes of the ground track, determined by the repeat-parameter N. For example, one should not attempt a spherical harmonic analysis of degree >N/2. Another limiting factor is the steep variation in scale of the tidal admittances $Z_\sigma(\theta,\phi)$ - e.g. Figure 2 (lower). In the middle of an ocean of roughly uniform depth, $Z(\theta,\phi)$ is usually fairly smooth, but near continents and large bathymetric features the tide admittances often rise or plunge steeply and therefore require high order spatial parameters for proper definition. There are, of course, discontinuities at every coast and across every land barrier separating distinct sea areas.

In a 1988 survey (published as Cartwright,1991) I described experiments in representing $Z(\theta,\phi)$ as continuous functions along narrow zonal bands, terminated by coasts, e.g. across the Pacific Ocean from New Zealand to Chile. Expansions in Fourier series and Chebyshev polynomials in ϕ were tried. In all cases, numerical instability was reached with increasing order of expansion, before adequate resolution of the steep variations of admittance near the continents. The same applies to double fourier expansions in (θ,ϕ) over an entire ocean, but with greater topological difficulties. My conclusion was that an equally good analysis was achieved by evaluating admittance functions **piecemeal** in grid-boxes of minimum size compatible with the ground-track spacing - about 1°x1.5° for the Geosat Exact Repeat Mission. Some mild smoothing over the grid sequence was also justifiable.

In theory, the most justifiable spatial definition is an expansion in Proudman Functions (§3). These embody all the physical requirements which depend only on basin shape and bathymetry, although their definition is also limited to a finite spatial grid for computational reasons. Such expansions were applied to the Pacific Ocean by Sanchez & Cartwright (1989), but the global set of Proudman Functions described by Sanchez (1991) has yet to be applied to altimetric analysis.

The altimetric tide

Finally, at some stage it is necessary to discriminate between the geocentric ocean tide sensed by an altimeter and the traditional 'ocean tide' as measured *in situ* by tide gauges. Denoting the 'altimetric tide' admittance by Z_a and the *in situ* admittance by Z_o, then

$$Z_a = Z_o + Z_b + Z_L$$

where the 'body tide' $\zeta_b = Z_b U_2 = h_2 U_2/g$ is usually subtracted from the raw data at a preliminary stage. Modern theory (Wahr, 1981) demands the use of a frequency-dependent h_2, especially near the frequency of K_1, in order to allow for the core-resonance.

(At present, a uniform value $h_2 = 0.609$ at all frequencies except the close neighborhood of K_1 ($h_2 = 0.52$), is considered to be a good compromise.)

Separation of Z_0 and Z_1 is more difficult. Given global definition of Z_0, the load tide admittance Z_1 can be computed as a spherical harmonic expansion from the loading Love numbers h_n' (equation 23). Deducing Z_0 from the measured $Z_e - Z_b = Z_0 + Z_1$ is not so straightforward, because over land, whereas $Z_0 = 0$ by definition, Z_1 is not zero and one has no direct measure of its value. Fortunately, an iterative procedure starting with $Z_1 = 0$ over land as a first approximation, converges rapidly and uses only moderate computing effort with a modern algorithm for spherical harmonic analysis. (Cartwright & Ray, 1991, Appendix A).

Research for and preparation of the Notes for this Course of lectures has been aided by a Grant to the author (David E. Cartwright) from the Leverhulme Foundation , Great Britain.

REFERENCES

Accad, Y. & C.L.Pekeris, Solution of the equations for the M_2 and S_2 tides in the world ocean from a knowledge of the tidal potential alone. Phil.Trans. R. Soc. London, A,290,235-266, 1978.

Bennett, A.F. & P.C.McIntosh, Open ocean modeling as an inverse problem : Tidal theory. J.Phys.Oceanog.,12,1004-1018, 1982.

Cartwright, D.E., A unified analysis of tides and surges round N and E Britain. Phil. Trans. R.Soc.London, A,263, 1-55, 1968.

Cartwright, D.E., Detection of tides from artificial satellites (Review).In:Tidal Hydrodynamics,(ed. Parker,B.B.- q.v.), 547-567, 1991

Cartwright, D.E. & A.C.Edden, Corrected tables of tidal harmonics, Geophys.J.R.astr.Soc. 33,253-264, 1973.

Cartwright, D.E., A.C.Edden, R.Spencer & J.M.Vassie, The tides of the northeast Atlantic Ocean, Phil.Trans.R.Soc.London, A,298, 87-139, 1980.

Cartwright, D.E. & R.D.Ray, New estimates of oceanic tidal dissipation from satellite altimetry. Geophys. Res.Lett., 16, 73-76, 1989.

Cartwright, D.E. & R.D.Ray, Oceanic tides from Geosat altimetry. J.Geophys.Res., 95,C3, 3069-3090, 1990.

Cartwright, D.E. & R.D.Ray, Energetics of global ocean tides from Geosat altimetry. J.Geophys.Res.,96,C9, 16897-16912, 1991.

Cartwright, D.E., R.Spencer, J.M.Vassie, & P.L.Woodworth, The tides of the Atlantic Ocean, 60°N - 30°S. Phil.Trans.R.Soc., A,324, 513-563, 1988.

Cartwright, D.E. & R.J.Tayler, New computations of the tide-generating potential. Geophys. J.R.astr.Soc., 23, 45-74, 1971.

Cazenave, A.C., Tidal friction parameters from satellite observations. In:Tidal friction and the Earth's rotation-II,(ed. P.Brosche & J.Sündermann),4-18, Springer-Verlag, 345pp, 1982.

Cazenave, A.C. & S.Daillet, Lunar tidal acceleration from Earth satellite orbit analysis. J.Geophys. Res., 86,B3, 1659-1663, 1981.

Darwin, G.H., The harmonic analysis of tidal observations. Pp.49-118 of: British Association for the Advancement of Science - Report for 1883.

Darwin, G.H., Tide. In: Encyclopedia Britannica, 11th edition, 26,938-961, 1910.

Dickey, J.O., J.G.Williams, & X X Newhall, The impact of Lunar Laser Ranging on geodynamics,(Abstract), EOS, Trans.Amer.Geophys.Union, 71,(17),475, 1990.

Doodson, A.T., The harmonic development of the tide-generating potential. Proc.R.Soc. London, A,100,305-329, 1921.

Doodson, A.T., Tides in oceans bounded by meridians, III - Semidiurnal tides. Phil.Trans. R.Soc.London,A,237,311-373, 1938.

Farrell, W.E., Deformation of the Earth by surface loads. Rev.Geophys. & Space Phys., 10,(3),761-737, 1972.

Gotlib, V.Yu., & B.A.Kagan. Resonance periods in the world ocean. Dokl.Akad.Nauk.SSSR, 252,725-728, 1980.

Groves, G.V. & R.W.Reynolds, An orthogonalised convolution method of tide prediction. J.Geophys.Res.,80,4131-4138, 1975.

Hendershott, M.C., The effects of solid Earth deformation on global ocean tides. Geophys. J.R.astr.Soc, 29,389-402, 1972.

Jourdin, F., O.Francis, P.Vincent, P.Mazzega, Some results of heterogeneous data inversions for oceanic tides. J.Geophys.Res., 96,B12,20267-20288, 1991.

Kaula, W.M., Theory of satellite geodesy. Blaisdell, Waltham, Mass. 124pp, 1966.

Lambeck, K., Tidal dissipation in the oceans : astronomical, geophysical and oceanographic consequences. Phil.Trans.R.Soc.London, A,287, 545-594, 1977.

Lambeck, K., The Earth's variable Rotation. Cambridge Univ.Press, 449pp, 1980.

Longuet-Higgins, M.S., Planetary waves on a rotating sphere, I,II. Proc.R.Soc.London, A,279, 446-473, and A,284,40-54, 1964.

Longuet-Higgins, M.S. & G.S.Pond, The free oscillations of a fluid on a hemisphere bounded by meridians of longitude. Phil.Trans.R.Soc.London, A,266,193-233, 1970.

Marchuk, G.I. & B.A.Kagan, Dynamics of ocean tides. Kluwer Academic Pubs, Dordrecht, 327pp, 1989.

Marsh, J.G. + 16 co-authors. The GEM-T2 gravitational model. J.Geophys.Res, 95,B 13, 22043-22071, 1990. (Ibid. Correction to Table 7, 96,B10,16651, 1991.)

Mazzega, P. The solar tides and the sun-synchronism of satellite altimetry. Geophys.Res.Lett. 16,(6),507-510, 1989.

Mazzega, P. & F.Jourdin, Inverting Seasat altimetry for tides in the northeast Atlantic; Preliminary results. In: Tidal Hydrodynamics,(ed. B.B.Parker - q.v.), 569-592, 1991.

Melchior, P., The tides of the planet Earth. Pergamon Press, 2nd Edn.,641pp, Oxford, 1983.

Miles, J.W., On Laplace's tidal equations. J.Fluid Mech.,66,241-260, 1974.

Miller, G.R., The flux of tidal energy out of the deep oceans. J.Geophys.Res.,71,(10), 2485- 2489, 1966.

Molodenskiy, S.M. Relation between Love numbers and load factors. Izvestiya, Earth Physics, (English edn.,pub. Amer. Geophys. Union),13,(3),147-149, 1977.

Morrison, L.V. & C.G.Ward, The analysis of the transits of Mercury. Mon.Not.R.astr.Soc., 173,183-206, 1975.

Munk, W.H. & D.E.Cartwright, Tidal spectroscopy and prediction. Phil.Trans.R.Soc.London, A,259,533-581, 1966.

Munk, W.H. & G.F.MacDonald, The rotation of the Earth; a geophysical discussion. Cambridge Univ.Press, 323pp, 1960.

Munk, W., F.Snodgrass & M.Wimbush, Tides offshore; transition from California coastal to deep waters. Geophys. Fluid Dynamics, 1,161-235, 1970.

Mysak, L.A., On the theory of continental shelf waves. J.Mar.Res.,25,205-227, 1967.

Newhall, X X , J.G.Williams & J.O.Dickey, Earth rotation from Lunar Laser Ranging. In: The Earth's rotation and reference frames for geodesy and geodynamics,(Ed.A.K.Babcock & G.A.Wilkins), Kluwer Academic Pubs., Dordrecht, 1988.

Pagiatakis, S.D., The response of a realistic earth to ocean tide loading. Geophys.J.Internat., 103,541-560, 1990.

Parke, M.E., O1,P1,N2 models of the global ocean tide on an elastic Earth, plus surface potential and spherical harmonic decomps. for M2,S2,K1. Mar.Geod.,6,35-8 1, 1982.

Parke, M.E. & M.C.Hendershott, M2,S2,K1 models of the global ocean tide on an elastic Earth. Mar.Geod.,3,379-407, 1980.

Parker, B.B. (Ed.) Tidal Hydrodynamics. John Wiley & Sons, New York, 883pp, 1991.

Platzman, G.W., World ocean tides synthesized from Normal Modes. Science, 220,602-604, 1983.

Platzman, G.W., Normal modes of the world ocean; III- A procedure for tidal synthesis; IV- Synthesis of diurnal and semidiurnal tides. J.Phys.Oceanog.14(10),1521-1550,1984a.

Platzman, G.W., Planetary energy balance for tidal dissipation. Rev.Geophys. & Space Physics, 22(1),73-84, 1984b.

Platzman, G.W., G.A.Curtis, K.S.Hansen, R.D.Slater, Normal modes of the world ocean; II- Description of modes in the period range 8 to 80 hours. J.Phys.Oceanog.11(5), 579-603, 1981.

Poincaré, H., Leçons de Mécanique Céleste, Tome 3: Théorie des Marées. Gauthier-Villars, Paris, 469pp, 1910.

Proudman, J., On the dynamic equations of the tides, I,II,III. Proc.London Math.Soc., 18, 1-68, 1917.

Proudman, J., On Laplace's differential equations for the tides. Proc.R.Soc.London, A,179, 261-288, 1941.

Sanchez, B.V., Proudman Functions and their application to tidal estimation in the world ocean. In: Tidal Hydrodynamics,(Ed. Parker,B.B. - q.v.), 27-39, 1991.

Sanchez, B.V. & D.E.Cartwright, Tidal estimation in the Pacific with application to satellite altimetry. Mar.Geod.,12(2),81-115, 1988.

Sanchez, B.V., D.B.Rao & P.G.Wolfson, Objective analysis for tides in an enclosed basin. Mar.Geod.,9(1),71-91, 1985.

Schwiderski, E.W., Ocean Tides; I : Global ocean tide equations, II : A hydrodynamic inter- polation model. Mar. Geod.,3,161-217, and 219-255, 1980a and b.

Schwiderski, E.W., Atlas of ocean tidal charts and maps, I : The semidiurnal principal lunar tide M_2. Mar.Geod.,6,(3-4),219-265, 1983.

Stephenson, F.R. & L.V.Morrison, Long term changes in the rotation of the Earth: 700 BC to AD 1980. Phil.Trans.R.Soc.London, A,313,47-70, 1984.

Sündermann, J., The semidiurnal principal lunar tide M_2 in the Bering Sea. Deutsche Hydrogr. Zeitschrift,30,91-101, 1977.

Taylor, G.I., Tidal friction in the Irish Sea. Phil.Trans.R.Soc.London,A,220,1-93, 1919.

Thomson, W., On gravitational oscillations of rotating water. Proc.R.Soc.Edinburgh,141-148, 1879.

Vincent, P. & C. Le Provost, Semidiurnal tides in the northeast Atlantic from a finite element numerical model. J.Geophys.Res.93,C1,543-555, 1988.

Wahr, J.M., Body tides on an elliptical, rotating, elastic and oceanless earth. Geophys.J. R.astr.Soc. 64,677-708, 1981.

Webb, D.J., Tides and tidal friction in a hemispherical ocean centred on the equator. Geophys.J.R.astr.Soc. 61,573-600, 1980.

Wunsch, C., The long-period tides. Rev.Geophys. 5,447-475, 1967.

Zahel, W., A global hydrodynamical-numerical 1° model of the ocean tide. Ann.Geophys. 33, 31-40, 1977.

Zahel, W., The influence of solid earth deformations on semidiurnal and diurnal oceanic tides. In: Tidal friction and the Earth's Rotation,(ed.J.Brosche & J.Sündermann), 98-124, Springer-Varlag, 243pp, 1978.

Zahel, W., Modeling ocean tides with and without assimilating data. J.Geophys. Res. 96,B12, 20379-20391, 1991.

Zschau, J., Tidal friction in the solid Earth: Loading tide versus Body tide. In: Tidal friction and the Earth's rotation,(ed. J.Brosche & J.Sündermann), 62-93, Springer-Verlag, 243pp, 1978.

Zschau, J., Tidal friction in the solid Earth: Constraints from the Chandler Wobble period. In: Space Geodesy & Geodynamics,(ed. A.J.Anderson & A.Cazenave),315-344, Academic Press, New York, 490pp, 1986.

Quantifying Time-Varying Oceanographic Signals with Altimetry

Victor Zlotnicki

Jet Propulsion Laboratory, MS 300-323
California Institute of Technology
4800 Oak Grove Drive
Pasadena CA 91109, U.S.A.

1. INTRODUCTION

Why altimetry? What is so special about this data type and what kind of information does it provide about the ocean? You probably know that tide gages have been measuring sea level for a long time, at selected locations, and sea level is the physical variable we want to extract from altimetry. As you will hear later, some of the properties we derive from sea level, like surface geostrophic velocities, can be measured from shipboard instrumentation that also supply information about deep oceanic properties which no satellite instrument can see.

The speed with which a satellite can sample the global ocean makes altimetry special. A ship steaming at 10 knots (18.52 km/h or about 5 m/s) and collecting data along the way (some oceanographic data require the ship to stop for several hours) samples the ocean about 1400 times slower than an altimetric satellite, whose footprint samples the ocean at about 7 km/s. Depending on the size of the gap between tracks one is willing to accept, uniform global sea level coverage from a satellite at 800 km altitude can be obtained in 3 days (8.4º between tracks), 10 days (2.5º between tracks), 17 days (1.5º between tracks), etc. Since typical satellite missions are expected to last 3 years or more, this gives us several realizations of global coverage to compute mean sea surfaces (for geoid and dynamic topography), and unique information about time-dependent flows in the world oceans, with about 1 month resolution. Because shorter repeat times are associated with larger gaps between adjacent tracks, oceanographers (e.g., Wunsch, 1989) have debated the somewhat complicated sampling characteristics of altimetry in terms of the signals they would like to retrieve. But the bottom line is that samping properties make altimetry unique.

I have chosen to discuss a few examples of the use of altimetry to detect and quantify specific time-varying ocean signals. I will use the examples to highlight the strengths and weaknesses of altimetric data, to discuss current problems, and to give the reader whose background is not Physical Oceanography a glimpse of the physics of oceanic motions. I had to make some assumptions about the audience in order to decide what to include. I have prepared these presentation for an audience of graduate students and researchers in Geodesy and Geophysics, but not Physical Oceanography. So I have mixed some basic oceanographic theory and facts without proof, but assumed a working knowledge of optimization methods.

A few words about what I will **not** touch upon here are in order. The details of what an altimeter measures are not covered, but any thoughtful user of the data must understand them. I recommend a technical report (Chelton, 1988) that summarizes a workshop on what was known at the time about the various instrumental and path corrections that convert the power spectrum of the difference between a received and a synthetic chirp (a signal whose frequency increases linearly in time) around 13.6 GHz into the sea level data we all use. I do want you to remember, however, that each altimetric pulse travels twice through some 800 km (1336 km for Topex) of atmospheric gases and free electrons, whose concentration changes daily, seasonally, and interannually -- while we try to retrieve similar time variability in the ocean. Also, the accuracy of the estimated sea level depends on the probability densities of heights and slopes associated with wind waves in the foortprint area, which also change geographically and with time, and which we have not yet been able to model fully. We will have something more to say about the water vapor correction and about the way Geosat sampled the tides below.

Other topics you need to know but I will not touch upon: the physics of large scale ocean motions and the permanent circulation are introduced in the chapter by Wunsch. Two crucial signals we wish to retrieve from altimetry are the surface expression of the time-averaged ocean circulation, addressed in the chapter by Wunsch from an oceanographic perspective and the one by Rummel from a geodetic perspective (see also the lectures in Wagner, 1989 and the now-classic paper by Wunsch and Gaposhkin, 1980), and the tides, highly predictable but not yet known with the desired accuracy, covered in the chapter by Cartwright. Finally, the largest error in altimetry and the one whose handling is most critical, residual orbit error, is discussed in the chapters by Balmino and Rummel.

2. SEA LEVEL and OCEAN CURRENTS.

In this section we will look at how and how well altimetry measures ocean surface currents, and constrains deep currents. Because the dynamics are somewhat different near the Equator, a subsection deals with that region. We will also look at a high energy region, the Gulf Stream, and a very low energy region, the Cape Verde area.

2.1 Basic altimetric data handling.

Geosat altimeter data (e.g., Cheney et al., 1989) will be used in all these examples. Many instrumental corrections and calibrations are applied before the data are written out as 'Geophysical Data Records' (GDRs, Cheney et al., 1988; a thorough description for Topex appears in a technical report by Callahan, 1990). In addition, the user applies environmental corrections for path delays through dry air, water vapor and free electrons, with suggested values given in the GDR. The user also applies 'corrections' for predictable oceanic signals, like the tides and the inverse barometer response. Unfortunately, these signals are not accurately known (an issue further discussed later). The user also removes out-of-range values, spikes, data over land, and other suspicious values (e.g., Rapp, 1983; Zlotnicki et al., 1990; Willebrand et al., 1990). The next step involves gridding the data *along* the track so that the various repeats of the same groundtrack have samples at approximately the same location.

To retrieve time-varying signals, we also remove the time-mean signals, containing geoid and various systematic errors. Because the notation will be needed later, this step is written out in detail here: let h (ϕ,λ,t) be the altimetric sea level at latitude ϕ, longitude λ, time t, corrected by all instrumental and path effects, including residual orbit error. Let h" be the time-average of h' at that position, averaged over all repeats during the mission (for example, 2 or 2.5 years for the Geosat ERM), and $\eta'=h-h''$ be the residual:

$$\eta'(\phi,\lambda,t) = h'(\phi,\lambda,t) - h''(\phi,\lambda) \qquad (2.1)$$

$$h''(\phi,\lambda) = (1/n) \sum_{i=1}^{n} h(\phi,\lambda,t_i) \qquad (2.2)$$

$$t_i = (i-1)*\Delta t + t_0, \quad i=1,2,\ldots,n \qquad (2.3)$$

(in the case of the Geosat ERM, $\Delta t=17.0505$ days, n=42 repeats in two years, $t_0 \approx$ Nov.8, 1986). This assumes we have interpolated the alongtrack (geoid) data to a set of positions (ϕ, λ) common to all repeats of the altimetric groundtracks, and we have done so with sufficient accuracy (<< 1 cm). This is actually not as trivial as it may appear, since departures of about 1 km from exact repeat patterns can cause errors due to the crosstrack geoid gradient of a few cm (Brenner et al, 1990).

Sea level differences at a fixed location with long alongtrack correlations are presumed to be due to residual orbit error, and removed by fitting a variety of simple alongtrack functions, ranging from straight lines over segments a few minutes long, to Fourier series over segments several days long (e.g., Tai, 1989; Sandwell et al., 1986; Zlotnicki et al., 1989).

After all this, one may wonder whether the residuals resemble the ocean. We know that the altimeter is measuring sea level to a good first approximation. For example, Tai et al (1988) compared altimetric sea level and equatorial tide gages, finding a median correlation (over 21 gages) of 0.60; however the best correlation was 0.79, seven correlations were below 0.20, and two of these were large negative correlations (-0.38, -0.43). They also compared the long wavelengths (actualy, the first Empirical Orthogonal Functions) of altimetry and dynamic topography in the upper 400 db from XBT data, and found that most of the area under study had correlations of 0.6 or higher. See also Wyrtki and Mitchum (1990). In summary, the altimetric sea level correlates well with dynamic topography and tide gages, but the differences exceed the ballpark error estimates of either data type. Are the differences due uncertain corrections, different physics (sea level and dynamic height relative to any depth are physically very different), or different wavenumber-frequency content?. Rather than review these results, altimetry will be compared to current data below.

Let us also assume that the residuals $\eta'(\phi, \lambda, t)$ have been interpolated onto a uniform ϕ, λ, t grid by a suitable space-time gridding and smoothing scheme from the equivalent values along the satellite track. We may use an optimal scheme in the minimum variance sense, (Bretherton et al, 1976) which is the space-time equivalent of collocation; or we may use a suboptimal but fast multiple scale scheme (Tripoli and Krishnamurti, 1975). One could have a whole lecture on this subject; suffice it to say that if the scheme is reasonable, the important thing to know is what smoothing parameters were used, i.e., what smoothing in space and time is imposed on the resulting grids by the interpolation length and time scales, and their assumptions about signal and noise variances. As we will see below, the smoothing parameters are crucial to correct use of the geostrophic equations and correct interpretation of any comparisons between altimetry and other in-situ data. It is also important to know what are the errors in the solution, both due to propagation of noise variance (commission errors) and those due to missing data (omission errors).

2.2 GULF STREAM time-varying surface currents.

The Gulf Stream is a very energetic area that will be addressed first. The standard deviation of sea level time series, averaged over the region, is about 41 cm (Zlotnicki et al., 1989).

Sea level slopes directly measure the geostrophic component of velocity at the sea surface. But if you measure water velocity at the sea surface it will include non geostrophic components. Let us assume the tides were correctly predicted and removed. Let us assume the effect of wind waves was correctly averaged out by processing under the altimeter footprint. We can remove some of the remaining energy by smoothing over a distance at least equal to the first baroclinic Rossby radius, the shortests length scale over which the rotation of the Earth affects the most energetic component of the motion. That length scale is about 40 km at mid-latitudes; in the Pacific it varies between 140 km at 5ºN to about 16 km at 50ºN (Emery et al., 1984). We also need to smooth the data over time scales much longer than 0.5 day /sin(latitude), the shortest time scale over which the rotation of the Earth affects the motion. In addition there is an Ekman component of motion balanced by wind stress, not water pressure gradients. To remove the Ekman component we must either a) model it given accurate estimates of the local wind or b) look deeper than about 100-200 m, the depth of the boundary layer (called the Ekman layer) over which the effect of the wind stress at the surface is felt. In this example, I have chosen the latter approach.

The geostrophic surface velocity components u' (eastward), v' (northward) with their time-mean removed are computed from the discretized forms of:

$$u' = -(g/f)\ \partial\eta'/\partial y \tag{2.4}$$

$$v' = (g/f)\ \partial\eta'/\partial x \tag{2.5}$$

where $x = R(\lambda - \lambda_0)\cos(\phi)$ is the local east coordinate, $y = R\phi$ is the northward coordinate, R=6371 km approximates the local radius of curvature, g=981 cm/s^2 approximates the local gravity acceleration, and f is the Coriolis parameter, twice the projection onto the local vertical of the Earth's rotation Ω.

$$f = 2\Omega\sin(\phi), \qquad \Omega=7.27e^{-5} \ s^{-1} \qquad\qquad (2.6)$$

The geostrophic approximation used in (2.4)-(2.5) is not just a steady state solution of the equations of motion, but is the zeroth order term in expansions of the time-varying velocity field in powers of the Rossby number ε (eq. 8 below; see Pedlosky, 1987, chapters 3 and 6) for a wide class of oceanic motions whose time scale is sufficiently slow that they are deflected by the Earth's rotation. Let U, L, T and D be characteristic velocity, length, time and depth scales of the oceanic motion under study. Then (2.4) and (2.5) are good approximations when conditions (2.7) through (2.9) hold:

$$1/f \approx 0.5 \ day/\sin(\phi) \quad << T \qquad\qquad (2.7)$$

$$\varepsilon \equiv U/(fL) \ << 1 \qquad\qquad (2.8)$$

$$Lo \equiv (g'D)^{1/2}/f < L \qquad\qquad (2.9)$$

T >> 1/f ensures that local accelerations are small compared to the Coriolis term, ε<< 1 ensures that convective accelerations (the non-linear terms in the full equations) are small relative to the Coriolis term; L > Lo (the Rossby radius of deformation) is about the same as T >> 1/f: it specifically refers to accelerations due to gravitational forces in the presence of buoyancy or sea level differences. For constant density the source of pressure differences is the horizontal gradient of $\rho\eta g$, where g is the gravity acceleration (~ 9.81 m s^{-1}), ρ is the density, so g'=g in (9) and D is the total water depth. For a stratified fluid, whose density ρ changes with depth, g is replaced by the 'reduced gravity', g'=g(D/ρ)($\partial\rho/\partial z$) which controls the buoyancy of fluid parcels (see Pedlosky, 1987, chapter 6; Gill, 1982, chapter 7), and D is a depth scale derived from the equation describing normal modes of oscillation associated with the profile of $\rho(z)$ (see Pedlosky, section 6.12). In the stratified case, (9) yields the 'internal' radius of deformation. Emery et al (1984) have mapped the distribution of these scales. Roughly speaking, the barotropic radius is about 2000 km and the first baroclinic radius about 30-50 km, except within a few degrees of the Equator.

In (4) and (5) u,v, η' can be considered functions of time if the data have been smoothed to satisfy (2.7), (2.8), so this argument about length and time scales affects both the validity of the equations to use and the practical decisions regarding smoothing lengths to be applied to the altimetric data. The length and time scales may also need to be empirically fine tuned to the local dynamics.

Figure 1 shows a comparison, made by Capotondi, Holland and Malanotte-Rizzoli (1992, pers.comm.), of current meter velocities and Geosat-derived velocities in the Gulf Stream. The current meters were averaged over 10 days, the altimetric data over about 85 km and 5 days (170 km and 17 day search radii). Even though the exact repeat period of Geosat is 17.0505 days, when data from a wide circle that includes neighboring tracks is considered, there are samples more frequently than once/17 days. The key issue to decide in each region is whether sea level values inside the wider circle are supposed to be correlated because the phenomenon we study is of a scale larger than the diameter of this circle. The velocities were computed from (2.4) and (2.5) with $\Delta x=\Delta y=28$ km.

Why does Geosat data work so well here, and does it work equally well everywhere? Let me go over some details since I provided them the fully corrected altimetric data. I used the old FNOC water vapor, which is very bad. I also modelled residual orbit error by a second degree polynomial over about 5000 km, which eliminates some oceanic signal, and due to the proximity to land (where the altimetric profiles are terminated for the purposes of this orbit correction), the Gulf Stream signal itself affects the orbit error estimate. But the signal is enormous, it is due to the meandering of a strong current, and it tends to mask errors in the orbit and environmental corrections. Furthermore, slopes over 28 km attenuate environmental errors which tend to vary over typical atmospheric scales of 1000 km.

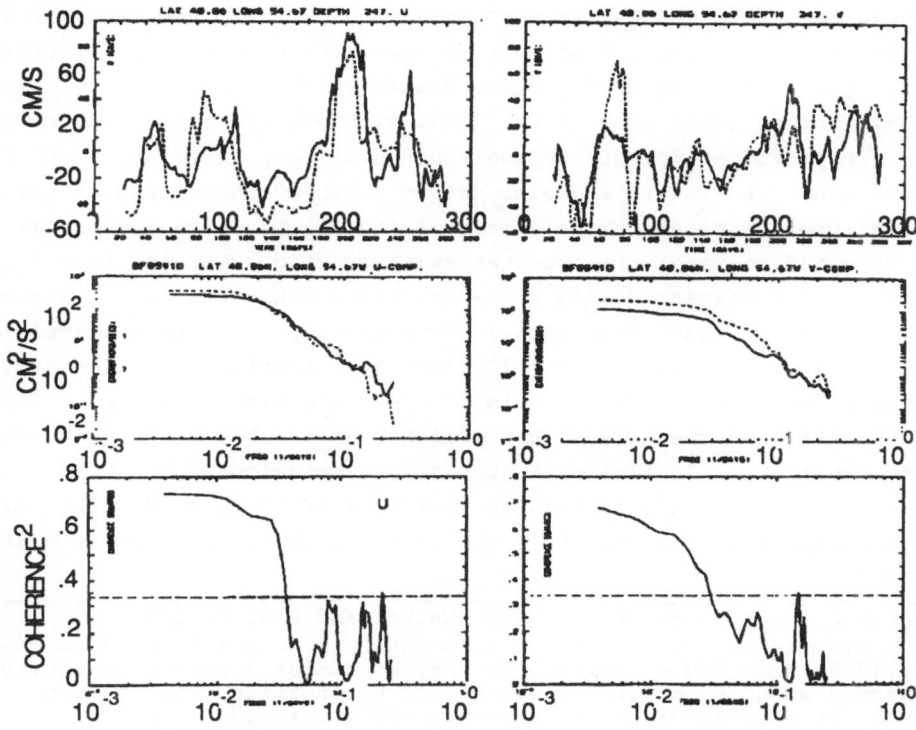

Figure 1. From A.Capotondi, W.Holland and P.Malanotte-Rizzoli (1992, pers. comm.). Comparison between surface altimetric geostrophic velocities (dashed) and current meter velocities (solid) at 247 m, at φ=40.86°N, λ=54.67°W. The right hand panels correspond to the East velocity, left hand panels are North velocity. Top panels have the time series over 300 days. Middle panels have the frequency spectra, in cycles per day. Lower panels show the amplitude of the coherence between current meter and altimetric measurements (the phase plot is omitted). Coherence is significant for periods longer than about 90 days, up to the 180 days half-length of the record.

2.3 GULF STREAM time-varying deep currents.

This is very encouraging. Can we infer the flow at depth in this region from the altimetric data? It is important to remember that the flow at depth can be not only weaker than the surface flow, but also in a different, even opposite direction, depending on the stratification and other factors. The moorings used by Capotondi et al. include current meters at various depths, but how to transfer the surface altimetric information to depth?

Before dealing with the issue of inferring the deep time-varying flow
from the surface flow, let us look at something more basic, a snapshot
of the ocean under the Gulf Stream. Figure 2A shows two profiles across
the Gulf Stream taken June 26-29, 1982, from R/V Endeavour. The numbers
along the profiles indicate hydrographic stations, where the ship stops
(a 'station') and lowers a device called a CTD/O2 that measures the
vertical profile of electrical conductivity, temperature and oxygen.
From the conductivity, one computes salinity (fig. 2C), and from the
salinity and temperature (fig 2B is the temperature that a water parcel
would have if brought adiabatically from its depth to the surface), one
computes the water density (fig 2E, see Gill, appendix 3). The arrows
in figure 2A indicate surface velocity measured with an acoustic Doppler
current profiler; the core of the Gulf Stream is the region with maximum
surface velocity. In figure 2E, the deep expression of the Gulf Stream
is the region where the isopicnals (surfaces of constant density) slope
down strongly with distance from the coast, with the current entering
the page.

Figure 2 (next page). From Joyce, Wunsch and Pierce, 1989.
A: positions of two profiles across the Gulf Stream taken June 26-
29, 1982, from R/V Endeavour. The arrows indicate surface velocity
measured with an acoustic Doppler current profiler. The numbers
along the profiles indicate hydrographic stations.
B: potential temperature in ºC vs depth along the southern section;
C,D,E: as B with salinity (parts per thousand), oxygen (parts per
thousand volume), potential density (1000*(potential density-1)).

If we have measurements of isopicnal slope at depth, then we can infer
the relation between surface and deep motions using the thermal wind
equation (e.g., Gill, section 7.7); for example, its y component relates
the x-component of density gradient on a surface of constant pressure
($(\partial \rho/\partial x)_p$) to the vertical shear of the y component of velocity, $\partial v/\partial z$.

$$f \, \partial v/\partial z = -(g/\rho) \, (\partial \rho/\partial x)_p \qquad (2.10)$$

However, while altimetry gives us about two global samples per month
of the time-varying surface flow, we have no practical way to measure
the interior isopicnals with comparable temporal and spatial coverage.

The next best way to infer the flow at depth is to have a numerical
model of the circulation in the area that includes both the physics and
thermodynamics that affect ocean motions, and properly describe the
stratification, realistic bathymetry and coastlines, and the effect of
the wind, heat and freshwater flux forcings in the more complex models,
and to assimilate the surface data into the model. Up to very recently,
oceanographers have dealt either with much simpler analytical models, or
with simplified versions of the 'complete' numerical model.

153

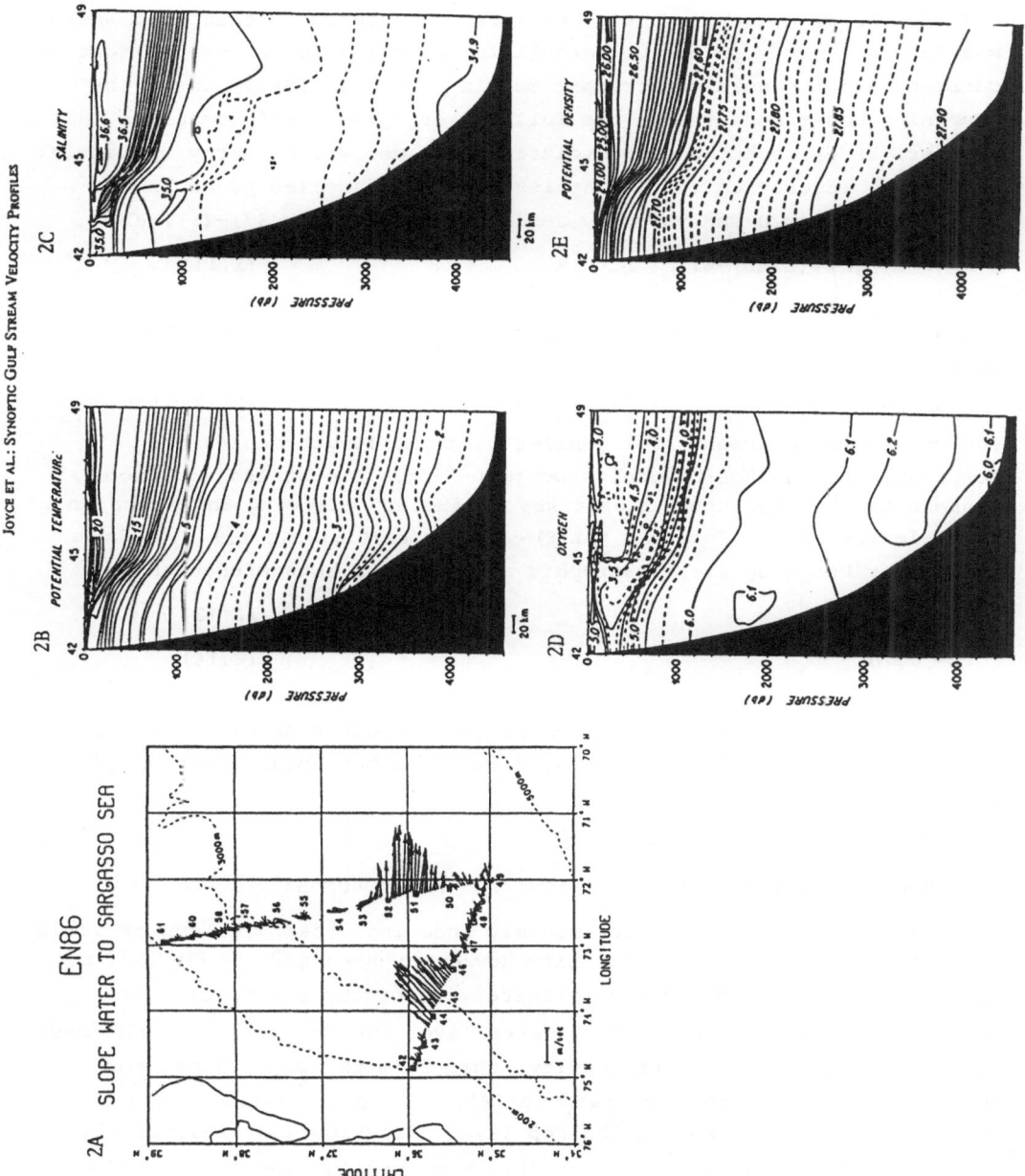

Figure 2. From Joyce, Wunsch and Pierce, 1989.

A simplified class of models, particularly useful in this area where most of the energy is due to baroclinic instabilities of the meandering current, are the quasigeostrophic models (Holland, 1986), where the governing equations are not the full set of 'primitive' equations (e.g., Semtner, 1986), but the time evolution of a derived property, potential vorticity, restricted to time scales T, length scales L, depth (or layer thickness) D and velocity scales U that satisfy (7)-(9), and

$$F=(f^2L^2/gD)=(L/Lo)^2 \approx 1, \quad L \ll R \qquad (2.11)$$

L<<R allows an expansion around a central latitude, so that the meridional change of f,

$$\beta=df/dy \qquad (2.12)$$

can be assumed constant; F≈1 implies that the length scales are comparable to the Rossby radius $Lo=(gD)^{1/2}/f$. The scaled equations contain the Rossby number ε as a key parameter, allowing an expansion of the velocity of the form $u(x,y,t,\varepsilon)= u_0(x,y,t)+\varepsilon.u_1(x,y,t)+...$, whose zero order terms u_0 are geostrophic and define a streamfunction ψ:

$$u_0=-\partial\psi/\partial y \qquad (2.13)$$
$$v_0= \partial\psi/\partial x \qquad (2.14)$$

The key derived equation is a form of the conservation of potential vorticity in a stratified rotating flow, on a beta plane, with applied stresses:

$$[\partial/\partial t + (\partial\psi/\partial x)\partial/\partial y - (\partial\psi/\partial y)\partial/\partial x]$$

$$[\partial^2\psi/\partial x^2 + \partial^2\psi/\partial y^2 +\partial/\partial z((1/S)\partial\psi/\partial z) + \beta y] = W_0+W_H+F \quad (2.15)$$

Simpler forms of (15) state, roughly speaking, that a column of fluid must change its relative vorticity ($\partial v/\partial x-\partial u/\partial y= \partial^2\psi/\partial x^2 + \partial^2\psi/\partial y^2$) to balance latitudinal changes of planetary vorticity f (the βy term), streching or contraction of each water layer due to changes in the depth of the surfaces of constant density that form the layer's top and bottom boundaries (the $\partial/\partial z$ term in 14), the curl of the stresses applied by the wind (W_0) on the top layer, friction (W_H) against the ocean floor, or lateral friction (F). The vertical density profile of the undisturbed state appears in the stratification parameter S:

$$S(z)=N^2(z)D^2/f^2L^2, \qquad (2.16)$$

$$N^2=(-(g/\rho)(\partial\rho/\partial z)] \qquad (2.17)$$

where N is the Brunt-Vaisälä (buoyancy) frequency at which a parcel of fluid at rest at depth z would oscillate if moved adiabatically to a

slightly different depth z+dz (2π/N usually ranges between 0.3 to 2
hours in the ocean). It is in looking for wave solutions of (14) and
separating variables, in such a way that the vertical dependence
multiplies functions indepent of z, that the concept of normal vertical
modes appear. They are the normal modes of $(1/\rho)(d/dz)((\rho/S) d\Phi/dz) =$
$-\lambda\Phi$ (see Pedlosky, section 6.12). Internal Rossby radii are associated
with each of the normal modes, shorter as the mode number increases;
they are the length scales that affect how we should smooth the data.
The first two are the barotropic (depth-independent flow) and first
baroclinic mode. Mention of the 'Rossby radius' without qualification
refers to the radius associated with the first baroclinic mode.

W. Holland has written several implementations of (15) with finite
differences in space and time and various conservation properties. Sea
level data can be 'assimilated' into such a model (Holland et al., 1991)
by adding to the vorticity equation for the upper layer a relaxation
('nudging') term of the form

$$\partial/\partial t \nabla^2\psi_1 = ... - (1/\tau)(\nabla^2\psi_1 - \nabla^2\psi_{obs}) \qquad (2.18)$$

where the index 1 refers to the top layer of the model (see Verron,
1992, for a good discussion of (2.18)), and ψ_{obs} is related to
altimetric sea level η' and the time averaged dynamic height $\bar{\eta}$ obtained
from classical hydrographic data by

$$\psi_{obs}(\phi,\lambda,t) = (g/f_0)(\eta'(\phi,\lambda,t) + \bar{\eta}(\phi,\lambda)) \qquad (2.19)$$

Capotondi et al. (1992, pers. comm.) used that formulation in the Gulf
Stream to assimilate the altimetric sea level data, discretizing the
ocean vertically into five layers whose lower level is at
300m, 750, 1500m, 2800m and 5000m, each with densities of approximately
1.0240, 1.0253, 1.0265, 1.0275, 1.0280 gm/cm^3.

Figure 3 shows one example of their findings for the deep velocity
generated in the model and those of an actual current meter. It is clear
that the good correlation between in situ current meter data and the one
derived from altimetry and downward continued with the model persists at
least to 1000 m.

There is a lot more to be said about modelling ocean currents,
assimilating altimetric data, or even simpler ways to estimate deep
velocities. For example, using simply the normal modes, without a time-
stepping numerical model, can yield a respectable estimate of the deep
flow based on the surface one. Nudging, while a computationally
efficient technique to assimilate, does not accept error estimates about
the data or model, nor does it give error bounds for its estimates.

Quasigeostrophic models do not take thermodynamic constraints, nor can
they handle steep topography well. These and other issues, while very
important, are beyond the scope of these lectures. My point here is that
Capotondi's results illustrate that surface altimetric data constrain
not only the surface flow but, in certain regions of the ocean and when
suitably modeled, much of the deep flow as well.

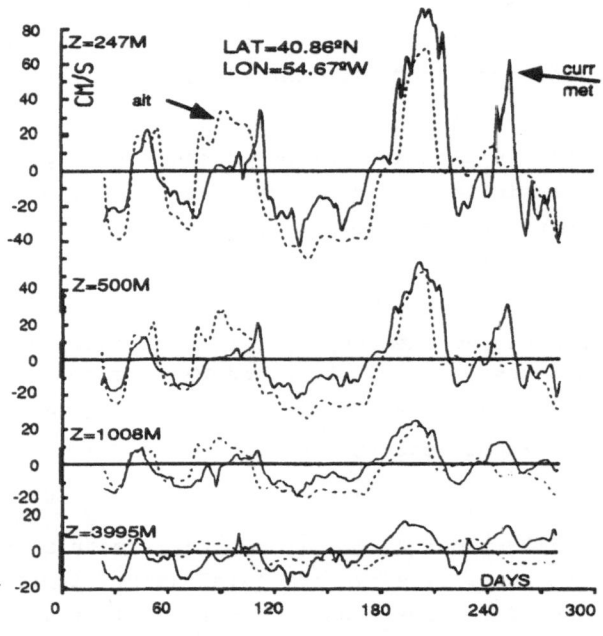

CAPOTONDI, HOLLAND, and MALANOTTE RIZZOLI

Figure 3. From A. Capotondi, W.Holland and P. Malanotte(1992,
pers. comm.). Current meter East velocities at various depths
(solid lines) and model-assimilated velocities at the same level
(dashed). Vertical axes in cm/s, horizontal in days.

2.4 Weak surface currents: the Cape Verde region.

When Geosat data first arrived most of us looked in the western
boundary currents, both because they are interesting and because the
signals are so large that, if altimetry didn't work here, it could
hardly be expected to work in more difficult areas. I looked at an area
of the eastern Atlantic where the sea level signals are very weak.

Even in a very weak current area, such as the Cape Verde region, where currents have an rms of 3 to 4 cm/s rms (Figure 4, upper panels), the effect of residual orbit error is negligible and significant correlations between altimetric and in-situ currents are found. However, honesty demands that figure 4 include the mooring (lower panels) where the comparison between altimetric and current meter velocities is very bad. In both cases the Geosat altimetric data were processed in the same way, and the current meter data belong to comparable instruments identically processed.

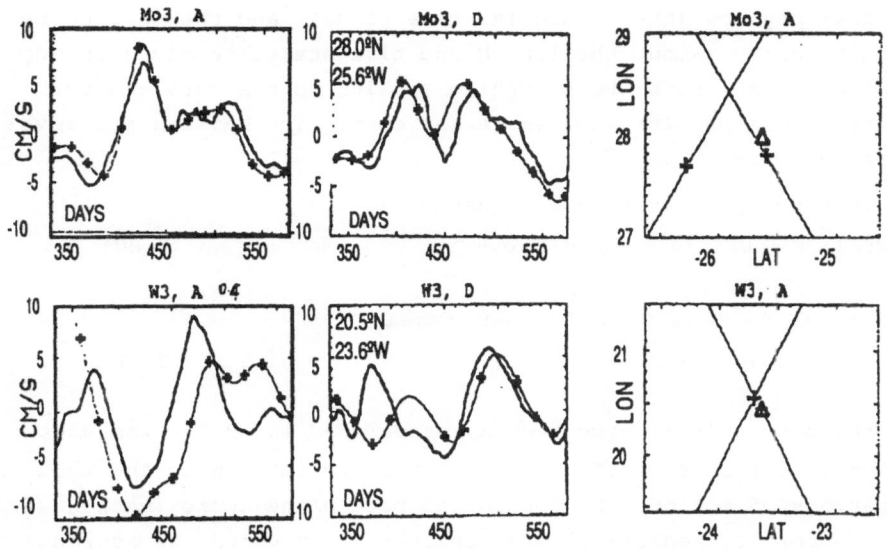

FIGURE 4. From Zlotnicki, Siedler and Klein (1992). Surface currents from altimetry (crosses) and current meter in the weak Cape Verde area. Left panels are currents normal to the ascending Geosat track, middle panels currents normal to the descending track, right panel location of the current meter and of the center of the altimetric slope used to compute current. Horizontal axis is time in days since 1/1/85. The correlations for the two top panels are 0.9 (A and D), those for the lower two are 0.4 (A) and 0.6 (D).

While the reason for the difference between the upper and lower panels is still being investigated, the student should keep such differences in mind when interpreting the data.

2.5 Equatorial surface currents.

Conversion of altimetric sea level changes η' to surface water velocities within one or two degrees of the Equator (the equatorial waveguide) cannot be accomplished with equations (4)-(5) since the

Coriolis parameter f→ 0 as the latitude φ→ 0. The equatorial region is very special as a consequence, and acts as a waveguide, a subject that we will pick up later. Can altimetry measure currents on the Equator?

It is possible to derive a modified geostrophic balance that allows meridional sea level curvature to measure zonal surface geostrophic currents. Picaut et al (1990) showed that good agreement existed between altimetrically derived geostrophic currents and current meter data in the Equator. I would like to show you what Picaut and his collaborators did and how it works because it introduces some nice additional physics (the equatorial beta plane), and the need to be specially careful about the length and time scales retained in any comparison, i.e., both the smoothing applied to the data and the validity of the physics implied in the conversion between sea level and current velocity.

Consider the y derivative of equation (2.4)

$$u' . \partial f / \partial y + f . \partial u' / \partial y = -g \ \partial^2 \eta' / \partial y^2 \qquad (2.20)$$

the limit as f→ 0 is well defined, namely

$$\beta . u' \ = -g \ \partial^2 \eta' / \partial y^2 \qquad (2.21)$$

where the conventional name β=df/dy as defined in (2.12) was used. Equation (2.21) is part of a consistent approximation to the whole set of equations of motion, the equatorial beta plane approximation (Gill, 1982, chapter 11; Pedlosky, 1987, chapter 8) in which the approximations cos(φ)≈1, sin(φ)≈1 are made, from which follows that f=βy (β=constant) and y=Rφ, for latitudes φ up to 30º from the Equator. Equation (21) *formally* allows us to compute geostrophic velocities on and near the Equator, but a) it forces us to compute second meridional derivatives from the data, a noise-enhancing process, and b) the formal application of L'Hôpital's rule to equation (2.4) does not provide much insight into the space and time scales for which (2.21) is valid, which requires both a better understanding of the equatorial beta plane approximation and actual data from which to estimate the characteristic scales.

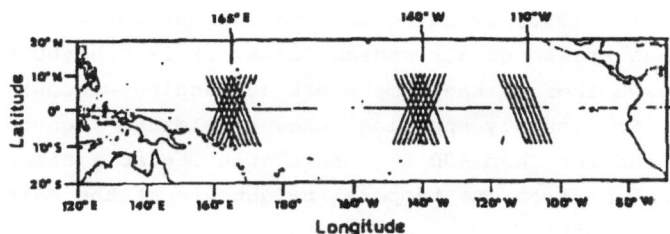

FIGURE 5. From Picaut et al.(1990, figure 1). Location of the Geosat tracks used to compute geostrophic velocities. The point to notice is that, by assuming that all tracks within a box around the current meter are co-located, thet authors both increased the temporal resolution from 17 days to 1.5 days and performed a zonal averaging.

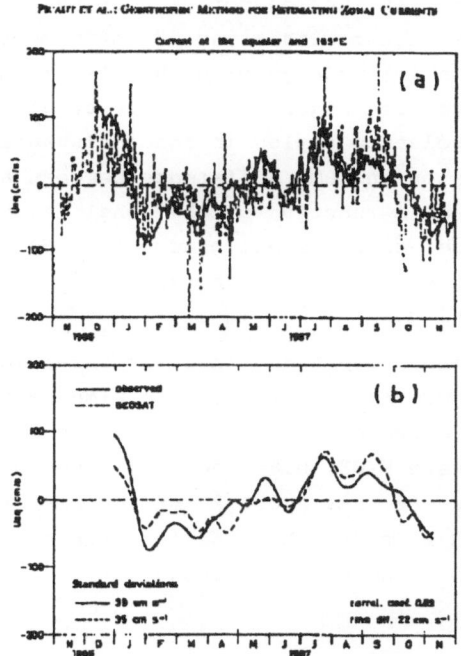

FIGURE 6. From Picaut et al.(1990, figure 4). Zonal velocities at φ=0, λ=165ºE from a current meter at z=50m and from Geosat altimetry. The upper panel shows the daily values, lower panel shows the time-smoothed values. The altimetry was smoothed over 400 km alongtrack. Data from a 9º zonal box centered at λ=165ºE were considered to be co-located, yielding altimetric samples every 1.5 days. The resulting current meter and altimetric time series were smoothed over 31 days.

Picaut et al. (1990) tried several alongtrack smoothing lengths and concluded that an alongtrack median filter of length 400 km, followed by a linear smoother of the same width (a Hanning window) yielded the best result. So, roughly speaking, they eliminated alongtrack wavelengths shorter than 400 km. Each such Geosat altimetric value is separated in time from its temporal neighbours at the same latitude-longitude by 17.0505 days.

Picaut et al. reasoned that *instantaneous* geostrophic currents did not make much sense so they needed to smooth in time. However, this could not be done unless data from neighboring passes was brought in, as Capotondi had done with the space-time gridding in the Gulf Stream. For the Geosat ERM, the pass to the East of a given pass and parallel to it occurs 3.00 days and, at the Equator, 1.475º away; the pass in the opposite direction can occur with a variety of time separations below 8.5 days, depending on latitude only; within 1º of the Equator the separation is 1.5 days (see Figure 3). Picaut et al. reviewed the literature for decorrelation scales in zonal distance and time, and then measured the correlation and RMS difference between altimetric and current meter values as a function of the zonal and temporal smoothing applied. They concluded that a zonal window 9º wide, centered at the current meter mooring yielded the best results, associated with a one month temporal smoothing.

After this processing, they found that for three locations they studied (shown in Figure 4; see also sample time series in Figure 5) the correlations between current meter and altimetric velocities, listed from West to East, were 0.83, 0.85 and 0.51, and the rms differences 22,18 and 28 cm/s, or 56%, 53% and 112% of the rms of the data which ranges between 39 and 25 cm/s (the location with higher error is the easternmost one, with no descending tracks).

2.6 Closing Comments.

In summary, the correlations are significant, clearly both techniques are measuring approximately the same phenomenon, but errors of 50% or higher would not usually be reasonable. In fact, these discrepancies exceed our rough error estimates for the data involved. While the finding in Picaut et al are very significant, as are those of Capotondi et al, and other authors who have compared geostrophic altimetric currents and directly measured ones, it is still not clear to what extent the discrepancies are due to errors in the altimetric or current meter data, or discrepancies in the physics measured by the two types of observations, i.e., whether equation (19) is all the physics we need to know in order to compare sea level and surface currents. The frequency spectra of current meters in general at periods shorter than about one month only contains 10%-20% of the total energy in the time series, so it would be reasonable to expect no more than this level of dicrepancy, plus data noise, between the two measurement types.

Let us discuss the effect of altimetric corrections. In all these cases in which slopes over 100 km or curvatures over 400 km are computed, the length scales involved are so much shorter than those of atmospheric phenomena and orbits, that environmental errors or even the huge orbit error are effectively attenuated. For example, a large orbit error looks along the track like $A.\sin(2\pi x/L)$, with A~1m (Geosat's NAG orbit), L=40,000km and x an alongtrack distance. So, the maximum error in the *slope* is 2.5×10^{-8} or, using equation (4), an error of 0.3 cm/s at mid-latitudes. So, to study these local, intense currents, even total failure to correct for residual orbit error would introduce only minimal noise in the results. The next largest environmental error in Geosat (but not Topex or ERS-1 if the onboard microwave radiometer is used), water vapor, also occurs over larger scales except just North and just South of the Equator, where the Intertropical Convergence Zone and the South Pacific COnvergence Zone cause sharp gradients in water vapor (see figure 6 within 15° of the Equator). Both Picaut et al. and Capotondi et al. used only the FNOC correction. Figure 6 illustrates the difference between the more accurate SSM/I correction and the FNOC correction.

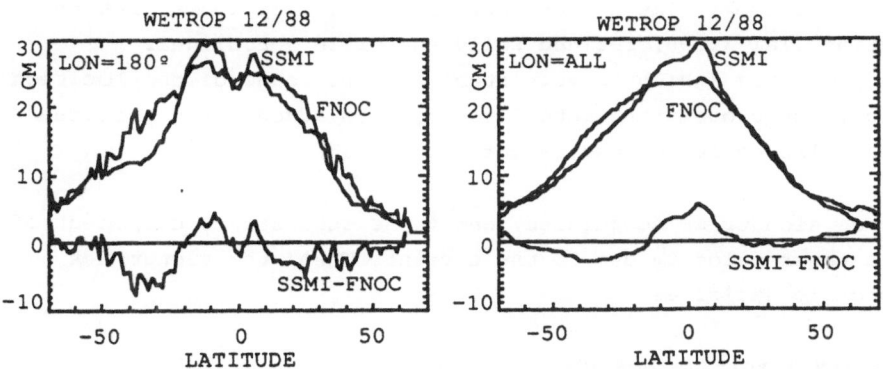

FIGURE 7. Wet tropospheric corrections for December 1988 from the surface FNOC fields included in the original Geosat GDR, from the SSMI data, and their difference, as a function of latitude. The left panel curves are along 180° longitude, averaged over 1°. The right hand panel curves are an average over all longitudes also every 1°.

The last source of large errors for Geosat, the tidal correction (more below) generally has much longer spatial scales.

3. TOTAL SURFACE CURRENTS IN THE GULF STREAM.

Almost all uses of altimetry without a geoid model lead to estimates of the time changes in the ocean, but there is one particular problem in which altimetry alone can define reasonably well the *time-averaged* surface currents: it is the case of strong *meandering* jets, like the Gulf Stream extension and the Kuroshio. The method may also work for portions of the Antarctic Circumpolar current, but it has not been tried yet.

3.1 Meandering. The shape of the current's sea level 'step'.

Kelly and Gille (1989) reasoned that the Gulf Stream extension basically looks like a 1 m sea level step, with the higher water on the South side. As the current meanders North and South of its mean position (see figure 7), the step moves, causing the variability we see. To the extent that we know the *shape* of the step that is moving, we only need fit its parameters (width, height, crossing angle, alongtrack position) to all the data and retrieve the full amplitude. But Kelly knew the shape of the current from transects measuring the profile of surface velocity with Doppler acoustic profiling (see figure 2 above): the velocity profile looks like a gaussian, hence the step looks like an error function. Tai (1990) refined Kelly and Gille's nonlinear estimation procedure (which is very stable if knowledge of the location where the current crosses the satellite track is added, from infrared imagery) and also computed results for the Kuroshio.

Let x be an alongtrack coordinate, and v the acrosstrack component of geostrophic water velocity due to the current. Then, the alongtrack velocity profile satisfies

$$v \approx V_0 . \exp(-\pi(x-x_0)^2/(w/\cos \vartheta)^2) \qquad (3.1)$$

where V_0 is the velocity maximum which occurs at alongtrack position x_0, w is a measure of the current's width, and ϑ corrects for the angle between the actual velocity vector and the normal to the track.

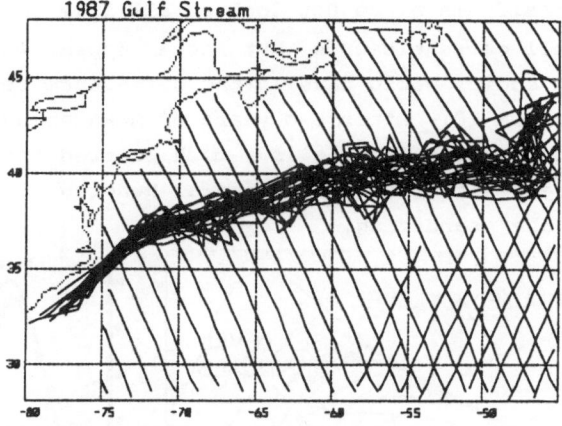

FIGURE 8. Weekly positions of the North Wall of the Gulf Stream during 1987, from NOAA AVHRR data, and ground tracks of Geosat altimetry. Most data from the descending tracks is missing (from Zlotnicki, 1990).

All the parameters ($V_0, x_0, w, \cos \vartheta$) are functions of both time t and the location λ_0 (roughly, the longitude at which the mean path intersect a Geosat ground track). The equivalent sea level profile is

$$\eta = (f/2g)(w/\cos \vartheta)V_0 \, \text{erf}(\pi^{1/2}(x-x_0)/(w/\cos \vartheta) \qquad (3.2)$$

(the hyperbolic secant and tangent will also do for the velocity and height profiles).

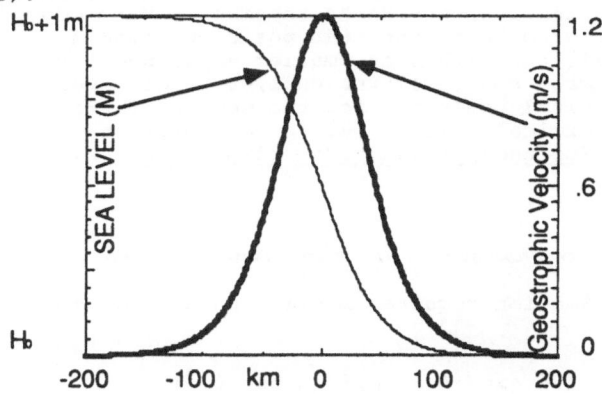

FIGURE 9. Approximate velocity profile and associated height profile with reasonable values for a normal crossing of the Gulf Stream extension (as found by Joyce et al., 1989). Horizontal axis is alongtrack distance, in km, with South to the left.

The sea level residuals η' do not look at all like the step in sea
level that the full current causes, but rather appear with the signature
that cold or warm eddies would have in the difference between sea level
and the geoid (figure 10). After two years of meandering back and forth,
the *time-averaged* current profile is a much blurred and streched
version of the step, with a width of several hundred km, depending on
the amplitude of the meandering.

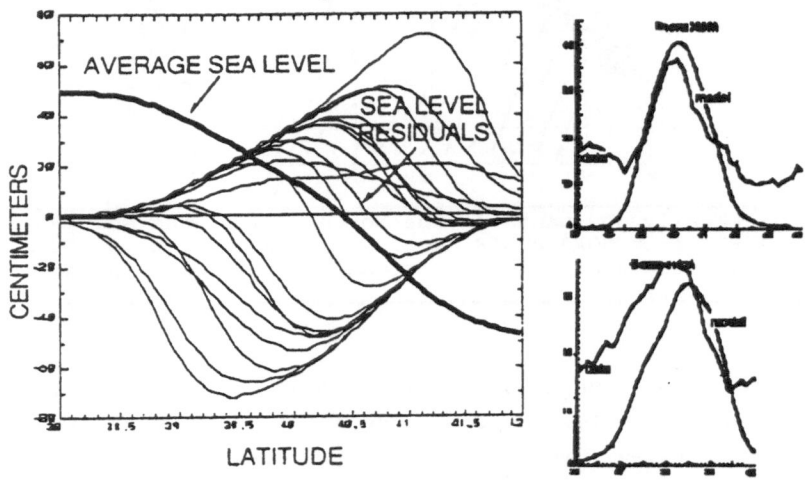

FIGURE 10. Left panel: simulated residuals from the yearly mean
along the ascending Geosat track that crosses the Equator at
322.6º. Horizontal axis is latitude from 38ºN to 42ºN, vertical
axis in cm between −80 and +80 cm. Notice that current positions
South of the mean yield a sea level residual that looks like a
cold eddy, currents to the North resemble a warm eddy, and only
currents within 20 km of the mean position cause a step in the sea
level residual, but with much smaller amplitude. Right panels:
Root mean squared sea level residual, averaged over 1 year at a
fixed geographical position, for two Geosat tracks and from the
tanh model; vertical axis range: 40 cm, horizontal axis range 7º
(top) and 4º (bottom) of latitude (from Zlotnicki, 1990).

3.2 Estimating parameters of the signal's shape.

The actual estimation requires minimizing sets of equations of the
form

$$(v(x)-v'(x)) - (V_0.\exp(-\pi(x-x_0)^2/(w/\cos \vartheta)^2)$$
$$- V'_0.\exp(-\pi(x-x'_0)^2/(w'/\cos \vartheta')^2)) \qquad (3.3)$$

where $v(x)-v'(x)$ are the alongtrack profiles of the difference between
altimetrically observed velocities along the same groundtrack
(identified by λ_0) at two different times, x is the common alongtrack
coordinate (or latitude), and the prime indicates data from two

different repeats (times, in integers of 17.05 days for the Geosat ERM).
The unknowns are V_0, x_0 and $w/\cos \vartheta$ at all times. If the set of
crossing positions $x_0(\lambda_0, t)$ is known from satellite infrared data, then
the crossing angles are computed from the $x_0(\lambda_0, t)$ and the coordinates
of the groundtracks, and the solution for V_0, x_0 is stable. Without the
infrared data the problem is somewhat unstable. There is no way to solve
for both w and $\cos \vartheta$ at the same time, so the minimization is performed
on the apparent width $w/\cos \vartheta$, and then the set of crossings $x_0(\lambda_0, t)$
is used to compute the path of the current and the consequent crossing
angles $\cos \vartheta$. Furthermore, the minimization is non-linear in x_0 which
requires a good initial estimate to avoid local minima. To further
complicate things, the profiles of $v(x)-v'(x)$ are not 'clean': eddies
may enter the picture and appear as false Gulf Stream crossings, or a
perverse meandering situation yielding two or three crossings of the
current on the same ground track may occur. Furthermore, even when
infrared data are available, they give a set of positions for the Gulf
Stream front that is the location of either the steepest gradient in
surface temperature or a particular isotherm; this 'front', called the
North Wall, is some 20 to 40 km to the North of the steepest part of the
sea level signal which responds to the gradient of vertically integrated
density. Given the difficulties, it is remarkable that any estimates
have been obtained at all, but the amount of data is so large, and each
set of repeats of one ground track yields results independent of other
sets (except for the angle ϑ), that internal consistency down the stream
provides a powerful check on the result. In addition, Doppler profiler
measurements of the surface velocity (Joyce et al., 1990) have confirmed
Kelly's approach.

Using variations of this approach on two year's worth of Geosat data,
Kelly concluded that V_0 changed from about 1.2 m/s between 75ºW and 60ºW
to about 0.8 m/s just East of that longitude, where the New England
seamount chain intercepts the current's path, assessed that downstream
changes in the water transport of the upper layer (equation (27) below)
were about the same as Fofonoff and Hall's (1983) changes in vertically
integrated transport obtained from hydrography, and computed seasonal
changes in V_0 that were in agreement with Zlotnicki's (1991) estimate of
seasonal changes obtained from the same data by very different methods.

3.3 The two layer model to estimate total transport.

Tai (1990) used a one year dataset for both Gulf Stream and Kuroshio
and estimated vertically integrated transports from the the sea level

jump, by assuming a simple two layer model (Gill, section 6.2), such that the topography η_2 of the density discontinuity between the layers satisfies

$$\eta_2 = -(\rho/(\rho_2-\rho_1)) \; \eta \qquad (3.4)$$

This reasonable assumption about the topography of the interface is essential, since that topography measures the relative velocity v_2-v_1 between top (1) and bottom (2) layers by a relation known as Margules equation (e.g., Gill, section 7.7).

$$v_2-v_1 = g' \; d\eta_2/dx \qquad (3.5)$$

$$g' = g(\rho_2-\rho_1)/\rho_2 \qquad (3.6)$$

g' is called the 'reduced gravity'. The transport in the upper layer is approximately

$$T = v_1H \; L = (g/f)H(\eta_H-\eta_C) \qquad (3.7)$$

where H is the mean depth of the top layer and L it cross-stream width. As pointed out by Tai (1990), a top layer with mean depth H= 500m, a 1 m jump $\eta_H-\eta_C$ in sea level at mid-latitudes $f=10^{-4} \; s^{-1}$ is associated with an upper layer transport of 50 Sv (where 1 Sv=$10^6 \; m^3 \; s^{-1}$, is the standard unit of volume transport in oceanograhy). Tai (1990) found that both Gulf Stream and Kuroshio left their respective coasts with transports of about 90 Sv, increased their transports to approximately 130 Sv (at 150°E for the Kuroshio, 63°W for the Gulf Stream), then decaying with the Kuroshio doing so over a much shorter distance (68 Sv at 165°E) than the Gulf Stream.

3.4 What is new here?

While oceanographers have had estimates of the surface velocity and vertically integrated transports of both currents for several decades, it wasn't until Geosat altimetry provided this dataset that estimates every 125 km along the currents, computed as a yearly average of data uniformly spaced in time during that year became possible. It was very difficult to ascertain whether the discrepancies between prior estimates at different times and positions was related to data quality, to time changes in the current, or to changes with downstream position. With Geosat altimetry it was also found that seasonal changes in both currents showed a surface transport maximum in the Fall, contrary to prior estimates with both hydrography and Seasat altimetry. It remains to be seen whether this discrepancy is due to interannual changes in the currents or to data errors whose nature was not understood at the time.

4. GLOBAL ANNUAL CYCLE.

4.1 Motivation: the simplest large scale signal.

Up to now the signals we discussed had short spatial scales, of the order of a few hundred km, and we saw that such retrievals from altimetry are relatively easy even in the presence of orbit and environmental errors. There are very interesting basin scale signals that we wish to retrieve, basin-wide wobbles or fast large scale barotropic responses to wind changes, which occur on the time scales of a few days. Such changes are extremely difficult to obtain with any other measurement technique. Wunsch (1991) gave a very nice example of such large scale retrievals, but since we do not know much about the signals to be retrieved, and Geosat was certainly not optimized for such observations, it is a bit early to assess the results (his methods are applicable to the more accurate data).

To illustrate the dangers and temptations, I would like to discuss the simplest large scale oceanic signal that we know something about from both theory and in-situ observation, the annual cycle of sea level, and point out the difficulties that arise.

Let me present the data first, then I will discuss theory and errors.

4.2 Computational scheme

Given the altimetric residuals $\eta'(\phi,\lambda,t)$ as defined in Equation (1) for 1987 and 1988, and corrected by all instrumental, orbit, and environmental effects as well as one can at the time (more about this later), let me just dive naïvely and estimate the coefficients A,C,S of the annual cycle by minimizing

$$\eta'(\phi,\lambda,t) - A(\phi,\lambda) - C(\phi,\lambda)\cos(\omega t) - S(\phi,\lambda)\sin(\omega t) \quad (4.1)$$

$$\omega = 2\pi/365. \ d^{-1} \quad (4.2)$$

at a set of positions (ϕ,λ) along the satellite track approximately every 7 km. Then let me average the A,C,S coefficients into 1º boxes in

PHASE

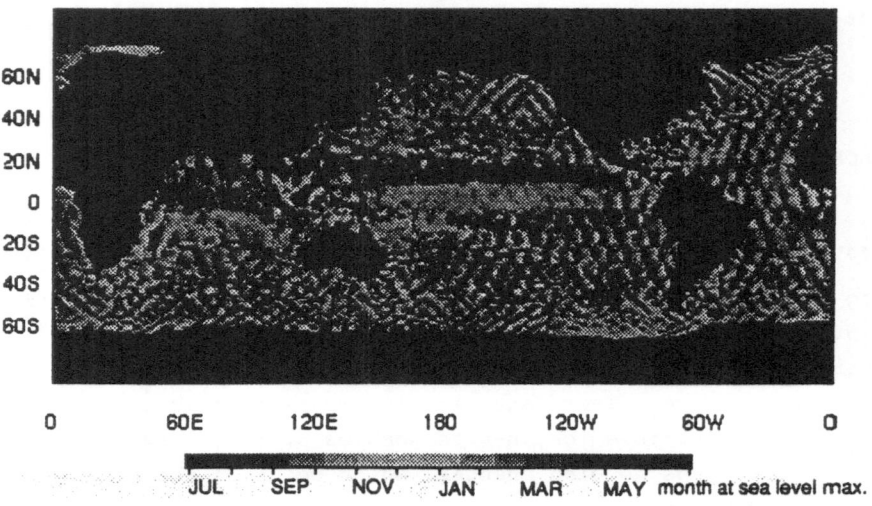

JUL   SEP   NOV   JAN   MAR   MAY   month at sea level max.

AMPLITUDE

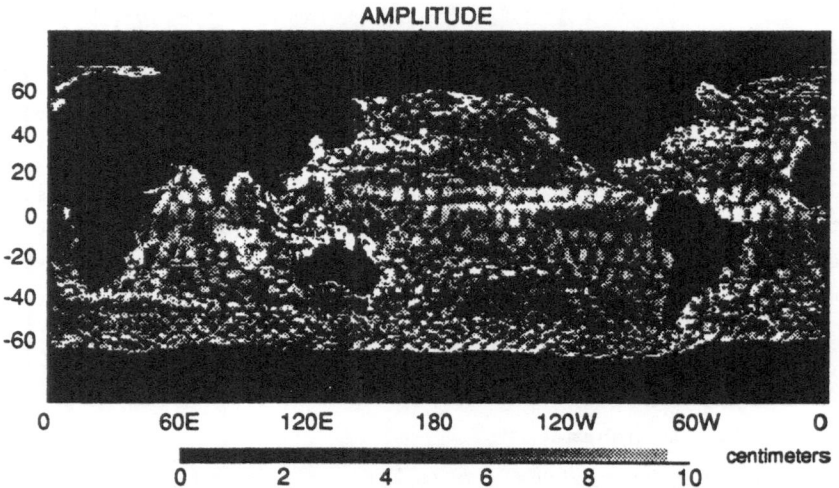

centimeters

0   2   4   6   8   10

FIGURE 11. Phase and Amplitude of the annual cycle of sea level from 2 years (11/86-12/88) of Geosat data, with GEM-T2 orbit, once/rev orbit correction, SSM/I water vapor after 7/87, including M2 tidal error.

both φ and λ, and convert the smoothed B and C coefficients to amplitude and phase (see Figure 11).

It is very clear in the phase plot (upper panel in figure 11) that there is a large error source that causes the 'trackiness', i.e., a pattern whose geographic distribution is strongly correlated along the data tracks. To see its spatial characteristics better, consider a simple profile of the phase along selected latitudes:

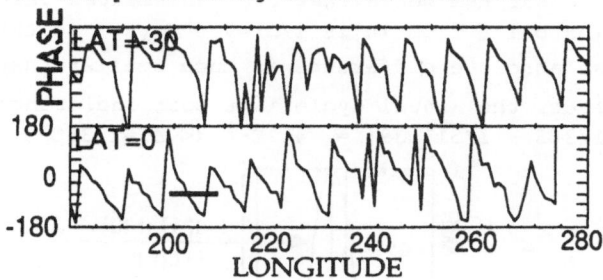

FIGURE 12. Phase of the annual cycle from figure 10 at latitudes 0º and 30ºS. Notice the periodicity of approx. 8º of longitude.

The zonal length scale of the error, about 8º, is suspiciously close to the 7.38º separation between closest tracks laid in one 3 day subcycle of the Geosat ERM orbit. So, the projection of the error on the annual cycle seems to progress slowly over 3 days.

What is more, suppose you are not working globally, but rather in a small piece of ocean with this data set, and so lack the big picture to ascertain that this signal is clearly an error correlated alongtrack. You compute the annual cycle, and it looks like a wave propagating to the West at approximately 2.5 km/day, a reasonable speed for Rossby waves around 25º to 30º of latitude (Gill, 1982, section 12.3). You can end up with a very interesting and completely wrong interpretation.

4.3 Tidal aliasing.

I actually sat on this problem for a while, incorrectly pursuing it as an orbit problem, until M. Parke (1991, pers. comm.) pointed out that it was an error in the M2 tide and explained to me how it comes about. Let us look at its time and space properties, since they illustrate some elementary time series analysis that we need to have handy. It was Jacobs et al.(1991) who proved that this trackiness was indeed the M2 error, although Cartwright and Ray (1990) had already estimated it from Geosat data.

To understand the error we need to remember a few facts . First, when a harmonic with the M2 period $1/f_m = 24/12.42058$ days is sampled only every $1/f_s = 17.0505$ days, it appears to have an alias period $1/f_a = 317.39$ days period

$$(fa = fm - fs*int(fm/fs + 0.5*(fm/abs(fm)))) \qquad (4.3)$$

That happens along a single pass and its repeats. The two year duration of the Geosat ERM means that the fundamental unit of frequency of any time series built from it is $1/\Delta f = 730$ days, so that two frequencies whose absolute difference is much smaller than $1/\Delta f$ are not separable. In fact, the annual cycle is almost indistinguishable from the M2 alias: $|1/365 - 1/317.39| = 4.1E-4 \ll \Delta f = 1.37E-3$.

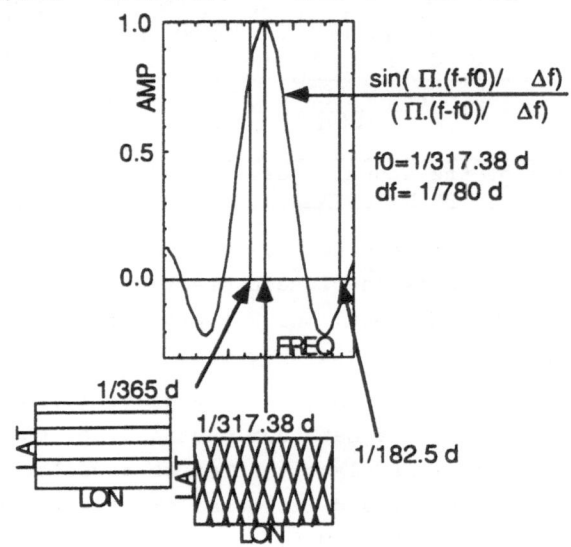

FIGURE 13. Because the annual cycle (1/365d) and the M2 alias (1/317.38d) are separated by less than 1/780d, they cannot be resolved in a time series from Geosat at a fixed latitude-longitude. However, these two signals so close in frequency have totally different geographic distribution: the anual cycle is predominantly zonal, while the M2 error is aligned with the tracks. By combining the space-time properties of the M2 error they can be separated.

Finally, to understand the zonal wavelength, one needs to know that for any given Geosat ERM pass, the nearest parallel track to the east is at a distance of 1.475° and is laid 3.00 days later. This means the M2 tide (assumed with constant coefficients over that distance) is sampled with a phase difference of ~ 73°, and if the coefficients remained constant the phase difference would reach 360° (one wavelength) at a zonal distance of 7.3° (about 5 tracks away). Of course, this simple

analysis is complicated by the spatial variability of the tidal
amplitudes, as well as by the interference caused by combining ascending
and descending tracks.

By assuming that both the M2 tides and the oceanic signals have
constant coefficients over an area some 5º to 10º in longitude, one can
compute the M2 correction, since its apparent phase changes between
neighbouring tracks in a predictable manner, while any oceanic signal
with a true near annual cycle and equally constant coefficients over the
area would have the same phase. See the chapter by Cartwright for
further details.

Remember, this error did not appear during our mesoscale and Gulf
Stream discussions. Let us see why below.

4.4 Orbit error, water vapor, sea state bias effects.

For mesoscale studies it is sufficient to model and remove residual
orbit error with a short (2000-5000 km) alongtrack quadratic polynomial,
but the M2 tide is itself a long oceanic wave whose characteristic
length scale is the barotropic Rossby radius, about 2000 km in the open
ocean, hence much of the tidal signal (error in this case) is removed by
short orbit corrections.

For this study, I used a once per revolution alongtrack harmonic
defined between the northernmost latitudes reached by the orbit (i.e.,
one set of once-per-revolution coefficients was computed for each
satellite period, with no constraints as to how much it could vary from
one period to the next), hence many long wavelength errors in the data
were retained, in addition to the M2 error just discussed. One way to
think about it is spectral: by removing a 2,000 km polynomial from the
data, all alongtrack signals with wavelengths longer than about 1000 km
are effectively removed, whether oceanic or erroneous. Notice however,
than cross-track signals are not so affected. Another way to think about
the effect of short orbit arc removals is this: in one period,
approximately 40,000 km along the track, there are 20 segments with
length 2,000 km; if a second degree polynomial is fit for each, 60
coefficients (degrees of freedom) are adjusted, while a once/rev sine
requires only 3 coefficients for the same data. As a consequence, 60
coefficients better adjust to both errors *and signals* along the track,
and remove more of *both* when substracted from the data; 3 coefficients
simply remove less energy.

Water vapor is the next big error source (after the orbit and the M2 errors) in Geosat. Again, this signal barely affects oceanic mesoscale (Jourdan et al., 1990) except near the Equator. While data from the DMSP8-SSM/I instrument exist from 7/87 onward, the instrument was turned off during 12/87 (to repair an overheating problem). The full water vapor error, with 5-10 cm amplitude at latitudes lower than 40º, would be retained for the better part of 1987 (see figure 6), and, of course, water vapor being related to heating and ocean temperatures, it has a strong annual signal. Let us assume for the time being that only data for 1988 are used, and that the SSM/I retrieval has errors below 1-2 cm. As I will show you below, there are strong traces of residual water vapor error.

Another error, sea state bias, can affect our results in a very important way. Sea state bias (Chelton, 1988) is a generic term for several errors associated with both the altimeter tracker (the component that estimates when the return from an altimeter pulse is expected at the antenna) and the properties of the interaction between electromagnetic waves from the altimeter and wind waves on the ocean surface. Tracker errors include skewness bias (discussed below); noise, because waveforms are averaged over only 1/20 to 1/10 of a second; coarse table lookups for the slope of the leading edge of the return pulse; effects of the satellite's attitude while the algorithms assume normal incidence; Doppler shifts due to the vertical component of the relative velocity between the satellite and the sea surface, etc.

The most important of these errors is the 'skewness bias', due to the assumption that the specular reflectors scattering the pulse back to the satellite have a gaussian probability distribution while the true distribution is skewed causing the median sea level to be below the mean sea level (Chelton, 1988; Rodriguez, 1988).

An 'electromagnetic' bias arises because the usually flatter troughs of ocean waves reflect more energy than the usually more pointed crests. As a consequence the mean reflecting surface is below the mean sea surface by an amount that is about 1% to 3%, for a radar wavelength of 2.2 cm, of significant wave height which is a measure of the distance between crests and troughs (Rodriguez et al, 1992).

To date, all these biases have been accounted for ('corrected') as a percentage of significant wave height, also measured by the altimeter. However, only recently has it been understood how this 'bias' depends on the degree of wave development that controls the shape of waves (Glazman and Pilorsz, 1990): when the waves come from afar (swell) they tend to

be longer, smoother and have a more 'gaussian' pdf; when the waves are
generated by local winds (seas) they are richer in shorter wavelengths
and their pdf is more non-gaussian.   Fu and Glazman (1991) showed that
this effect implied that the sea state bias correction associated with 4
m waves could range between 3 and 11 cm depending on the wave age (the
ratio of the phase speed of the dominant wave to the wind speed. Wave
age is a measure of wave development, which increases slowly with fetch,
a measure of the distance traveled by the waves). There are as yet no
large scale studies of this effect, or the accuracy of a correction
based on wave age, but the generation of waves by winds varies
seasonally so there is bound to be a several cm signal associated with
this effect in the annual cycle of sea level measured by altimetry. When
we talk about the Antarctic Circumpolar Current, I will bring up these
biases again.

In summary, when we chose a longer wavelength model for the orbit
error, all these other errors became more prominent.

4.5 N.Equatorial countercurrent. Somali current.

The maps in figure 13 look oceanographically 'plausible' if you
mentally remove the tidal trackiness, so it is tempting to describe what
we see, at least qualitatively for now.   First, the broad patterns are
very zonal, with the strongest signal, both in amplitude and in phase
contrast, in the equatorial Pacific.   The sharp phase discontinuity
between 5°N and 10°N appears to indicate the seasonality of the
eastward-flowing N. equatorial countercurrent (Wyrtki, 1974), which
flows downhill the slope caused by the N.E. trade winds and has maximum
transport in September (Lolk, 1992).   One must haste to point out that
this band is also the location of the InterTropical Convergence Zone,
where the N.E. trade winds meet with the northward branch of the S.E.
trade winds, causing the highest gradient of atmospheric water vapor
concentration of the oceans.   The discontinuity extending ESE from New
Guinea to about 140°W is a bit more suspicious, since it is the
approximate position of the South Pacific Convergence Zone (SPCZ) which.
like its better known counterpart, the ITCZ, is a zone of intense water
evaporation; in other words, this line looks suspiciously like residual
water vapor error and possibly casts doubt on any NECC calculations.

The next strongest phase contrast is in the Indian ocean, the area of
the most famous seasonal wind variations, the Monsoon. During January-
February atmospheric pressure is greater over S.E. Asia than over the
ocean, driving the dry, cool, N.E. Monsoon which converges with the

trade winds at around 10-20ºS. During May-June low atmospheric pressure over S.E. Asia leads to a southwesterly wind, the S.W. Monsoon, which is laden with moisture from the Arabian Sea and discharges over India (Brown et al., 1989). The ocean signals in the Bay of Bengal and the powerful Somali current are both visible in the amplitude of the annual cycle in figure 13, but the phase signal of the Somali current is not, perhaps because it so closely hugs the coastline. The signal around 5ºN looks just like the winter N. Equatorial current and its summer counterpart, the S.W. Monsoon current. Now for the suspicious part: the biggest phase signal in the Indian Ocean, the southeastward line from about 5ºS, 60ºE to Australia is the location of -- you guessed it, strongest gradient in water vapor over that ocean.

4.6 Heating and cooling.

Now, let us look at the broad lines around 40ºN, and 40ºS in the phase plot, which roughly separate the subtropical from the subpolar gyres. North of 40ºN, sea level is high around January; between 40ºN and the Equator it is high around September/October; essentially the pattern holds for the southern hemisphere with a 6 month offset from the Northern one.

This broad pattern for the subtropical gyres is the result of the seasonal heating and cooling of the upper ocean (above the seasonal thermocline, roughly the upper 200 m), identified by Patullo et al (1955) in tide gage data and explained by Gill and Niiler (1973) as the result of local storage of heat since on the annual time scale involved, large scale surface currents are too slow by an order of magnitude to advect this surface heat. This pattern does not seem affected by any clearly defined altimetric errors.

4.7 The ACC

Finally, the region of the Antarctic Circumpolar current shows strong variability at the annual cycle, most likely associated with current instabilities (Chelton et al., 1990). This is also the region where sea state errors can be expected to be largest (since it has the highest winds and waves), where our uncertainty about the precise scales of validity of the ocean's 'static' response to atmospheric pressure changes (inverse barometer, about 1 cm per mbar) is greater because of the highly variable atmospheric pressure, and an inaccessible region with few measurements of atmospheric pressure to constrain models from which the dry tropospheric correction and the inverse barometer correction are derived. The signal in the subpolar gyres is strongly affected by the wind, as opposed to the subtropical gyres where heating was dominant.

4.8 Summary.

The main point of this whole section is not how to compute accurate tidal estimates, or the description and meaning of the global annual cycle of sea level, but to warn you of system errors that appear when we get more ambitious and eliminate fewer signals through the orbit correction. It is meant to counter any excessive optimism the reader may have developed from seeing that no matter what we did in section 2, oceanographically reasonable results appeared. So, when reading various researchers' results, it is crucial to understand how their stated corrections can affect their results.

5. EQUATORIAL SEA LEVEL ASSIMILATION and WINDS.

5.1 Getting rigorous about errors in data and models.

Up to now I did some hand-waving about the errors in the data and used either contemporary data of another nature, or whatever we knew from historical oceanographic data to argue that Geosat altimetry did indeed measure oceanic phenomena -- much of the time. This may be reasonable when a new data type appears, as was the case with altimetry, but in order to truly quantify oceanic processes whose details we did not know before, it is necessary to get rigorous about errors in the data, errors in the equations we use to describe the ocean motions, and the propagation of those errors through all the computations.

I am sure you are familiar with minimum variance estimation, and derived methods such as least squares collocation (Moritz, 1980) used in Geodesy, inverse methods (Backus and Gilbert, 1970) used in Geophysics, Kalman filter methods (Gelb, 1970) used for time-evolving linear dynamic systems, and adjoint schemes (Thacker and Long, 1988) computationally tailored to time-stepping numerical models. They have much in common: they minimize an L_2 norm (as opposed to linear programming which minimizes an L_1 norm, or schemes that minimize the largest discrepancy, an $L\infty$ norm) for linear or linearized systems, they use information on the covariance of the signal (its power spectrum when it is a continuous function), of the errors in the data and in the equations, and they tend to be tailored to particular applications, especially where the number of unknowns is huge and larger than the rank of the data covariance matrix, applications that transcend disciplines. Their application in Oceanography is rather recent (Wunsch, 1978; Thacker and Long, 1988; Miller and Cane, 1989; Gaspar and Wunsch, 1989).

I will describe the work of Fu, Fukumori and Miller (1992) in using a Kalman filter to assimilate Geosat data in the equatorial Pacific region which both shows you some nice equatorial physics, the assimilation scheme, and an example of the very tight connection between winds and sea level in this region, so tight in fact that if one of them is known accurately you can improve your knowledge of the other one.

5.2 The equatorial beta plane.

The equatorial region is very nice for a linear optimization scheme because linear physics give a very good approximation. Gill's chapter 11 and Pedlosky's section 8.5 are excellent introductions to the dynamics. This very brief summary follows Gill. Within 30º of the Equator one can write

$$\sin(\varphi) \approx \varphi, \quad \cos(\varphi) \approx 1, \quad y = R\varphi \qquad (5.1)$$

$$f(\varphi) = \beta y, \quad \beta \approx 2.3 \times 10^{-11} \text{ m}^{-1}\text{s}^{-1} \approx 2.0 \times 10^{-3} \text{ km}^{-1}\text{day}^{-1} \quad (5.2)$$

(the error of 5.1 is no greater than 14%). If you consider a frictionless ocean of uniform depth H and uniform density ρ, the forced equations of motion for the east velocity u, the north velocity v, and the sea level η, all with horizontal length scales L >> H are

$$\partial u/\partial t - \beta y v = -g\, \partial \eta/\partial x + X/\rho H \qquad (5.3)$$

$$\partial v/\partial t + \beta y v = -g\, \partial \eta/\partial y + Y/\rho H \qquad (5.4)$$

$$\partial \eta/\partial t + \partial(Hu)/\partial x + \partial(Hv)/\partial y = -E/\rho \qquad (5.5)$$

where X and Y are forcings that we should think of as wind stresses. Equation 5.5 states mass conservation, and its forcing E is the difference between evaporation and precipitation, but this forcing is not be used below, E=0. The potential vorticity equation that follows from (5.3-5.5) is

$$\partial(\zeta - f\eta/H)/\partial t + \beta v = (\partial Y/\partial x - \partial X/\partial y)/\rho H \qquad (5.6)$$

which is linear thanks to the absence of vorticity advection terms. The right hand side of 5.6 has the wind stress curl, and $\zeta=\partial v/\partial x-\partial u/\partial y$ is the usual relative vorticity.

For a stratified ocean with $\rho(z)$ the equations (5.3-5.6) are more complex, but separation of variables allows us to express $u(x,y,z,t)=\hat{u}(x,y,t)\xi(z)$ if such terms as $2\Omega u$, $2\Omega w$ and $\partial w/\partial t$ in the stratified equations can be neglected. The $2\Omega u$ term (geodesists would call it the Eötvos acceleration, but not oceanographers) in the equivalent of 5.5, and a related term $2\Omega w$ in the equivalent of 5.3 are negligible if the Brunt-Väisälä frequency $N(z) >> 2\Omega$ (Gill, section 11.9), which is true almost everywhere in the oceans (1/1 hr >> 1/12 hrs). The $\partial w/\partial t$ term in the z equation is negligible if the horizontal scale of the motions is much larger than the vertical scale over which $\rho(z)$ varies, which is also true almost everywhere in the oceans. The exceptions are some very important and very small regions of upwelling and subduction at high latitudes. Then, the separation of variables works if $\xi(z)$ satisfies (Gill, section 6.11; Pedlosky, section 6.12),

$$(1/\rho)d(\rho d\xi/dz)/dz + (N^2/c)\xi = 0 \qquad (5.7)$$

where N is again the Brunt-Väisälä frequency. The normal modes of 5.7 have eigenvalues associated with permitted values of velocities

$$c_m, \quad m=0,1,2,\ldots,\infty \qquad (5.8)$$

and vertical eigenfunctions ξ_m, both ordered by decreasing values of the c_m (this separation of variables works at all latitudes, not just the equatorial region). The 0-th mode is always associated with depth-independent functions, the barotropic mode. All modes $1,2,\ldots,\infty$ are the baroclinic modes, associated with increasingly shorter vertical scales of variation of $N(z)$, and its associated $\rho(z)$, for which surfaces of constant density are not parallel to surfaces of constant pressure.

Associated with each vertical normal mode c_m there are horizontal structure functions $u_m(x,y,t)$, etc for each of u,v,η, in the equations, which satisfy *the same* equations (5.3-5.6) provided H is interpreted not as the depth of the ocean but as a depth scale associated with the mth mode of (5.10). This depth scale is not closely related to any layer thickness (unless one uses the reduced gravity g' instead of g): typical values for c_1 are 2 to 3 m/s, with equivalent depth c^2/g less than 1 meter.

Also associated with each vertical normal mode there are characteristic scales for length a_m, time τ_m, and velocity c_m. They are,

$$c_m = \sqrt{gH_m} \qquad\qquad\qquad (5.9)$$

$$a_m = \sqrt{c_m/2\beta} \qquad\qquad\qquad (5.10)$$

$$\tau_m = 1/\sqrt{2c_m\beta} \qquad\qquad\qquad (5.11)$$

c_m is the velocity of gravity waves with length scales $\gg$ H for which rotation is negligible (gravity is the key restoring force and their frequency is higher than the inertial period $1/f$; more about it later), a_m is the equatorial Rossby radius corresponding to that mode, and is also the distance from the Equator within which wave solutions to 5.3–5.6 are trapped. (I apologize for using Lo in section 2 for the Rossby radius, called a_m here). While in mid-latitudes the two most energetic modes are the barotropic and first baroclinic (m=0 and 1) modes, in the equatorial region the barotropic mode has scales too long for the equatorial β plane approximation to remain valid, hence it is not used. Typical scales are

	m=0	m=1	m=2
c_m(km/day)	17,000	250	154
a_m(km)	2,085	251	196
τ_m	2.9 hrs	1.0 d	1.3 d
τ_{PAC}	1.05 d	72 d	117 d

where τ_{PAC}=18,000 km/c_m is an estimate of the time a wave with velocity c_m would take to cross the equatorial Pacific.

Now you 'know' that (5.3–5.6) apply to each mode m. But those linear equations have an infinite set of wave solutions: one Kelvin wave, infinitely many gravity waves, infinitely many Rossby waves, and one mixed gravity-Rossby wave (more about them below). So, the solution to the stratified version (5.3–5.6) for the velocity u is a sum over the barotropic and all baroclinic normal modes m, and all wave modes n of the $u_{m,n}$, and similarly with the the $v_{m,n}$, $w_{m,n}$, $\eta_{m,n}$. As usual, not all normal modes and all wave modes are needed. For example, friction kills the short Rossby waves, and the first two or three normal modes have most of the variance.

5.3 Equatorially-trapped waves.

There are many properties of the various waves that satisfy 5.3–5.6 that are both instructive and interesting, since the waves carry information about local disturbances (such as wind changes) across the ocean. With the time we have here, only their dispersion properties and some characteristic scales can be briefly discussed.

The wave solutions of 5.3-5.6 have frequencies ω_{nm} and zonal (+ eastward) wavenumbers k_{nm} that satisfy the dispersion relation (Gill, section 11.6),

$$(\omega_{nm}/c_m)^2 - k_{nm}^2 - \beta k_{nm}/\omega_{nm} = (2n+1)\beta/c_m \qquad (5.12)$$

where again, the index m refers to the normal vertical mode, and the index n identifies eigenmodes (waves) that are solutions of 5.3-5.6.

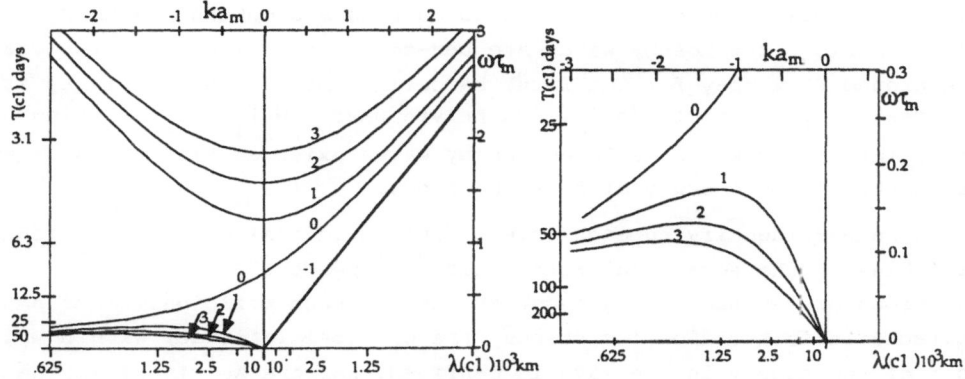

Figure 14: Dispersion of equatorial waves.
LEFT PANEL: Top axis=nondimensional zonal wavenumber, right axis= nondim. frequency. Bottom axis=wavelength in km, left axis=periods in days, for the first baroclinic mode, assuming c_1=2.8 m/s. Right quadrant is Eastward.
RIGHT PANEL: expands the Rossby wave region of the left-hand panel

Figure 14 shows the dispersion curves that satisfy 5.12. The straight line at 45° in the positive quadrant is the Kelvin wave. Related Kelvin waves propagate along coasts, sometimes a consequence of the water pileup caused by Ekman transport toward the coast, sometimes the poleward radiation of energy brought to eastern boundaries by equatorial Kelvin waves, sometimes associated with tides. In one case the coast, in the other the Equator, provides the boundary. Here they propagate along the Equator, their sea level signal symmetric across the Equator, with a meridional 'half-width' equal to the Rossby radius a_m. As figure 14 shows, they are non-dispersive, with group velocity $c_m=\sqrt{gH_m}$ at all frequencies, approximately 250 km/day for the first baroclinic mode.

Each hyperbola is a gravity wave. For the first baroclinic mode all their periods are shorter than about 6 days. They can propagate energy in either zonal direction.

The curves in the lower left hand corner, expanded in the lower panel, are the Rossby waves, the first baroclinic waves with periods always longer than about 1 month. All Rossby waves have westward *phase* propagation, but those with with long wavelengths also have westward energy propagation (group velocity), while those with short wavelengths propagate energy eastward (and do so 8 times slower). For the values in figure 14, the boundary between these two regimes in the first baroclinic Rossby wave of order 1 is at about 1200 km wavelength.

Since friction tends to kill the shorter Rossby waves (Cane and Sarachik, 1981) most discussions assume the long wavelength portion and make appropriate approximations based on this assumption. In this approximation, the Rossby waves are non-dispersive. Typical velocities are around 60 km/day for the first baroclinic mode. While gravity is the restoring force of gravity and Kelvin waves, β is the restoring force for Rossby waves. Related Rossby waves exist at all latitudes and are central to the study of rotating flows.

The number identifying each wave is its meridional mode. Separation of variables in the equatorial zone yields independent factors for functions of latitude, longitude, and depth. Each meridional mode has a characteristic meridional distance, the critical latitude, within which most of the energy in the wave is confined. For the Kelvin wave, the critical latitude is the equatorial Rossby radius. For both gravity and Rossby waves it increases as $\sqrt{n}$.

5.4 The Miller and Cane formulation.

Miller and Cane (1989), Cane (1984) used this long wavelength approximation, which allowed them to express the sea level η and zonal velocity u in the equatorial area as

$$\begin{Bmatrix} \eta_m \\ u_m \end{Bmatrix} = (2^{-1/2}) \, [r_{0m}(x,t) \, \Psi_0(y) \, + $$

$$\sum_{n=1}^{N} (2^{-1} \, r_{nm}(x,t)) \, (n+1)^{-1/2} \, \Psi_{n+1}(y) \begin{Bmatrix} + \\ - \end{Bmatrix} n^{-1/2}\Psi_{n-1}(y)] \quad (5.13)$$

where $\Psi_n(y)$ is the product of the nth Hermite polynomial of meridional distance with $e^{(-(y/y_0)^2)}$, $r_{0m}(x,t)$ is the local amplitude of the Kelvin wave in the mth baroclinic mode, $r_{nm}(x,t)$ is the local amplitude of the nth meridional mode Rossby wave in the mth baroclinic mode. Miller and Cane (1989) then numerically solved the equations satisfied by the r_{nm},

$$\partial r_{0m(x,t)}/\partial t + c_m \, \partial r_{0m(x,t)}/\partial x = T_0(x,t) \qquad (5.14)$$

$$\partial r_{nm(x,t)}/\partial t - (2n+1)^{-1}c_m \, \partial r_{nm(x,t)}/\partial x = T_n(x,t) \qquad (5.15)$$

where T are projections of the wind stress onto the appropriate meridional mode. The boundary conditions, discussed in Cane and Sarachik (1981), are

$$u=0 \text{ at } x=X_E \qquad (5.16)$$

$$\int u \, dy = 0 \text{ at } x=0 \qquad (5.17)$$

Notice that the only driving force is the wind. These equations 'convert' wind to sea level and water velocity. This important point will come up again later.

In the work of Fu et al (1992), the equations are sampled every 5º of longitude between 125ºE an 80ºW, with two baroclinic modes and five meridional mode Rossby wave as well as the Kelvin wave.

5.5 Kalman filter and smoother data assimilation

After formulating the model as stated above, Miller and Cane (1989) included in the code a Kalman filter scheme to assimilate sea level data in order to estimate the amplitudes of the waves. Fu et al. (1992) added a Kalman smoother with asympotic steady state covariance matrix to the filter. I recommend Gelb (1974) as an excellent introduction to Kalman filtering, Goodwin and Sin (1984) for more recent results.

Given the number of modes they used, the assimilation scheme needs to compute, at each time step t, 384 parameters: the coefficients $r_{nm}(x,t)$ for 2 vertical modes x 6 meridional modes x 32 longitudes. Their time stepping was chosen as 3 days, because of the Geosat subcycle.

In the notation of Fu et al.(1989), equations (5.14-15) are equivalent to

$$\mathbf{q}_f(j) = \mathbf{A} \, \mathbf{q}_a(j-1) + \mathbf{B} \, \tau(j-1) \qquad (5.18)$$

which states that the model computes the vector of unknown parameters at time j as a linear combination of the 'analyzed' set of parameters $\mathbf{q}_a(j-1)$ at the previous time j-1 and another linear combination of the wind stress forcing $\tau(j-1)$ also at the previous time step. By 'analyzed' it is meant that $\mathbf{q}_a(j-1)$ are the result of data assimilation at the previous step, rather than the simple model forecast. In Kalman filter parlance, $\mathbf{q}$ are the state variables, τ the control variables.

Let $\mathbf{P}_a$ be the error covariance of the $\mathbf{q}_a$, and Γ the error covariance of the wind stress τ. Then the error covariance $\mathbf{P}_f$ of $\mathbf{q}_f$ is

$$P_f(j) = A P_a(j-1) A^T + B \Gamma B^T \qquad (5.19)$$

In the traditional formulations of Kalman filter there is a term describing the error of the model itself, i.e., the error of the equation (5.18) or its equivalent (5.14-15). In this application, all the blame for misfit is assigned to the wind, since experience shows that the inaccuracy of wind data dominates the error of (5.18) (Miller and Cane (1989)). However, it is also true that it is seldom easy to quantify the errors of the neglected terms in the model equations.

The vector of forecast sea level η_f is also linearly related to the model parameters q by

$$\eta_f(j) = C(j) q_a(j) \qquad (5.20)$$

but we have a data vector η_o of observed sea level, whose error covariance is N.

So, the Kalman filter estimate of $q_a(j)$ is a blend of the model forecast $q_f(j)$ with the projection of the data η_o onto the parameter space q,

$$q_a(j) = q_f(j) + G(j) [\eta_o(j) - C(j) q_f(j)] \qquad (5.21)$$

$$G(j) = P_f(j) C(j)^T [C(j) P_f(j) C(j)^T + N(j)]^{-1} \qquad (5.22)$$

$$P_a(j) = [I - G(j) C(j)] P_f(j) \qquad (5.23)$$

$G(j)$ is the all-important Kalman gain matrix, whose role is to project the difference between observed and predicted data (here sea level) onto the parameter space, weighing such difference by the relative reliability of model forecast and data. The computational burden arises because at each time step all of (5.18-19, 5.21-23) must be computed.

Fukumori et al (1992) took advantage of the fact that, when the A, B, and C matrices are time-invariant, $P_f(j)$ sometimes has an asymptotic limit which can be computed by some efficient algorithms. If the assimilation reaches the steady limit relatively quickly, then it is unnecessary to time-step $P_f(j)$ as (5.19) requires, and it can be replaced by its asymptotic limit. This insight was included in Fu et al (1992). When they compared the full Kalman computation against the steady state version, the computer time savings were a factor of 20, and the q_a accounted for 49% of the data variance, vs. 52% when P_f was recomputed at each step. Care had to be taken to wait for the steady state limit to be reached, which for this model takes about 100 days, the time for a Kelvin wave to cross the Pacific.

The traditional Kalman filter designed to control a spacecraft uses only *past* observations to predict the next time step. However, in this case we have all the data, past and future. The scheme to use future data is called an 'optimal smoother' in filtering parlance, and it is equivalent to time-stepping the model backward from its final state after completing the forward time-stepping Kalman filter run. Again, an asymptotic steady-state version of G, P_a, P_f made the computation feasible.

5.6   Error covariances of the altimetric data

All this formalism would do no good if realistic errors for the data and the wind forcing were not used. Fu et al (1992) used the following description of errors:

orbit: they generated a once per rev 'signal' with 30 cm rms plus 10 cm correlated over 100 km (mesoscale), then estimated the once per rev signal. The covariance of the difference between the known and estimated once per rev was their orbit error model.

wet tropospheric correction: covariance of the difference between FNOC and SSMI estimates, after removing a once per rev component.

ocean tides: they used the covariance of 10% of the actual Schwiderski tides in the Geosat GDR, after removing a once per rev. The assumption is that the error in the tides has the same spatial characteristic as the tides themselves.

inverse barometer: the covariance of the full correction, since we do not know how accurate it is, after removing a once per rev.

sea state bias: the covariance of 2% of the time varying part of significant wave height, after removing a once per rev. Again, this assumes that the correction is entirely unreliable.

wind forcing: a gaussian covariance with 10º zonal and 2º meridional scales. Their 32.9 m^2/s^2 wind error variance was estimated from the assimilation itself, as the value that gave the best fit to the altimetry. The spatial scales had been calibrated in Miller and Cane's (1989) assimilation to tide gage data.

While one may argue with the details, this is a very realistic and thorough attempt at modeling the residual errors in all the quantities entering the calculation.

5.7 Match to tide gage data and withheld altimeter data.

To have a baseline, Fu et al (1992) computed the match of the model to the altimeter data in the absence of assimilation: 0.25 correlation, just at the 95% confidence limit for the number of independent pieces of information they estimated (2 degrees of freedom per 2400 km altimeter pass). The subsequent 52% correlation after filtering, and 59% correlation after Kalman smoothing must be measured against this initial value.

However, there is always an external way to ascertain whether the formal error estimates of any large-scale least squares adjustment are close to reality, and it involves witholding some data from the computation altogether, to be used for checking purposes. With a time-stepping scheme such as the Kalman filter, the time history of the mismatch between predicted (not analyzed or smoothed) and observed data, $\eta_f(j)-\eta_o(j)$, can also provides an ongoing check.

Fu et al (1992) tried withholding every other pass of altimeter data. Result: 52% correlation with the *withheld* data.

They also completely withheld tide gage data they had. Result: the 66% correlation between forecast sea level and tide gages increased to 81% after the altimeter assimilation step. Notice that the tide gages match either the forecast or analyzed values better than altimetry (52%) because they are more accurate at the few locations where they exist; the altimeter data is less accurate on a point by point basis, but its spatial and temporal coverage better constrain the model.

5.8 Can anything be said about the wind?

To close this example, I would like to tell you that an oceanographer jokingly told another one 'if you give me the wind, I'll give you the sea level', implying that known the main driving force, the rest could be modelled, to which the other oceanographer, well versed in inverse methods, answered 'give me the sea level and I'll give you the wind'. Actually, both sides know that over the global oceans, not only wind but also heating and cooling, and evaporation minus precipitation are the forcings on the ocean circulation, so the joke is an oversimplification.

But the equations (5.14-15) only relate sea level to wind, so Fu et al (1992) got tempted: could the altimetric sea level tell us anything about the wind?

They obtained wind data from 14 buoys. Result: the average correlation between the gridded wind data they used (from Florida State University) and the bouys was 60.4%, and the average correlation between model-corrected wind and the buoys was 60.8%. Of the 14 buoys, 10 showed increased correlation with the model-derived wind correction.

Clearly, and they state so, the improvement is not statistically significant. However, it is remarkable how tight a link there is between wind and sea level in the equatorial oceans, allowing this model to estimate a correction to the wind, and that applying such a correction does not worsen its correlation to external data. Remember that all model errors, including neglected physics, were blamed on the poor wind data.

6. ACKNOWLEDGEMENTS

This work was supported by NASA's Climate and Hydrologic Systems Program and by the Topex/Poseidon Project, and performed at the Jet Propulsion Laboratory, California Institute of Technology under contract with the National Aeronautics and Space Administration. I am very grateful to Fernando Sansò and Reiner Rummel for inviting me to participate in the Summer School, for what I learned there, and for their encouragement.

7. REFERENCES

Backus, G.E. and F. Gilbert   Uniqueness in the inversion of inaccurate gross Earth data   Phil. Trans. R. Soc. London ser. A, 266, 123-192   1970

Brenner, A.C., C.J.Koblinsky and B.D. Beckley,   A preliminary Estimate of Geoid-Induced Variations in Repeat Orbit Satellite Altimeter Observations, J.Geophys.Res., 95(c3), 3033-3040, 1990

Bretherton F., R. Davis and C.Fandry.,   A technique for objective analysis and design of oceanographic experiments applied to MODE-73., Deep Sea Res., 23, 559-582. , 1976.

Brown, J., A.Colling, D.Park, J.Phillips, D.Rothery, and J.Wright, Ocean Circulation, The Open University, Pergamon Press, 238 pp, 1989

Callahan, P.S.,   TOPEX Ground System Science Algorithm Specification, JPL D-7075, available from Document Copies Desk, Jet Propulsion Laboratory, Pasadena, CA 91109, USA, 1990

Cane, M.A. and E.S. Sarachik,   The response of a linear baroclinic equatorial ocean to periodic forcing, J.Marine Res., 39, 651-693, 1981

Cane, M.A.,   Modeling Sea Level during El Nino, J.Phys.Oceanog., 14, 1864-1874, 1984

Chelton, D.B., M.G. Schlax, D. L. Witter and J.G. Richman,   Geosat Altimeter Observations of the Surface Circulation of the Southern Ocean, J.Geophys. Res., 95, 17877-17904, 1990

Chelton, Dudley B., WOCE/NASA Altimeter Algorithm Workshop, U.S. WOCE Technical Report No.2, 70pp., U.S. Planning Office for WOCE, COllege Station, TX, 1988

Cheney R.E., B.C. Douglas, R.W. Agreen, L. Miller and N. Doyle, The NOAA Geosat Geophysical Data Records: Summary of the first year of the Exact Repeat Mission, NOAA Technical Memor. NOS-NGS-48, 1988

Cheney R.E., B.C. Douglas and L. Miller, Evaluation of Geosat altimeter data with application to tropical pacific sea level variability, J. Geophys. Res., 94 (C4),4737-4747., 1989

Emery, W. J., G. H. Born, D. G. Baldwin and C. Norris, Satellite Derived Water Vapor Corrections for GEOSAT Altimetry, J. Geophys. Res. 95,2953:2964, 1990

Emery, W.J., W.G. Lee and L. Magaard, Geographic and seasonal distribution of Brunt-Väisälä frequency and Rossby radii in the North Pacific and North Atlantic., J.Phys. Oceanog., 14(2), 294-317, 1984

Fofonoff, N. and M.M. Hall, Estimates of mass, momentum, and kinetic energy fluxes of the Gulf Stream., J.Phys.Oceanog., 13, 1868-1877, 1983

Fu, L.-L. and R. Glazman, The effect of the degree of wave development on the sea state bias in radar altimetric measurement, J.Geophys. Res.,96(C1), 829-834, 1991

Fu, L.-L., I.Fukumori, and R.N. Miller, Fitting Dynamic Models to the Geosat Sea Level Observations in the Tropical Pacific Ocean. Part II: A linear, Wind-Driven Model, submitt. to J.Phys. Oceanog., 1992

Fukumori, I., J.Benveniste, C. Wunsch and D.B. Haidvogel, Assimilation of sea surface topography into an ocean circulation model using a steady-state smoother, submitt. to J.Phys.Oceanog., , 1992

Gaspar P., and C. Wunsch, Estimates from Altimeter data of barotropic Rossby waves in the Northwerstern Atlantic Ocean, J. Phys.Oceanog, 19, 1821-1844, 1989

Gelb, A., Applied Optimal Estimation, The MIT Press, 374 pp, Cambridge, MA., 1974

Gill A.E. and P.P. Niiler, A theory of the seasonal variability in the ocean, Deep Sea Research, 20, 141-177, 1973

Gill A.E., Atmosphere-Ocean Dynamics, Academic Press, 662 pp., 1982

Glazman R.E. and S.H. Pilorz, Effects of Sea Maturity on Satellite Altimeter Measurements, J. Geophys. Res. 95:C3, 2857-2870, 1990

Goodwin, G.C. and K.S. Sin, Adaptive Filtering Prediction and Control, Prentice-Hall, Inc., Englewood CLiffs, N.J., 540pp, 1984

Halpern, D., V.Zlotnicki, J.Newman, O.Brown, and F. Wentz, An Atlas of Monthly Mean distributions of GEOSAT Sea Surface Height, SSMI Surface Wind Speed, AVHRR/2 Sea Surface Temperature, and ECMWF Surface Wind Components during 1988, JPL Publication 91-8, available from D. Halpern, Jet Propulsion Lab., Pasadena, CA 91109, 1991

Holland, W.R, Quasigeostrophic modelling of eddy resolved ocean circulation, in J.J.O'Brien (ed), Advanced Physical Oceanographic Numerical Modelling, 203-231, D. Reidel Pub.Co, 1986

Holland, W.R., V.Zlotnicki, and L.-L. Fu, Modelled Time Dependent Flow in the Agulhas Retroflection Region as deduced from Altimeter Data Assimilation, S. African J. of Marine Science, 10:407-427., 1991

Jacobs, G.A., G.H.Born, P.C. Allen, and M. E. Parke, The global structure of the annual and semi-annual sea surface height variability from Geosat Altimetry data, J.Geophys. Res., submitted, 1992

Jourdan, D., C.Boissier, A.Braun, and J.-F. MInster, Influence of the Wet Tropospheric Correction in Mesoscale Dynamic Topography as derived from Satellite Altimetery, J.Geophys.Res., 95 (C10), 17-993-18004, 1990

Joyce T.M., C. Wunsch and S.D. Pierce, Synoptic Gulf Stream Velocity Profiles through simultaneous inversion of Hydrographic and acoustic doppler data, J. Geophys. Res., 91 (C6), 7573-7585, 1989

Joyce, T.M., K.A. Kelly, D.M.Schubert and M.J.Caruso, Shipboard and altimetric studies of rapid Gulf Stream variability between Cape Cod and Bermuda, Dee-Sea Res., 37(6)897-910, 1990

Kelly, K. A. and S. T. Gille, Gulf Stream Surface Transport and Statistics at 69ºW from the Geosat Altimeter, J. Geophys. Res., 95(c3), 3149-3161, 1989

Kelly, K.A., The meandering Gulf Stream as seen by the Geosat altimeter: surface transport, position and velocity variance from 73ºW to 46ºW, J.Geophys.Res., 96, 16721-16738, 1991

Lolk, N.K., Annual and longitudinal variations of the Pacific North Equatorial Countercurrent, JPL Pub. 92-8, available from D. Halpern, Jet Propulsion Lab., Pasadena, CA 91109, 1992

Miller, R. N. and M. A. Cane, A Kalman filter analysis of sea level height in the tropical Pacific, J.Phys.Oceanog., 19, 773-790, 1989

Moritz, H. Advanced Physical Geodesy Herbert Wichmann Verlag, Karlsruhe, Germany, 500 pp 1980

Pedlosky, J., Geophysical Fluid Dynamics, Springer Verlag, 710 pp, 1987

Picaut, J., A.J.Busalacchi, M.J.McPhaden, and B.Camusat, Validation of the Geostrophic Method for Estimating Zonal Currents at the Equator from Geosat Altimeter Data, J.Geophys.Res., 95(C3), 3015-3024, 1990

Rapp R.H., The determination of Geoid Undulations and Gravity Anomalies from SEASAT altimeter Data., J.Geophys.Res. 88, C3, 1552:1562, 1983

Rodriguez, E., Altimetry for non-gaussian oceans: height biases and estimation of parameters, J.Geophys.Res., 93, 14107-14120, 1988

Rodriguez, E., Y. Kim and J.M. Martin, The effect of small-wave modulation on the electromagnetic bias, J.eophys.Res. 97(C2), 2379-2389, 1992

Sandwell D.T., Milbert D.G., Douglas B.C., Global non-dynamic orbit improvement for altimetric satellites, J.Geophys. Res, 91 9447-9451, 1986

Semtner, A.J. Jr., Finite Difference Formulation of a World Ocean Model, In: Advanced Physical Oceanographic Numerical Modelling, J.J. O'Brien, ed., 187-202, D.Reidel Publishing Co., 1986

Stammer, D., H.-H. Hinrichsen, R.H. Käse, Can Meddies be detected by Satellite Altimetry, J.Geophys.Res., 96(C4) 7005-7014, 1991

Tai, C.-K., Estimating the surface transport of Meandering Jet Streams from Satellite Altimetry: Surface Transport Estimates for the Gulf Stream and Kuroshio extension, J.Phys.Oceanog., 20(6), 860-879, 1990

Tai, C.-K. , Accuracy assessment of widely used orbit error approximations in satellite altimetry and its oceanographic implications, J.Atm Oceanic Techn., 6, 147-150, 1989

Tai, C.-K., W.White and S.E. Pazan, Geosat crossover analysis in the Tropical Pacific, part 2. Verification analysis of altimetric sea level maps with XBT and island sea level data, J. Geophys.Res., 94 (C1), 897-908, 1988

Thacker, W.C. and R.B. Long, Fitting dynamics to data., J. Geophys. Res. 93, 1227-1240, 1988

Tripoli, Gregory J. and T.N. Krishnamurti, Low-Level Flows over the GATE Area during Summer 1972, Monthly Weather Review, 103: 197-216, 1975

Verron, J., Nudging Satellite Altimeter Data into Quasi-Geostrophic Ocean Models, J.Geophys.Res., 97(C5), 7479-7491, 1992

Wagner, C. A., Summer School Lectures on Satellite Altimetry, in: F. Sansò and R. Rummel (eds), Theory of Satellite Geodesy and Gravity Field Determination, LEcture Notes in Earth Sciences 25, pp285-334, Springer Verlag, , 1989

Willebrand, J., R.H. Käse, D. Stammer, H.-H. Hinrichsen and W. Krauss, Verification of Geosat sea surface topography in the Gulf Stream extension with surface drifting buoys and hydrographic measurements., J. Geophys. Res., 95(c3), 3007-3014, 1990

Wunsch, C. and E.M.Gaposhkin, On using satellite altimetry to determine the general circulaiton of the oceans, with application to geoid improvement, Rev.Geophys. Space Phys., 18, 725-745, 1980

Wunsch, C., Large -Scale response of the ocean to atmospheric forcing at low frequencies, J.Geophys.Res., 96(C8), 15083-15092, 1991

Wunsch, C., Sampling characteristics of satellite orbits, J.Oceanic and Atmosp. Tech., 6, 891-907, 1989

Wunsch, C., The North Atlantic General Circulation west of of 50ºW determined by inverse methods, Rev. Geophys. and Space Physics, 16, 583-620, 1978

Wyrtki K., Equatorial currents in the Pacific 1950 to 1970 and their relations to the trade winds, J. Phys. Oceanog., 4, 372-380, 1974

Wyrtki, K. and G. Mitchum, Interannual differences of Geosat altimeter heights and sea level: the importance of a datum, J. Geophys. Res., 95(c3), 2969-2976, 1990

Zlotnicki, V., G.Siedler and B.Klein, Can the weak surface currents of the Cape Verde Frontal Zone be measured with Altimetry?, J. Geophys. Res., in press.

Zlotnicki, V., Sea Level Differences across the Gulf Stream and Kuroshio extension, J. Physical Oceanog., 21(4), 599-609., 1991

Zlotnicki, V., The Mean Sea Level of the Gulf Stream estimated from Satellite Altimetric and Infrared Data, in: H.Sünkel and T.Baker (ed.), Sea Surface Topography and the Geoid, 108-116. IAG Symposia, Springer Verlag, 1990

Zlotnicki, V, A. Hayashi and L.-L. Fu, The JPL-Oceans 8902 version of Geosat Altimetry data, JPL Internal Document D-6939, Jet Propulsion Lab., California Institute of Technology, Pasadena, CA 91109, 1990

Zlotnicki V., L.-L. Fu, and W. Patzert, Seasonal Variability in Global Sea Level observed with Geosat altimetry, J. Geophys. Res. 94 (C12), 17,959-17,970, 1989

Principle of Satellite Altimetry and Elimination of Radial Orbit Errors

Reiner Rummel

Delft University of Technology
Faculty of Geodetic Engineering
Thijsseweg 11, 2629 JA Delft,
The Netherlands

1. Introduction.

Satellite altimetry is but a measurement technique, one out of several of the geodetic arsenal. Nevertheless did altimetry change the face of geodesy. Before space age the oceans were scarcely accessible to any geodetic activity. Only relatively few shipborne gravity measurements were available. Not until the 1970's, with the launch of the first altimeter satellites, did the situation change profoundly. Since the world's ocean are almost in equilibrium, the deviations of its surface from a level surface being only one or two meters, the altimeter measures with this precision the geometric shape of one of the equipotential surfaces of the earth's gravity field. As a consequence nowadays the gravity field in ocean areas is known more completely than the continental one. Many geophysical investigations of the past two decades are based on it.

But why then still devoting an entire summerschool to altimetry? With the launch of SEASAT in 1978 and with the results of three years of GEOSAT the enormous potential of altimetry for ocean research became visible and ERS-1, launched in 1991, as well as TOPEX-POSEIDON, to be launched in 1992, focus on oceanographic applications. The potential of altimetry for ocean research shall be subject of this school. The problem with the oceanographic signal is that it is rather small, hardly exceeding various sources of noise. Thus, if altimetry is to be used quantitatively e.g., for data assimilation, the data must be precise and reliable. It is - in good geodetic tradition - the objective of these lectures to look into the value of the adjusted altimeter measurements, in particular into the elimination of the radial orbit error, the definition and precision of the adjusted altimetric heights, the dependence of the outcomes on the mathematical model and the selected area and the consistency of the outcomes as a function of time. In short, our discussions want to provide guidelines concerning the usefulness of altimetric data for oceanography.

Before doing so I would like to introduce into the peculiarities of altimetry with a little comparison. In many ways one could consider altimetry as an extension of tide gauge measurements to the open ocean. In many places tide gauge measurements are carried out for more than a century. They serve the modeling of tidal variations, monitoring of coastal water changes and in recent years more and more, due to the growing concern about global warming, the analysis of sea level rise. Tide gauges measure with a certain sample rate the variation of the instantaneous sea level, usually freed from high frequency fluctuations. The measurements refer to the solid ground of the instrument site.

Besides giving the temporal variation at an individual site, the regional or global changes of sea level are of interest. They are derived from a connection of gauges into one network. Tidal records are affected by a long list of local and regional disturbances (Wemelsfelder, 1970), ranging from errors in the registration itself, via local meteorological effects, to regional land subsidence. Hence supplementary, e.g. meteorological measurements and correction models are required. The connection of gauges by geodetic leveling or modern space techniques permits separation of sea level variation from land uplift or subsidence.

A satellite altimeter provides so-to-say hundreds of thousands of "tide gauges" all over the oceans. They measure the variable ocean surface at a variety of temporal and spatial scales. Via tracking from a number of ground stations the space-borne "gauge" is tied to the solid earth and to a well-defined coordinate frame. The altimetric measurements are disturbed by a number of instrument and meteorological errors. What is land subsidence for gauge records is the orbit error for altimeter measurements. However the latter is rather regular, following the laws of mechanics. Since in the case of altimetry all "sites" have so-to-say to share the same radar source (the satellite) sampling in space and time is closely connected and determined by the choice of the orbit elements of the spacecraft. Thus, in contrast to real tide gauge records, in case of satellite altimetry undersampling of periodic phenomena is more the rule than the exception, leading to rather complicated aliasing problems. Continuity is difficult to achieve, because it implies that consecutive satellite missions, and inside each mission, consecutive adjusted orbits, have to be connected without any off-set or drift. So far this example. The fact, that at certain stations comparison of tide gauge records and altimetry show good agreement, (Miller, Cheney, Douglas, 1988) or (Cheney & Miller, 1990), show that the above example is more than an experiment of thought. Also the hope that in the future long term sea level monitoring could be done by altimetry points in this direction, (Tapley, 1990).

2. Altimetric Model.

In this chapter the basic model of satellite altimetry shall be introduced. We can then address the question of how the main systematic error source, the radial orbit can be eliminated, in principle. For this purpose also the fundamentals of the measurements have to be understood, the elements that lead to the stochastic model. Only then, in the following chapters, we shall discuss various approaches to the actual orbit adjustment.

2.1. Altimeter Principle.

Let us depart from the following idealized situation:

- Using all available tracking data the best possible orbit of the altimeter satellite has been computed. For its computation the altimetric heights themselves were not used. From the computed ephemerides the radial distance r_s of the satellite has been determined. The coordinate system, in which the ephemerides are given, is well-defined and (hopefully) geocentric.
- After applying a number of corrections to the raw altimeter measurements, precise ranges ρ from the center of mass of the spacecraft to the ocean surface are available.

- A mean earth ellipsoid with semi-major axis a_e and flattening f_e is chosen as reference surface (for example a_e = 6378 137 m and f_e = 1:298.257 222 101). With the satellite ephemerides defining the sub-satellite points on the ellipsoid, the radial distance r_e of these points is determined. The ellipsoid is concentric with the satellite coordinate system.

Then the basic equation of satellite altimetry becomes

$$h = r_s - \rho - r_e \qquad (2.1)$$

where h is the geometric height of the ocean surface above the reference ellipsoid. The situation is displayed in Figure 2.1. More correctly r_s, ρ and $(r_e + h)$ form a triangle and a correction term $C = \dfrac{r_e}{8}\left(1 - \dfrac{r_e}{r_s}\right)e^4 \sin^2 2B$ has to be applied to (2.1): $h = r_s - \rho - r_e - C$, with e the eccentricity and B the geographic latitude; C is between 0 and 5 m. See (Gopalapillai, 1974).

Would the ocean surface be at rest, with no external forces acting on it, such as tides, winds, or variations in air pressure, it would be an equipotential surface, purely determined by the gravitational field of the earth and by its angular rate. The equipotential surface at sea level is called the geoid. Its deviations from the reference ellipsoid, the geoid heights N, are in the range from +85 m to -100 m. [If a hydrostatic equilibrium figure with flattening f = 1:299.63 would be chosen as reference body, compare (Hager & Richards, 1989), the geoid heights would reach the 200 m level.]

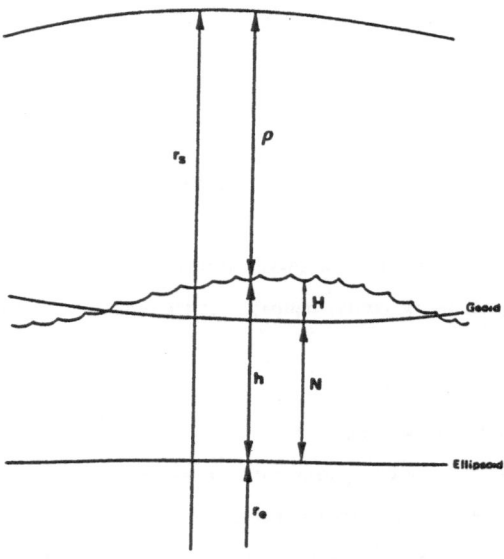

FIGURE 2.1: Geometry of satellite altimetry.

In agreement with the definition of land topography the separation of the actual sea surface from the geoid is denoted sea surface topography (SST), H. It reflects the dynamics of the ocean and is therefore a key unknown in oceanography. Thus, the geometric sea height h is split into geoid height N and sea surface topography H (see Figure 2.2):

$$h = N + H \quad . \tag{2.2}$$

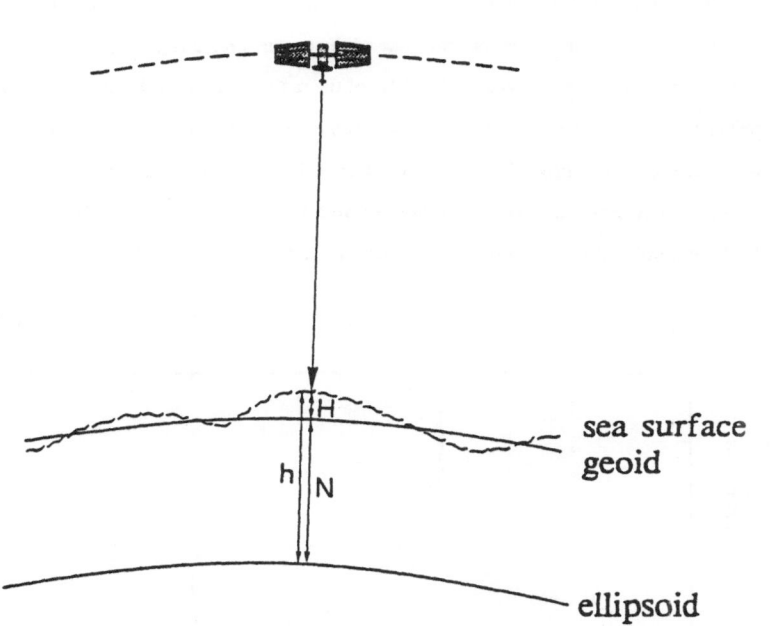

FIGURE 2.2: Sea surface height h, geoid height N and sea surface topography H.

Unfortunately altimetry can only deliver h and not N and H separately. [Later we shall check whether this is really true.] Sometimes this is considered as the fundamental dilemma of altimetry, since oceanographers would hope geodesists could provide the geoid by independent gravimetric means, whereas geodesists expect oceanographers to offer an estimate of SST. Actually, however, geodesists are in a more favorable position because the SST can be modeled to 10 or 30 cm basically everywhere, (Roemmich & Wunsch, 1982), whereas the geoid uncertainty is easily of the order of 1 to 2 m. In the near future satellite gradiometry may resolve the geoid uncertainty, (Rummel, 1991).

As this point the separate introduction of N and H is little more than a formal statement. Combination of (2.1) with (2.2) yields:

$$N + H = h = r_s - \rho - r_e \quad . \tag{2.3}$$

The sea surface topography is composed of a constant or almost stationary part H_0 and a time variable part ζ:

$$H = H_0 + \zeta \quad . \tag{2.4}$$

In conclusion r_e, N and H_0 are functions only of their geographic location, whereas r_s, ρ, and ζ depend on time and location. Consequently comparison of two altimeter measurements taken at the same location at times t_1 and t_2 gives:

$$\Delta \zeta = \zeta(t_2) - \zeta(t_1)$$

$$= h(t_2) - h(t_1) = r_s(t_2) - r_s(t_1) - (\rho(t_2) - \rho(t_1)) \tag{2.5}$$

Hence today altimetry in combination with the computed orbit can, in principle, provide either the geometric sea surface height h, eq. (2.3), or the variation with time of the ocean surface $\Delta \zeta$, eq. (2.5).

The main obstacle towards this goal is the effect of the radial orbit error. Even with optimal tracking and using the most recent geopotential models, its size remains somewhere between 30 cm and 2 m, exceeding the altimeter precision by a factor of ten to fifty. However the orbit error is rather systematic. It is concentrated around the zero and one cpr frequency (cpr = cycles per revolution). Above 2.3 cpr, corresponding to 17400 km, the error contribution is negligible. An example of the radial orbit error spectrum is shown in Figure 2.3.

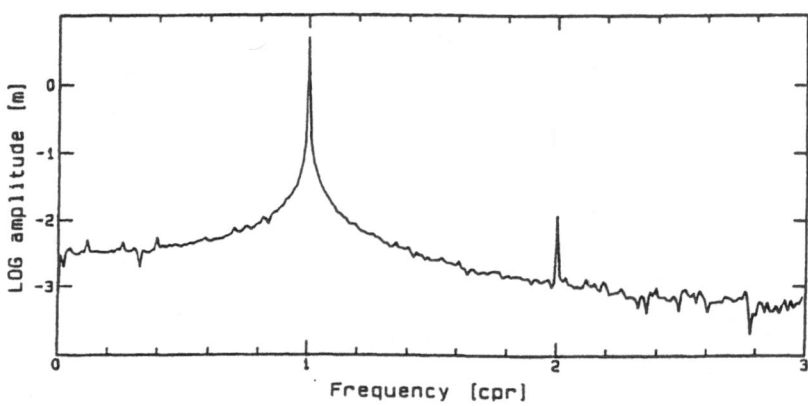

FIGURE 2.3: Spectrum of simulated radial orbit error about 1 cpr (with zero frequencies excluded).

How do we know at all that such a significant error is still present in the determined sea heights? Repeat arcs and arc intersections display systematic differences that can only be explained as orbit errors. Thus these lectures shall focus on realistic ways of a mathematical description of the radial orbit uncertainty and its elimination in a least-squares adjustment. Thereby basic conditions are that the orbit error can be distinguished and separated from the ocean variation $\Delta \zeta$ and the altimeter error, see eq. (2.5). We shall first look into these effects, in particular discuss the measurement process.

2.2. Measurement Process and Error Budget.

In this section we give a short description of the principle of radar altimetry and discuss the error sources that have to be taken into account. One could also say we try to establish a stochastic model for our adjustment problem. Basically we follow (Stewart, 1985) and (Robinson, 1985) with input from (Bomford, 1980), (Lorell, Parke and Scott, 1980), (Maul, 1985), (Chelton, 1988), (Lambeck, 1988) and (Seeber, 1989).

A radar altimeter operates typically at 13.5 GHz (1 GHz = 10^9 s^{-1}). For height determination a transmitter is required generating pulses that are reflected at the ocean surface and returned to a receiver. The pulse travel time is measured by a clock. A precision of 1 cm in height would require a clock resolution of τ = 30 ps (1 ps = 10^{-12} s; c = 3 · 10^8 ms^{-1}). The generation of a rectangular pulse of 30 ps duration would occupy an unacceptably broad frequency bandwidth (bandwidth $\approx \frac{1}{\tau}$ = 30 GHz). Actually the relatively narrow band of 300 MHz (3 ns) centered at 13.5 GH is used. Instead of a sharp pulse the shape of a much longer pulse is analyzed by curve fitting. The necessary signal-to-noise ratio is achieved by averaging up to 1000 of the reflected pulses, the approximate data rate per second. For example for SEASAT, 50 consecutive pulse returns (0.2 s) were averaged. At this rate the automatic gain control (AGC) loop is up-dated. The AGC loop determines the attenuation that must be applied to the reflected signal to keep the returned power at about constant level (Chelton, 1988). Delays in the adoption to a jump in the signature of the return pulse, e.g. at the boundary of land to sea or ice to sea can lead to unreliable outcomes or data loss. Two consecutive of these averaged curves were used to determine a height each 0.1 s. The shape of a smooth curve is fitted through the averages. Depending of the significant wave height (SWH), (= four times the rms value of sea surface elevation), its shape varies being rather steep for calm sea (sharp return pulses) and flat for rough sea. In Figure 2.4 the result of the averaging process is displayed in relation to the employed number of pulses. Figure 2.5 shows the average curve fitting for three sea states.

Sufficient signal power is transmitted by pulse compression, working with chirped pulses, (Maul, 1985), (Chelton, 1988) or (Cantafio, 1989). The returned signal represents a certain average over the illuminated area. As the beam is relatively wide, in order to confine the signal to a small area, so-called beam limited geometry is used. The emitted radar pulse propagates as a spherical wavefront. The leading edge strikes the ocean surface at the shortest distance from the satellite. The illuminated area shall first be a disk with growing radius. After τ (the pulse width) the trailing edge of the pulse reaches the water surface. The disk becomes an annulus of growing radius sin ψ (ψ opening angle) and radial thickness sin^{-1} ψ with constant area A = 2π sin ψ sin^{-1} ψ, see Figure 2.6. Consequently the return signal is first increasing from zero level to its maximum at the point where the disk changes into an annulus. It is then slowly decreasing. See again Figure 2.5. The maximum radius of the disk, or in other words the maximum foot print of the radar signal is (compare (Robinson, 1985)):

$$r = \sqrt{2\rho c\tau} \qquad \text{, calm sea} \qquad\qquad (2.6a)$$

$$r = \sqrt{2\rho c\tau'} \qquad \text{, rough sea} \qquad\qquad (2.6b)$$

where the effective pulse width is increased by the significant wave height (SWH) via

$$\tau' = \tau + \frac{2SWH}{c} \qquad . \qquad\qquad (2.7)$$

For SEASAT with ρ = 800 km and τ = 3 ns the foot print radius is somewhere between 1 km (calm sea) and 6 km (rough sea).

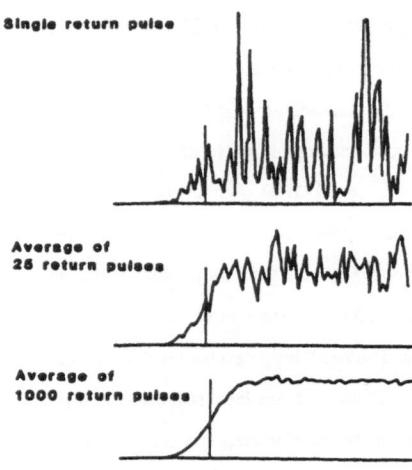

FIGURE 2.4: Averaging of return pulses, from (Robinson, 1985).

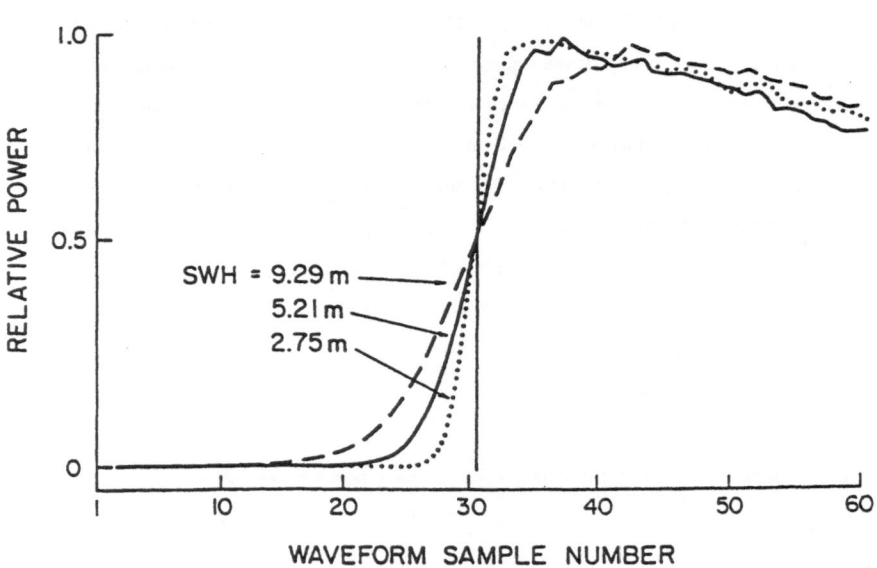

FIGURE 2.5: Averaged return pulse curves for three sea states, from (Chelton, 1988).

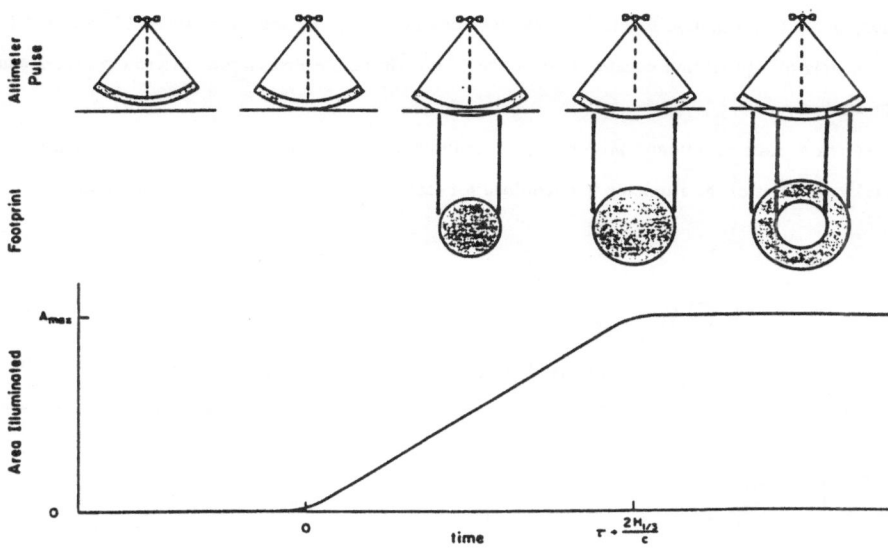

FIGURE 2.6: Altimeter pulse reflection, from (Chelton, 1988).

ERROR BUDGET:

The real instrument noise can - at a rate of 1 Hz - be considered purely uncorrelated with a standard deviation (sd) of a few centimeters. There exist, however, additional error sources and uncertainties, that need to be taken into account. The radar signal propagates through a medium, it is scattered at the ocean surface and the ocean surface is continuously changing. It shall now be looked into the implication of these effects on the altimetric measurement.

A. Radio Wave Scattering: For calm sea about 60% to 70% of the emitted radiation is reflected. Thus a large fraction is reflected back in the direction of the receiver. With increasing sea surface roughness a larger portion is reflected away from the satellite. The power of the received signal decreases.

First there is the tracker bias. It is caused by systematic errors of the on-board tracker algorithm, which is designed to determine the half-power point of the return wavefront. A discussion is given in (Chelton, 1988). The bias is mainly influenced by the sea state. Second, an electromagnetic bias occurs, due to the systematic difference between mean scattering surface and mean sea level. Wave throughs are generally flattened, wave crests are peaked. The result is a higher reflection rate of throughs than of crests. The altimeter height ρ gets over determined, or the sea surface height h underestimated. The total effect of tracker and electromagnetic bias is estimated to be 7% of the SWH with an uncertainty of 2% to 3%. Some improvement in modelling these biases seems possible. An intrinsic limit is the uncertainty in the estimated SWH. The above numbers (7% $\pm$ 3%) imply 70 cm $\pm$ 30 cm for SWH = 10 m (rough sea) and 3.5 cm $\pm$ 1 cm for SWH = 0.5 m (calm sea). In order to demonstrate that in more than 50% of the cases the SWH is higher than 2 m we give a histogram of SWH values of 3 years of GEOSAT, Figure 2.7 and a world map of its geographical distribution, Figure 2.8. Typical values are 6 m in the Southern Ocean, 1-2 m in the tropics, and 3-4 m in the mid-to high-latitude oceans, see (Chelton, 1988). Thus these biases represent a source for systematic errors in ocean studies.

For other errors such as antenna mispointing or calibration biases it is again referred to (Chelton, 1988). The "absolute" length bias of the altimeter is determined for each mission by a calibration experiment.

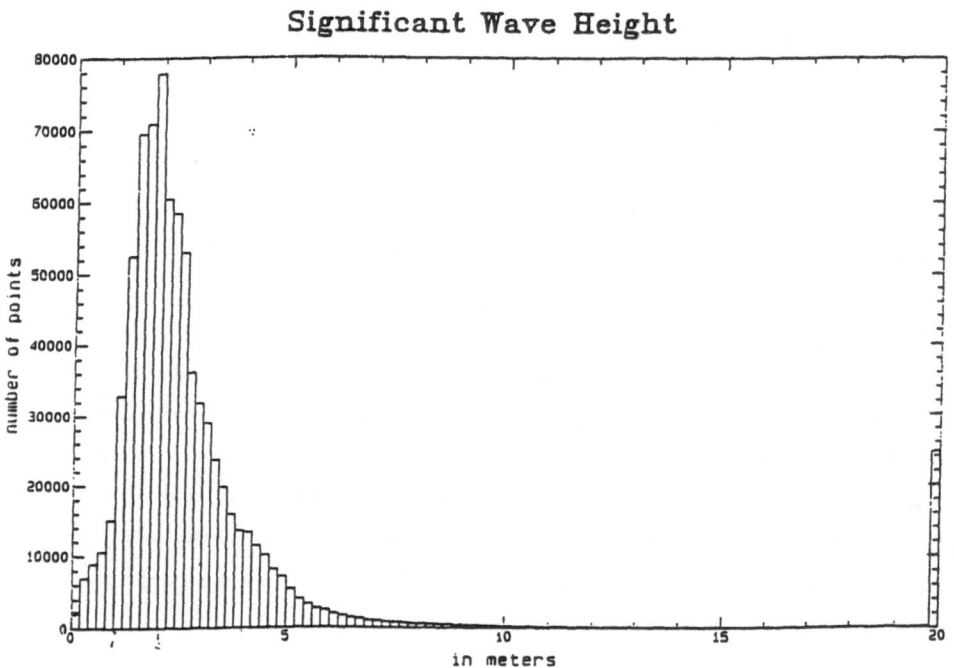

FIGURE 2.7: Histogram of significant wave heights (SWH) of Geosat.

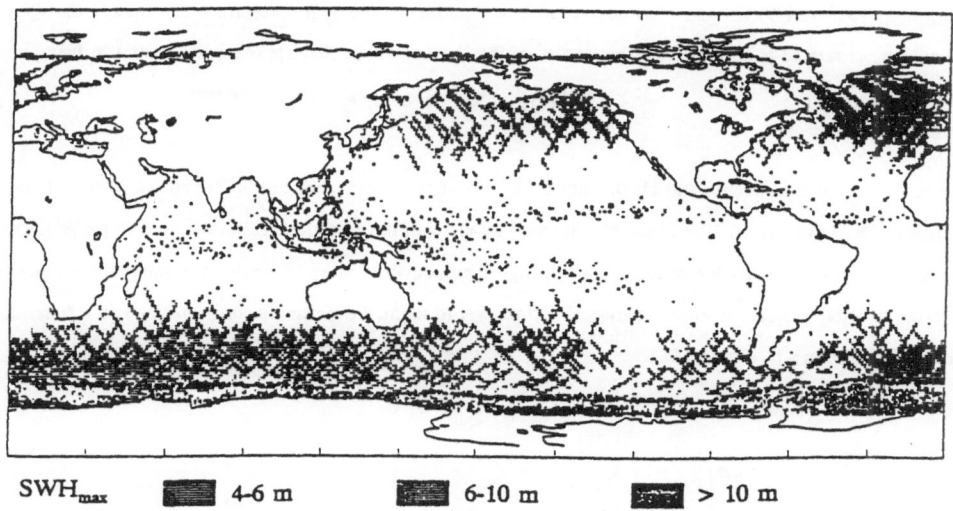

SWH$_{max}$ ▨ 4-6 m   ▨ 6-10 m   ▨ > 10 m

FIGURE 2.8: Geographical distributions of high SWH's.

B.Propagation Effect: Propagation of electromagnetic waves is affected by ionospheric and tropospheric delay. The problem is alleviated to a considerable degree by the fact that the measurements are made along a vertical path. Main difficulty is to collect representative ancillary data for correction computation. Since certain propagation effects vary with sunlight intensity they have the unfortunate property of getting aliased into variations at orbital period and precession rate (Williamstown, 1970).

In vacuum the altimeter height follows from

$$\rho = \frac{1}{2} ct \tag{2.8}$$

with c the velocity of light and t the pulse travel time. If c_m is the pulse propagation velocity in a medium and therefore $\rho_m = \frac{1}{2} c_m t$, the correction to the estimated distance becomes

$$\Delta \rho = \rho - \rho_m = \frac{1}{2} (c - c_m)t = \frac{1}{2} c_m t \left(\frac{c - c_m}{c_m} \right) \tag{2.9}$$
$$= \rho_m(n-1) \quad .$$

Thereby $n = c/c_m$ is the refractive index. In a medium with variable refractivity we find

$$\Delta \rho = \int_0^\rho (n-1)dz \quad . \tag{2.10}$$

where z is the vertical direction upwards.

<u>Ionosphere</u> (between 60 km and 1000 km altitude): Refraction caused by free electrons and ions in the upper atmosphere is related to the dielectric properties of the ionosphere. In first approximation one can take

$$(n-1) = \frac{40.2}{f^2} N_e + O(f^{-3}) \qquad (2.11)$$

where f is the signal frequency and N_e the number of electrons per m^3. Terms of $O(f^{-3})$ and $O(f^{-4})$ are only marginally negligible, (Bomford, 1980). The factor 40.2 is equal to $\alpha/2$ with the empirical constant $\alpha = 80.4 \ [m^3/s^2]$. The height correction term (after insertion of (2.11) into (2.10)) is kept small because f is squared in the de-nominator, but the uncertainty of N_e is large. It varies considerably from day to night, from summer to winter and as a function of the solar cycle, as shown in Figure 2.9.

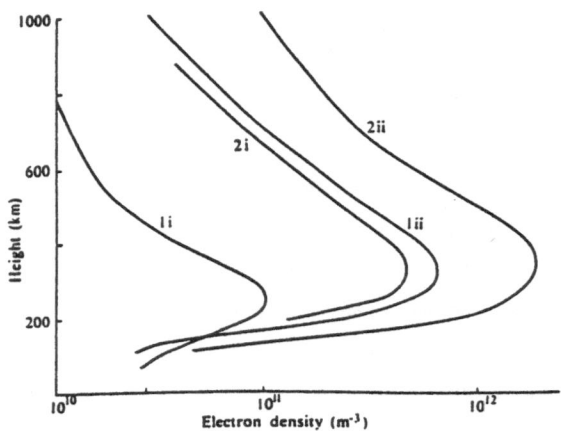

FIGURE 2.9:   Typical electron density profiles: quiet solar activity (curves 1); high solar activity (curves 2); average night time (i); average day time (ii), from (Lambeck, 1988).

Example: Take the columnar value E of free electrons, $E = \int_0^p N_e \, dz$, as one time $3.5 \cdot 10^{10}$ x $8 \cdot 10^5$ (el/m²) = $28 \cdot 10^{15}$ (el/m²) and a second time as $7 \cdot 10^{11}$ x $8 \cdot 10^5$ (el/m²) = $56 \cdot 10^{16}$ (el/m²) we find at f = 13.5 GHz:

$\Delta \rho = 6$ mm   (average night-time minimum)

and

$\Delta \rho = 12$ cm   (average day-time maximum).

Since the direct measurement of the electron content is difficult, a real remedy would only come from using <u>two-frequencies altimeters</u>.

<u>Troposphere</u> (between 0 and 40 km altitude): As many examples in geodesy show the computation of the troposphere correction, in particular of its wet component, forms a problem. On the other hand, the situation is less critical over ocean areas than on land where the topographic relief complicates the situation.

A highly simplified, but common model is, (Bomford, 1980), (Lorell et al., 1980), (Robinson, 1985):

$$\Delta\rho = 2.277 \cdot 10^{-5} \left[(1 + 0.0026 \cos 2B)P_s + (\frac{1255}{T_s} + 0.05)e_s \right] \tag{2.12}$$

with P_s sea level air pressure (in Pa), T_s surface atmospheric temperature (in K) and e_s partial water vapor pressure (in Pa). In can be split into

$$\Delta\rho_{DTD} = 2.277 \cdot 10^{-5} (1 + 0.0026 \cos 2B)P_s \quad \text{(dry tropospheric delay)} \tag{2.13a}$$

and into

$$\Delta\rho_{WTD} = 2.277 \cdot 10^{-5} (\frac{1255}{T_s} + 0.05)e_s \quad \text{(wet tropospheric delay)}. \tag{2.13b}$$

Filling in the surface pressure ($\approx 10^5$ Pa) we see that $\Delta\rho_{DTR} \approx 2.3$ m. The refractive index is proportional to the atmospheric density d_{ATM}: $n-1 \sim c d_{ATM}$. In good approximation d_{ATM} decreases exponentially with altitude: $d_{ATM} = e^{\alpha z + \beta}$. Hence it can easily be linked to the surface conditions. This explains why the dry part can be modeled with high accuracy. The error caused by the uncertainty in the sea level atmospheric pressure is of the order of 0.5 cm but is certainly higher in the southern hemisphere. Relatively large errors occur e.g. on the boundaries of large depressions. For seasonal or tidal pressure variations see (Lambeck, 1988).

Surface values of air temperature T_s, pressure P_s and partial water vapor pressure e_s are supplied by the Fleet Numerical Oceanography Center (FNOC). However, their model does not resolve spatial scales shorter than about 2000 km whereas the water vapor content varies rather irregularly. This makes modeling of $\Delta\rho_{WTD}$ difficult. Better results are achieved using a satellite passive microwave sounder, a so-called scanning multichannel microwave radiometer (SMMR), mounted onboard of the altimeter satellite. Thereby the radiation W emitted by water vapor is measured at a number of frequencies. These data are calibrated by means of radiosonde measurements. The following empirical formula is often adopted:

$$\Delta\rho_{WTR} = 6.36 \cdot 10^{-3} \, W \tag{2.14}$$

with W in kg $\cdot$ m^{-2}. See (Tapley, Lundberg & Born, 1984). See also (Emery, 1990) and (Ray et al., 1991).

Rain and Liquid Cloud Droplets: At altimeter frequencies rain presents an obstacle for the propagation of electromagnetic waves. Light rain causes rapid changes in signal strength, heavy rain leads to complete data loss.

In the case of non-raining clouds an empirical formula for the delay (in cm) is $\Delta\rho = 0.15 \cdot 10^{-6} \, d_{cl} h_{cl}$ with d_{cl} the cloud density in kg/m^3 and h_{cl} the cloud thickness in m. For a cloud thickness of 1 km the delay is on the order of 0.15 cm. Therefore even uncertainty by a factor of three would be tolerable.

C. Ocean Variations: Naturally ocean time variations $\Delta\zeta$ are signal rather than noise. However in the context of orbit adjustment where very long wavelength orbit uncertainties are modeled, $\Delta\zeta$ becomes part of the stochastic model. The various ocean phenomena vary considerably in amplitude, surface extension and time period. They range from tides, the barotropic effect, equatorial currents, western and eastern boundary currents, rings, mesoscale eddies to large gyros. It is referred to the lectures by Cartwright and Wunsch (this issue).

As some of these effects, in particular the tides, get easily aliased into orbit frequency a priori correction is advisable. The tidal signal in open ocean areas is typically below 1 m with length scales longer than 1000 km. Both for the contribution of the solid earth and the ocean tides good correction models exist. The uncertainty of the ocean tide model of Schwiderki (1980) is estimated to be typically 5-10 cm globally. In large shelf sea areas the model does not apply or has high uncertainty. The result of inaccurate tidal modeling is pseudo-ocean-variability, see e.g. (Oskam, 1990).

Another factor of uncertainty is the effect of atmospheric pressure on the ocean surface. In the case of perfect isostatic response of the ocean to surface pressure loading the response would be $h = -(d_{WA}g)^{-1}P$, with d_{WA} water density, g gravity, and P atmospheric density. This is the inverse barometer effect. It implies one-to-one response of the ocean surface to atmospheric pressure. Actually it is known that such a response can only be expected over basin scale and after uniform pressure change, (Trupin & Wahr, 1990) or (Chelton, 1988).

CONCLUSION.

Radio wave scattering, propagation effects and ocean variations easily exceed the size of the altimeter noise. Partly these influences can be corrected for, but also after correction their size reach or exceed that of the actual measurement error. The quality of the corrections largely depends on the availability of ancillary data, e.g., from complementary sensors onboard of the spacecraft. The described effects are correlated with environmental effects such as pressure, temperature, wind and sunlight activity and have typical variations in space and time, partly interfering with the spectrum of the orbit error. A list of effects is given in Table 2.1. Certain influences like atmospheric pressure affect the altimeter height in several ways (troposphere, barotropic effect) and result in correlation. For high precision studies it can be advisable to delete certain classes of data, e.g., data with SWH above a certain threshold, etc.

Table 2.1: Typical Altimeter Error Budget				
Type of effect	Source	Amplitude (cm)	Residual (cm)	Wavelength (km)
altimeter	noise	< 5		
altimeter	bias	20	2	
sea state	electromagn. + tracker bias	7% SWH	3% SWH	100-1000
ionosphere	free electrons	2-20	3	50-10,000
dry troposphere	mass of air	230	1	1000
wet troposphere	water vapor	6-30	3	50-1000
liquid water	rain	10-100		30-50
ocean surface	tides	100	5	aliased
currents	equatorial	30	?	5000
	western boundary	130	?	100-1000
	eastern boundary	30	?	100-1000
mesoscale	eddies	25		100
	rings	100		100
	gyres	50		3

3. Regional Cross-Over Adjustment.

We return to the problem of the elimination of the radial orbit error. In this chapter the regional cross-over adjustment is discussed. It is the oldest and simplest method of orbit error elimination. In view of very advanced, global and integrated adjustment approaches, meanwhile applied by various groups, the local cross-over adjustment method seems almost obsolete. But it maintains its value, for even after global adjustment, a second phase regional one is often required. In addition, for groups with limited facilities it is the easiest manner to get into altimeter processing.

The satellite orbit, when projected onto the earth's surface, produces a regular pattern of ground tracks, see the example of Figure 3.1. Depending on the scientific goals of the altimeter mission, the orbit parameters can be chosen either so as to produce a rather dense grid of ground tracks measured with a low repetition rate in time, or a coarse grid with high repetition rate. See lectures Balmino (this issue). The orbit inclination defines the most northern and most southern latitude of the ground track pattern. Ascending arcs are those along which the satellite moves from south to north, descending arcs those along which the satellite moves from north to south. Thus each full revolution consists of one ascending and one descending arc. Due to the steadiness of the motion of the satellite and of the rotation of the earth also the pattern of cross-over points, the intersections of ascending and descending arcs, is very regular. Figure 3.1 shows that the equatorial crossings are equidistant and the cross-over points are arranged along latitude circles.

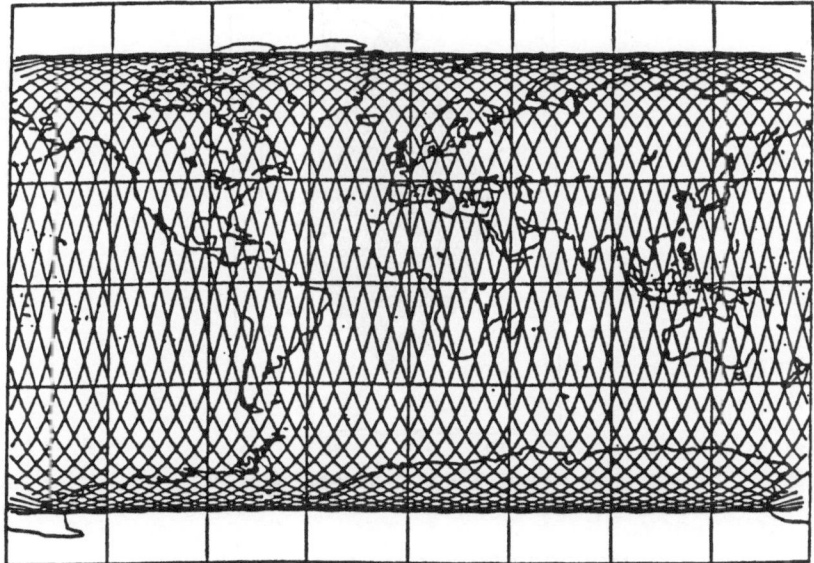

ERS-1: Projected 3-day repeat groundtrack

FIGURE 3.1: Global ground track pattern.

Let us assume the orbit error is to be eliminated in a certain region on basis of the height differences at cross-over points, a so-called <u>cross-over adjustment</u>. First it is advisable to select a diamond shaped region, bounded by two ascending and two descending arcs, compare Figure 3.2. This choice has the advantage of all ascending and all descending arcs, respectively, being almost of the same length. The adjustment model is derived from eq. (2.5), which becomes after linearization:

$$\Delta \tilde{h}_{ad} = \Delta r_a - \Delta r_d + \Delta \tilde{\zeta} + \tilde{v}_\rho \qquad . \qquad (3.1)$$

Thereby is $\Delta \tilde{h}_{ad}$ the observed, stochastic ("~") cross-over difference, the sub-indices indicating ascending and descending, Δr_a and Δr_d are the radial orbit errors of the ascending and descending arc, $\Delta \zeta$ the sea surface variation between times t_a and t_d and $\tilde{v}_\rho$ the altimetric measurement error, being roughly $\sqrt{2}$ times that of the individual measurement. The considerations that lead to an appropriate definition of the stochastic model of $\Delta \zeta$ and v_ρ were discussed in ch. 2.2. In particular the change in signature of $\Delta \zeta$ and the space/time correlation of $\Delta \zeta$ and v_ρ need attention. In the context of cross-over orbit adjustment often the sum of $\Delta \tilde{\zeta} + \tilde{v}_\rho = \tilde{e}$ is considered to be one uncorrelated random variable with $E\{\tilde{e}\} = 0$ and $D_y = E\{\tilde{e} \ \tilde{e}^T\} = \sigma_0^2 I$ (I = unit matrix). See however (Wunsch & Zlotnicki, 1984), (Knudsen, 1987).

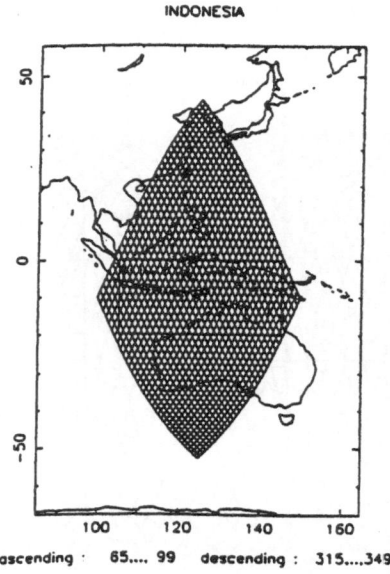

FIGURE 3.2: Diamond shaped adjustment region.

The model of the radial orbit error is considered non-stochastic. At least for shorter intervals a satellite orbit can be described as Keplerian orbit. Its geocentric radial distance is then

$$r_s = a(1 - e \cos E)$$

with a the semi-major axis, e the eccentricity (the orbits of altimeter satellites are almost circular, i.e. $e \approx 0$) and E the eccentric anomaly. After linearization the radial correction becomes

$$\Delta r = \Delta a + e\Delta Ma \sin M - \Delta ea \cos M \qquad (3.2)$$

where we assume the mean anomaly $M = E$. In the context of altimetry this model is discussed in detail in (Colombo, 1984) and (Wagner, 1985). With $M = M_0 + \dot{M}t$ and $\dot{M} = \bar{n} = \dfrac{2\pi}{T}$ ($\bar{n}$ = mean angular velocity; T = orbit period), we observe that Δr describes a function with orbit period. At least for intervals up to one revolution the perturbations Δa, ΔM, and Δe can be considered constant. Thus a mathematical model for the description of the radial orbit error is available.

With time tag $\mu \overset{def.}{=} \dfrac{2\pi}{T}(t-t_0)$ and t_0 a reference time, model (3.2) can be written as

$$\Delta r = x_0 + x_1 \sin \mu + x_2 \cos \mu \qquad [long\ segments] \qquad (3.3a)$$

which can be reduced for relatively short arc segments (< 2000 km) ($\sin \mu = \mu$; $\cos \mu = 1$) to

$$\Delta r = x_0 + x_1\mu \qquad [medium\ length] \qquad (3.3b)$$

and for very short segments to

$$\Delta r = x_0 \qquad . \qquad [short\ segments] \qquad (3.3c)$$

In Figure 3.3 a diamond shaped region is shown bounded by ascending arcs 1 and 2 and descending arcs 3 and 4. An additional ascending arc segment i and descending arc segment j intersect in X. The reference time t_0 of μ is chosen such that μ is zero when the tracks enter the area (intersection of all ascending arcs with descending arc 3 and of all descending arcs with ascending arc 1). Hence intersection time tag of segment i with segment j (and of segment j with i, respectively) becomes:

$$\mu_{ij} = \frac{t_{ij} - t_{i3}}{T} \qquad \left(\mu_{ji} = \frac{t_{ji} - t_{j1}}{T}\right) \qquad . \qquad (3.4)$$

Since for not all too large regions, all segments of one kind (asc. or desc.) are of almost of the same length, the whole configuration can be mapped with little distortion onto a rectangular line configuration, see Figure 3.4. Thereby it holds

$$\mu_{1j} = \mu_{2j} = \dots + \mu_{ij} = \dots \overset{def.}{=} \mu_{aj} \qquad (3.5a)$$

and

$$\mu_{1i} = \mu_{2i} = \dots = \mu_{ji} = \dots \overset{def.}{=} \mu_{di} \qquad . \qquad (3.5b)$$

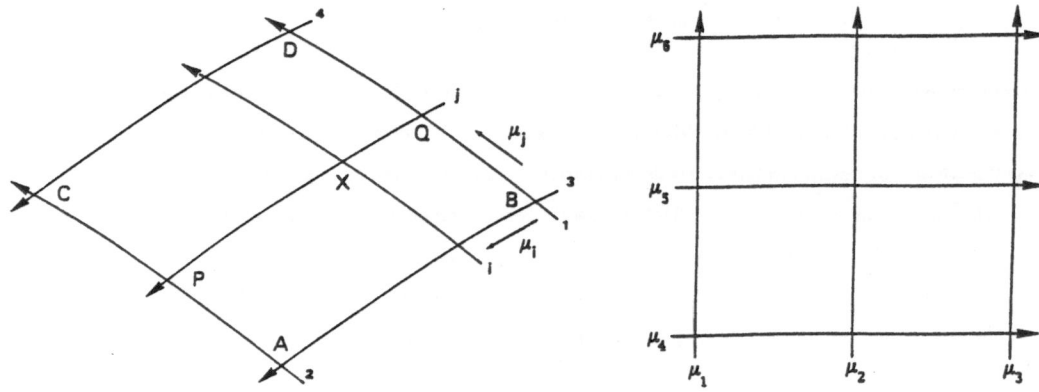

FIGURE 3.3: Regional cross-over scheme.   FIGURE 3.4: Mapping of Figure 3.3 into rectangular scheme.

After these preparations the linear model of the least-squares adjustment is written as

$$\bar{y} = Ax + \bar{e} \tag{3.6}$$

with $E\{\bar{e}\} = 0$ and $D_y = \sigma_0^2$ I (or something more adequate, see ch. 2.2). The vector of observations $\bar{y}$ contains all cross-over discrepancies ordered e.g. along all ascending arcs. Let us assume the arc segment configuration is very regular (no gaps) with k ascending arcs, and ℓ descending arcs. Thus $\dim(\bar{y}) = n\times 1 = (k\cdot\ell) \times 1$. The coefficient matrix is divided into two parts

$$\underset{n\times m}{A} = [\underset{n\times kp}{A_a} \mid \underset{n\times\ell p}{A_d}] \tag{3.7}$$

where A_a and A_d refer to the parameters describing the ascending and descending orbit error, respectively; p is the number of parameters per segment, for example, for a two-parameter model, (3.4), p = 2, and m = (k+ℓ)2. The structure of A is shown in Table 3.1. The parameter vector x has dimension m×1.

The formal least-squares solution is

$$\hat{x} = (A^T D_y^{-1} A)^{-1} A^T D_y^{-1} \bar{y} \tag{3.8}$$

with the a posteriori variance-covariance matrix of the estimated parameters

$$D_x = (A^T D_y A)^{-1} \quad . \tag{3.9}$$

With the prior variance-covariance matrix $D_y = \sigma_0^2 I$ the normal matrix becomes

TABLE 3.1 : Structure of the coefficient matrix A for the case of a two-parameter model

		A_a				A_d			
		1	2	...	k	1	2	...	l
1	1	$1 \quad \mu_{a1}$				$-1 \quad {}^{-}\mu_{d1}$			
	2	$1 \quad \mu_{a2}$					$-1 \quad {}^{-}\mu_{d1}$		
	⋮	$\cdot \quad \cdot$						$\cdot \quad \cdot$	
	l	$1 \quad \mu_{al}$							$-1 \quad {}^{-}\mu_{d1}$
2	1		$1 \quad \mu_{a1}$			$-1 \quad {}^{-}\mu_{d2}$			
	2		$1 \quad \mu_{a2}$				$-1 \quad {}^{-}\mu_{d2}$		
	⋮		$\cdot \quad \cdot$					$\cdot \quad \cdot$	
	l		$1 \quad \mu_{al}$						$-1 \quad {}^{-}\mu_{d2}$
k	1				$1 \quad \mu_{a1}$	$-1 \quad {}^{-}\mu_{dk}$			
	2				$1 \quad \mu_{a2}$		$-1 \quad {}^{-}\mu_{dk}$		
	⋮				$\cdot \quad \cdot$			$\cdot \quad \cdot$	
	l				$1 \quad \mu_{al}$				$-1 \quad {}^{-}\mu_{dk}$

$$A^T D_y^{-1} A = \sigma_0^2 \begin{pmatrix} A_a^T A_a & A_a^T A_d \\ A_d^T A_a & A_d^T A_d \end{pmatrix} \qquad . \tag{3.10}$$

For our example it is given in Table 3.2.

From Table 3.2 we see that for very regular situations the normal matrix can be generated analytically. It has a peculiar structure that can be treated very efficiently, see also (Rummel, 1986).

In order the adjustment to be overdetermined a necessary condition is

$$n = k \cdot \ell > m = (k + \ell)p \qquad . \tag{3.11}$$

Furtheron the rank r of A must be analyzed, and it is easily conceivable that r < m. This leads to the datum problem[1] of cross-over adjustment. It is evident, for example, that in the case of a one-parameter model (p = 1) a common shift of the entire cross-over configuration would not affect the cross-over discrepancies. Thus this shift cannot be determined from the latter either. The situation is comparable to that of a leveling network. Measured are height increments. Heights can only be determined by fixing the height of one arbitrary point. The same holds true here, although it is less evident what the general rule is for a p parameter adjustment. Schrama (1989) could derive the following rule:

> In a regional cross-over adjustment with p parameters per arc the datum defect is p^2. A regular adjustment is attained after fixing p parameters of p non-intersecting arcs.

$$\tag{3.12}$$

The means, for example, that for the two-parameter model two arbitrary "parallel" (ascending or descending) arc segments must be fixed. Technically this is done, either by eliminating the columns of the coefficient matrix that refer to the p parameters of these arcs or by adding to the linear system (3.6) constraints, that give for example fixed values of these parameters, with variance zero:

$$c = Bx \qquad . \tag{3.13}$$

Eq. (3.12) says that the rank defect of A is

$$m - r = p^2 \qquad . \tag{3.14}$$

In a real cross-over adjustment with data gaps and interruptions due to islands and coast lines and in addition in one where a different number of parameters p is selected for different arcs (long ones and short ones) careful prior analysis of the rank of the system is required.

---

[1]In geodesy a datum is defined as the minimum number of parameters that need to be fixed in order to obtain a well-defined system. For example, in a leveling network the height value attributed to one fundamental bench mark defines the datum.

Table 3.2 : Structure of normal matrix $A^T A$ for the case of a two-parameter model

k = number of ascending arcs

ℓ = number of descending arcs

$M = \sum_{i=1}^{\ell} \mu_{ai}$

$MM = \sum_{i=1}^{\ell} \mu_{ai}^2$

$M' = \sum_{j=1}^{k} \mu_{dj}$

$M'M' = \sum_{j=1}^{k} \mu_{dj}^2$

Although by (3.12) the datum problem is solved, a second thought is in order. The rank defect of A, (3.14), implies the existence of a null space of A of dimension p^2, i.e. the exist p^2 linearly independent vectors x_i for which

$$Ax_i = 0 \qquad i = 1,...,p^2 \quad . \qquad (3.15)$$

They are also referred to as homogeneous solution of the problem. Any linear combination of the x_i may be added to the solution of the (full rank) least-squares problem without affecting the cross-over differences. In order to make the character of the null space better visible the linear combination is written as

$$x_0 = q_1 x_1 + q_2 x_2 \cdots + q_{p^2} x_{p^2} \qquad (3.16)$$
$$= Xq$$

where the solution vectors x_i of (3.15) form the columns of X and q is a $p^2 \times 1$ vector. Each x_i, and therefore also X itself, can be separated into a part referring to the ascending and one referring to the descending arc segments: $X = \begin{pmatrix} X_a \\ X_d \end{pmatrix}$ with dimension: dim $X_a = kp \times p^2$ and dim $X_d = \ell p \times p^2$. Then one can visualize the non-estimable radial orbit error by the expression

$$\Delta r_a = A_a\, X_a\, q \qquad (\text{"ascending" invariance surface}) \qquad (3.17a)$$

and

$$\Delta r_d = A_d\, X_d\, q \qquad (\text{"descending" invariance surface}) \qquad (3.17b)$$

where q can be chosen freely, and (resulting from (3.15)) with

$$\Delta r_a + \Delta r_d = (A_a\, X_a + A_d\, X_d)q = AXq = 0 \quad . \qquad (3.18)$$

In other words Δr_a and Δr_d represent that portion of the radial orbit error that cannot be detected on basis of the given cross-over differences. The most elementary - and easiest conceivable - case of Δr_a and Δr_d is that of a common radial shift, which never could be detected from cross-over differences alone.

In general it is difficult to find X. For the case of the regular cross-over adjustment discussed here X is given in Table 3.3. See (Schrama, 1989). For the two-parameter case the invariance surface for one arbitrary choice of q is shown in Figure 3.5.

What has been achieved so far? The linear model for the least-squares adjustment of cross-over differences has been derived. The resulting system of normal equations has a favorable structure, compare Table 3.2, that can easily be solved. The definition of the datum has been discussed as well. Eq. (3.9) gives the posteriori variance of the adjusted orbit parameters. Thereby one has to keep in mind, that D_x is given in a certain datum. Hence the variances of the fixed p arcs are zero and variances of arc parameters shall grow with increasing distance from them. Transformation of D_x from one datum to another is done by means of an S-transformation, see (Baarda, 1973), (Teunissen, 1986), (Strang van Hees, 1982), or (Schrama, 1989). There remains an undetectable part of the orbit error, expressed by the invariance surfaces (null space).

Table 3.3 : Structure of X-matrices (null space) for the case of a one-, two-, and three-parameter cross-over adjustment

$(sd1 = \sin \mu_{d1} \; ; \; ca1 = \cos \mu_{a1})$

	one (p=1)	two (p=4)				three (p=9)								
X_a ($p \times p^2$)	1	1	μ_{d1}	0	0	1	sd1	cd1	0	0	0	0	0	0
	1	0	0	1	μ_{d1}	0	0	0	1	sd1	cd1	0	0	0
	1	1	μ_{d2}	0	0	0	0	0	0	0	0	1	sd1	cd1
	·	0	0	1	μ_{d2}	1	sd2	cd2	0	0	0	0	0	0
	·	·	·	·	·	0	0	0	1	sd2	cd2	0	0	0
	·	·	·	·	·	0	0	0	0	0	0	1	sd2	cd2
	·	1	μ_{dk}	0	0	·	·	·	·	·	·	·	·	·
	1	0	0	1	μ_{dk}	0	0	0	0	0	0	1	sdk	sdk
X_d ($p \times p^2$)	1	1	0	0	μ_{a1}	1	0	0	0	0	0	0	sa1	ca1
	1	0	1	μ_{a1}	0	0	1	0	0	0	sa1	ca1	0	0
	1	1	0	0	μ_{a2}	0	0	1	sa1	ca1	0	0	0	0
	·	0	1	μ_{a2}	0	1	0	0	0	0	0	0	sa2	ca2
	·	·	·	·	·	0	1	0	0	0	sa2	ca2	0	0
	·	·	·	·	·	0	0	1	sa2	ca2	0	0	0	0
	·	1	0	0	μ_{al}	·	·	·	·	·	·	·	·	·
	1	0	1	μ_{al}	0	0	0	1	sal	cal	0	0	0	0

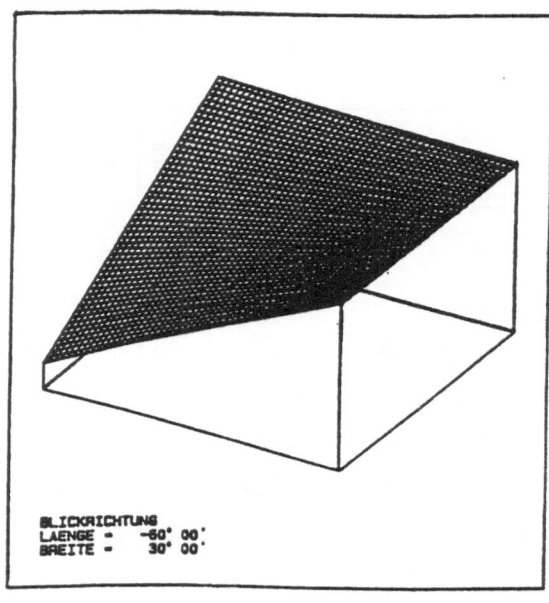

FIGURE 3.5: Invariance surface for the case of a two-parameter cross-over adjustment.

The choice of the p master arcs that define the datum is arbitrary. What is a wise choice? In principle one choice is as good as any other. However those arcs should be held fixed which are (1) representative, in the sense of being long and rather central and (2) accurate, if there are indications that the error of certain arcs is smaller than that of others. However, it is always possible to change from one datum to any other by means of (3.16). Let in the old datum the p^2 parameters of the master arcs have the values $c_1, c_2, ..., c_{p^2}$ and in the new datum $d_1, d_2, ..., d_{p^2}$, e.g. zero, then the transformation parameters q are derived from the $p^2 \times p^2$ system

$$d = c + \bar{X}q \tag{3.19}$$

where $\bar{X}$ is that part of X referring to the master arcs. With q derived from (3.19) a correction X_q can be applied to all other arcs. By the same transformation neighboring adjustment areas can be merged.

The cross-over adjustment reduces the differences at the intersections significantly. A typical histogram of the prior and posteriori differences is shown in Figure 3.6. The root mean square (rms) values decrease in this example from 1.08 m to 0.11 m. Some main points concerning the regional cross-over adjustment are summarized in Table 3.4.

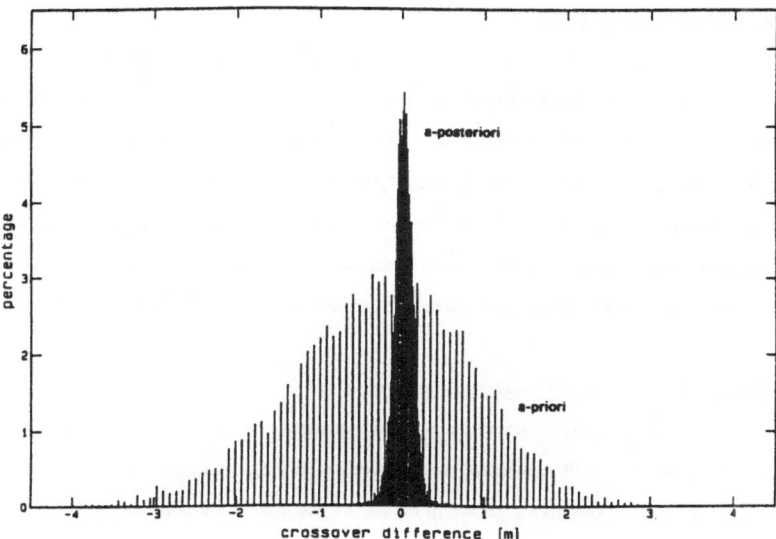

FIGURE 3.6: A-priori/a posteriori histogram of cross-over differences.

TABLE 3.4: Conclusions of Regional Cross-Over Adjustment.

- It is recommended to define the adjustment area by two ascending and two descending arcs.
- Depending on the size of the area the radial orbit error can be described by a one (bias), two (bias + tilt), or three (bias + sinus (orbit period) + cos (orbit period)) model. Local cross-over adjustment leads to a significant reduction of the radial orbit error in the considered region.
- The linear system has a rank defect of p^2 (datum defect). It is eliminated by fixing the p parameters of p non-intersecting arc segments.
- Part of the orbit error cannot be detected on basis of the cross-over differences. This part can be visualized as an "invariance" or null-space surface.
- The datum can be transferred to an adjacent area or redefined by datum transformation.
- The error variances refer to a datum.

Critique:

The commonly applied stochastic models are hardly adequate. The internal and external accuracy should be evaluated in terms of precision and reliability.

- The estimated sea surface depends on the size and shape of the chosen area and on the chosen parameter model.
- The rank defect depends on the chosen parameter model.
- Main complaint is that the orbit correction model does not take into account the (mechanics) time history of an orbit (see however (Wunsch & Zlotnicki, 1984), (Knudsen, 1987)).

4. Global Cross-Over Adjustment.

The main complaint against the procedure of regional cross-over adjustment, just described, is that it does not take into account the actual time history of the radial orbit error. Each arc segment is treated independently, whereas we know that the orbit error increases and decreases in a periodic manner over the entire length of the computed orbit (typically 3 to 6 days). This almost certainly leads to unrealistical bias, tilt or amplitude values of the arc parameters. A first step towards an improvement is the so-called chronological, segmented cross-over adjustment. Thereby the chosen arc parameters are still considered independent for each ascending and descending arc and it remains to be seen if the chronological sequence of adjusted arc corrections can reproduce the actual radial orbit error.

4.1. Chronological, Segmented Cross-Over Adjustment.

This type of radial orbit adjustment is global, i.e. covering the entire altimetric surface, but each ascending and descending arc segment is still considered independently. In our discussion we assume a globally homogeneous distribution of arc intersections, as e.g. shown in Figure 3.1. No distinction between land and ocean areas is made. The radial orbit error model for each arc is that derived in eq. (3.3a):

$$\Delta r = x_0 + x_1 \sin \mu + x_2 \cos \mu \quad .$$

However the previous definition of the relative time tags μ does not apply anymore. The time tags are now referred to the equator crossing time t_{ie} of each segment:

$$\mu_{ij} = \frac{2\pi}{T} (t_{ij} - t_{ie}) \quad . \tag{4.1}$$

Hence μ runs from $-\frac{\pi}{2}$ to $+\frac{\pi}{2}$. As discussed in (Schrama, 1989) each crossing ascending and descending arc forms with an equator segment the symmetric curved triangle, shown in Figure 4.1, with the advantageous property

$$\mu_{ij} \ (ascending) = -\mu_{ji} \ (descending) \quad . \tag{4.2}$$

The interrelation of equator longitude difference $\Delta\lambda$ between ascending and descending arc, latitude ϕ of crossing, and time tag μ_{ij} is given in Figure 4.2 for a realistic set of orbit parameters, (Schrama, 1989). It displays an increasing interval in elapsed time and longitude difference versus a decreasing step size with higher latitude (more cross-overs towards maximum latitudes).

Due to (4.2) the adjustment model of sea surface differences becomes in this case

$$\begin{aligned}
\Delta \bar{h}_{ij} &= \Delta r_i - \Delta r_j + \bar{e} \\
&= x_{i0} + x_{i1} \sin \mu_{ij} + x_{i2} \cos \mu_{ij} - x_{j0} - x_{j1} \sin \mu_{ji} - x_{j2} \cos \mu_{ji} + \bar{e} \\
&= (x_{i0} - x_{j0}) + (x_{i1} + x_{j1}) \sin \mu_{ij} + (x_{i2} - x_{j2}) \cos \mu_{ij} + \bar{e} \quad .
\end{aligned} \tag{4.3}$$

The structure of coefficient and normal matrix are given in Tables 4.1 and 4.2. Thereby two types of ground track configurations are considered (see symbolical maps): Table 4.1 is based upon a prograde orbit, in which case the rate of precession $\dot{\Omega}$ adds to the earth's angular rate ($I < 90^0$); Table 4.2 is based on a retrograde orbit ($I > 90^0$). In both

Table 4.1 :
Ground track pattern, coefficient matrix and normal matrix of global (segmented), chronological cross-over adjustment (prograde orbit)

Symbolic map of ground tracks

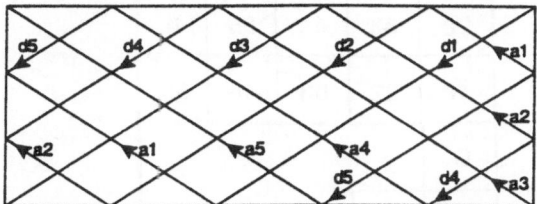

Coefficient matrix

a	d	a					d				
		1	2	3	4	5	1	2	3	4	5
1	2	a1						-d1			
	3	a2							-d2		
	4	a3								-d3	
	5	a4									-d4
	1	a5					-d5				
2	3		a1						-d1		
	4		a2							-d2	
	5		a3								-d3
	1		a4				-d4				
	2		a5					-d5			
3	4			a1						-d1	
	5			a2							-d2
	1			a3			-d3				
	2			a4				-d4			
	3			a5					-d5		
4	5				a1						-d1
	1				a2		-d2				
	2				a3			-d3			
	3				a4				-d4		
	4				a5					-d5	
5	1					a1	-d1				
	2					a2		-d2			
	3					a3			-d3		
	4					a4				-d4	
	5					a5					-d5

Normal matrix

A					55	11	22	33	44
	A		0		44	55	11	22	33
		A			33	44	55	11	22
0			A		22	33	44	55	11
				A	11	22	33	44	55
					D			0	
		*				D			
							D		
						0		D	
									D

Table 4.2 :
Ground track pattern, coefficient matrix and normal matrix of global (segmented), chronological cross-over adjustment (retrograde orbit)

Symbolic map of ground tracks

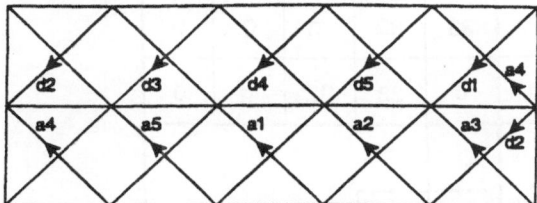

Coefficient matrix

a	d	a					d				
		1	2	3	4	5	1	2	3	4	5
1	5	a1									-d1
	4	a2								-d2	
	3	a3							-d3		
2	1		a1				-d1				
	5		a2								-d2
	4		a3						-d3		
3	2			a1				-d1			
	1			a2			-d2				
	5			a3							-d3
4	3				a1				-d1		
	2				a2			-d2			
	1				a3		-d3				
5	4					a1				-d1	
	3					a2		-d2			
	2					a3	-d3				

Normal matrix

A					0	0	33	22	11
	A		0		11	0	0	33	22
		A			22	11	0	0	33
0			A		33	22	11	0	0
				A	0	33	22	11	0
					D				
						D		0	
	★						D		
						0		D	
									D

cases a_i denotes the set of all three coefficients (1, sin μ, cos μ) of the ascending arc and taken at time tag i; d_j is the analogous expression of the descending arc. It is seen that the normal matrices consist of two block-diagonal submatrices with constant 3x3 blocks A and D and two circulant Töplitz off-diagonal submatrices, in the retrograde case partly filled with zeros. This structure

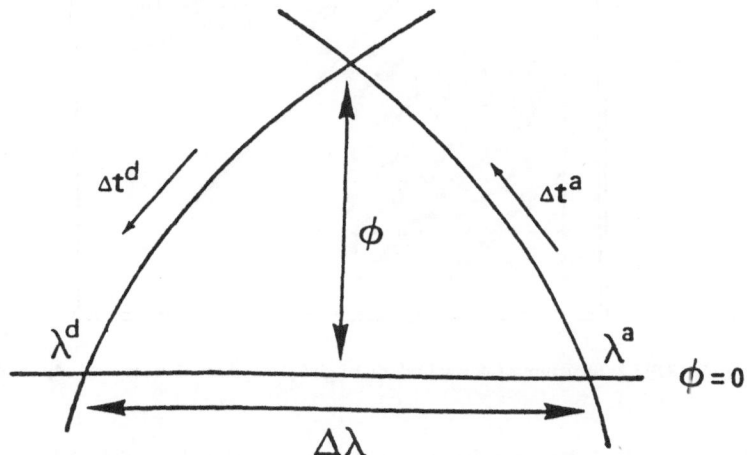

FIGURE 4.1: Equator, ascending and descending arc configurations.

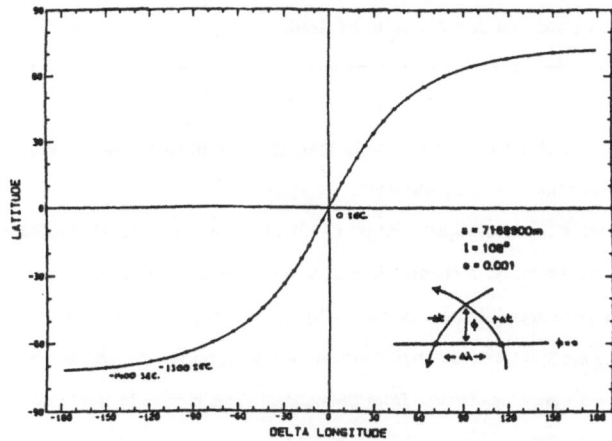

FIGURE 4.2:   Longitude separation, latitude and time relative to the equator crossing of cross-overs, from (Schrama, 1989).

allows an easy solution and analysis of the system of normal equations. In a real world situation, with land and island interruptions, the situation is disturbed but the general character of this structure is still maintained, see Figure 4.3.

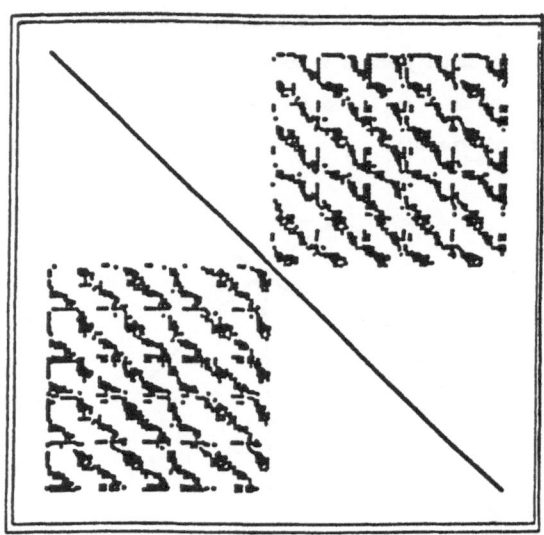

FIGURE 4.3: Structure of normal matrix of global, segmented cross-over adjustment.

Also in this case the coefficient matrix A is deficient. It has a rank defect of three, $rg(A) = r = m-3$. The rank defect can be eliminated by fixing one arc, compare also (Rowlands, 1981).

In a global chronological and segmented cross-over adjustment with $p = 3$ parameters per arc, the p parameters of one master arc have to be fixed.

(4.4)

It should be noted that fixing more parameters than those implied by the rank defect leads to an <u>over-constraint</u>, i.e. a deformation of the adjusted parameters.

Why is the rank defect of the local adjustment p^2, (3.12), and that of the global chronological one only p? The regional cross-over adjustment, described in chapter 3, was so-to-say carried out in plane approximation. Going back to the diamond shaped adjustment area (Figures 3.3a and 3.3b), the time tags of e.g. the first cross-over of all ascending arcs after crossing descending arc 3, were considered identical. In reality, on the curved surface of the earth these time tags get shorter and shorter with growing distance from the equator. This change in time intervals decreases the rank defect from p^2 to p (with one arc fixed). However only in really global or almost global situations a stable adjustment can be performed. In all other cases only the theoretical rank defect is p whereas the actual (numerical) one is p^2. Then fixing p^2-parameters is required and therefore a slight over-constraint is unavoidable.

This result exhibits a nice analogy with geodetic network adjustment on a two-axial ellipsoid. The network is only invariant with respect to rotations about the z-axis (semi-minor axis), corresponding to a rank defect of one. It can be eliminated by fixing one longitude. Actually, however, even for networks of the size of Europe the numerical rank-defect is four. Hence in practice, besides the scale, the two coordinates of a central point and one azimuth are fixed resulting in a slight over-constraint, compare (Teunissen, 1985; ch. 2.2).

The homogeneous part of the solution is again expressed by

$$x_0 = X q \qquad (3.16)$$

where the null space is spanned by the columns of X

$$X = \begin{pmatrix} X_a \\ X_d \end{pmatrix} = \begin{pmatrix} 1 & 0 & 0 \\ 0 & 1 & 0 \\ 0 & 0 & 1 \\ & \cdots & \\ 1 & 0 & 0 \\ 0 & -1 & 0 \\ 0 & 0 & 1 \\ & \cdots & \end{pmatrix} \qquad (4.5)$$

The invariance surface is displayed in Figure 4.4. It can be derived either from

$$\Delta r_a = A_a X_a q = q_1 + q_2 \sin \mu + q_3 \cos \mu$$

or from

$$\qquad (4.6a,b)$$

$$\Delta r_d = A_d X_d q = q_1 - q_2 \sin \mu + q_3 \cos \mu \qquad .$$

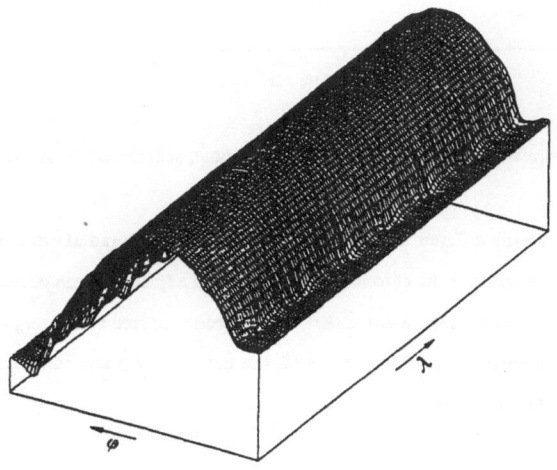

FIGURE 4.4: Invariance surface for the case of a global, segmented cross-over adjustment.

The purpose of the chronological cross-over adjustment is to take into account the time history of the radial orbit error. Let us first consider the unobservable part, expressed by (4.6). When merging Δr_a and Δr_d in a chronological sequence and mapping it onto a sphere, the constant and sine apart can form a continuous surface function whereas this is impossible with the cosine part. Thus we are led with the conclusion that only the constant plus sine function remains unobservable.

In a simulated 3-day example, the adjusted radial orbit error is shown in Figure 4.5, taken from (Schrama, 1989). It shows the sequence of ascending and descending arc segments. One master arc had been fixed (at t = 0.71 days). It shows up as a horizontal bar. We observe the typical once/rev. behavior of the orbit error with an envelope expressing modulation by additional, e.g. resonant, effects. However, there also appear discontinuities (in the derivative) in the upper envelope. In other words, the applied method leads to unnatural features. The nature of this folding down of the envelope can be influenced by the choice of the datum (reference arc). So far no conclusive study has been carried out on the exact interrelation of reconstructed orbit error and datum reference arc.

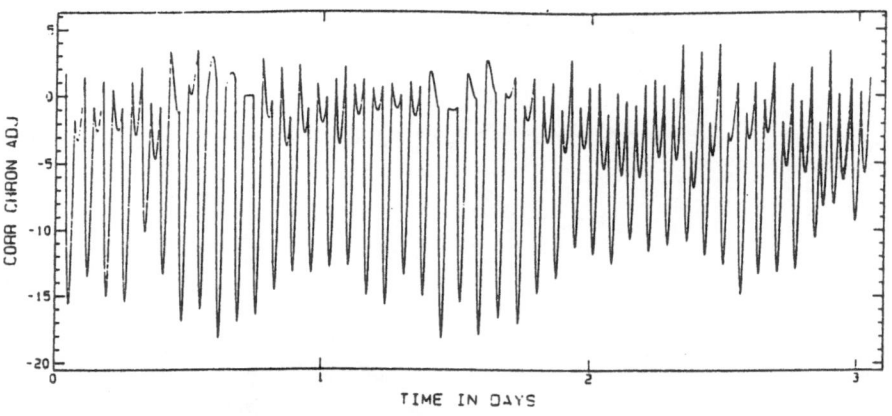

FIGURE 4.5: Adjusted radial orbit resulting from global, segmented cross-over adjustment.

Schrama (ibid) analysed the difference between the actual (simulated) radial orbit and the one determined by the cross-over adjustment to see whether the chronological segmented adjustment can correctly reconstruct the radial orbit error. Spectral analysis of the residuals revealed that they are concentrated at $2\omega_e$, $\omega_0-\omega_e$, $\omega_0+\omega_e$, and $2(\omega_0-\omega_e)$ with $\omega_e = (\Omega-\theta)$, the apparent earth rate, and $\omega_0 = \omega+M$, the orbit rate (Ω ascending node, θ Greenwich siderial time, ω argument of perigee, M mean anomaly).

4.2. Discussion of Rank Defect and Null Space.

Already during the first cross-over adjustments of GEOS-3 data attention was given to the rank defect, see

(Rummel & Rapp, 1977). But no thought was given to the null space and its interpretation in terms of the non-detectable part of the radial orbit error. Goad et al. (1980) pointed out the problem and later it has been discussed at various places, e.g. (Sandwell et al., 1986), (Rosborough, 1986), (Engelis, 1987), (Schrama, 1989), (Barzaghi et al., 1991), or (Schrama, 1992).

Let us give, in addition to the more mathematical derivation of the previous sections, a more intuitive discussion of this problem, see also (Schrama et al., 1990). When we follow a ground track from its most southern point first up (ascending arc) and then down again (descending arc) the first intersection occurs at time Δ, at somewhat less than the orbit period T in the case of a retrograde orbit. This can be verified from Figure 4.6. The intersection of the original ascending arc with the next descending arc takes place at almost exactly 2Δ and so on. The cross-over points are numbered in sequential order in Figure 4.6. We also see that the cross-over points slowly move northwards. The radial orbit error is dominated by a once/rev. oscillation. Let us assume it is a perfect once/rev. sinus, as displayed in Figure 4.7, with zero value at the most southern and northern latitude and period T. The time tags from 1 to 6 of the first ascending arc segment and of all following descending arcs refer to the cross-overs of Figure 4.6. They indicate the radial orbit error at the time tags. Observable is the difference in orbit error at corresponding time tags, the cross-over difference. This difference would be zero if the elapsed time would be T. In this case a once/rev. periodical error would be unobservable. The configuration of Figure 4.7 is invariant with respect to a constant (a vertical shift of the curve does not affect the differences) and a sine with period Δ (it has the same period between corresponding ascending and descending points). These two are not observable. They establish the invariance surface, the null space. By the fact that Δ slightly changes from maximum latitudes towards the equator a somewhat distorted situation emerges.

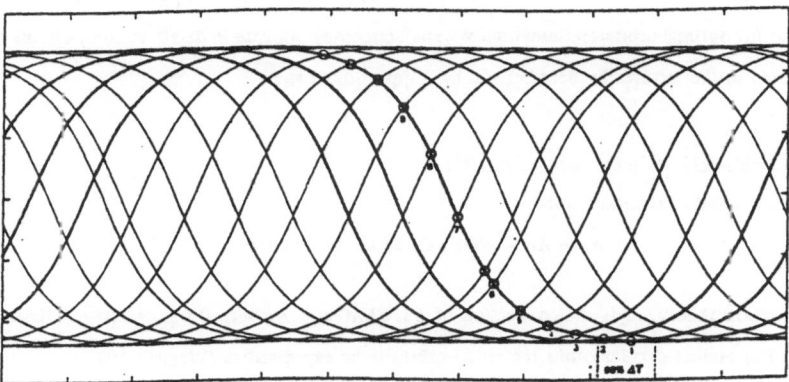

FIGURE 4.6: Cross-over occurence after one, two, ... revolutions.

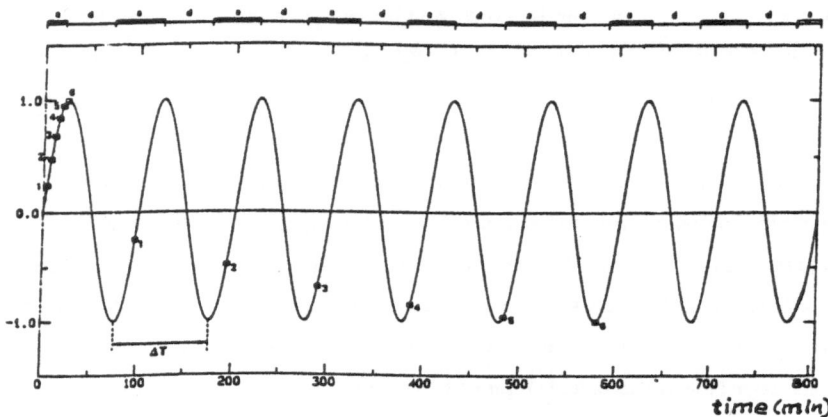

FIGURE 4.7: Cross-over occurence and idealized radial orbit error.

4.3. Global Continuous Cross-Over Adjustment.

For many geophysical and oceanographic investigations the regional cross-over adjustment is an easy and appropriate tool for radial orbit error elimination. Global chronological and segmented adjustment is the next logical step, still rather easily implemented but with remaining discontinuities between consecutive segments. Since it is known that the main cause of the radial orbit error is due to our insufficient knowledge of the earth's gravitational field the next logical step is to consider complete computed orbits (3 to 6 days length) and express the radial orbit error in terms of corrections $\Delta \bar{C}_{\ell m}$ and $\Delta \bar{S}_{\ell m}$ of the spherical harmonic expansion of the earth's gravitational field. This is saying that the cross-over discrepancies $\Delta \bar{n}_{ij}$ are considered a (so far not used) information source from which the gravity field can be improved, either simultaneously with all other (tracking) data or in a second phase after all other sources have been used already for optimal orbit determination. We shall not cover this case in detail, but only discuss some aspects, as it will be treated in the lectures by Balmino and by Rapp of this issue.

4.3.1. Analytical Expression of the Radial Orbit Error.

The linear model of the radial orbit error is

$$\Delta r = \Delta a + e \Delta M a \sin M - \Delta e a \cos M \qquad (3.2)$$

Lagrange perturbation theory yields the connection of Δa, $e \Delta M$, and Δe with the potential coefficient corrections $\Delta \bar{C}_{\ell m}$ and $\Delta \bar{S}_{\ell m}$. For almost circular orbits the radial orbit can be expressed as (Wagner, 1985):

$$\Delta r(t) = \sum_{m=0}^{\ell max} \sum_{k=-\ell max}^{\ell max} [A_{km} \cos \psi_{km} t + B_{km} \sin \psi_{km} t] \qquad (4.7)$$

with

$$(A_{km}, B_{km}) = \sum_{t=m}^{tmax} H_{tmk} \left[\begin{Bmatrix} \Delta \bar{C}_{tm} & \Delta \bar{S}_{tm} \\ -\Delta \bar{S}_{tm} & \Delta \bar{C}_{tm} \end{Bmatrix}_{t-m:odd}^{t-m:even} \cos \bar{\Psi}_{km} + \begin{Bmatrix} \Delta \bar{S}_{tm} & -\Delta \bar{C}_{tm} \\ \Delta \bar{C}_{tm} & \Delta \bar{S}_{tm} \end{Bmatrix}_{t-m:0}^{t-m:e} \sin \bar{\Psi}_{km} \right] \tag{4.8}$$

and

$$H_{tmk} = \frac{\mu(a_e/a)^t}{\bar{n}^2 a^2} \bar{F}_{tm}^k (I) \begin{bmatrix} \beta_{km}(t+1) - 2k \\ \beta_{km}(\beta_{km}^2 - 1) \end{bmatrix} \tag{4.9}$$

Thereby it is

$$\begin{aligned}
\beta_{km} &= \dot{\Psi}_{km} / \bar{n} \\
\Psi_{km} &= \bar{\Psi}_{km} + \dot{\Psi}_{km} t \\
\bar{\Psi}_{km} &= k(\bar{\omega} + \bar{M}) + m(\bar{\Omega} - \bar{\theta}) = k\omega_0 + m\omega_e \\
\dot{\Psi}_{km} &= k(\dot{\omega} + \dot{M}) + m(\dot{\Omega} - \dot{\theta}) = k\dot{\omega}_0 + m\dot{\omega}_e
\end{aligned} \tag{4.10a-d}$$

The quantities are described in Table 4.3. The orbit error is now a continuous function of time over the complete length of the computed arc. At cross-over points the difference of (4.7) at times t_{ij} and t_{ji} is taken.

TABLE 4.3: Quantities of Equations (4.7) to (4.10)	
$\mu = GM$	gravitational constant times mass
a_e	semi-major axis reference ellipsoid
a	semi-major axis orbit
$\bar{F}_{tm}^k$	inclination functions
$\bar{n}$	mean angular velocity of satellite
ω	argument of perigee
M	mean anomaly
$\omega_0 = \omega + M$	orbit rate
Ω	longitude of node
θ	Greenwich siderial time ($\dot{\theta}$ earth rate)
$\omega_e = \Omega - \theta$	earth rate
$\Delta \bar{C}_{tm}$ $\Delta \bar{S}_{tm}$	fully normalized gravity potential coefficients
$tmax$	maximum degree of series expansion
ℓ, m	spherical harmonic degree and order

The adjustment process can be divided into two steps. First the coefficients A_{km} and B_{km} are determined from the cross-over differences. They are called <u>lumped coefficients</u>, because as seen from (4.8) they consist of a linear combination of $\Delta \bar{C}_{lm}$ and $\Delta \bar{S}_{lm}$. See (Wagner & Klosko, 1981). Then, in the second step, the $\Delta \bar{C}_{lm}$ and $\Delta \bar{S}_{lm}$ are derived from the lumped coefficients. In total there are $4(tmax)^2$ lumped coefficients and $(tmax)^2$ unknown potential coefficients. The actual separability of the individual $\Delta \bar{C}_{lm}$ or $\Delta \bar{S}_{lm}$ coefficients depends very much on the sensitivity coefficients $H_{\ell m k}$. We see that at the zero frequency of the ψ_{km} spectrum and at once/rev. ($\psi_{km} = \bar{n}$) resonance occurs. Coefficients close to the resonance bands are well estimable; far off the resonance bands discrimination is difficult or even impossible. In this case regularization methods are applied. In other words prior information on $\Delta \bar{C}_{lm}$ and $\Delta \bar{S}_{lm}$ has to be introduced.

4.3.2. Null Space of Global Adjustment.

Let us look into the datum defect of this adjustment. Any orbit effect that has the same signature at intersections of ascending and descending arcs cannot be determined. In other words any part of the orbit error that can be represented purely as a function of location is invariant with respect to cross-over differences. Sandwell et al. (1986) were able to show that there exists such an expression:

$$X(\phi,\lambda) = \sum_{l=0}^{l_{max}} \sum_{m=0}^{l} \sum_{k=-l(2)}^{+l} \bar{F}_{lm}^{k}(I) \left[\begin{bmatrix} a_{lm} \\ -b_{lm} \end{bmatrix}_{l-m:0}^{l-m:e} \cos \psi_{km} + \begin{bmatrix} b_{lm} \\ a_{lm} \end{bmatrix}_{l-m:0}^{l-m:e} \sin \psi_{km} \right] \quad . \tag{4.11}$$

We take (4.7) as linear model with $\Delta h_{ij} = \Delta r_i - \Delta r_j$ as observable and the vector of unknowns x containing all lumped coefficients A_{km} and B_{km}. It is also reasonable to truncate the series at 2.3 cpr considering the error spectrum of the radial orbit error, see Figure 2.3. Eigenvalue analysis of the coefficient matrix reveals a rank defect of nine: p = 9. Consequently nine linear independent vectors exist for which $Ax_i = 0$. Again, with x_i, i = 1,...,9 arranged as columns of a matrix X the homogeneous solution can be written as

$$x_0 = X q \quad . \tag{4.12}$$

With (4.11) serving as a basis of a description of the invariance surface, the nine elements of q are (a_{00}, a_{10}, a_{11}, b_{11}, a_{20}, a_{21}, b_{21}, a_{22}, b_{22}} and the non-zero elements of the columns of X are those given in Table 4.4. An example of the invariance surface is shown in Figure 4.8. The non-zero elements are given together with the corresponding frequencies (sine or cosine term) which are those we met already in section 4.1. In addition the constant, once per rev. and, new, a twice per rev. term are part of the null space. A higher truncation frequency of the time series would lead to a larger null-space. We observe that all columns of X are orthogonal to each other except those belonging to the coefficients a_{00} and a_{20}. The notation of the elements of the q-vector was chosen so that no confusion can arise between the a_{lm} and b_{lm} of q on one hand and the potential coefficient correction $\Delta \bar{C}_{lm}$ and $\Delta \bar{S}_{lm}$ on the other hand. Again the nine components of q have to be constrained to adopted values in order to arrive at a regular system of equations.

Function $X(\phi,\lambda)$ represents a geographically correlated error, that is not detectable from cross-over differences. What is its relation to the geographically correlated orbit error as introduced in (Tapley & Rosborough, 1985) or (Rosborough, 1986) and discussed also in (Engelis, 1987)? Although Sandwell's null space analysis is correct it is not in the line of considerations of the previous chapters. It is applied to the ψ_{km}-spectrum, or in other words, unknowns are the lumped coefficients A_{km} and B_{km}. The transformation from the vectors spanning the null space to an invariance surface on the sphere is performed by means of eq. (4.11). Alternatively the rank defect analysis could be applied directly to (4.7)-(4.9) with the vector of potential coefficients $\Delta \bar{C}_{lm}$ and $\Delta \bar{S}_{lm}$ as vector of unknowns. In this case the sensitivity coefficients H_{lmk} would be part of the coefficient matrix. This is basically what Rosborough (1986) did. In this case the vectors x_i of the homogeneous solution $Ax_i = 0$ (spanning the null space) also consists of potential coefficients. If the radial orbit error spectrum is limited to 2.3 cpr the same null space should result as in Sandwell's case. Or more planely, a contour line representation of the invariance surface, Figure 4.8, should yield the representation of the geographically correlated orbit error, shown in Figure 22 of Balminos lectures (this issue). A thorough study is needed to prove that the two approaches are consistent. The indeterminancy can be eliminated regionally or globally by dedicated high precision tracking stations as discussed e.g. in (Wagner & Melchioni, 1989).

Readers interested in the details of processing of altimeter data (preprocessing, regional and global cross-over adjustment, orbit error) are referred to (Colombo, 1984), (Denker, 1990), (Engelis, 1987), (Rowlands, 1981), (Schrama, 1989), (Shum et al. 1990), (Tai, 1989), (Tai & Fu, 1989), (Wagner, 1989) or (Zandbergen, 1991) or to the special issues on altimetry in some journals e.g. (Bernstein, 1982), (Born, 1980), (Born, 1984), (Seasat Special Issue II, 1983), (Seeber & Apel, 1984), (Special Section: Geosat part II, 1990).

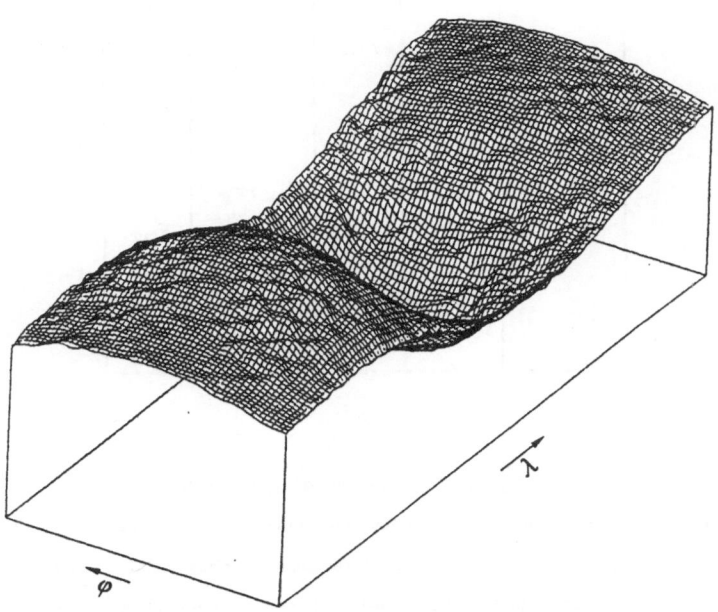

FIGURE 4.8: Invariance surface of global adjustment (orbit error truncated at 2.3 cpr).

TABLE 4.4: The Null Space Generating Matrix X of the Global Cross-Over Adjustment (truncation frequency of the orbit representation at 2.3 cpr).

	a_{00}	a_{10}	a_{11}	b_{11}	a_{20}	a_{21}	b_{21}	a_{22}	b_{22}
1	F_{000}				F_{200}				
$\sin \omega_e$						F_{210}			
$\cos \omega_e$							$-F_{210}$		
$\sin 2\omega_e$									F_{220}
$\cos 2\omega_e$								F_{220}	
$\sin(\omega_0-\omega_e)$				$-F_{11-1}$					
$\cos(\omega_0-\omega_e)$			F_{11-1}						
$\sin \omega_0$		$F_{101}-F_{10-1}$							
$\sin(\omega_0+\omega_e)$				F_{111}					
$\cos(\omega_0+\omega_e)$			F_{111}						$-F_{22-2}$
$\sin(2\omega_0-2\omega_e)$									
$\cos(2\omega_0-2\omega_e)$								F_{22-2}	
$\sin(2\omega_0-\omega_e)$						$-F_{21-2}$			
$\cos(2\omega_0-\omega_e)$							$-F_{21-2}$		
$\cos 2\omega_0$					$F_{202}+F_{20-2}$				
$\sin(2\omega_0+\omega_e)$						F_{212}			
$\cos(2\omega_0+\omega_e)$							$-F_{212}$		F_{222}
$\sin(2\omega_0+2\omega_e)$									
$\cos(2\omega_0+2\omega_e)$								F_{222}	

5. On the Separation of Sea Surface Topography and Geoid or the "Munchhausen Problem".

All preceeding chapters were concerned with the determination of the ocean surface, free of orbit errors. Oceanographers and geodesists want to know the separation of the sea surface height h into geoid height N and sea surface topography (SST) H. However altimetric measurements alone seem to provide no means for such a separation. Altimetry provides h, the sum of N and H. How should the identification of the parts be possible with only the sum being measured? This is why I refer to this issue as "Munchhausen problem[2]".

The problem posed here is: Given the sea surface height h and

---

[2]Baron van Munchhausen, *1720 in Hannover, †1797 Bodenwerder, famous for his enormously exaggerated adventure tales. They were published in 1793 by R.E. Raspe: The Adventures of Baron Münchhausen, with additional tales very likely not from Munchhausens pen. In one of the tales Munchhausen gets trapped in a mud hole and is on the verge to drown. He saves his life by taking his own hair and pulling himself out of the hole.

$$h = N + H \qquad , \qquad \text{(2.2)}$$

how can N and H be determined individually. Unfortunately even the most advanced available geoid models, derived independently of altimetry, are typically inaccurate to one or more meters, see Figure 5.1 and (Marsh et al., 1990) and (Haagmans & Van Gelderen, 1991). As an example one could use for N the geoid computed from GEM-T1. It is complete up to degree and order 36. Truncation at a lower degree would help to produce the very long wavelengths part with higher precision (10-20 cm), but would still not satisfy oceanographic needs and, even worse, increase the chance of aliasing higher frequency gravity effects into SST. Another approach that helps to separate N and H is to allow SST-models only that meet the laws of fluid dynamics, i.e. the equations of motions. In other words H would have to follow geostrophic or quasi-geostrophic models. This idea has been put forward in (Wunsch & Gaposchkin, 1980), see also (Mather et al., 1979). At several places work is currently undertaken in this direction. A third approach tries to solve the "Munchhausen problem" directly. It makes use of the fact that the unmodeled part of the geoid is connected to the unmodeled part of the radial orbit whereas the SST is not. Quite successful attempts to employ this difference for the separation of N and H were undertaken by Tapley et al. (1988) and Marsh et al. (1990). See also (Wagner, 1986), (Wagner, 1989), (Engelis & Rapp, 1984), (Engelis, 1985), (Engelis, 1987), (Schrama, 1989), (Denker & Rapp, 1990), (Rapp, Wang & Pavlis, 1991) and (Visser, 1992). It is not our intention to describe these methods. The problem shall be treated in Rapp's lectures (this issue). We only like to address the question of to what extent a separation is possible in principle and let oceanographers decide on the relevance of the outcome.

Standard deviation of geoid heights (cm)

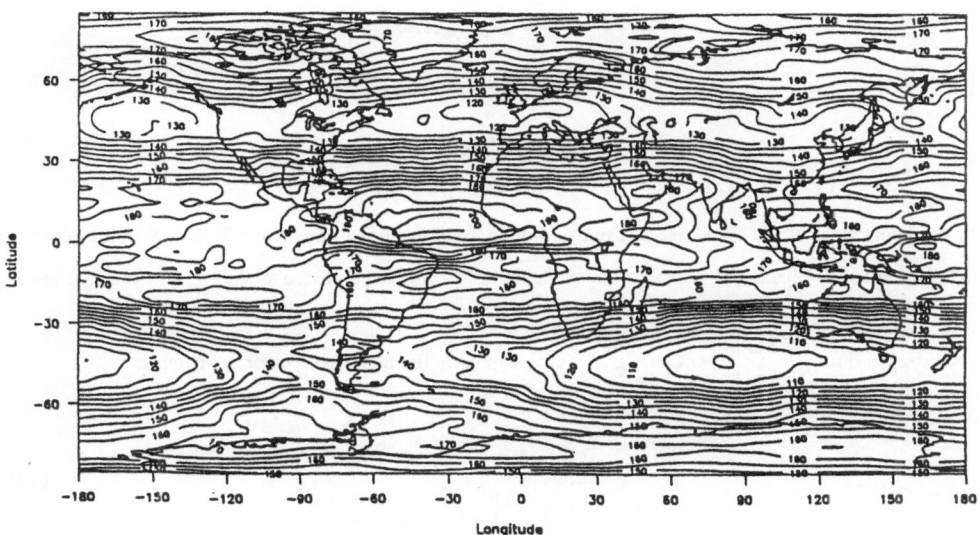

FIGURE 5.1: Geoid accuracy of GEM-T1, from (Haagmans & Van Gelderen, 1991).

Let us assume the unknown part of the earth's gravitational field is expressed in a series of spherical harmonics. (For reasons of compactness we choose here to use complex and normalized spherical harmonic functions. Hopefully this does not cause confusion. Their relation to the usual expansion is given e.g. in (Kaula, 1967), (Ilk, 1983), (Sneeuw, 1992) or in Balmino's lectures (this issue). With complex spherical harmonic coefficients c_{lm} (m ranging now from - l to $+l$) the geoid part becomes

$$N(\varphi,\lambda) = R \sum_l \sum_m c_{lm} \, Y_{lm} \, (\varphi,\lambda) \qquad . \tag{5.1}$$

In an analogous way, the permanent part of SST is expanded into the spherical harmonic series

$$H_0(\varphi,\lambda) = R \sum_l \sum_m a_{lm} \, Y_{lm}(\varphi,\lambda) \qquad . \tag{5.2}$$

In both expressions we leave it open whether an a priori model (geoid or SST) has been subtracted. Eqs. (5.1) and (5.2) underline that given $h = N + H_0$ a separation into N and H_0 (or c_{lm} and a_{lm}) were impossible, would the gravity potential coefficients c_{lm} not also enter into the description of the radial orbit error Δr.

Transformation of N and H_0 from the earth-fixed equatorial system into a coordinate system defined by the orbit trajectory yields for the geoid height

$$N(\varphi,\lambda) = R \sum_l \sum_m c_{lm} \, Y_{lm}(\varphi,\lambda)$$
$$= R \sum_l \sum_m \sum_k \bar{F}_{lm}^k \, (I)c_{lm} \, e^{i\psi_{km}}$$
$$= R \sum_l \sum_m N_{lm} \tag{5.3}$$

and analogously for SST

$$H_0(\varphi,\lambda) = R \sum_l \sum_m a_{lm} \, Y_{lm}(\varphi,\lambda)$$
$$= R \sum_l \sum_m \sum_k \bar{F}_{lm}^k \, (I)a_{lm} \, e^{i\psi_{km}}$$
$$= R \sum_l \sum_m H_{0lm} \tag{5.4}$$

Thereby one could write N and H_0 either as a function of (φ,λ) or as one of time. (In (5.3) and (5.4) the advantage of the complex notation becomes obvious: Before, in ch. 4, $e^{i\psi_{km}}$ had to be written as sin ψ_{km} and cos ψ_{km} and c_{lm} had to be grouped into $\begin{pmatrix} \Delta \bar{C}_{lm} \\ -\Delta \bar{S}_{lm} \end{pmatrix}_0^e$ and $\begin{pmatrix} \Delta \bar{S}_{lm} \\ \Delta \bar{C}_{lm} \end{pmatrix}_0^e$ etc.). Similarly the radial orbit error is written as

$$\Delta r(\varphi,\lambda) = R \sum_l \sum_m \Delta r_{lm}$$
$$= R \sum_l \sum_m \sum_k \left(\frac{R}{a}\right)^{l-1} B_{lmk} \, \bar{F}_{lm}^k(I)c_{lm} \, e^{i\psi_{km}} \tag{5.5}$$

with

$$B_{\ell m k} = \frac{\beta_{\ell m}(\ell+1) - 2k}{\beta_{\ell m}(\beta_{\ell m}^2 - 1)} \qquad . \tag{5.6}$$

We observe that (compare eq. (4.9)):

$$H_{\ell m k} = \frac{\mu(a_e/a)^\ell}{\bar{n}^2\, a^2}\, B_{\ell m k}\, \bar{F}^k_{\ell m}(I)$$

$$\dot{=} R\left(\frac{R}{a}\right)^{\ell-1} B_{\ell m k}\, \bar{F}^k_{\ell m}(I) \qquad .$$

In (5.5) initial condition errors are omitted, they can be introduced with (3.2).

We now assume $\bar{h}$ to be given. Thus for the first time we do not work anymore with cross-over differences. The adjustment model becomes for each given point

$$\bar{h} = N + H_0 - \Delta r + \bar{e}$$

or after insertion of (5.3), (5.4), and (5.5):

$$\bar{h} = R \sum_\ell \sum_m \left\{ \sum_k (1 - \left(\frac{R}{a}\right)^{\ell-1} B_{\ell m k})\, \bar{F}^k_{\ell m}(I)\, e^{i\psi_{km}} \right\} c_{\ell m}$$

$$+ R \sum_\ell \sum_m \left\{ \sum_k \bar{F}^k_{\ell m}(I)\, e^{i\psi_{km}} \right\} a_{\ell m} + \bar{e} \tag{5.7}$$

with unknowns $c_{\ell m}$ and $a_{\ell m}$. The elements of the coefficient matrix are, respectively

$$A_{\ell m}\ (GRAVITY) = R \sum_k (1 - \left(\frac{R}{a}\right)^{\ell-1} B_{\ell m k})\, \bar{F}^k_{\ell m}(I)\, e^{i\psi_{km}} \tag{5.8a}$$

and

$$A_{\ell m}\ (SST) = R \sum_k \bar{F}^k_{\ell m}(I)\, e^{i\psi_{km}} \qquad . \tag{5.8b}$$

NOTE: We restrict ourselves to the absolutely essential in this derivation: N and H_0 are expressed in spherical approximation (R = const.), Δr contains only the gravity part, no distinction is made between land and sea, the maximal degree ℓmax has not been defined so far!

The separability of the coefficients $a_{\ell m}$ from the $c_{\ell m}$ is determined by the degree of linear independence of corresponding columns (per ℓ and m) of $A_{\ell m}$ (GRAVITY) and $A_{\ell m}$ (SST). Inspection of (5.8) shows that they only differ through the presence of $(1 - \left(\frac{R}{a}\right)^{\ell-1} B_{\ell m k})$. As mentioned above, only because of the presence of Δr the separation can be possible. On the other hand if this orbit error Δr were zero also the geoid would be known. A closer look at (5.6) shows that the separability solely depends on the size of the product $\left(\frac{R}{a}\right)^{\ell-1} B_{\ell m k}$. We take a look at $\left(\frac{R}{a}\right)^{\ell-1}$ and $B_{\ell m k}$ individually:

- The ratio $\lambda_{\ell} = (R/a)^{\ell+1}$, the so-called <u>upward continuation</u> factor causes a significant damping with increasing altitude and degree.

  Example: $h = 1000$ km $\rightarrow$ $R/(R+h) = 0.864$;

ℓ	10	30	50
λ_{ℓ}	0.23	0.012	0.0007

- The $B_{\ell m k}$ get singular for $\beta_{km} = 0$ or $\beta_{km} = \pm 1$. These are <u>resonance</u> cases. Close to the resonance frequencies $B_{\ell m k}$ gets large and able to compensate damping. From

$$\beta_{km} = k + m \, \frac{\dot{\omega}_e}{\dot{\omega}_0} \qquad (5.9)$$

one sees that the resonance frequency, i.e. the degree and order of its occurrence, depends on the choice of the orbit.

In order to get more insight a closer look on the size of the coefficients $A_{\ell m}$ is required. With Schwarz' inequality we derive

$$A_{\ell m} (GRAVITY) = R \sum_{k} (1 - \left(\frac{R}{a}\right)^{\ell-1} B_{\ell m k}) \bar{F}_{\ell m}^{k}(I) \, e^{i\psi_{km}} \leq \sqrt{\sum_{k} [R(1 - \left(\frac{R}{a}\right)^{\ell-1} B_{\ell m k} \, \bar{F}_{\ell m}^{k}]^2}$$

$$= |A_{\ell m}(GRAVITY)|$$

and $\qquad\qquad\qquad\qquad\qquad\qquad\qquad\qquad\qquad\qquad\qquad\qquad\qquad\qquad\qquad$ (5.10a,b)

$$A_{\ell m} (SST) = R \sum_{k} \bar{F}_{\ell m}^{k} (I) \, e^{i\psi_{km}} \leq \sqrt{\sum_{k} (R \, \bar{F}_{\ell m}^{k})^2} = |A_{\ell m}(SST)|$$

Thus we can define as a <u>measure of separability</u> the ratio

$$P_{\ell m} = \frac{|A_{\ell m}(GRAVITY)|}{|A_{\ell m}(SST)|} \qquad (5.11)$$

A ratio $P_{\ell m} = 1$ implies complete dependence of SST and gravity part (no separation possible), $P_{\ell m} > 1$ indicates separability. Separability plots are given in Figures 5.2, 5.3, and 5.4 for the altimeter missions GEOSAT or SEASAT, ERS-1, and TOPEX/POSEIDON. We see that inherently SST and geoid can only be separated close to some resonance lines. Since these depend solely on the choice of the orbit and are not related in any way to the dynamics of the oceans the estimated SST models could be very misleading. In case an apriori ocean dynamic topography is employed, e.g. the Levitus model, this model can only be checked or improved in these resonance areas. In any event, never could wavelengths be identified that would enable us to adequately represent narrow features such as current boundaries.

233

We conclude that any approach of this type is perhaps valuable to demonstrate qualitatively the separability of geoid and SST at certain wavelengths. Actually, the precise computation of SST from altimetry - and thus of ocean surface circulation - has to wait till independent high quality geopotential models become available. A high resolution satellite gravity field mission, like ARISTOTELES, would be the solution to this problem, (Rummel, 1991).

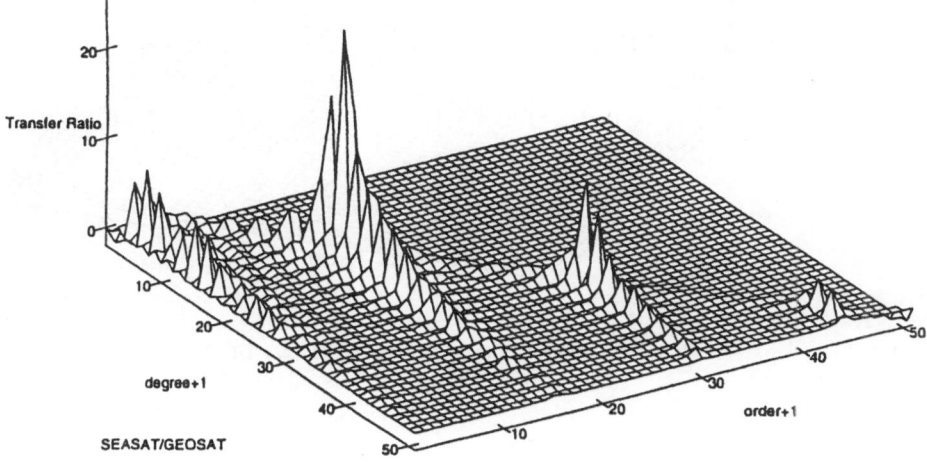

FIGURE 5.2: Separability of SST and geoid, case GEOSAT or SEASAT.

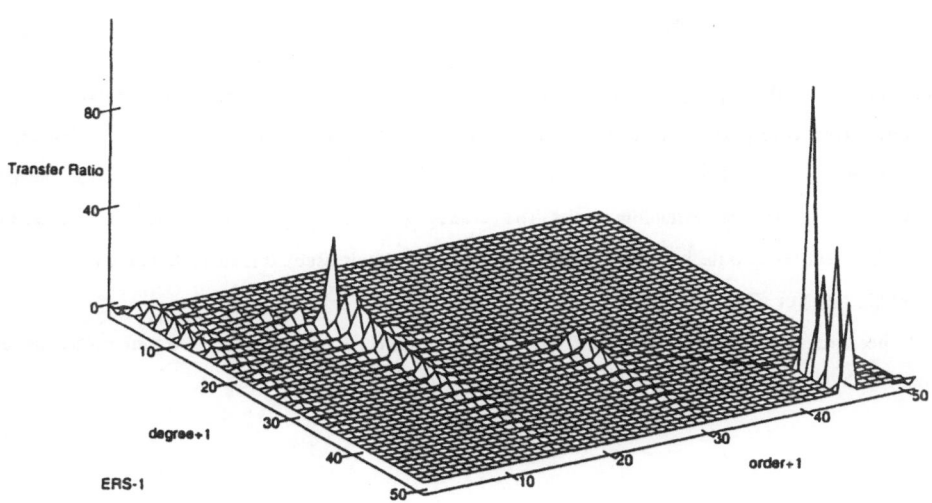

FIGURE 5.3: Separability of SST and geoid, case ERS-1.

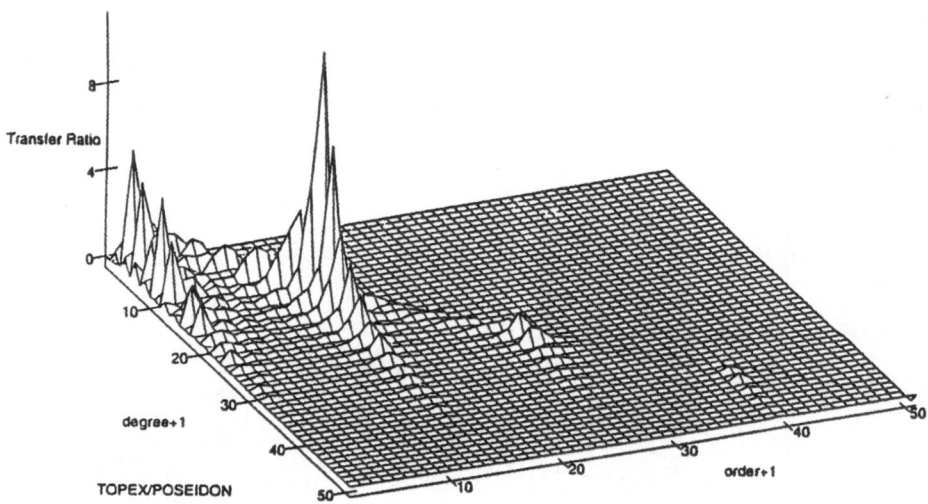

FIGURE 5.4: Separability of SST and geoid, case T/P.

6. Ocean variability.

So far only the determination of the ocean surface from altimetry has been considered. As long as the separation of the ocean topography from the geoid fails, due to the uncertainties in the earth's gravity field this path is only important for geodetic and geophysical studies but of limited value for oceanography. The uncertainty of the geoid plays no role when analysing the time variation of the ocean surface. This leads back to eq. (2.5)

$$\Delta \zeta = \zeta(t_2) - \zeta(t_1)$$
$$= h(t_2) - h(t_1) + r_s(t_2) - r_s(t_1) - (\rho(t_2) - \rho(t_1)) \qquad . \tag{2.5}$$

Basic requirement for the application of (2.5) is the repetition of altimeter measurements at one location. For this purpose either cross-over points can be considered or so-called <u>co-linear</u> or <u>repeat arcs</u>. In the latter case altimetric analysis is based on the periodic repetition of the complete ground track configuration. This makes the method particularly attractive for ocean variability studies. The oceanographic use of repeat measurements is discussed in the lectures by Zlotnicki, see also the parts by Cartwright and Wunsch (all this issue). It is also referred to (Fu et al., 1988), (Cheney et al., 1991) or (Zlotnicki, 1991).

Repeat arcs require that the orbit rate $\dot{\omega}_0 = \dot{M} + \dot{\omega}$ and earth rate $\dot{\omega}_e = \dot{\Omega} - \dot{\theta}$ are chosen as relative primes:

$$\frac{\dot{\omega}_0}{\dot{\omega}_e} = \frac{N_R}{N_D} \tag{6.1}$$

with N_R the full number of satellite revolutions and N_D that of full days, see (Colombo, 1984) or lectures Balmino. In addition the orbit must be almost spherical. Basically one wants an orbit that "bites its own tail" after a certain number of revolutions (and days). Ultimately any orbit (remaining at the same altitude) would do this. However in the case of altimetry the repeat period is very important and determined by oceanographic arguments. In order really to meet a chosen repeat requirement with high accuracy orbit corrections are regularly applied to the spacecraft. The above ratio was -43/3 for SEASAT and -244/17 for GEOSAT. This means, the SEASAT ground track pattern repeated itself every three days.

Again, as seen from (2.5), the radial orbit error is a major obstacle and has to be eliminated. The model is in this case

$$\Delta \bar{h}(t_2 - t_1) = \Delta r(t_2) - \Delta r(t_1) + \Delta \zeta(t_2 - t_1) + \vartheta_\rho(t_2 - t_1) \tag{6.2}$$

where the altimeter noise is, for example, assumed to be random and $\sqrt{2}$ times that of the individual altimeter measurement. We first look into the difference of radial orbit errors. From (6.1) it follows for the repeat period

$$\Delta t = N_R \frac{2\pi}{\dot{\omega}_0} = N_D \frac{2\pi}{\dot{\omega}_e} \quad . \tag{6.3}$$

The complete analytical model of the gravitational part of the radial orbit error is, see (4.7) and (5.5) (we keep the compact complex notation):

$$\Delta r(t) = \sum_\ell \sum_m \sum_k H_{\ell m k} c_{\ell m} e^{\kappa(\psi_{\ell m} + \dot{\psi}_{\ell m} t)} \tag{5.5}$$

We had

$$\dot{\psi}_{km} = k \dot{\omega}_0 + m \dot{\omega}_e \quad . \tag{6.4a}$$

If also a slight non-eccentricity of the orbit is taken into account a more accurate formula is

$$\dot{\psi}_{kmq} = k \dot{\omega}_0 + m \dot{\omega}_e - q \dot{\omega} \quad . \tag{6.4b}$$

In the case of repeat arcs it is possible to take the difference of $\Delta r(t)$ exactly after the repeat period. We find with (6.3) (see Schrama, 1989):

$$\dot{\psi}_{kmq}(t + \Delta t) = k \dot{\omega}_0 (t + N_R \frac{2\pi}{\dot{\omega}_0}) + m \dot{\omega}_e (t + N_D \frac{2\pi}{\dot{\omega}_e}) - q \dot{\omega}(t + \Delta t)$$

$$= \dot{\psi}_{kmq} t + 2\pi(kN_R + mN_D) - q \dot{\omega} \Delta t \quad . \tag{6.5}$$

Eq. (6.5) says that all gravitational orbit perturbations cancel, because $2\pi(kN_R + mN_D)$ is a shift by an integer number of full periods, except for a long periodic effect related to $-\dot{\omega}\Delta t$. With q being +1 or -1 the remaining term is of precession of perigee period which is small, $\dot{\omega} \approx 0$, (Colombo, 1984). Hence except for this remaining effect repeat arc differences act as an ideal filter of gravitational orbit errors.

There remains a second radial orbit error part, $\Delta r_0(t)$, related to uncertainties in the initial conditions. It can be modeled by (3.2) with Δa, Δe, $e\Delta M$ the errors in the Keplerian mean elements, or, following Hill's theory, as

$$\Delta r_0(t) = (-3\Delta z_0 - \frac{2}{n} \Delta \dot{x}_0)\cos \bar{n}t + \frac{\Delta \dot{z}_0}{n} \sin \bar{n}t +$$
$$+ (4\Delta z_0 + \frac{2}{n} \Delta \dot{x}_0) \quad . \tag{6.6}$$

In (6.6) Δz_0, $\Delta \dot{z}_0$ are the initial position and velocity errors in radial direction, and Δx_0, $\Delta \dot{x}_0$ those in along-track direction. This part does not filter out over the repeat period, as can easily be verified. For e.g. a 17 day repeat period the initial elements belong anyway to different computed arcs. However Δr_0 can be eliminated, on an arc by arc basis, by a simple least-squares fit with one of the three models of eqs. (3.3).

The processing of repeat or co-linear tracks is easiest demonstrated by Figure 6.1. It shows the same profile measured by SEASAT eight times at intervals of three days. First from all profiles a reference profile is subtracted, e.g. the average of all nine or simply the most complete one. Then the radial orbit error Δr_0 is eliminated by a least-squares fit. A time series of profiles results, that exhibits the slow motion of an oceanographic signal, e.g. an ocean eddy. If the procedure is terminated at this point a global or regional time series of the ocean surface can be constructed from all arcs. Alternatively the root mean square (rms.) value at each point along the profile over all arcs can be determined. It expresses the ocean variability. See (Cheney et al., 1983), (Fu, 1983), (Oskam, 1990), or (Wakker et al., 1990).

Variability from Cross-Overs: A completely different approach uses cross-overs. The time differences at cross-overs are very different from those between collinear tracks, the shortest interval being of the order of the orbit period, see ch. 4.2. Hence this technique can nicely complement the repeat arc method. It can also be employed to detect aliasing effects of the latter. Naturally the space sampling of the cross-over technique is much sparser.

In short the procedure runs as follows. First the radial orbit error is eliminated from the data by cross-over adjustment, either globally or in a diamond shaped region as described earlier. Then the area of investigation is subdivided into small diamond shaped blocks, typically of size 200 km x 200 km. In each block the remaining cross-over discrepancies down to the noise level are attributed to time variations of the ocean surface. Inside each block the variation is now modeled (compare eq. (6.2)) as

$$\Delta \bar{h}_{ad} = \zeta(t_a) - \zeta(t_d) + \vartheta_p$$

with a constant x_0 expressing each sea state

$$\zeta(t_a) = x_0(t_a) \quad and \quad \zeta(t_d) = x_0(t_d) \quad .$$

One reference sea level has to be fixed inside each block. This way, taking all cross-overs inside the block into account, a time series is established for the particular block. Smooth interpolation in time is produced by smoothing splines or least-squares prediction. Comparing the time tags of the reference levels of all blocks a time series can be established for the entire area under investigations. See (Fu & Chelton, 1985), (Milbert et al., 1988), (Feron, 1989).

A description of the collinear track and cross-over method and an analysis of the consistency of the two approaches is given in (Feron et al., 1992).

STEP 0: Elimination of spurious data, elimination of small data gaps by interpolation and synchronisation of all profiles by interpolation (e.g. at standard one second intervals relative to the time of equator crossing).

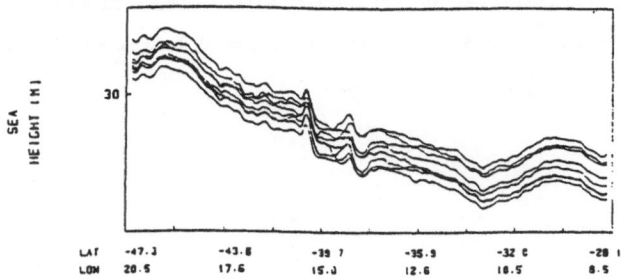

STEP 1: Subtraction of one arbitrary profile (e.g. the most complete one) or of the mean over all profiles.

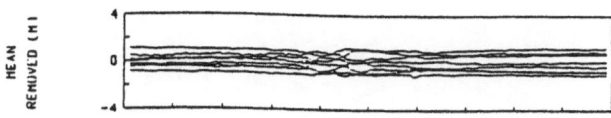

STEP 2: Removal of radial orbit error by least squares fit from the difference between each individual profile and the reference one. The chosen model (bias, bias + tilt, or bias + sine + cosine) depends on length of profile.

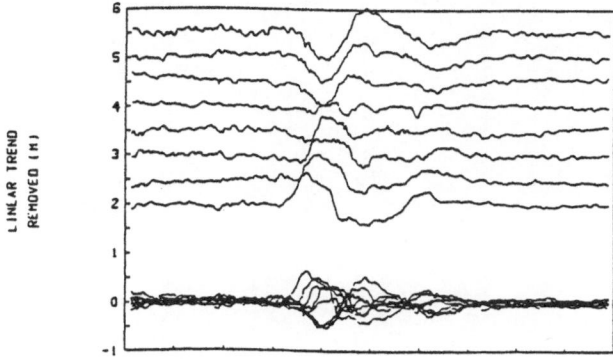

STEP 3: Computation of the ocean variability (rms. value over all profiles).

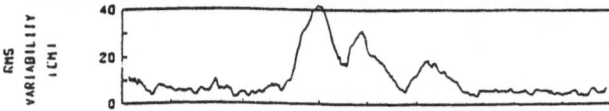

FIGURE 6.1: Computation steps of ocean variability determination

Acknowledgement.

These lectures benefit from the work of the Dutch Altimeter Team, coordinated by K. Wakker (faculty of Aerospace Engineering, TU Delft). Input from Dinos Danas, Raymond Feron, Roger Haagmans, Dick Oskam, Nico Sneeuw and in particular from E.J.O. Schrama is greatly appreciated. I thank Wil Coops, M.G.G.J. Jutte and D. Vukotic for their help in the preparation of these notes.

References.

Baarda, W.: S-transformation and Criterion Matrices, Netherlands Geodetic Commission, New Series, 5, 1, Delft, 1973.

Barzaghi, R., M. Brovelli, F.Sansò: Altimetry Rank Deficiency in Crossover Adjustment, in: Determination of the Geoid, eds. R.H. Rapp & F. Sansò, IAG Symposia 106, 108-118, Springer, New York 1991.

Bernstein, R.L. (ed): SEASAT Special Issue I, J.Geophys. Res, 87(05) 1982.

Bomford, G.: Geodesy, 4th edition, 1980.

Born, G.H. (ed): Special Issue on Seasat Ephemeris Analysis, Astronautical Sciences, 28(4) 1980.

Born, G.H. (ed.) : Theme Issue: Satellite Altimetry, Marine Geodesy, 8(1-4) 1984.

Cantafio, L.J.: Space-Based Radar Handbook, Artech House, Norwood, 1989.

Chelton, D.B.: WOCE/NASA Altimeter Algorithm Workshop, U.S. WOCE Technical Report, No. 2, U.S. Planning Office for WOCE, College Station, TX, 1988.

Cheney, R.E., J.G. Marsh, B.D. Beckley: Global Mesoscale Variability from Collinear Tracks of Seasat Altimeter Data, J. Geophys. Res., 88, 4343-4354, 1983.

Cheney, R.E., L. Miller: Recovery of the Sea Level Signal in the Western Tropical Pacific from Geosat Altimetry, J. Geophys. Res., 95(C3), 2977-2984, 1990.

Cheney, R.E., W.J. Emery, B.J. Haines, F. Wentz : Recent Improvements in Geosat Altimeter Data, EOS Transactions AGU, 72(51), 577-579, 1991.

Colombo, O.L.: Altimetry, Orbits and Tides, NASA Technical Memorandum 86180, Greenbelt, 1984.

Denker, H., R.H. Rapp: Geodetic and Oceanographic Results from the Analysis of 1 year of Geosat Data, J. Geophys. Res., C8, 13151-13168, 1990.

Emery, W.J., G.H. Born, D.G. Baldwin, Ch.L. Norris: Ionospheric Effects on Geosat Altimeter Observations, J. Geophys. Res., 95, G3, 2953-2964, 1990.

Engelis, T., R.H. Rapp: Global Ocean Circulation Patterns Based on Seasat Altimeter Data and the GEM L2 Gravity Field, Marine Geophys. Res., 7, 55-67, 1984.

Engelis, T.: Global Circulation from Seasat Altimeter Data, Marine Geodesy, 9, 45-69, 1985.

Engelis, T.: Radial Orbit Error Reduction and Sea Surface Topography Determination Using Satellite Altimetry, Dept. Geodetic Science and Surv., 377, Ohio State University, Colubus, 1987.

Feron, R.C.V., M.C. Naeije, D. Oskam: Quality of Ocean Variability Results from Satellite Altimetry, Mar. Geod., (accepted for publication), 1992.

Fu, L.-L.: Recent Progress in the Application of Satellite Altimetry to Observing the Mesoscale Variability and General Circulation of the Oceans, Rev. Geophys., 21, 1657-1666, 1983.

Fu, L.-L., D.B. Chelton: Observing Large-Scale Temporal Variability of Ocean Curents by Satellite Altimetry with Application to the Antartic Circumpolar Current, J. Geophys. Res., 90, C3, 4721-4739, 1985.

Fu, L.-L., D.B. Chelton, V. Zlotnicki: Satellite Altimetry: Observing Ocean Variability from Space, Oceanography, 1(2), 4-12, 1988.

Goad, C.C., B.C.Douglas, R.W. Agreen: On the Use of Satellite Altimeter Data for Radial Ephemeris Improvement, in: (Born, 1980), 419-428, 1980.

Gopalapillai, S.: Non-Global Recovery of Gravity Anomalies from a Combination of Terrestrial and Satellite Altimetry Data, Dept. Geodetic Science, 210, Ohio State University, Columbus, 1974.

Haagmans, R.H.N., M. van Gelderen: Error Variances-Covariances of GEM-T1: Their Characteristics and Implications in Geoid Computation, J. Geophys. Res., 96, B12, 20 011-20 022, 1991.

Hager, B.H., M.A. Richards: Long-wavelength Variations in Earth's Geoid: Physical Models and Dynamical Implications, In "Seismic Tomography and Mantle Circulation "(R.K. O'Nions and B. Parsons, eds.), 309-327 Royal Society London, 1989.

Ilk, K.H.: Ein Beitrag zur Dynamik ausgedehnter Körper - Gravitationswechselwirkung, DGK, C, 288, 1983.

Kaula, W.M.: Theory of Statistical Analysis of Data Distributed over a Sphere, Reviews of Geophysics, 5, 1, 83-107, 1967.

Knudsen, P.: Adjustment of Satellite Altimetry Data from Cross-Over Differenes Using Covariance Relations for the Time Varying Components Represented by Gaussian Functions, proc. IAG Symposia, 617-628, 1987.

Lambeck, K.: Geophysical Geodesy, Clarendon Press, Oxford, 1988.

Lorell, J., M.E. Parke, J.F. Scott: Geophysical Data Record (GDR) Users Handbook, Jet Propulsion Laboratory, 622-97 Pasadena, 1980.

Marsh, J.G., C.J. Koblinsky, F.J. Lerch, S.M. Klosko, J.W. Robbins, R.G. Williamson, G.B. Patel: Dynamic Sea Surface Topography, Gravity, and Improved Orbit Accuracies from the Direct Evaluations of Seasat Altimeter Data, J. Geophys. Res., 95(C8), 13129-13150, 1990.

Mather, R.S., C. Rizos, R. Coleman: Remote Sensing of Surface Ocean Circulation with Satellite Altimetry, Science, 205, 4401, 11-17, 1979.

Maul, G.A.: Introduction into Satellite Oceanography, Martinus Nijhof Publ., 1985.

Milbert, D., B. Douglas, R. Cheney, L. Miller, R. Agreen: Calculation of Sea Level Time Series from Noncollinear Geosat Altimeter Data, Mar. Geod., 12, 287-302, 1988.

Miller, L., R.E. Cheney, B.C. Douglas: Geosat Altimeter Observations of Kelvin Waves and the 1986-87 El Niño, Science, 239, 52-54, 1988.

Oskam, D.: Sea Surface Variability in the North Sea as Derived from Seasat Altimetry, Geophys. J. Int., 100, 1-7, 1990.

Rapp, R.H., Y.M. Wang, N.K. Pavlis: The Ohio State 1991 Geopotential and Sea Surface Topography Harmonic Coefficients Models, Dept. Geod. Sci. and Surveying, 332, The Ohio State University, Columbus, 1991.

Ray, R.D., C.J. Koblinsky, B.D. Beckley: On the Effectiveness of Geosat Altimeter Corrections, (unpublished manuscript) 1990. Robinson, I.S.: Satellite Oceanography, Ellis Horwood, Chichester, 1985.

Roemmich, D., C. Wunsch: On Combining Satellite Altimetry with Hydrographic Data, J. Marine Res. 40 Supplement 605-619, 1982.

Rosborough, G.W.: Satellite Orbit Perturbations Due to the Geopotential, Center for Space Research, Univ.Texas at Austin, Austin, 1986.

Rowlands, D.: The Adjustment of Seasat Altimeter Data on a Global Basis for Geoid and Sea Surface Height Determinations, Dept. Geodetic Science and Surv., 325, Ohio State University, Columbus, 1981.

Rummel, R., R.H. Rapp: Undulation and Anomaly Estimation Using GEOS-3 Altimeter Data Without Precise Satellite Orbits, Bull. Geod., 51, 73-88, 1977.

Rummel, R.: Satellite Altimetry as Part of a Geodetic Model, in proc.: 1st Hotine-Marussi Symposium on Mathematical Geodesy, vol.2, 757-786, Milano 1986.

Rummel, R.: On the Principle of Aristoteles, Proc. International Workshop: Solid-Earth Mission ARISTOTELES, Anacapri, 11-15, ESA Publications Division, ESTEC, Noordwijk, 1991.

Sandwell, D.T., D.G. Milbert, B.C. Douglas: Global Nondynamic Orbit Improvement for Altimetric Satellites, J. Geophys. Res., 91(B9), 9447-9451, 1986.

Schrama, E.J.O.: The Role of Orbit Errors in Processing of Satellite Altimeter Data, Neth. Geodetic Commission, New Series, 33, 1989.

Schrama, E.J.O., D. Oskam, R. Rummel: Geodätische Aspekte bei der Verarbeitung von Satelliten-altimetriedaten, ZfV, 1, 13-23, 1990.

Schrama, E.J.O.: Some Remarks on Several Definitions of Geographically Correlated Orbit Errors: Consequences for Satellite Altimetry, Manuscripta Geodaetica, 1992 (in print).

Schwiderski, E.W.: Ocean Tides, Part I: Global Ocean Tidal Equations, & Part II: A Hydrodynamical Interpolation Model, Mar. Geod., 3, 161-218, 1980.

Seasat Special Issue II : Scientific Results, J. Geophys. Res, 88 (C3) 1529-1952, 1983.

Seeber, G., J.R. Apel (eds.): Geodetic Features of the Ocean Surface and their Implications, (reprinted from Marine Geophysical Researches, 7(1/2)), Reidel, Dordrecht, 1984.

Seeber, G.: Satellitengeodäsie, de Gruyter, Berlin, 1989.

Shum, C.K., B.H. Zhang, B.E. Schutz, B.D. Tapley: Altimeter Crossover Methods for Precision Orbit Determination and the Mapping of Geophysical Parameters, J. Astronautical Sci., 38(3), 355-368, 1990.

Sneeuw, N.J.: Representation Coefficients and their Use in Satellite Geodesy, manuscripta geodaetica, 17, 117-123, 1992.

Special Section: Geosat Part II, J. Geophys. Res., 95(C3), 17865-18027, 1990.

Stewart, R.H.: Methods of Satellite Oceanography, Univ. of California Press, Berkeley, 1985.

Strang van Hees, G.L.: Variance-Covariance Transformations of Geodetic Networks, manuscripta geodaetica, 7, 120, 1982.

Tai, Ch.-K., C. Wunsch: An Estimate of Global Absolute Dynamic Topography, J. Physical Ocean., 14, 457-463, 1984.

Tai, Ch-K., L.-L. Fu: On Crossover Adjustment in Satellite Altimetry and Its Oceanographic Implications, J. Geophys.Res., 91 (C2), 2549-2554, 1986.

Tai, Ch.-K.: Accuracy Assessment of Widely Used Orbit Error Aproximations in Satellite Altimetry, J. Atmospheric and Oceanic Technology, G(1), 147-150, 1989.

Tapley, B.D., J.B. Lundberg, G.H. Born: The Seasat Altimeter Wet Tropospheric Range Correction Revisited, Mar. Geod., 8, 221-248. 1984.

Tapley, B.D., G. Rosborough: Geographically Correlated Orbit Error and its Effect on Satellite Altimetry Missions, J. Geophys. Res., 909 (C6), 11817-11831, 1985.

Tapley, B.D., R.S. Nerem, C.K. Shum, J.C.Ries, D.N. Yuan: Determination of the General Ocean Circulation from a Joint Gravity Field Solution, Geophys, Res. Lett., 15 (10), 1109-1112, 1988.

Tapley, B.D.: The Monitoring of Changes in Global Mean Sea Level Using Satellite Altimetry, Pres. at: Topex/Poseidon Science Working Team Meeting, Washington D.C., 1990.

Teunissen, P.J.G.: The Geometry of Geodetic Inverse Linear Mapping and Non-linear Adjustment, Netherlands Geodetic Commission, New Series, 8, 1, Delft, 1985.

Trupin, A., J. Wahr: Spectroscopic Analysis of Global Tide Gauge Sea Level Data, Geophys. J. Int., 100, 441-453, 1990.

Visser, P.N.A.M.: The Use of Satellites in Gravity Field Determination and Model Adjustment, Delft University Press, 1992.

Wagner, C.A., S.M. Klosko: Spherical Harmonics Representation of the Gravity Field from Dynamic Satellite Data, Planetary and Space Sciences, 30, 1, 5-28, 1981.

Wagner, C.A.: Radial Variations of a Satellite Orbit Due to Gravitational Errors: Implications for Satellite Altimetry, J. Geophys. Res., 90 (B4) 3027-3036, 1985.

Wagner, C.A.: Accuracy Estimate of Geoid and Ocean Topography Recovered Jointly from Satellite Altimetry, J. Geophys. Res. 91, B1, 453-461, 1986.

Wagner, C.A.: Comment on "Determination of the General Ocean Circulation from a Joint Gravity Field Solution", Geophys. Res. Lett., 16 (4) 335-336, 1989.

Wagner, C.A., E. Melchioni: On Using Precise Laser Ranges to Provide Vertical Control for Satellite Altimetric Surfaces, Manuscripta geodaetica, 14, 339-344, 1989.

Wagner, C.A.: Summer School Lectures on Satellite Altimetry, in: Theory of Satellite Geodesy and Gravity Field Determination, ed. F. Sansò & R. Rummel, 285-334, Springer, Berlin, 1989.

Wakker, K.F., R.C.A. Zandbergen, R. Scharroo, B.A.C. Ambrosius: Geosat Altimeter Data Analysis for the Oceans around South Africa, J. Geophys. Res. 95, C3, 2991-3006, 1990.

Wemelsfelder, P.J.: Sea Level Observations as a Fact and as an Illusion in: Report on the Symposium on Coastal Geodesy (ed. R. Sigl) 65-80, Munich, 1970.

Williamstown: The Terrestrial Environment: Solid Earth and Ocean Physics NASA CR-1579, 1979.

Wunsch, C., E.M. Gaposchkin: On Using Satellite Altimetry to Observing the Mesoscale Variability and General Circulation of the Oceans, Reviews of Geophysics and Space Physics, 18, 4, 725-745, 1980.

Wunsch, C., V. Zlotnicki: The Accuracy of Altimetric Surfaces, Geophys. J.R. astr. Soc., 78, 795-808, 1984.

Zandbergen, R.C.A.: Satellite Altimeter Data Processing: From Theory to Practice, Ph.-D. Thesis, Delft Univ. Technology, Delft, 1991.

Zlotnicki V.: Sea Level Differences Across the Gulf Stream and Kuroshio Extension, J. Phys. Oceanogr. 21(4), 599-609, 1991.

Orbit Choice and the Theory of Radial Orbit Error for Altimetry

George Balmino

Département de Géodésie Terrestre et Planétaire
Groupe de Recherches de Géodésie Spatiale
18, Avenue Edouard Belin, Toulouse, France

Our goal is to give the minimum of what is necessary to fully understand the gross evolution of current altimeter satellite orbits around the Earth, what are the major gravitational perturbations, how global geopotential models are determined, and how large are the radial errors induced by uncertainties in such models. This is certainly the most critical concern when designing or running a mission, and that is why quite detailed developments and explanations will be given on the spectrum of these errors, on their relationship with geopotential coefficient covariances, and finally on their geographically correlated characteristics.

This does not pretend to be a course of celestial mechanics, and we will use the most simple mathematical tools whenever possible, still preserving the rigour of the proofs.

1 INTRODUCTION AND BASIC CONCEPTS

The motion of an Earth artificial satellite is the motion of a body with very small mass and negligible dimensions with respect to the planet. For most forces acting on the satellite, and especially the gravitational forces, it is sufficient to approximate it by a point mass. Surface forces, which are more complex due to their nature, require special treatments in which the shapes and surface properties of all the spacecraft elements are modelled ; they will not be treated in detail in this course.

Other forces, such as the attraction of the Sun and Moon, the solid and fluid tidal effects will also be considered as being very small with respect to the main gravitational ones. Also, the equations of motion will be written, with sufficient approximations, in a reference system assumed to be fixed in space, corrections to this hypothesis being suitable of a treatment in terms of small perturbations. Finally, the actual law of forces will be written according to the classical mechanics ; the framework of general relativity is the rigourous one but differences with the classical approach are negligible for our purpose here.

Therefore, we consider that a massive point O, with mass M, exerts on a point mass S, of mass m, a force $\overline{F}_{O \to S} = -GmM\overline{OS}/OS^3$, (where G is the constant of gravitation and the overbar denotes a vector), and that the acceleration acquired by S is proportional to the force. This principle must however be written in a galilean reference system $(\mathcal{R})$. If we take a system of axis (Σ) centered at O, with fixed directions in space, we write :

$$m\left[\frac{d^2\overline{OS}}{dt^2} + \overline{\Gamma}_{O/(\mathcal{R})}\right] = -GmM\frac{\overline{OS}}{OS^3}$$

and :

$$M\overline{\Gamma}_{O/(\mathcal{R})} = GMm\overline{OS}/OS^3$$

Therefore :

$$\frac{d^2\overline{OS}}{dt^2} = -G(M+m)\frac{\overline{OS}}{OS^3}$$ (1)

Of course, in the case of an artificial satellite $m \ll M$ and $G\,(M+m)$ is replaced by GM.

1.1 THE UNPERTURBED SATELLITE ORBIT (two-body problem)

This is the orbit of S around a perfectly spherical Earth, of center of mass O. This point is one focus of the conic on which S moves ; in our case it will always be an ellipse (1.st Kepler's law). The closest point to O is the perigee P, the farthest is the apogee A, $\overline{OS}$ is the radius vector $\overline{r}$ of modulus r, the ellipse semi-major axis is a, its eccentricity is e, $p = a(1-e^2)$ is the parameter. S is positioned : either by the true anomaly $v = (\overline{OP},\overline{OS})$, or by the eccentric anomaly $E = (\overline{OP},\overline{OS'})$, where S' is on the principal circle (of which the ellipse is the affine transformed), or by the mean anomaly $M = (\overline{OS},\overline{OS''})$ where S'' moves on the principal circle with a uniform velocity at the motion period, T. In the plane of the orbit (fig. 1), we have for the radius r and coordinates x, y of S :

$$x = r\cos v = a(\cos E - e)$$
$$y = r\sin v = a\sqrt{1-e^2}\sin E$$ (2)
$$r = \frac{a(1-e^2)}{1+e\cos v} = a(1-e\cos E)$$

(from which $\sin v$, $\cos v$ can be expressed in terms of $\sin E, \cos E$, and vice versa). From these, other useful formulas are derived :

$$\text{tg}\frac{v}{2} = \sqrt{\frac{1+e}{1-e}}\,\text{tg}\frac{E}{2}$$
$$\text{tg}\frac{v-E}{2} = \frac{\beta\sin E}{1-\beta\cos E} = \frac{\beta\sin v}{1+\beta\cos v}$$ (3)
$$\beta = e/(1+\sqrt{1-e^2})$$
$$\sin v - \sin E = \beta\sin(v+E)$$

The main motion n is defined by $n = 2\pi/T$ and we have :

$$M = n(t - t_o)$$ (4)

t_o being an epoch when S passes through perigee.

Finally we have the second Kepler's law :

$$r^2\frac{dv}{dt} = na^2\sqrt{1-e^2} \qquad , \tag{5}$$

the kinetic energy integral from which the velocity V is such that :

$$V^2 = GM\left(\frac{2}{r}-\frac{1}{a}\right) \tag{6}$$

and Kepler's third law :

$$n^2a^3 = GM = \mu \tag{7}$$

Practically, in order to compute x, y from t, it is necessary to compute M, then E from the Kepler equation :

$$E - e\sin E = M \tag{8}$$

then to apply (2).

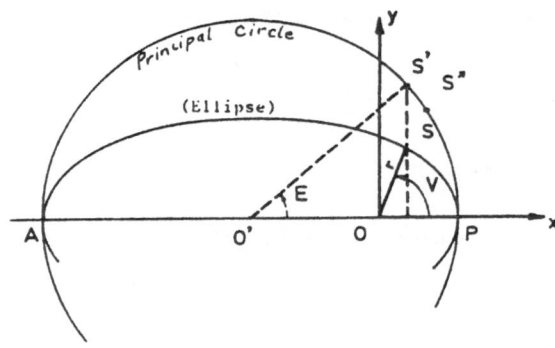

Fig. 1. The elliptical motion of S around O

Actually, the orbit lies in a plane which position in space depends on the initial conditions. The plane intersection with the (X, Y) plane of (Σ) is the line of nodes, the ascending node N being the point where S crosses the equatorial plane with Z increasing ; the longitude of N is $\Omega = (\overline{OX}, \overline{ON})$. The angle between the equatorial plane and the orbital plane is the inclination I, counted from $0°$ to $90°$ for direct motion and from $90°$ to $180°$ for retrograde motion. In the orbital plane, the direction of the perigee P is counted from N : $\omega = (\overline{ON}, \overline{OP})$ is the argument of perigee. These angles are shown on figure 2.

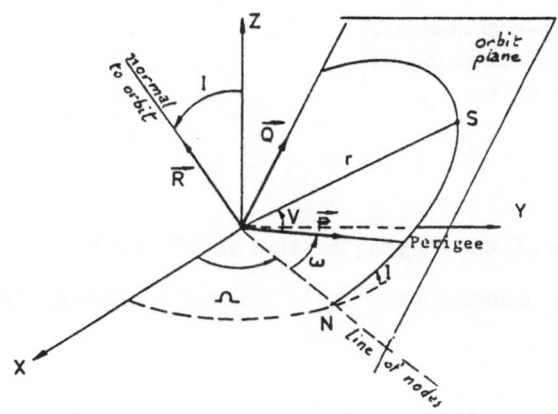

Fig. 2. The orbit in space

Let us call T the transformation : $(a, e, I, \Omega, \omega, M) \rightarrow (\bar{r}, \dot{\bar{r}})$.

With :

$$\bar{r} = [X_1 X_2 X_3]^+ = [XYZ]^+$$

we have :

$$\bar{r} = R_3(\Omega) R_1(I) R_3(\omega)[xyo] \qquad (9)$$

where :

$$R_1(\alpha) = \begin{pmatrix} 1 & 0 & 0 \\ 0 & c & -s \\ 0 & s & c \end{pmatrix}, \quad R_2(\alpha) = \begin{pmatrix} c & 0 & s \\ 0 & 1 & 0 \\ -s & 0 & c \end{pmatrix}, \quad R_3(\alpha) = \begin{pmatrix} c & -s & 0 \\ s & c & 0 \\ 0 & 0 & 1 \end{pmatrix}$$

$(c = \cos \alpha, s = \sin \alpha)$

x, y are of course given by (2).

The second part of the transformation is given by :

$$\dot{\bar{r}} = n \frac{a^2}{r}(-\sin E \bar{P} + \cos E \sqrt{1-e^2} \bar{Q}) \qquad (10)$$

where :

$$\overline{P} = \begin{pmatrix} \cos\Omega\cos\omega - \cos I \sin\omega\sin\Omega \\ \sin\Omega\cos\omega + \cos I \sin\omega\cos\Omega \\ \sin I \sin\omega \end{pmatrix}$$

and :

$$\overline{Q} = \partial\overline{P}/\partial\omega$$

(to derive (10) from (9), one uses the fact that $dE/dt = n\,a/r$).

The inverse transformation T^{-1} can be achieved in different ways. One is the following :

. $a = (2/r - \dot{r}^2/\mu)^{-1}$

. compute $\overline{C} = \overline{r} \times \dot{\overline{r}}$ (angular momentum) ; $|\overline{C}| = C$ is constant

. $I = \cos^{-1}(C_3/C)$

. $k\sin\Omega = C_1, k\cos\Omega = -C_2$ with $k = (C_1^2 + C_2^2)^{1/2}$: define Ω.

. compute $p = C\sqrt{\mu a} \Rightarrow e = (1 - p/a)^{1/2}$

. compute $\mathrm{tg}\,\gamma = \overline{r}.\dot{\overline{r}}/[\mu a(1-e^2)]^{1/2}$, then E is defined by : $e\sin E = \mathrm{tg}\,\gamma(1-e^2)^{1/2}$, $e\cos E = 1 - r/a$, and $M = E - e\sin E$

. compute v from : $re\sin v = p\,\mathrm{tg}\,\gamma, re\cos v = p - r$
  then $(\omega + v)$ from : $r\sin I \sin(\omega + v) = X_3, r\sin I \cos(\omega + v) = (X_1\cos\Omega + X_2\sin\Omega)\sin I \Rightarrow \omega$.

For future usage, we now need some series expansions of a few functions of the two-body problem. Such expansions may be viewed as power series in e, or as Fourier series in M depending on the utilization. By an application of a theorem by Lagrange applied to Kepler equation, it can be shown that any function $F(E)$ of the eccentric anomaly E may be expanded as :

$$F(E) = F(M) + \sum_{n=1}^{\infty} \frac{e^n}{n!} \frac{d^{n-1}}{dM^{n-1}}[F'(M)\sin^n M] \tag{11}$$

which is convergent for $e < e_0$, e_0 being defined by $1 + (1 + e_0^2)^{1/2} = e_0\exp(1 + e_0^2)^{1/2}$, that is $e_0 = 0.6627...$

We then find, for example :

$$E = M + \sum_{n=1}^{\infty} \frac{e^n}{2^{n-1}n!} \sum_{j=0}^{(n-1)/2}(-1)^j\binom{n}{j}(n-2j)^{n-1}\sin(n-2j)M \tag{12}$$

$$\frac{r}{a} = 1 - e\cos M - \frac{e^2}{2}(\cos 2M - 1) - \sum_{n=3}^{\infty} \frac{e^n}{2^{n-1}(n-1)!} \sum_{j=0}^{(n-1)/2}(-1)^j\binom{n}{j}(n-2j)^{n-2}\cos(n-2j)M \tag{13}$$

It is obvious that, since the motion is periodic, any function of coordinates is 2π-periodic in M. For example, $\cos E$ can be expanded as a cosine series in M :

$$\cos E = a_o + \sum_{p=1}^{\infty} a_p \cos pM$$

with :

$$a_o = \frac{1}{2\pi} \int_o^{2\pi} \cos E \, dM \; , \; a_p = \frac{1}{\pi} \int_o^{2\pi} \cos E \cos pM \, dM$$

From $E - e \sin E = M$ we derive $dM = (1 - e \cos E)dE$; therefore : $a_o = -e/2$. For a_p we note that $\cos pM = [d(\sin pM)/dM]/p$ and we integrate by part, finding :

$$a_p = -\frac{1}{p\pi} \int_o^{2\pi} \sin pM \, d(\cos E)$$

Replacing M by $E - e \sin E$ noting the invariance of the integral limits, we find :

$$a_p = \frac{1}{p} \frac{1}{2\pi} \int_o^{2\pi} [\cos(p-1)E - pe \sin E] dE - \frac{1}{p} \frac{1}{2\pi} \int_o^{2\pi} [\cos(p+1)E - pe \sin E] dE$$

We define the Bessel function of the first kind and of order n as :

$$J_n(x) = \frac{1}{2\pi} \int_o^{2\pi} \cos(nt - x \sin t) dt \tag{14}$$

So that :

$$a_p = [J_{p-1}(pe) - J_{p+1}(pe)]/p$$

and finally :

$$\cos E = -\frac{e}{2} + \sum_{p=1}^{\infty} \frac{1}{p} [J_{p-1}(pe) - J_{p+1}(pe)] \cos pM \tag{15}$$

A property of the Bessel functions is that :

$$J_{p-1}(x) - J_{p+1}(x) = 2J'_p(x) \tag{16}$$

which simplifies a little the form of (15).

From this it is easy to find that :

$$\frac{r}{a} = 1 + \frac{e^2}{2} - 2e \sum_{p=1}^{\infty} \frac{1}{p^2} \frac{d}{de} [J_p(pe)] \cos pM \tag{17}$$

Other useful formulas are :

$$v = M + 2 \sum_{p=1}^{\infty} \frac{1}{p} \left\{ J_p(pe) + \sum_{n=1}^{\infty} \beta^n [J_{p-n}(pe) + J_{p+n}(pe)] \right\} \sin pM \tag{18}$$

(β : as given in (3)).

$$\cos v = -e + 2\left(\frac{1-e^2}{e}\right) \sum_{p=1}^{\infty} J_p(pe) \cos pM \tag{19}$$

$$\sin v = 2\sqrt{1-e^2} \sum_{p=1}^{\infty} \frac{1}{p} \frac{d[J_p(pe)]}{de} \sin pM \tag{20}$$

More general formulas will be needed in the expression of the gravity field perturbations, for functions of the form $(r/a)^n \cos mv$ and $(r/a)^n \sin mv$ for any n and m, positive or negative. In complex form we write :

$$\left(\frac{r}{a}\right)^n \exp(imv) = \sum_{k=-\infty}^{+\infty} X_k^{nm} \exp(ikM) \tag{21}$$

This defines the Hansen coefficients. They are real functions of e, and can be evaluated in many different ways. Since X_k^{nm} is a Fourier coefficient, we have :

$$X_k^{nm} = \frac{1}{2\pi} \int_{-\pi}^{+\pi} \left(\frac{r}{a}\right)^n \exp(imv) \exp(-ikM) dM$$

Introducing $z = \exp(iE)$, we find that $\exp(iM) = z \exp[-e(z-1/z)/2]$, $dM = -i[1-(e/2)(z+1/z)]dz/z$, $r/a = (1-\beta/z)(1-\beta z)/(1+\beta^2)$, and $\exp(iv) = z(1-\beta/z)/(1-\beta z)$.

Finally :

$$X_k^{nm} = \frac{1}{(1+\beta^2)^{n+1}} \frac{1}{2i\pi} \oint z^{m-k-1} \left(1-\frac{\beta}{z}\right)^{n+m+1} (1-\beta z)^{n-m+1} \cdot \exp\left[\frac{ke}{2}\left(z-\frac{1}{z}\right)\right] dz$$

From this, the following series expansion can be found (after some laborious algebra !) : for $k = m+s$, with $s \geq o$

$$X_{m+s}^{n,m}(e) = (-1)^s \left(\frac{e}{2}\right)^s \sum_{t=o}^{\infty} \left\{ \sum_{j=o}^{t} \sum_{p=o}^{\infty} \binom{n+m+1}{j-p} \frac{(m+s)^p}{p!} \sum_{q=o}^{s+j} \binom{n-m+1}{s+j-q} \frac{(m+s)^q}{q!} (-1)^q \right.$$

$$\left. \left[2\binom{2t-n+s-p-q-2}{t-j} - \binom{2t-n+s-p-q-1}{t-j} \right] \right\} \left(\frac{e}{2}\right)^{2t} \tag{22}$$

If $s < o$, we compute $X_{m+s}^{nm} = X_{-m-s}^{n,-m}$ by the same formula, using the property of symmetry of the Hansen functions ($X_k^{nm} = X_{-k}^{n,-m}$).

In the above formula, binomial coefficients $\binom{-\mu}{p}$, where $\mu \geq o$, must be computed as being equal to $(-1)^p \binom{\mu+p-1}{p}$, p being always positive.

Another expression for the Hansen coefficients is :

$$X_k^{nm}(e) = \frac{1}{(1+\beta^2)^{n+1}} \sum_{p=0}^{\infty} \sum_{q=0}^{\infty} \binom{n+m+1}{p} \binom{n-m+1}{q} (-\beta)^{p+q} J_{k-m+p-q}(ke) \qquad (23)$$

which can be more economical to evaluate than (22). From (22) and from the symmetry property, it is obvious that $X_k^{nm}(e) = o(e^{|k-m|})$.

For large values of the n, m, k indices, it is numerically more efficient and precise to compute the Hansen coefficients by Fourier transform, from their definition (formula 21).

1.2 DISTURBING FORCES ON AN ARTIFICIAL SATELLITE

These forces are of different types :

- gravitational forces : first, and most important for our purpose here, are the forces due to the non-sphericity of the Earth in the general sense (geometrical form, internal density distribution). The main term is related to the flattening of our planet, the others describe all lateral density variations. This will be the whole subject of chapter 2. We then have :

  . the third-body perturbations due to the Moon, the Sun, and the closest and/or biggest planets,

  . the tidal forces of various origins : solid tides due to the global yielding of the elastic Earth to the disturbing forces of the Sun and Moon ; ocean tides with numerous frequencies and varied amplitudes.

- surface forces :

  . atmospheric drag, which acts in a very complex way due to the variations of the atmosphere density under the action of the sun (solar cycle, yearly, seasonal, monthly, daily and hourly variations do exist due to the sun activity, geomagnetic effects and induced chemical reactions), also due to the complex shapes of satellites and the nature of their surface elements,

  . radiation pressures : there is the direct solar pressure but also the one coming from the reradiation of the sun light by the Earth (albedo effect, the most complex since it is related to the cloud coverage), plus the infrared radiation of the Earth (considered as a black body), all requiring a careful modeling of the spacecraft components.

Finally, correction terms to the total acceleration of the satellite must be added to account for the correct relativistic description of the equations of motion and, if the reference frame in which these equations are written is moving, apparent accelerations have to be included.

1.3 EQUATIONS OF PERTURBED MOTION (LAGRANGE, GAUSS, HILL)

We have already written the cartesian equations of motion, (1), in the reference frame (Σ) in the case of the two body problem. With initial conditions $(\bar{r}_o, \dot{\bar{r}}_o)$ at t_o, this is a system of ordinary differential equations which solution is uniquely defined. Actually, (1) is equivalent to the system :

$$\frac{d\bar{\alpha}}{dt} = A(\bar{\alpha}, t)$$

with :

$$\bar{\alpha} = [a, e, I, \Omega, \omega, M]^+$$

$$A = [o, o, o, o, o, n_o]^+ \quad , \quad n_o = (\mu/a_o^3)^{1/2}$$

a_o being computed from $(\bar{r}_o, \dot{\bar{r}}_o)$

This simply reflects the fact that, if one transforms the system (1) by $(\mathcal{T})$ - given by (9) and (10), one finds the system for $\bar{\alpha}$. It is therefore quite natural, when introducing disturbing accelerations which are very small with respect to μ/r^2, to use the same transformation, hoping that the solution of the transformed system will be expressed as small variations around the solution of the two body problem, that is around

$$\bar{\alpha}_o = [a_o, e_o, I_o, \Omega_o, \omega_o, M_o + n_o(t - t_o)]^+ \quad .$$

Let us write the equations of motion including the disturbing accelerations $\bar{\gamma}$ (one will often say disturbing "forces") as :

$$\ddot{\bar{r}} = -\mu\bar{r}/r^3 + \bar{\gamma} \tag{24}$$

with $\bar{r}(t_o) = \bar{r}_o$, $\dot{\bar{r}}(t_o) = \dot{\bar{r}}_o$.

We assume that the right-hand side member satisfies conditions such that (24) has always one and only one solution $[\bar{r}(t), \dot{\bar{r}}(t)]$ for $|t - t_o|$ large enough for our application (Cauchy-Arzela conditions).

At any time t in this interval, we can therefore apply the transformation $\mathcal{T}^{-1}$ to the solution of (24) and we get quantities $a(t), e(t), I(t), \Omega(t), \omega(t), M(t)$ which are no longer constant (or linear in time for M). These are called osculating elements. Their physical meaning is simple ; if, for $t' > t$ we suppress $\bar{\gamma}$, then the satellite motion obeys the system :

$$t' > t : \ddot{\bar{r}} = -\mu\bar{r}/r^3, \text{ with } \bar{r}(t) = \bar{r} \quad , \quad \dot{\bar{r}}(t) = \dot{\bar{r}} \quad ,$$

of which the solution is : $a(t') = a(t), \ e(t') = e(t), \ I(t') = I(t), \ \Omega(t') = \Omega(t), \ \omega(t') = \omega(t),$ $M(t') = M(t) + [\mu/a(t)^3]^{1/2}(t - t')$, that is a keplerian ellipse. This ellipse passes through the point

$S(t)$ of radius vector $\bar{r}$ and a mobile on it has the same velocity vector $\dot{\bar{r}}$, but the acceleration is different by construction (that is the term "osculating" is improper from the geometrical viewpoint).

Thus, using the variables $a\,(t)$, $e\,(t)$,... $M\,(t)$ allows to visualize the trajectory evolution (e.g. rotation of the plane, of the line of apside : apogee-perigee,...). Now, we want to deduce from (24) and from the formulas for $\mathcal{T}$ and $\mathcal{T}^{-1}$ the system verified by the osculating elements, which must be of the form :

$$\frac{d\bar{\alpha}}{dt} = \text{function } of\ \bar{\alpha} \text{ and } \bar{\gamma}$$

with :

$$\bar{\alpha}(t_o) = \bar{\alpha}_o = \mathcal{T}^{-1}(\bar{r}_o, \dot{\bar{r}}_o)$$

The perturbing acceleration is projected on the mobile reference system axis : $(\hat{r}, \hat{s}, \hat{w})$ defined by : $\hat{r} = \bar{r}/r$, $\hat{s}$: unit vector orthogonal to $\hat{r}$ in the osculating plane and in the direction of $\dot{\bar{r}}$, $\hat{w} = \hat{r} \times \hat{s}$; that is $\hat{w} = (\bar{r} \times \dot{\bar{r}})/|\bar{r} \times \dot{\bar{r}}|$, and $\hat{s} = \hat{w} \times \hat{r}$ (fig. 3).

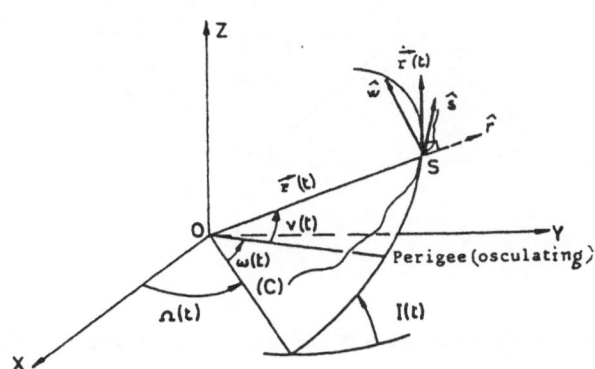

Fig. 3. The Gauss mobile system $(\hat{r}, \hat{s}, \hat{w})$

So : $\bar{\gamma} = R\hat{r} + S\hat{s} + W\hat{w}$. Now, by derivation of (6) with respect to time and since $r\dot{r} = \bar{r}.\dot{\bar{r}}$ (from $r^2 = \bar{r}.\bar{r}$), we readily find :

$$(\mu/a^2)\dot{a} = 2\dot{\bar{r}}.\bar{\gamma}$$

Rewriting (10) as $\dot{\bar{r}} = na(1-e^2)^{-1/2}[e\sin v\hat{r} + (1+e\cos v)\hat{s}]$, obtained by a rotation around $\hat{w}$ with angle v, we obtain :

$$\dot{a} = (2/n)(1-e^2)^{-1/2}[e\sin v\mathbf{R} + (1+e\cos v)\mathbf{S}]$$

We then use the angular momentum vector $\overline{C} = C\hat{w}$, with $C = [\mu a(1-e^2)]^{1/2} = (\mu p)^{1/2}$; $\overline{C}$ verifies $d\overline{C}/dt = \bar{r} \times (-\mu\bar{r}/r^3 + \bar{\gamma}) = \bar{r} \times \bar{\gamma} = 1/2(\mu/p)^{1/2}\dot{p}\hat{w} + (\mu p)^{1/2}\dot{\hat{w}}$.

We define the following unit vectors : $\overline{N}$ in the ascending node direction, $\overline{N}'$ orthogonal to $\overline{N}$ in the equatorial plane, $\overline{M}$: orthogonal to $\overline{N}$ in the (osculating) orbit plane, and $\bar{i}, \bar{j}, \bar{k}$: unit vectors of the (Σ) frame (fig. 4).

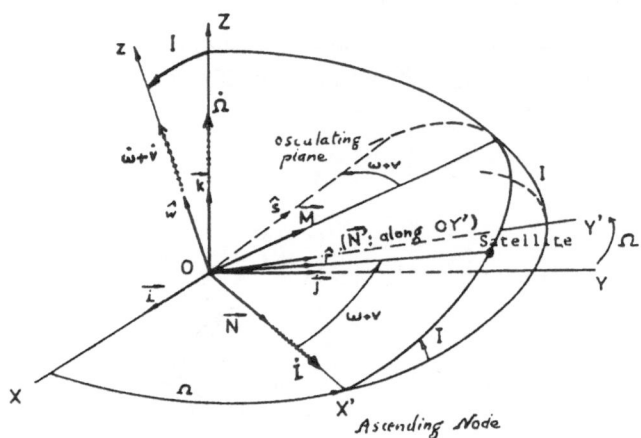

Fig. 4. The intermediate vectors $\overline{N}, \overline{N}', \overline{M}$ introduced for the equations for e, I, Ω

Writing $\bar{r} \times \bar{\gamma} = -r\mathbf{W}\hat{s} + r\mathbf{S}\hat{w}$, $\hat{w} = \overline{N} \times \overline{M}$, $\overline{M} = \overline{N}'\cos I + \bar{k}\sin I$, noting that $\overline{N}$ depends only on Ω, that $\dot{\overline{N}}' = \dot{\Omega}d\overline{N}'/d\Omega = -\overline{N}\dot{\Omega}$, also that $\dot{\overline{N}} = \overline{N}'\dot{\Omega}$, and taking account that $\overline{N}' \times \bar{k} = \overline{N}$, $\overline{N} \times \overline{N}' = \bar{k}$, $\overline{N} \times \bar{k} = -\overline{N}'$, we obtain :

$$\dot{\hat{w}} = \dot{\Omega}\sin I\overline{N} - \dot{I}\overline{M}$$

Noting also that $\hat{r} = \overline{N}\cos(\omega+v) + \overline{M}\sin(\omega+v)$, $\hat{s} = -\overline{N}\sin(\omega+v) + \overline{M}\cos(\omega+v)$, and equating the components of $d\overline{C}/dt$ on $\hat{r}, \hat{s}, \hat{w}$, we find three equations for $\dot{p} = \dot{a}(1-e^2) - 2ae\dot{e}$, $\dot{I}$ and $\dot{\Omega}$.

The equation for $\dot{M}$ is obtained through : $r = r(a,e,M)$ which implies :

$\dot{r} = \dot{a} \partial r / \partial a + \dot{e} \partial r / \partial e + \dot{M} \partial r / \partial M$

It is easy to find $\dot{r} = nae \sin v / (1 - e^2)^{1/2}$, $\partial r / \partial a = r/a$, $\partial r / \partial E = ae \sin E$, $\partial E / \partial e = a \sin E / r$, $\partial E / \partial M = a/r$, $\partial r / \partial e = -a \cos v$, $\partial r / \partial M = ae \sin v / (1 - e^2)^{1/2}$, and this yields the equation for $\dot{M}$.

The last equation, for $\dot{\omega}$, is more tricky. We start from $\psi = \omega + v$, and from the second Kepler's law : $\dot{\psi} = na^2 (1 - e^2)^{1/2} / r^2$ which is valid in the osculating motion if ψ is counted from a fixed direction. But, in the real motion, all elements vary and the direction ON from which one would like to count ψ varies too ! Therefore, we cannot write $\dot{\psi} = \dot{\omega} + \dot{v}$ if we apply Kepler's law. We derive $\dot{\psi}$ directly from $\mathrm{tg}\,\psi = \eta / \xi$, that is $d\psi = (\xi d\eta - \eta d\xi) / (\xi^2 + \eta^2)$, where $\overline{OS} = \xi \overline{N} + \eta \overline{M}$. For getting $d\xi$, $d\eta$, we compute $d\overline{OS}/dt$ when Ω, $\omega + v$ and I vary ; in this case :

$$d\overline{OS}/dt = [\dot{\Omega}\overline{k} + \dot{I}\overline{N} + (\dot{\omega} + \dot{v})\hat{w}] \times \overline{OS}$$

Writing this equality in $(\overline{N}, \overline{M}, \hat{w})$ with $\overline{k} = \sin I \overline{M} + \cos I \hat{w}$, we arrive at :

$$\dot{\xi} = -r \sin(\omega + v)(\dot{\omega} + \dot{v} + \dot{\Omega} \cos I)$$

$$\dot{\eta} = r \cos(\omega + v)(\dot{\omega} + \dot{v} + \dot{\Omega} \cos I)$$

(the 3rd equation would give $\dot{\zeta} = $ (component of $d\overline{OS}/dt$ on $\hat{w}$) $= \eta \dot{I} - \xi \sin I \dot{\Omega}$) ; we then find :

$$d\psi = d\omega + dv + d\Omega \cos I$$

Consequently :

$$\dot{\omega} = na^2 (1 - e^2)^{1/2} / r^2 - \dot{v} - \dot{\Omega} \cos I$$

$\dot{v}$ is computed as $dv(e, M)/dt = \dot{e} \partial v / \partial e + \dot{M} \partial v / \partial M$, with $\partial v / \partial e = \sin v [a/r + 1/(1 - e^2)]$, $\partial v / \partial M = (1 - e^2)^{1/2} a^2 / r^2$. The equation for $\dot{\omega}$ follows...

We now summarize the six equations, known as Gauss equations, which are obtained by the elementary manipulations shown above :

$$\dot{a} = 2[Re \sin v + S(1 + e \cos v)] / (nf)$$

$$\dot{e} = f[R \sin v + S(\cos E + \cos v)] / (na)$$

$$\dot{I} = Wr \cos(\omega + v) / (na^2 f \sin I) \tag{25}$$

$$\dot{\Omega} = Wr \sin(\omega + v) / (na^2 f)$$

$$\dot{\omega} = f[-R \cos v + S(1 + (1 + e \cos v)^{-1}) \sin v$$

$$- Wr \cos I \sin(\omega + v) / (na^2 f \sin I)] / (nae)$$

$$\dot{M} = n + f^2 \{R[-2e/(1 + e \cos v) + \cos v] - S[1 + (1 + e \cos v)^{-1}] \sin v\} / (nae)$$

(here : $f = \sqrt{1 - e^2}$).

Next, we will derive the Lagrange equations. They are a particular case of the Gauss equations when the disturbing acceleration $\bar{\gamma}$ is the gradient of a function $\mathcal{R}$(force function) : $\bar{\gamma} = \overline{\nabla}\mathcal{R}$. This is the case of all forces of gravitational origin and this leads to a simpler differential system. In the $(\hat{r}, \hat{s}, \hat{w})$ system we can write :

$$d\mathcal{R} = \overline{\nabla}\mathcal{R} \cdot d\overline{OS} = Rdr + Srd\psi + Wd\zeta$$

or, for any orbital element α :

$$\frac{\partial \mathcal{R}}{\partial \alpha} = R\frac{\partial r}{\partial \alpha} + Sr\frac{\partial \psi}{\partial \alpha} + W\frac{\partial \zeta}{\partial \alpha},$$

where $(dr, rd\psi, d\zeta)$ are the orthogonal components of $d\overline{OS}$. Clearly, dr can be due to changes da, de, dM only and we have : $\partial r/\partial a = r/a$, $\partial r/\partial e = -a\cos v$, $\partial r/\partial M = ae\sin v/(1-e^2)^{1/2}$, already used above. Similarly, $d\psi$ can result from changes in $\Omega, \omega,$ and v as shown in the derivation of the Gauss equation for ω, and we have : $\partial\psi/\partial\omega = 1$, $\partial\psi/\partial\Omega = \cos I$, $\partial\psi/\partial v = 1$; and since v is function of e and M, $\partial\psi/\partial e = \sin v[a/r + (1-e^2)^{-1}]$, $\partial\psi/\partial M = (a/r)^2(1-e^2)^{1/2}$. Finally, we already obtained $d\zeta = \eta dI - \xi \sin I d\Omega$, from which $\partial\zeta/\partial I = \eta = r\sin(\omega+v)$ and $\partial\zeta/\partial\Omega = -\xi\sin I = -r\cos(\omega+v)\sin I$.

All the other partial derivatives are equal to zero. Therefore we have found :

$$\frac{\partial \mathcal{R}}{\partial a} = \frac{r}{a}R$$

$$\frac{\partial \mathcal{R}}{\partial e} = -Ra\cos v + S\sin v\left(a + \frac{r}{1-e^2}\right)$$

$$\frac{\partial \mathcal{R}}{\partial I} = r\sin(\omega+v)W \tag{26}$$

$$\frac{\partial \mathcal{R}}{\partial \Omega} = rS\cos I - rW\cos(\omega+v)\sin I$$

$$\frac{\partial \mathcal{R}}{\partial \omega} = rS$$

$$\frac{\partial \mathcal{R}}{\partial M} = R\frac{ae\sin v}{\sqrt{1-e^2}} + S\frac{a^2}{r}\sqrt{1-e^2}$$

We now transform the Gauss equations one by one. For $\dot{a}$, we replace $1 + e\cos v$ by $a(1-e^2)/r$ and relate the right-hand side to $\partial\mathcal{R}/\partial M$. For $\dot{e}$, we note that $\cos E + \cos v = [a(1-e^2)/r - r/a]/e$, we have R $\sin v$ in terms of $\partial\mathcal{R}/\partial M$ and S, which we replace by $(1/r)\partial\mathcal{R}/\partial\omega$. $\dot{I}$ is obtained from $\partial\mathcal{R}/\partial\Omega$ and $\partial\mathcal{R}/\partial\omega$. $\dot{\Omega}$ is immediately written in terms of $\partial\mathcal{R}/\partial I$. The first two terms in the bracket for $\dot{\omega}$ equals $(1/a)\partial\mathcal{R}/\partial e$ and the last one is proportional to $\partial\mathcal{R}/\partial I$. For $\dot{M}$, we first express S and its factor in terms of R and $\partial\mathcal{R}/\partial e$ and then replace $(1-e^2)R/(1+e\cos v)$ by $(r/a)R = \partial\mathcal{R}/\partial a$.

Finally, the six Lagrange equations are :

$$\frac{da}{dt} = \frac{2}{na}\frac{\partial \mathcal{R}}{\partial M}$$

$$\frac{de}{dt} = \frac{1-e^2}{na^2e}\frac{\partial \mathcal{R}}{\partial M} - \frac{\sqrt{1-e^2}}{na^2e}\frac{\partial \mathcal{R}}{\partial \omega}$$

$$\frac{dI}{dt} = \frac{\cos I}{na^2\sqrt{1-e^2}\sin I}\frac{\partial \mathcal{R}}{\partial \omega} - \frac{1}{na^2\sqrt{1-e^2}\sin I}\frac{\partial \mathcal{R}}{\partial \Omega}$$

$$\frac{d\Omega}{dt} = \frac{1}{na^2\sqrt{1-e^2}\sin I}\frac{\partial \mathcal{R}}{\partial I} \qquad\qquad (27)$$

$$\frac{d\omega}{dt} = \frac{\sqrt{1-e^2}}{na^2e}\frac{\partial \mathcal{R}}{\partial e} - \frac{\cos I}{na^2\sqrt{1-e^2}\sin I}\frac{\partial \mathcal{R}}{\partial I}$$

$$\frac{dM}{dt} = n - \frac{2}{na}\frac{\partial \mathcal{R}}{\partial a} - \frac{1-e^2}{na^2e}\frac{\partial \mathcal{R}}{\partial e}$$

The form of this system is remarquable. If we take $\sigma = M - nt$ instead of M, we have $\partial \mathcal{R}/\partial M = \partial \mathcal{R}/\partial \sigma$ and :

$$\frac{d}{dt}[a,e,I,\Omega,\omega,\sigma]^+ = \mathcal{M}(a,e,I)[\mathcal{R}'_a, \mathcal{R}'_e, \mathcal{R}'_I, \mathcal{R}'_\Omega, \mathcal{R}'_\omega, \mathcal{R}'_M]^+$$

where $\mathcal{R}'_a = \partial \mathcal{R}/\partial \alpha$ and where n is replaced by $(\mu/a^3)^{1/2}$. $\mathcal{M}$ is an antisymmetric matrix with only ten non-zero elements. The system may be simplified further if one adopts the so-called Delaunay variables :

$$L = \sqrt{\mu a}$$
$$G = \sqrt{\mu a(1-e^2)}$$
$$H = \sqrt{\mu a(1-e^2)}\cos I \qquad\qquad (28)$$
$$l = M$$
$$g = \omega$$
$$h = \Omega$$

In this case, we simply have, with $\mathcal{F} = \mathcal{R} + \mu^2/(2L^2)$

$$\frac{dL}{dt} = \frac{\partial \mathcal{F}}{\partial l} \quad , \quad \frac{dG}{dt} = \frac{\partial \mathcal{F}}{\partial g} \quad , \quad \frac{dH}{dt} = \frac{\partial \mathcal{F}}{\partial h}$$

$$\frac{dl}{dt} = -\frac{\partial \mathcal{F}}{\partial L} \quad , \quad \frac{dg}{dt} = -\frac{\partial \mathcal{F}}{\partial G} \quad , \quad \frac{dh}{dt} = -\frac{\partial \mathcal{F}}{\partial H} \qquad (29)$$

This system is said to be canonical, with the hamiltonian $\mathcal{F}$. It is the best suited one for some sophisticated techniques of deriving analytical solutions.

In the case of quasi-circular orbits, it may be of interest to describe the real motion in terms of discrepancies with respect to a reference circular trajectory whose plane is fixed in (Σ) and defined by its mean motion $\tilde{n}$, the radius of the orbit, $\tilde{r}$, satisfying Kepler's 3rd. law : $\tilde{n}^2\tilde{r}^3 = \mu$.

The true position S of the satellite will be given by its three coordinates $(u,\ v,\ w)$ in the mobile system rotating with the fictitious reference point $\tilde{S}$ (fig. 5).

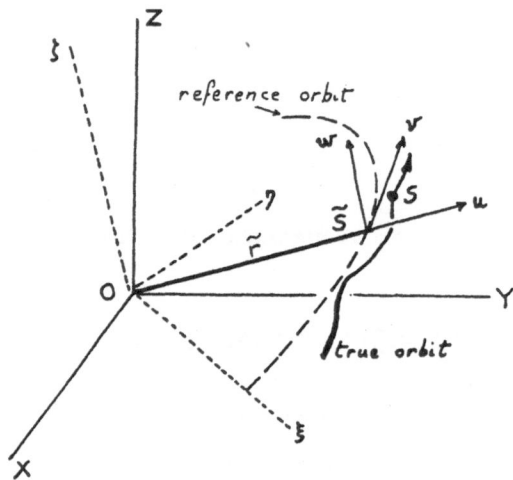

Fig. 5. The Hill reference orbit and rotating system

We here restrict ourselves to conservative forces, that is $\overline{\gamma} = \overline{\nabla}\mathcal{R}$. In the rotating system $\tilde{\Sigma} = \{uvw\}$, which rotation vector $\overline{\rho}$ with respect to (Σ) is $\tilde{n}\overline{w}$, we have :

$$-\mu\overline{r}/r^3 + \overline{\gamma} = [d^2\overline{r}/dt^2]_\Sigma = [d^2\overline{r}/dt^2]_{\tilde{\Sigma}} + 2\overline{\rho} \times [d\overline{r}/dt]_{\tilde{\Sigma}} + \overline{\rho} \times (\overline{\rho} \times \overline{r}) + \dot{\overline{\rho}} \times \overline{r}$$

The last term is equal to zero since $\tilde{n}$ and therefore $\overline{\rho}$ are constant. This equation is projected on $\tilde{\Sigma}$, in which the coordinates of S are $\tilde{r}+u,v,w$. We find :

$$\ddot{u} - 2\tilde{n}\dot{v} - \tilde{n}^2(\tilde{r}+u) = -\frac{\mu}{r^3}(\tilde{r}+u) + \frac{\partial\mathcal{R}}{\partial u}$$

$$\ddot{v} + 2\tilde{n}\dot{u} - \tilde{n}^2 v = -\frac{\mu}{r^3}v + \frac{\partial\mathcal{R}}{\partial v}$$

$$\ddot{w} = -\frac{\mu}{r^3}w + \frac{\partial \mathcal{R}}{\partial w}$$

Hill equations are finally obtained by linearizing this system around $u = v = w = o$. We first write : $r^2 = (\bar{r} + u)^2 + v^2 + w^2 \approx \bar{r}^2 + 2u\bar{r} = \bar{r}^2(1 + 2u/\bar{r})$, from which : $r^{-3} \approx \bar{r}^{-3}(1 - 3u/\bar{r})$. Hence, the first term in the right hand side members of the above equations become $-\mu(\bar{r} - 2u)/\bar{r}^3$, $-\mu v/\bar{r}^3$, $-\mu w/\bar{r}^3$. Replacing μ by $\bar{n}^2\bar{r}^3$ yields the final Hill system :

$$\ddot{u} - 2\bar{n}\dot{v} - 3\bar{n}^2 u = \partial \mathcal{R}/\partial u$$

$$\ddot{v} + 2\bar{n}\dot{u} = \partial \mathcal{R}/\partial v \tag{30}$$

$$\ddot{w} + \bar{n}^2 w = \partial \mathcal{R}/\partial w$$

Note that the last equation is decoupled from the others, allowing a separate treatment.

1.4 APPROXIMATE ANALYTICAL SOLUTIONS OF THE EQUATIONS OF MOTION

For further use in this course, it is sufficient to consider only the case of the Lagrange equations with a disturbing force function $\mathcal{R}$. However, much of what follows may be applied to other cases treated with the Gauss system.

$\mathcal{R}$ is a function of the position, hence of the six osculating elements, and also of the coordinates of the disturbing bodies (Moon, Sun). It is a 2π-periodic function in Ω, ω, M since it must have the same numerical value when these arguments change by 2π, the others being constant. On the other hand, the position of a disturbing body may be expressed via the orbital elements of its trajectory : $a^*, e^*, I^*, \Omega^*, \omega^*, M^*$, with respect to the reference frame (Σ) or to an intermediate reference frame with given (slowly varying) Euler angles $\varepsilon = (\varepsilon_1, \varepsilon_2, \varepsilon_3)$. It will be assumed, with enough accuracy here, that (a^*, e^*, I^*) are constant and that Ω^*, ω^*, M^* and ε are linear functions of time. Obviously, $\mathcal{R}$ must also be 2π-periodic in Ω^*, ω^*, M^* and in these Euler angles. Finally, and as concerned all Earth gravity direct and tidal effects, $\mathcal{R}$ must be 2π-periodic in θ, the sidereal time (we here assume that the equatorial plane of (Σ) is the Earth mean equator and that θ is the mean sidereal time, discrepancies from this hypothesis being treated as very small corrections to the solution).

Therefore, $\mathcal{R}$ may be expanded in a Fourier series of the form :

$$\mathcal{R} = \Sigma B_{jklj^*k^*l^*pq}(a, e, I, a^*, e^*, I^*)\cos(j\Omega + k\omega + lM + j^*\Omega^* + k^*\omega^* + l^*M^* + p\theta + q\varepsilon + \Phi)$$

The summation runs on all indices, for all disturbing bodies, and the phase Φ is a function of these indices in general. We write $\mathcal{R}$ in a more compact form, as :

$$\mathcal{R} = \Sigma B_{ii^*h}(m, m^*)\cos(iA + i^*A^* + hH) \tag{31}$$

where i stands for (j,k,l), i^* for (j^*,k^*,l^*), h for (p,q). m is the triplet (a,e,I) of the satellite metric[1] elements, A the triplet (Ω,ω,M) of its angular elements ; m^*,A^* designate the metric and angular elements of a disturbing body ; and H stands for all other angular parameters (and the phase is distributed among all pertinent arguments and indices).

The form of the Lagrange equations is such that :

$$\frac{dm}{dt} = \Sigma C_{ii^*h}(m,m^*)\sin(iA + i^*A^* + hH) \tag{32}$$

$$\frac{dA}{dt} = \Sigma D_{ii^*h}(m,m^*)\cos(iA + i^*A^* + hH) \tag{33}$$

In (33) there generally exist terms with all indices equal to zero, that is terms which are independent of the angular elements : $D_{ooo}(m,m^*)$.

Many different methods exist for obtaining the solution of these equations. Modern approaches all use algebric manipulators, but it is usely not too difficult to obtain by hand calculations a first good idea of the solution characteristics by retaining only the most significant terms, in particular by neglecting all the terms that are of the order of the square of small quantities characterizing the disturbing function. Such a procedure is called a first order solution and is simple to apply once the equations are written as in (32) and (33).

Let us note $m_o = (a_o,e_o,I_o)$ the mean values of (a, e, I), which are obtained if one neglects all the terms in (32). These are substituted in (33) in which we also provisionally neglect the periodic terms, keeping only the D_{ooo}'s. We find : $dA/dt = D_{ooo}$, or : $\dot{\Omega}^{(o)} = n_\Omega(m_o)$, $\dot{\omega}^{(o)} = n_\omega(m_o)$, $\dot{M}^{(o)} = n_M(m_o)$. The superscript (o) indicates that this is the beginning of the process of successive approximations. Actually, n_M consists of $n_o = (\mu/a_o^3)^{1/2}$ and of the term coming from the development. Integrating these equations, we obtain :

$$\overline{\Omega} = n_\Omega(t - t_o) + \Omega_o$$

$$\overline{\omega} = n_\omega(t - t_o) + \omega_o \qquad \Leftrightarrow \qquad \overline{A} = n_A(t - t_o) + A_o \tag{34}$$

$$\overline{M} = n_M(t - t_o) + M_o$$

These are linear, hence unbounded functions of time ; they are called secular terms and are the largest perturbations.

The next step of the process is to substitute m_o and $\overline{A}$ in the right hand sides of (32), (33), taking also into account that $m^*(t) \approx m_o^*$, $A^*(t) \approx \overline{A}^* = n_{A^*}(t - t_o) + A_o^*$, $H(t) \approx \overline{H} = n_H(t - t_o) + H_o$.

After integration, we obtain :

---

[1] Although I is an angle, it is called a "metric" element similar to a, e, because of the type of equation which governs its behaviour.

$$m = m_o - \Sigma \frac{C_{ii^*h}}{in_A + i^* n_{A^*} + h n_H} \cos(i\overline{A} + i^* \overrightarrow{A} + h\overline{H}) \tag{35}$$

$$A = \overline{A} + \Sigma \frac{D_{ii^*h}}{in_A + i^* n_{A^*} + h n_H} \sin(i\overline{A} + i^* \overrightarrow{A} + h\overline{H}) \tag{36}$$

Of course, the coefficients C_{ii^*h}, D_{ii^*h} are different for each of the metric or angular elements.

In this procedure, we have overlooked the fact that the first term of dM/dt is n and not n_o. We have $n^2 a^3 = \mu$, from which $2\Delta n/n + 3\Delta a/a = o$, hence to first order : $n = n_o[1 - 3/2(\Delta a/a)]$. From (35) we get :

$$\Delta a = -\Sigma \frac{C_{ii^*h}^{(a)}}{in_A + i^* n_{A^*} + h n_H} \cos(i\overline{A} + i^* \overrightarrow{A} + h\overline{H})$$

Then :

$$n = n_o + \frac{3n_o}{2a_o} \Sigma \frac{C_{ii^*h}^{(a)}}{in_A + i^* n_{A^*} + h n_H} \cos(i\overline{A} + i^* \overrightarrow{A} + h\overline{H})$$

So we must add to the solution in M given by (36), with $D_{ii^*h}^{(M)}$, the integral of these additional terms, that is :

$$\Delta' M = \frac{3n_o}{2a_o} \Sigma \frac{C_{ii^*h}^{(a)}}{\left(in_A + i^* n_{A^*} + h n_H \right)^2} \sin(i\overline{A} + i^* \overrightarrow{A} + h\overline{H}) \tag{37}$$

All periodic terms in (35), (36), (37) look similar. They are usually grouped in short, middle, long period terms depending on their period $2\pi/\left(in_A + i^* n_{A^*} + h n_H\right)$ with respect to the mean period of the satellite $2\pi/n_o$. It is interesting to note that there may exist combinations of the indices i, i^*, h such that, for n_A, n_{A^*}, n_H being given, the divisor $in_A + i^* n_{A^*} + h n_H$ becomes very small with respect to C_{ii^*h} or D_{ii^*h}, thus enhancing greatly the perturbation. This is called a resonance phenomenon. When the divisor becomes too small, the linear theory outlined above becomes meaningless and other techniques are required.

If one stops the procedure at the stage of the last equations, we usually do not have a full 1st. order theory with respect to the small parameter (s) of $\mathcal{R}$. There exist additional first order terms coming from the next step, that is when one substitutes m and A as given by (35), (36) and also (37), in the Lagrange equations and integrate again.

Finally, the form of the Lagrange (or Gauss) equations is such that orbits with small eccentricity and/or with small $\sin I$ cannot be properly treated without care. Either one must adopt another set of variables, such as ($e \sin \omega$, $e \cos \omega$) instead of (e, ω) - for which there exist an equivalent system of equations, or one must expand the solution in the vicinity of $e = o$, or of $\sin I = o$, and properly re-arrange or group some terms (... which we will do later when dealing with quasi-circular orbits).

2 THE GEOPOTENTIAL AND ITS REPRESENTATION

2.1 SPHERICAL HARMONIC REPRESENTATION OF THE GEOPOTENTIAL

Let us consider the Earth (E) with its actual shape (grossly approximated by an ellipsoid of revolution) and its density distribution such that, at the current point P', the mass element is $dM' = \rho(P')dV'$ in the elementary volume dV'. Let (Σ_o) be a reference system fixed in (E). The gravitational force at any point S outside (E) derives from the force function (called geopotential) :

$$U = G \int_{(E)} dM'/\Delta \tag{38}$$

where Δ is the distance SP' (fig. 6).

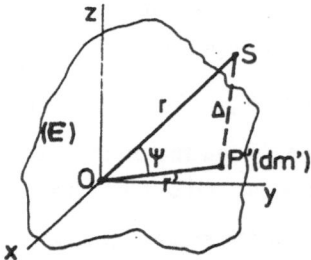

Fig. 6. The Earth and satellite point S

$1/\Delta$ is written as $r^{-1}[1-2(r'/r)\cos\psi+(r'/r)^2]^{-1/2}$. Now, if $r' < r$ for all P', the term $[1-2t\cos\psi+t^2]^{-1/2}$ with $r'/r = t < 1$, can be expanded in a convergent Legendre series :

$$[1-2t\cos\psi+t^2]^{-1/2} = \sum_{l=o}^{\infty} t^l P_l(\cos\psi) \tag{39}$$

where $P_l(x) = \{d^l[(x^2-1)^l]/dx^l\}/(2^l l!)$ is the usual Legendre polynomial of degree l. Then, $P_l(\cos\psi)$ can be transformed as follows. Denoting by (ϕ',λ') the latitude and longitude of P' and by (ϕ,λ) those of S, we have :

$$\cos\psi = \sin\phi\sin\phi' + \cos\phi\cos\phi'\cos(\lambda-\lambda')$$

which is transformed by the operator P_l as (Legendre addition formula) :

$$P_l(\cos\psi) = \sum_{m=-l}^{+l} (-1)^m P_{lm}(\sin\phi)P_{l,-m}(\sin\phi')\exp[im(\lambda-\lambda')] \tag{40}$$

In this formula, $P_{lm}(x)$ is the Legendre associated function of the first kind, of degree l and order m, and is defined by :

$$m = o : P_{lo}(x) = P_l(x)$$

$$m > o : P_{lm}(x) = (1-x^2)^{m/2}d^m[P_l(x)]/dx^m$$

$$P_{l,-m}(x) = (-1)^m \frac{(l-m)!}{(l+m)!} P_{lm}(x)$$

M being the mass of the Earth (and since $P_o(x) = 1$), we obtain :

$$U = \frac{GM}{r} + \frac{G}{r}\sum_{l=1}^{\infty}\left(\frac{1}{r}\right)^l \sum_{m=-l}^{+l}\left[(-1)^m \int_{(E)} r'^l P_{l,-m}(\sin\phi')\exp(-im\lambda')dM'\right]P_{lm}(\sin\phi)\exp(im\lambda)$$

This expansion requires that S be exterior to the smallest sphere containing (E), let us say a sphere of radius R. Introducing it as a factor of homogeneity, we obtain :

$$U = \frac{GM}{r} + \mathcal{R}$$

$$\mathcal{R} = \frac{GM}{r}\sum_{l=1}^{\infty}\left(\frac{R}{r}\right)^l \sum_{m=-l}^{+l} K_{lm}P_{lm}(\sin\phi)\exp(im\lambda) \tag{41}$$

with :

$$K_{lm} = \frac{(-1)^m}{MR^l}\int_{(E)} r'^l P_{l,-m}(\sin\phi')\exp(-im\lambda')\rho(r',\phi',\lambda')dV'$$

The K_{lm} coefficients depend on the shape and density function of the Earth. They are called harmonics of the geopotential (for U, and $\mathcal{R}$ are harmonic functions), of degree l and order m. In practice, noting that K_{l0} is real, we define real coefficients C_{lm}, S_{lm} for any $m > 0$ by :

$$K_{lm} = \frac{1+\delta_{0m}}{2}(C_{lm} - iS_{lm})$$

$$K_{l,-m} = \frac{1+\delta_{0m}}{2}(C_{lm} + iS_{lm})\frac{(l+m)!}{(l-m)!}(-1)^m \tag{42}$$

(where $\delta_{0m} = 0$ if $m \neq 0$, $\delta_{00} = 1$). When $m = 0$, $K_{l0} = C_{l0}$ and $S_{l0} = 0$ and it is then easy to verify that $\mathcal{R}$ can be written as :

$$\mathcal{R} = \frac{GM}{r}\sum_{l=1}^{\infty}\left(\frac{R}{r}\right)^l\left[C_{l0}P_l(\sin\phi) + \sum_{m=1}^{l}(C_{lm}\cos m\lambda + S_{lm}\sin m\lambda)P_{lm}(\sin\phi)\right] \tag{43}$$

The C_{l0} coefficients are sometimes denoted as $-J_l$, and are called zonal harmonics, since they characterize variations of U which are independent of the longitude. The other harmonics (C_{lm}, S_{lm}) are called tesseral ; a peculiar case is when $l = m$ and the (C_{ll}, S_{ll}) are named sectorial harmonics. Practically, the origin of (Σ_e) is taken at the Earth's center of mass and the Z axis along the mean Earth axis of rotation, assumed to be a principal axis of inertia. These hypothesis implies that $C_{10} = C_{11} = S_{11} = C_{21} = S_{21} = 0$. Furthermore, we have the important relations :

$$C_{20} = -\frac{1}{MR^2}\left(C - \frac{A+B}{2}\right)$$

$$C_{22} = \frac{1}{MR^2}\frac{B-A}{4}$$

where A, B, C are the moments of inertia of (E) in (Σ_e).

In the following, equation (43) which gives the expression of $\mathcal{R}$ in terms of the spherical harmonics of the geopotential, will start at $l = 2$. It will also be used with normalized Legendre functions $\overline{P}_{lm}(x)$ and normalized harmonics $(\overline{C}_{lm}, \overline{S}_{lm})$ such that :

$$\overline{P}_{lm}(\overline{C}_{lm}, \overline{S}_{lm}) = P_{lm}(C_{lm}, S_{lm})$$

$$\overline{P}_{lm}(x) = \left[(2-\delta_{0m})(2l+1)\frac{(l-m)!}{(l+m)!}\right]^{1/2}P_{lm}(x) = v_{lm}P_{lm}(x) \tag{44}$$

This normalization is such that :

$$\frac{1}{4\pi}\int\int_{unitsphere}\overline{P}_{lm}^2(\sin\phi)\begin{bmatrix}\cos^2 m\lambda \\ \sin^2 m\lambda\end{bmatrix}\cos\phi\, d\phi\, d\lambda = 1$$

Hence :

$$\mathcal{R} = \frac{GM}{r} \sum_{l=2}^{\infty} \left(\frac{R}{r}\right)^l \sum_{m=0}^{l} (\overline{C}_{lm} \cos m\lambda + \overline{S}_{lm} \sin m\lambda) \overline{P}_{lm}(\sin\phi)$$

$$= \frac{GM}{r} \sum_{m=0}^{\infty} \sum_{l=\sup(m,2)}^{\infty} \left(\frac{R}{r}\right)^l (\overline{C}_{lm} \cos m\lambda + \overline{S}_{lm} \sin m\lambda) \overline{P}_{lm}(\sin\phi)$$

$$= \Sigma_{l,m} \mathcal{R}_{lm}$$

2.2 THE REPRESENTATION OF THE GEOID SHAPE

The geoid is a conventional equipotential surface of the total potential $W = U + C$, where U is the gravitational potential, and C is the centrifugal potential of the rotating Earth ($C = [\theta^2 r^2 \cos^2\phi]/2$ with θ = sidereal rotation rate). This equipotential, in the oceanic areas, is the surface the sea would have if there was no motion of the sea water, even averaged over an infinite time (this assumes that mass movements, such as those due to tectonic motions or internal convection, are neglected in the "solid" Earth) ; this geoid physical definition is implicity extended (mathematically valid) over the continental areas. If the Earth was fluid and composed of (for instance) homogeneous confocal layers, its surface would be a perfect ellipsoid of revolution. Besides the observed fact that the Earth's surface may actually be approximated by such an ellipsoid flattened at the poles, this is why the shape of the geoid is described with respect to an ellipsoid of revolution, called a dynamical ellipsoid. It is defined as having the same mass, center of mass and mean rotation axis as the Earth's ; it has a prescribed semi-major axis a_e and a flattening $\alpha = (a_e - a_p)/a_e$ (a_p : semi-minor (polar) axis) ; it rotates with the Earth with the same sidereal rate θ and its surface is an equipotential of its own total potential $W_E = U_E + C$ (U_E = gravitational part) ; conventionally, the value of W_E on its surface is taken equal to the value of the real potential W on the geoid surface.

Under these assumptions, the height, usually denoted by N in physical geodesy, of the geoid with respect to the ellipsoid, counted positively along the outward normal $\overline{n}$ to the ellipsoid (fig. 7) is given by (Brun's formula) :

$$N = \frac{W - W_E}{\gamma} = \frac{U - U_E}{\gamma} \tag{45}$$

with γ being the gravity on the ellipsoid : $\gamma = |\partial W_E/\partial n|$. As a result, and since the ellipsoid gravitational potential expansion involves even degree zonal terms only, we have :

$$N = \frac{GM}{r\gamma} \sum_{l=2}^{\infty} \left(\frac{R}{r}\right)^l \sum_{m=0}^{l} (\overline{C}_{lm}^* \cos m\lambda + \overline{S}_{lm} \sin m\lambda) \overline{P}_{lm}(\sin\phi) \tag{46}$$

with :

$$\overline{C}_{lm}^{\ \prime} = \overline{C}_{lo} - \overline{C}_{lo}(ellipsoid)\dots \quad \text{if} \quad l \quad \text{is even}$$

$$= \overline{C}_{lo}\dots \quad \text{if} \quad l \quad \text{is odd}$$

$$\overline{C}_{lm}^{\ \prime} = \overline{C}_{lm} \text{ if } m > 0.$$

This expression is often used in the simplified form (taking $\gamma = GM/r^2$ and $r = R = a_e$ instead of their mathematical expressions at the surface of the reference ellipsoid) :

$$N \approx R \sum_{l,m} \left(\overline{C}_{lm}^{\ \prime} \cos m\lambda + \overline{S}_{lm} \sin m\lambda \right) \overline{P}_{lm}(\sin \phi) \tag{47}$$

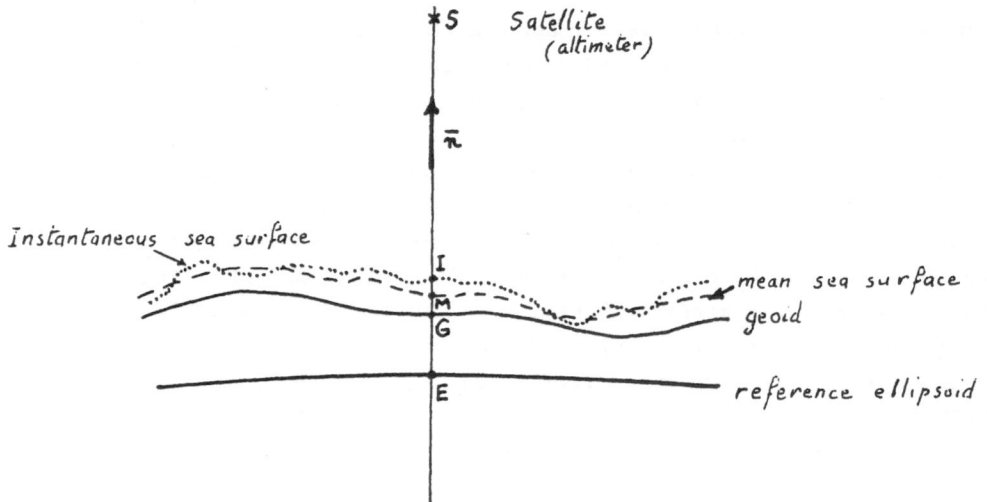

Fig. 7. Surfaces to be considered in satellite altimetry

$N = \overline{EG}$ measures the departure of the geoid shape from the ellipsoid.

$\overline{GI}$ is the dynamic topography (instantaneous), $\overline{GM}$ its mean value.
An altimeter on board satellite S measures $\overline{SI}$.
If S is known from ground tracking observations and a posteriori orbit determination, then $\overline{ES}$ is known ; $\overline{SI}$ being measured, $\overline{EI}$ is known.

2.3 TRANSFORMATION IN ORBITAL ELEMENTS ; KAULA'S SOLUTION

Our goal is now to use Lagrange equations to derive the main geopotential perturbations on a satellite orbit. It is therefore necessary to transform $\mathcal{R}$, as given by (44) and expressed in (Σ_o), in a function of all six orbital elements. It is clear that r will involve the elements (a, e, M), whereas ϕ and λ will involve I, Ω, ω and M. The transformation is therefore splitted into two parts.

2.3.1 Transformation of $\overline{P}_{lm}(\sin\phi)\cos m\lambda$ and $\overline{P}_{lm}(\sin\phi)\sin m\lambda$

There are several ways of achieving it. One, originally due to Kaula (hence the name of Kaula's solution) starts from the exact expression of $\overline{P}_{lm}(\sin\phi)$ in terms of powers of $\sin\phi$, divided by $\cos^m\phi$, transforms $\cos m\lambda$ and $\sin m\lambda$ in terms of powers of $\cos(\omega+v)$, $\sin(\omega+v)$, $\cos I$, with the factor $\exp[im(\Omega-\theta)]/\cos^m\phi$. There remains a triple summation which gives the quantities in terms of cosines and sines of the argument $(l-2p)(\omega+v)+m(\Omega-\theta)$ with the so-called Kaula's inclination functions $\overline{F}_{lmp}(I)$ in factor (Kaula, 1966).

Another derivation starts from the theorem on the rotation of the spherical harmonic functions $Y_{lm}(\phi,\lambda) = P_{lm}(\sin\phi)\exp(im\lambda)$, when going from a reference system (σ) to another one (σ') by three rotations according to the three usual Euler angles Ψ, Θ, Φ; this theorem states that :

$$(l-m)!Y_{lm}(\phi,\lambda) = \sum_{m'=-l}^{+l} (l-m')!Y_{lm'}(\phi',\lambda')E_{lm}^{m'}(\Psi,\Theta,\Phi) \tag{48}$$

The Euler functions $E_{lm}^{m'}$ are defined as :

$$E_{lm}^{m'}(\Psi,\Theta,\Phi) = (-1)^{l-m}\exp\left[i(m'-m)\frac{\pi}{2}\right]\exp\left[i(m\Psi+m'\Phi)\right]C_{lm}^{m'}\left(\frac{\Theta}{2}\right)$$

where the $C_{lm}^{m'}$ are the Clifford trigonometric polynomials :

$$C_{lm}^{m'}\left(\frac{\Theta}{2}\right) = \sum_{j=j_{inf}}^{j_{sup}} (-1)^j \binom{l-m}{j}\binom{l+m}{m+m'+j}\cos^v\frac{\Theta}{2}\sin^{2l-v}\frac{\Theta}{2}$$

with :

$$j_{inf} = \max(0, -m-m')$$

$$j_{sup} = \min(l-m, l-m')$$

$$v = 2j+m+m'$$

We apply this transformation to $\sigma = (\overline{X}_o, \overline{Y}_o, \overline{Z}_o) = (\Sigma_o)$ and $\sigma' = (\overline{r}, \overline{s}, \overline{w})$ - fig. 8.

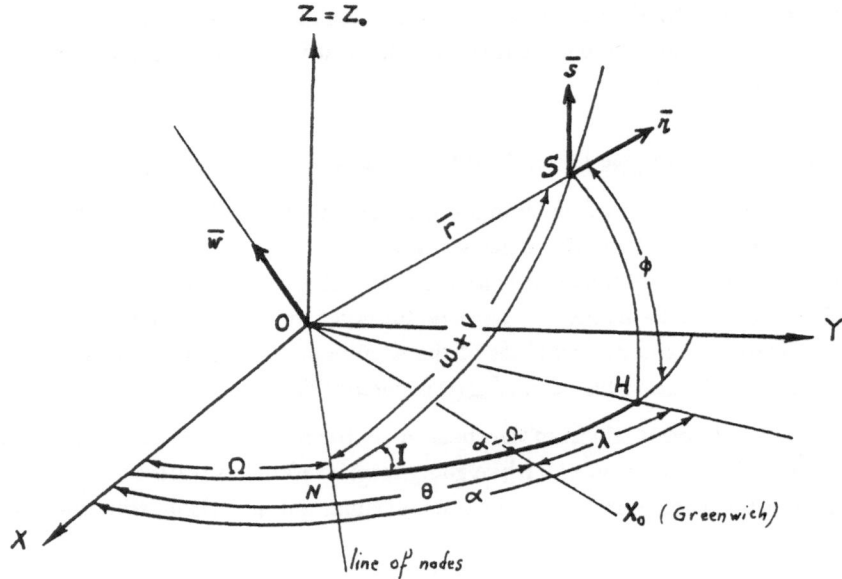

Fig. 8. Angles encountered in Kaula's transformation

Hence : $\Psi = \Omega - \theta$, $\Theta = I$, $\Phi = \omega + \nu$. In (σ'), we have $\lambda' = 0$, $\phi' = 0$. We take advantage of the fact that $P_{lm'}(o) = o$ if $l - m'$ is odd and $P_{lm'}(o) = (-1)^{(l - m')/2}(l + m')!/\{2^l[((l - m')/2)!][((l + m')/2)!]\}$ if $l - m'$ is even. The result is :

$$P_{lm}(\sin\phi)\exp(im\lambda) = i^{l-m} \sum_{p=o}^{l} D_{lmp}(I)\exp[i(l-2p)(\omega+\nu) + m(\Omega - \theta)] \tag{49}$$

with :

$$D_{lmp}(I) = (-1)^{l-m}\frac{(l+m)!}{2^l l!}(lp)\sum_{j=j_1}^{j_2}(-1)^j\binom{2p}{j}\binom{2l-2p}{l-m-j}\left(\cos\frac{I}{2}\right)^{l-m-2p+2j}\left(\sin\frac{I}{2}\right)^{l-m+2p-2j} \tag{50}$$

($j_1 = \max(o, 2p - l - m)$, $j_2 = \min(l - m, 2p)$).

These expressions are easier to evaluate than the Kaula's original ones. They are related to the classical functions $F_{lmp}(I)$ by ($m \ge o$) :

$$D_{lmp}(I) = (-1)^{l-m+[(l-m)/2]}F_{lmp}(I)$$

$$D_{l,-m,p}(I) = (-1)^{[(l-m)/2]}\frac{(l-m)!}{(l+m)}F_{l,m,l-p}(I) \tag{51}$$

There exist numerous recursive relations between the F_{lmp}, or the D_{lmp} functions, which are more efficient for numerical evaluations, especially for large values of l, m, p.

2.3.2 Transformation of terms containing r and v

We take again formula (41) for $\mathcal{R}_s$ with $P_{lm}(\sin\phi)\exp(im\lambda)$ being replaced by (49). We have to transform $r^{-l-1}\exp[i(l-2p)(\omega+v)] = r^{-l-1}\exp[i(l-2p)v]\exp[i(l-2p)\omega]$. From the definition of the Hansen coefficients we immediately write :

$$\frac{1}{r^{l+1}}\exp[i(l-2p)v] = \sum_{k=-\infty}^{+\infty} X_k^{-l-1,l-2p}\exp(ikM)$$

$$= \sum_{q=-\infty}^{+\infty} X_{l-2p+q}^{-l-1,l-2p}\exp[i(l-2p+q)M] \tag{52}$$

where the second expression is obtained by a change of index : $k = l - 2p + q$. Kaula introduced the notation :

$$G_{lpq}(e) = X_{l-2p+q}^{-l-1,l-2p} \tag{53}$$

From (22), it is clear that $G_{lpq}(e) = o(e^{|q|})$. For most geodetic satellites, e is small ($< 10^{-2}$), and only terms with $q = o$, $q = \pm 1$ and sometimes $q = \pm 2$ need to be taken into account in (52) for sufficient accuracy in the analytical solution.

Using (22) and (53), it is easy to get :

. if $q > o$:

$$G_{lpq} = \left(-\frac{e}{2}\right)^q \sum_{t=o}^{q}\binom{2p-2l}{q-t}\frac{(-1)^t}{t!}(l-2p+q)^t + o(e^{q+2})$$

When $p < l$:

$$\binom{2p-2l}{q-t} = (-1)^{q-t}\binom{2l-2p+q-t-1}{q-t}$$

When $p = l$:

$$G_{llq} = \left(-\frac{e}{2}\right)^q\frac{(-1)^q}{q!}(q-l)^q + o(e^{q+2})$$

. if $q < o$:

$$G_{lpq} = \left(-\frac{e}{2}\right)^{-q} \sum_{t=o}^{-q}\binom{-2p}{-q-t}\frac{(l-2p+q)^t}{t!} + o(e^{-q+2})$$

When $p > o$:

$$\binom{-2p}{-q-t} = (-1)^{-q-t}\binom{2p-q-t-1}{-q-t}$$

When $p = o$:

$$G_{loq} = \left(-\frac{e}{2}\right)^{-q} \frac{(q+l)^{-q}}{(-q)!} + o(e^{-q+2})$$

. if $q = o$, $G_{lpo} = 1 + o(e^2)$. One sometimes needs the term in e^2 ; it is computed as ($t = 1, s = o$ in formula (22), and $n + m + 1 = -2p$, $n - m + 1 = 2p - 2l$, $m = l - 2p$) :

$$\sum_{j=o}^{1} \sum_{h=o}^{1} \binom{-2p}{j-h} \frac{(l-2p)^h}{h!} \sum_{k=o}^{l} \binom{2p-2l}{j-k} \frac{(l-2p)^k}{h!} (-1)^k$$

$$\left[2\binom{l+1-h-k}{1-j} - \binom{l+2-h-k}{1-j}\right]\frac{e^2}{4} = [l + (4p - 3l)(l - 4p)]\frac{e^2}{4}$$

To summarize, we have :

$$G_{lpo} = 1 + g_{lpo}\frac{e^2}{2} + o(e^4)$$

$$G_{lp \pm 1} = g_{lp \pm 1}e + o(e^3)$$ (54)

$$G_{lp \pm 2} = g_{lp \pm 2}\frac{e^2}{2} + o(e^4)$$

so that :

$$G'_{lpo} = g_{lpo}e + o(e^3)$$

$$G'_{lp \pm 1} = g_{lp \pm 1} + o(e^2)$$ (55)

$$G'_{lp \pm 2} = g_{lp \pm 2}e + o(e^3)$$

where :

$$g_{lpo} = [l + (4p - 3l)(l - 4p)]/2$$

$$g_{lp1} = (3l - 4p + 1)/2$$

$$g_{lp-1} = (4p - l + 1)/2$$ (56)

$$g_{lp2} = (l - p)(2l - 3p + 5/2) + (l - 2p + 2)^2/4$$

$$g_{lp-2} = p(3p - l + 5/2) + (l - 2p - 2)^2/4$$

2.3.3 Final form of the geopotential disturbing function

Putting together (49) and (52), we obtain :

$$\mathcal{R} = \frac{\mu}{a} \sum_{l=2}^{\infty} \left(\frac{R}{a}\right)^l \sum_{m=-l}^{+l} i^{l-m} K_{lm} \sum_{p=o}^{l} D_{lmp}(I) \sum_{q=-\infty}^{+\infty} G_{lpq}(e) \exp i \psi_{lmpq}$$ (57)

where :

$$\psi_{lmpq} = (l-2p)\omega + (l-2p+q)M + m(\Omega-\theta) \tag{58}$$

Actually $\mathcal{R}$ is a real function and, taking account of (42) and (51), adopting normalized coefficients $(\overline{C}_{lm}, \overline{S}_{lm})$, we find :

$$\mathcal{R} = \frac{\mu}{a} \sum_{l=2}^{\infty} \left(\frac{R}{a}\right)^l \sum_{m=0}^{l} \sum_{p=0}^{l} \overline{F}_{lmp}(I) \sum_{q=-\infty}^{+\infty} G_{lpq}(e) S_{lmpq}(\Omega, \omega, M, \theta)$$

$$= \sum_{l,m} \mathcal{R}_{lm} = \sum_{l,m,p,q} \mathcal{R}_{lmpq} \tag{59}$$

with :

$$\overline{F}_{lmp}(I) = v_{lm} F_{lmp}(I)$$

(cf. formula (44))

$$S_{lmpq} = \hat{C}_{lm} \cos \psi_{lmpq} + \hat{S}_{lm} \sin \psi_{lmpq}$$

and :

$$\hat{C}_{lm} = \overline{C}_{lm}, S_{lm} = \overline{S}_{lm} \quad if \quad l-m \quad \text{is even},$$

$$\hat{C}_{lm} = -\overline{S}_{lm}, S_{lm} = \overline{C}_{lm} \quad if \quad l-m \quad \text{is odd}$$

2.3.4 First order perturbations in the elements

The term of the geopotential which dominates the disturbing function is $\mathcal{R}_{20}$ since $\overline{C}_{20}$ (or $-\overline{J}_2$) is at least hundred times larger than any other $\overline{C}_{lm}$ or $\overline{S}_{lm}$ (it has been found empirically that the magnitude of these coefficients decrease approximately as $10^{-5}/l^2$ - Kaula's rule). We can therefore get a fairly good estimation of the major perturbations by restricting ourselves to :

$$\mathcal{R}_{20} = \frac{\mu}{a} C_{20} \left(\frac{R}{a}\right)^2 \sum_{p,q} F_{20p}(I) G_{2pq}(e) \cos[(2-2p)\omega + (2-2p+q)M] ,$$

(written here in non normalized form with $C_{20} \approx 1.082628 \, 10^{-3}$).

Following the successive approximation technique described in 1.4, we first get the (main) secular terms on Ω, ω, M from the above, with $p = 1$ and $q=o$, that is :

$$\mathcal{R}_{2010} = \frac{\mu}{a} C_{20} \left(\frac{R}{a}\right)^2 F_{201}(I) G_{210}(e)$$

where :

$$F_{201}(I) = 3/4 \sin^2 I - 1/2$$

$$G_{210}(e) = (1-e^2)^{-3/2}, \quad \text{exactly.}$$

From the last three Lagrange equations, we find :

$$\dot{\Omega}_{sec} = n_\Omega = \frac{3}{2} n C_{20} \left(\frac{R}{a}\right)^2 \frac{1}{(1-e^2)^2} \cos I$$

$$\dot{\omega}_{sec} = n_\omega = \frac{3}{4} n C_{20} \left(\frac{R}{a}\right)^2 \frac{1}{(1-e^2)^2} (1 - 5 \cos^2 I) \tag{60}$$

$$\dot{M}_{sec} = n_M = n - \frac{3}{4} n C_{20} \left(\frac{R}{a}\right)^2 \frac{1}{(1-e^2)^{3/2}} (3 \cos^2 I - 1)$$

These are very important formulas. They show that the mean orbital plane has a precession motion which is retrograde if $0 \leq I < 90$ (since $C_{20} < 0$) and prograde if $90° < I \leq 180°$ (at this degree of approximation, it is extremely small for $I = 90°$) ; the line of apsides rotates (motion of ω) in this orbital plane clockwise if $I > I_c$, counterclockwise if $I < I_c$, with $I_c \approx 63°26'$; I_c is the critical inclination and the periapsis undergoes a peculiar libration motion when $I \approx I_c$ (in what follows, we will always assume that $I \neq I_c$) ; finally, with respect to the mean motion n (in the absence of perturbations), the satellite goes faster on its orbit if $I < I_o$, and slower if $I > I_o$, with $3 \cos^2 I_o - 1 = o$ ($I_o \approx 35°16'$).

In reality, it is easy to see that other terms, namely the zonal harmonics $C_{2k,o}$ (of even degree), for $k > 2$, also give secular perturbations which can be computed as above in a first approximation ; in the following, we will assume that $\dot{\Omega}_{sec}, \dot{\omega}_{sec}, \dot{M}_{sec}$ (denoted simply $\dot{\overline{\Omega}}, \dot{\overline{\omega}}, \dot{\overline{M}}$) contain these perturbations.

One must realize that there is no mean to have secular perturbations on a, e, I with this type of disturbing function.

To finish with, we apply the remaining of the procedure described in 1.4, and we obtain (including the variations in M resulting from changes in the mean motion n, arising from perturbations of the semi-major axis) :

$$\Delta\alpha = \sum_{lmpq} \Delta\alpha_{lmpq} \tag{61}$$

where α represents anyone of the orbital elements, and the $(lmpq)$ set of indices is such that it does not produce any secular effect (already included in $\dot{\Omega}, \dot{\omega}, \dot{M} = n$), that is $\dot{\psi}_{lmpq} \neq o$.

The $\Delta\alpha_{lmpq}$ for the metric elements a, e, I are of the form :

$$\Delta\alpha_{lmpq} = C^\alpha_{lmpq}(\overline{a}, \overline{e}, \overline{I}, \dot{\overline{\Omega}}, \dot{\overline{\omega}}, \dot{\overline{M}}) S_{lmpq}(\Omega, \omega, M, \theta) \tag{62}$$

and for the angular elements Ω, ω, M :

$$\Delta\alpha_{lmpq} = C^\alpha_{lmpq}(\overline{a}, \overline{e}, \overline{I}, \dot{\overline{\Omega}}, \dot{\overline{\omega}}, \dot{\overline{M}}) S^*_{lmpq}(\Omega, \omega, M, \theta) \tag{63}$$

where : $\overline{a},\overline{e},\overline{I}$ are the mean values of a, e, I, as opposed to their osculating values $a = \overline{a} + \Sigma(\Delta a_{lmpq})$, etc..., and $\dot{\overline{\Omega}},\dot{\overline{\omega}},\dot{\overline{M}}$ are the mean rates of Ω, ω, M evaluated with the secular terms as said above ; that is for instance, $\Omega(osculating) = \Omega_o + \dot{\overline{\Omega}}(t - t_o) + \Sigma \Delta \Omega_{lmpq}$.

S_{lmpq} is as in (59), and $S^*_{lmpq} = \dot{C}_{lm} \sin \psi_{lmpq} - S_{lm} \cos \psi_{lmpq}$. ψ_{lmpq} itself is evaluated with the mean angular elements $\Omega_o + \dot{\overline{\Omega}}(t - t_o)$, $\omega_o + \dot{\overline{\omega}}(t - t_o)$, $M_o + \overline{n}(t - t_o)$ and with $\theta = \theta_o + \dot{\theta}(t - t_o)$. In the remaining of this course, we will drop all the overbars to simplify the notations since there should be no confusion : 1.st order perturbations are evaluated with the values of the mean elements.

The $\dot{C}_{lmpq}$ coefficients are the following :

$$C^a_{lmpq} = 2A\, G_{lpq}(l - 2p + q)/\dot{\psi}_{lmpq}$$

$$C^e_{lmpq} = \frac{A}{a}\frac{\sqrt{1-e^2}}{e} G_{lpq}[\sqrt{1-e^2}(l-2p+q) - (l-2p)]/\dot{\psi}_{lmpq}$$

$$C^I_{lmpq} = \frac{A}{a}\frac{1}{\sin I \sqrt{1-e^2}} G_{lpq}[(l-2p)\cos I - m]/\dot{\psi}_{lmpq} \tag{64}$$

$$C^\Omega_{lmpq} = \frac{A'}{a}\frac{1}{\sin I \sqrt{1-e^2}} G_{lpq}/\dot{\psi}_{lmpq}$$

$$C^\omega_{lmpq} = \frac{1}{a}\left[\frac{\sqrt{1-e^2}}{e} A\, G'_{lpq} - \frac{\cos I}{\sin I \sqrt{1-e^2}} A'\, G_{lpq}\right]/\dot{\psi}_{lmpq}$$

$$C^M_{lmpq} = \frac{A}{a}\left[2(l+1)G_{lpq} - \frac{1-e^2}{e} G'_{lpq} - 3G_{lpq}\frac{(l-2p+q)n}{\dot{\psi}_{lmpq}}\right]/\dot{\psi}_{lmpq}$$

with :

$$A = na\left(\frac{R}{a}\right)^l \overline{F}_{lmp}(I)$$

$$A' = na\left(\frac{R}{a}\right)^l \overline{F}'_{lmp}(I)$$

As an example, we have computed the perturbations for the TOPEX-POSEIDON satellite with the following mean elements $a = 7714410$ m, $e = 9.3\ 10^{-5}$, $I = 66°02$. They have been converted to rectangular coordinate perturbations in the Gauss system by the method which is the subject of chapter 3 for all l, m, p, q's.

R. Rapp's 1991 global geopotential model truncated at degree and order 60 has been used, and $|q|$ limited at 2. Then, since the perturbations for given $(\overline{C}_{lm}, \overline{S}_{lm})$ are composed of many frequencies, the r.m.s. has been computed. The diagram on figure 9 shows the r.m.s. perturbation in position, in meters, for each couple of harmonics (for low degrees and orders, the perturbations are quite large and their graphic representation was truncated...).

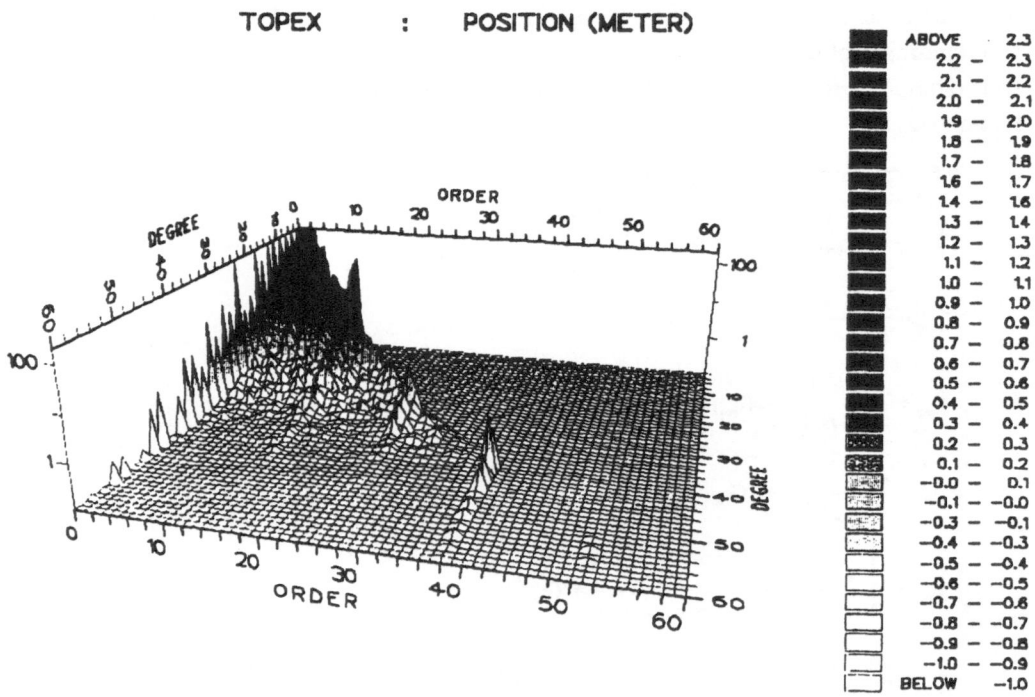

Fig. 9. Diagram of r.m.s. perturbations in the position of the Topex-Poseidon satellite.

2.3.5 Choosing the orbit of a satellite

It all depends on the usage of the satellite, of the on board sensors and their operational constraints.

The mean motion is quite important for it is the major angular parameter which very directly interacts with the sidereal time rate $\dot\theta$ and it conditions greatly the overall coverage. The mean semi-major axis which corresponds to it immediately places the spacecraft far enough from the Earth's upper atmosphere or directly in it (e.g. from 200 to 1000 km) which may entail problems as concerns the mission life-time, the proper operation of some sensors, the attitude and orbit controls of the satellite... ; also, one must note the decrease of the geopotential perturbations as $(R/a)^\ell$, (apart from sharp resonance cases), of which one may take advantage, for instance in the case of geodynamic satellites (e.g. LAGEOS). The mean eccentricity will usually be rather small, so as to operate at more or less constant altitude, apart from variations due to the radial orbit perturbations and due to the Earth's flattening. The inclination is a very important parameter since it is through it that the orbital plane precesses and, for many sensors of geodetic and Earth observation missions, it governs the coverage one finally obtains throughout the mission.

Important cases are : the polar inclination by which the orbital plane is practically fixed in space (if the altitude is sufficient to neglect the effects of drag) ; the heliosynchronous case in which the orbit plane follows (approximately) the motion of the sun with respect to Earth, that is $\dot{\Omega} = 360°/365.2422 \ d = 0.98565°/day$ (it cannot follow the sun exactly since the right ascension of the sun does not vary linearly but has additional periodic terms which depend on the Earth mean anomaly, eccentricity and obliquity), which requires an inclination generally in the range 96° to 100° (I (helios.) $= \cos^{-1}[-4.784204.10^{-15}a^{7/2}]$, with a in km).

In all cases, figure 10 illustrates how successive tracks are placed with respect to the Earth, from which one can derive algorithms to compute the coverage of the ground tracks or to determine repeat orbits. The algorithms are based on the value of the longitude interval, $\Delta\lambda$, between two successive tracks, with respect to Earth.

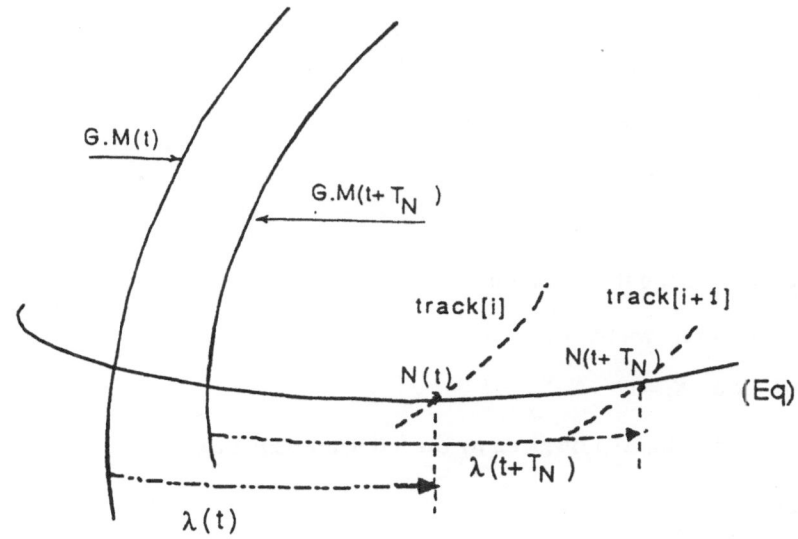

Fig. 10. Geometry of successive tracks

N : ascending node
T_N : nodal period
G.M. : Greenwich meridian
Eq : Earth equator

We have :

$$\Delta\lambda = \lambda(t + T_N) - \lambda(t) \quad ,$$

$$= (\dot\Omega - \dot\theta)T_N$$

where $\dot\Omega$ is the secular drift of the ascending node and T_N the nodal (draconitic period), given by :

$$T_N = 2\pi/(\dot M + \dot\omega)$$

with, as before :

$\dot M =$ "mean" mean motion $(\dot M > o)$

$\dot\omega =$ secular drift of argument of periapsis $(|\dot\omega| \quad < \quad \dot M)$

$\dot\theta =$ sidereal time rate

Two types of problems can then be solved :

(a) Resolution at the equator versus time :

Let us call ρ_j the resolution on the equator after a time interval ΔT_j counted from the beginning of the mission (with $\Delta T_j < \Delta T_{j+1}...$), that is after an integer number of revolutions, K_j.

We have :

$$\Delta T_j = K_j T_N$$

and we write :

$$\rho_j = \overline{R}\Delta_j$$

where $\overline{R}$ is the Earth equatorial radius.

At the beginning, we have :

$$\rho_o = 2\pi\overline{R} \quad \text{and} \quad K_o = 1$$

Then the series $\{\rho_j, \Delta T_j\}_j$ is given by the following sequence :

- let us define :

. if $j = 1$: $A_1 = 2\pi,\quad \Delta_1 = |\Delta\lambda|$

. if $j > 1$: $B_j = q_{j-1}\Delta_{j-1} - A_{j-1}$

$\qquad A_j = \Delta_{j-1}$

$\qquad \Delta_j = \inf(B_j, A_j - B_j)$

- then, for all $j \geq 1$: $q_j = [A_j/\Delta_j + 1]$, $[...]$ = integer part

$$K_j = [2\pi/\Delta j + 1/2]$$

$$\rho_j = \overline{R}\Delta j$$

$$\Delta T_j = K_j T_N$$

This algorithm takes account only of ascending or descending passes. If both types of passes are considered, which is reasonable if the orbital eccentricity is small - that is the spacecraft altitude will be almost the same at the descending and ascending nodes and sensor "operation" conditions may be similar too, the actual resolution will be between $\rho_i/2$ and ρ_i. Finally, if one is interested in the mean resolution at some latitude, resolution numbers must be multiplied by the cosine of that angle.

(b) Determination of repeat orbits of given repeat period :

A repeat orbit is characterized by the existence of integer solutions $\{h,k\}, h \in N, k \in Z$, to the following equations :

$$hT_N = T_{r\varphi}$$

$$h(\dot{\Omega} - \dot{\theta})T_N = 2k\pi \text{ , or } \frac{\dot{\omega} + \dot{M}}{\dot{\Omega} - \dot{\theta}} = \frac{h}{k} \qquad (65)$$

where notations are as before and the given repeat period is $T_{r\varphi}$. Being given a_o, e_o, I_o (usual metric elements), and allowed intervals of their variations : $[a_o - \Delta a, a_o + \Delta a] = \mathcal{A}$, $[e_o - \Delta e, e_o + \Delta e] = \mathcal{E}$, $[I_o - \Delta I, I_o + \Delta I] = \mathcal{I}$, one searches the possible values $h_1, h_2, ... h_p$ and associated values (k) which may satisfy the equations for $a \in \mathcal{A}, e \in \mathcal{E}, I \in \mathcal{I}$. Actually, the h_i's are all consecutive, that is $h_{min} \leq h \leq h_{max}$ (and $h_i = h_{min} + i - 1$) and, for a given h, possible values of k are found to be between $k_{min}(h)$ and $k_{max}(h)$. There may be no such value for a given $T_{r\varphi}$.

For any couple of values (h,k) and a given value of e in $\mathcal{E}$, one then tries to find a and I so that $\dot{\omega}(a,e,I), \dot{M}(a,e,I), \dot{\Omega} = (2\pi/T_N)(k/h) + \dot{\theta}$ satisfy exactly the system.

There may be no solution, or sometimes solutions outside $\mathcal{A}$ and $\mathcal{I}$.

The physical interpretation of the repeat orbit is that the ground track repeat itself after h revolutions of the satellite in the orbital plane and after k revolutions of the orbital plane itself about the Earth's mean rotation axis and with respect to the Earth's surface : $T_{r\varphi} = hT_N = kT_{\Omega-\theta}$.

In reality nearly every circular trajectory resembles a repeating one since any real value of $(\dot{\omega} + \dot{M})/(\dot{\Omega} - \dot{\theta})$ may be approximated by a ratio of two integers. A practical problem might be that the integer values become quite large for an accurate approximation of this ratio ; therefore one usually limits oneself to repeat periods which are less than a few months or so.

The longitude spacing of the ground tracks is obviously $360°/h$. For example, SEASAT had in its last month a repeat orbit at the mean altitude of 790 km with $I = 108°$, resulting in $h/k = -43/3$, hence a longitude spacing of the ground tracks at the equator equal to $360°/43 = 8.37°$, TOPEX-POSEIDON, with $a = 7714.5$ km, $e = 9.5 \ 10^{-5}$ and $I = 66.039°$, is such that $(\dot{\omega} + \dot{M})/(\dot{\Omega} - \dot{\theta}) = -12.7 = -127/10$, hence a longitude spacing of $2.83°$ for a repeat period of 9.92 days.

It is interesting to look at the spectral characteristics of repeat arc differences in this case. All orbital elements being expressed, as in (62) and (63), as Fourier series with co-efficients which are functions of the mean (fixed) metric elements, $\alpha(t + T_{rep}) - \alpha(t)$ is the product of such a coefficient (independent of t) by a sine or cosine of :

$$\psi_{lmpq}(t + T_{rep}) - \psi_{lmpq}(t)$$

Writing $\psi_{lmpq} = (l - 2p + q)(\omega + M) + m(\Omega - \theta) - q\omega$,   taking   account   of
$T_{rep} = h2\pi/(\dot{\omega} + \dot{M}) = k2\pi/(\dot{\Omega} - \dot{\theta})$, and then of $\Omega(t + T_{rep}) = \Omega(t) + \dot{\Omega}T_{rep}$, $\omega(t + T_{rep}) = \omega(t) + \dot{\omega}T_{rep}$, $M(t + T_{rep}) = M(t) + \dot{M}T_{rep}$, we find :

$$\psi_{lmpq}(t + T_{rep}) - \psi_{lmpq}(t) = [(l - 2p + q)h + mk]2\pi - q\dot{\omega}T_{rep}$$

To the order o in eccentricity ($q = o$), we find that the argument after T_{rep} differs by a multiple of 2π ; therefore the differences of any two elements are equal to zero. In particular, the radial perturbations are the same on any ascending *or* descending arc... but not necessarily at a cross-over between an ascending <u>and</u> a descending arc (it will be shown in chapter 5 how they actually differ). The term $q\dot{\omega}T_{rep}$ causes this result to be approximate : we can only say that all short periodic perturbations due to geopotential model errors are eliminated in repeat arc differences. In the case of a frozen repeat orbit, we have $\dot{\omega} \approx o$ and we can expect the effect of $q\dot{\omega}T_{rep}$ to be negligible.

2.4 THE DETERMINATION OF A GEOPOTENTIAL MODEL-OVERVIEW

Global modeling of the Earth's gravity field has been a concern since the beginning of the artificial satellites era. Observing the trajectories in space of such proof-masses allows in principle to determine the forces which act upon them and compute the coefficients inherent to their parameterization ; this is the oldest inverse problem of celestial mechanics. In practice, however, trajectories are observed from ground stations (sometimes from another satellite) by means of ranging devices (radars, laser system which now reach centimeter precision), range-rate measurement apparatus (measuring the Doppler effect), even tracking cameras which observed, in the old days, the directions to the satellites on the sky background. All these instruments have limitations (biases and noise) and, since satellites must be flown at a minimum altitude H if we want them to live long enough (say above 350 km for a life-time of a few months - without manoeuvering the orbit), the attenuation factor $[R/(R+H)]^l$ ultimately limits the degree l (and order $m \le l$) to which we can determine the spherical harmonics of the geopotential.

Another important fact lies in the frequency spectrum of the geopotential orbital perturbations, which comes from the decomposition of the disturbing function $\mathcal{R}$ as given by (59). We can re-arrange the quadruple summation over (l,m,p,q) as follows : for a model to be determined up to degree and order L, we first write that :

$$\sum_{l=2m=o}^{L}\sum_{}^{L} \ldots = \sum_{m=o}^{L}\sum_{l=\max(m,2)}^{L} \ldots$$

Then, changing p into $k = l - 2p$, q into $s = l - 2p + q$, interchanging the summations over l and k and finally limiting the series in $G_{lpq}(e)$ to $|q| \le Q$, we readily find :

$$\mathcal{R} = \frac{\mu}{a}\sum_{m=o}^{L}\sum_{k=-L}^{+L}\sum_{s=-k-Q}^{k+Q}\left\{\left(\sum_{\substack{l=\max(m,2,|k|)\\l-k:even}}^{L} i^{l-m}\overline{K}_{lm}\left(\frac{R}{a}\right)^l D_{l,m,(q-k)/2}(I)G_{l,(q-k)/2,s-k}(e)\right)\right\}\exp(i\psi_{ksm}) \tag{66}$$

with : $\psi_{ksm} = k\omega + sM + m(\Omega - \theta)$.

From (66) it is obvious that several harmonics (an infinity when $L \to \infty$) give rise to perturbations of the same frequency. By varying s, one goes from the so-called m-daily perturbations (period $\sim 2\pi/(m\dot\theta)$ when $s = o$, with $\dot\omega$ and $\dot\Omega \ll \dot\theta$) to short period perturbations ($s \ne o$) which all involve the same $\overline{K}_{lm}$ harmonics. For a given distribution of tracking stations it is usually not possible to observe well enough the orbit, that is to sample well enough the perturbations it undergoes, and the resulting observation equations are then insufficient to separate the different harmonics. That is why, with this approach of geopotential determination, it is necessary to have several satellites with varied altitudes and especially inclinations so as to get very different coefficients in the bracketted term of (66), hence independent observation equations for the harmonics. A very favourable situation is also when the orbit is in shallow resonance, that is when there exist (l, m, p, q) or equivalently (k, s, m) such that $\dot\psi_{ksm} \ll n$ ($\dot\psi_{ksm}$ may eventually come to zero in cases of sharp resonance, but these are transient phenomena). Neglecting $\dot\omega$ and $\dot\Omega$ with respect to n and θ, such a situation occurs when $sn \sim m\dot\theta$. If n is expressed in revolution per day, we have approximately $n \sim m/s$. When n is an integer, the main resonant perturbations are with coefficients $\overline{K}_{ln}$ ($s = 1$), then with $\overline{K}_{l,2n}(s = 2)$, $\overline{K}_{l,3n}(s = 3)$, ... ; if $n = r/2$ (r : integer), resonance occurs with $\overline{K}_{lr}(s = 2)$, $\overline{K}_{l,2r}(s = 4)$, and so on... These enhanced perturbations allow to better determine the corresponding class (es) of harmonics. Finally, we remark that, if we have a polar satellite mission which results, after some time, in a ground track pattern with equatorial inter-track distance Δ, and if observations are made along the orbit at least every $(\Delta/R)/n$ seconds (n is here in rd/sec), then the data sample allows, in the perfect case, to recover all harmonics up to degree and order $L \sim [\pi R/(2\Delta)]$.

Most accurate geopotential models have also been determined by combining satellite data (from which satellite only solutions may be computed) with surface gravity measurements and also satellite derived geoid heights from past altimetry missions (Geos 3, Seasat, Geosat) - after correction for sea surface topography (the difference between the ocean surface and the geoid), from a model, or by simultaneously determining it. Equation (46) is the basis for performing this combination as far as the geoid height is concerned. A similar equation exists for the gravity anomalies Δg derived from surface measurements : $\Delta g \approx g_{measured}$(reduced on the geoid) $- \gamma_{ref.ellipsoid}$(on the ellipsoid) is the basic quantity used in this case (γ is the theoretical gravity - it is called "normal gravity") ; the relationship between Δg and the geopotential harmonics is :

$$\Delta g = \frac{\mu}{r^2} \sum_{l=2}^{\infty} (l-1) \left(\frac{R}{r}\right)^l \sum_{m=0}^{l} \left(\overrightarrow{C}_{lm} \cos m\lambda + \overline{S}_{lm} \sin m\lambda\right) \overline{P}_{lm}(\sin\phi) \tag{67}$$

where $\overrightarrow{C}_{lm}$ is as in (46). Such equations are in general rewritten for mean values which are derived from real measurements (or sometimes from predicted anomalies - with a larger uncertainty, to avoid artefacts in the poorly covered areas).

As an example of what can be obtained from the above described techniques, figures 11 and 12 show contour maps of the geoid height and predicted errors of one GRIM 4 (combined) solution (Reigber et al., 1992). Usual features in the geoidal surface are visible, and the errors accumulate over land areas not well covered by gravity data (altimetry data were used over the oceans, thus providing a much better control). A full account on how the recent GRIM models were derived can be found in the CNES-DARA 1991 report.

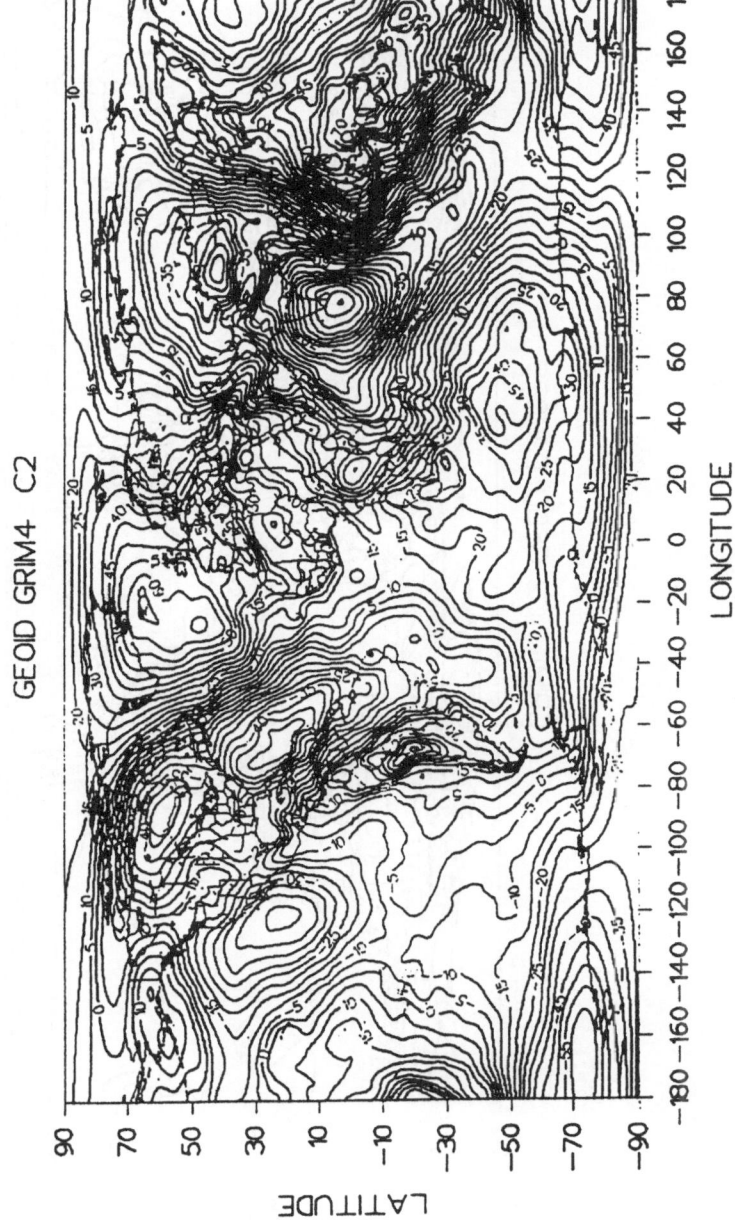

Fig.11.

GEOID GRIM4 C2

Fig. 12.

GRIM4 C2 GEOID ERRORS

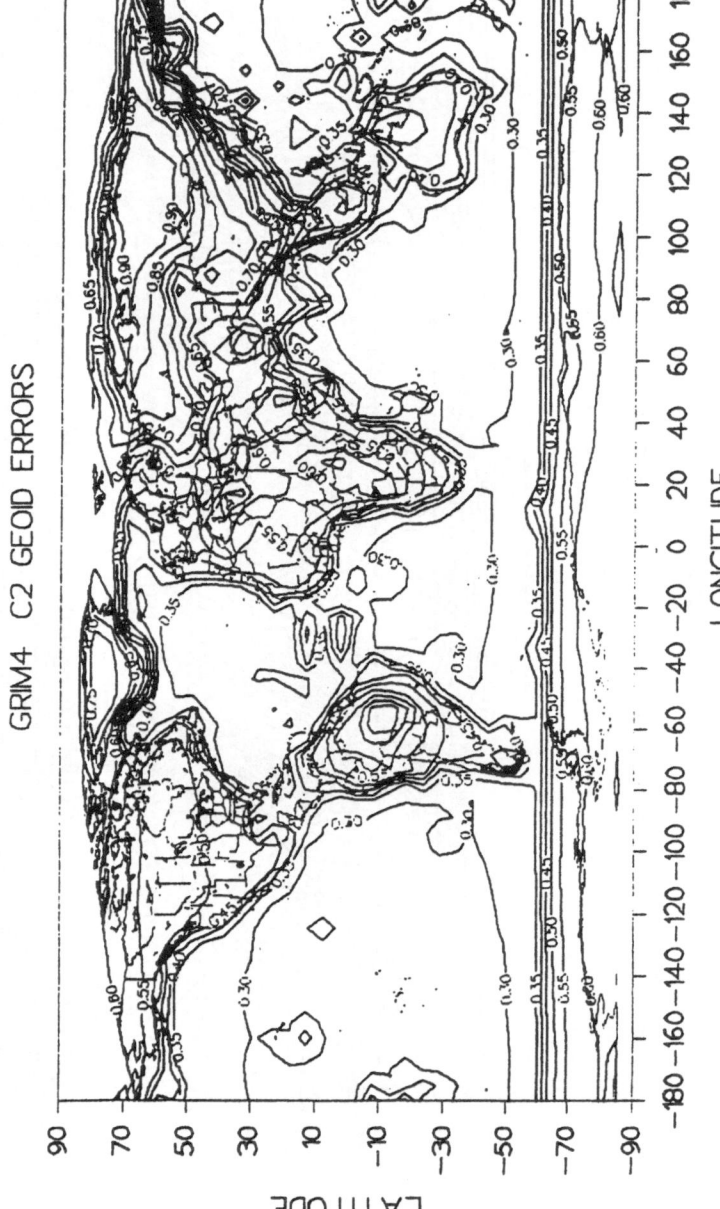

3 TEMPORAL REPRESENTATION OF THE RADIAL PERTURBATIONS DUE TO THE GEOPOTENTIAL

The radial perturbation, Δr, on a satellite orbit due to the geopotential may be derived in various ways, and with various approximations. One approach uses the Hill equations (Schrama, 1989), others start from the Lagrange equations and Kaula's formulation of the solution (Wagner, 1985 ; Rosborough, 1986, Engelis, 1987). Here, we will also start from the analytical expressions of the orbital perturbations derived from the Lagrange equations (formulas 64) and from $r(a,e,M)$ as given by (13) or (17). We will first derive the perturbations of order o and order 1 in an elementary fashion, to show some of the traps which are encountered when working with almost singular elements ($e \sim o$ usually), also to correct some mistakes made in earlier works in the terms of order 1.

We will therefore write :

$$\Delta r = \sum_{l,m,p} (\Delta r_{lmp}^{o} + \Delta r_{lmp}^{1} + \dots \Delta r_{lmp}^{k} \dots) \tag{68}$$

where the superscript stands for the order in eccentricity.

Perturbations in the two other orthogonal directions (transverse and normal) may be derived in a similar way. The transverse component is $\Delta \tau$, derived from $\Delta \tau = r \Delta \psi = r(\Delta v + \Delta \omega + \Delta \Omega \cos I)$, can be evaluated from (18) and from the perturbations in M, ω, and Ω. The normal component, $\Delta \zeta$, comes from $\Delta \zeta = r[\Delta I \sin(\omega + v) - \Delta \Omega \sin I \cos(\omega + v)]$, requires to expand $\cos v$ and $\sin v$ by (19) and (20)[1]. Results can be found in Rosborough (ibid).

3.1 RADIAL PERTURBATIONS OF ZERO AND FIRST ORDER IN ECCENTRICITY

We start from (13) truncated at e^2 : $r = a[1 - e \cos M - e^2(\cos 2M - 1)/2]$ from which, by differentiation, we obtain :

$$\Delta r = (1 - e \cos M)\Delta a + a(e - \cos M - e \cos 2M)\Delta e + a \sin M e \Delta M$$

To obtain Δr^o and Δr^1, we rewrite Δr as :

$\Delta r =$	Δa .... with terms of order	0,1 in e
	$- e \Delta a \cos M$	0
	$- a \Delta e \cos M$	0,1

---

1 $\Delta \tau$ and $\Delta \zeta$ come from $\dot{\xi}, \dot{\eta}, \dot{\zeta}$ derived prior to the Gauss equation for ω...

$$+ \, ea\Delta e \, ... \qquad\qquad 0$$

$$- \, ea\Delta e \cos 2M \, ... \qquad\qquad 0$$

$$+ \, ae\Delta M \sin M \, ... \qquad\qquad 0,1$$

Now we use (64). We take advantage of (54), (55) and (56) and we keep terms with :

. $q = o, +1, -1$ for Δa

. $q = o, +1, -1$ for $e\Delta M$ and retain g_{lpo} for G'_{lpo} (and $1 - e^2 \approx 1$)

. $q = o, \pm 1, \pm 2$ for $a\Delta e$ expanding $\sqrt{1 - e^2}(l - 2p + q) - (l - 2p)$ as $\approx -(l - 2p)\frac{e^2}{2} + q$; therefore

the singularity disappears when $q = o$; also, there is no singularity in e for $q = \pm 1$ due to $e g_{lp \pm 1}$; finally the remaining $\sqrt{1 - e^2}$ term is replaced by 1, and $q = \pm 2$ generates a term in e.

Hence, dropping other e^2 terms which are generated by these choices of q and following the prescribed orders per term in Δr :

$$\Delta a = 2A \left\{ \frac{l - 2p}{\dot{\psi}_{lmpo}} S_{lmpo} + e \left[\frac{l - 2p + 1}{\dot{\psi}_{lmp1}} g_{lp1} S_{lmp1} + \frac{l - 2p + 1}{\dot{\psi}_{lmp-1}} g_{lp-1} S_{lmp-1} \right] \right\}$$

$$-e\Delta a \cos M = -2A \left\{ e \left[\frac{l - 2p}{\dot{\psi}_{lmpo}} S_{lmpo} \cos M \right] \right\}$$

$$-a\Delta e \cos M = -A \left\{ \frac{g_{lp1}}{\dot{\psi}_{lmp1}} S_{lmp1} \cos M - \frac{g_{lp-1}}{\dot{\psi}_{lmp-1}} S_{lmp-1} \cos M \right.$$
$$\left. + e \left[-\frac{l - 2p}{2\dot{\psi}_{lmpo}} S_{lmpo} \cos M + \frac{g_{lp2}}{\dot{\psi}_{lmp2}} S_{lmp2} \cos M + \frac{g_{lp-2}}{\dot{\psi}_{lmp-2}} S_{lmp-2} \cos M \right] \right\}$$

$$ea\Delta e = A \left\{ e \left[\frac{g_{lp1}}{\dot{\psi}_{lmp1}} S_{lmp1} - \frac{g_{lp-1}}{\dot{\psi}_{lmp-1}} S_{lmp-1} \right] \right\}$$

$$-ea\Delta e \cos 2M = -A \left\{ e \left[\frac{g_{lp1}}{\dot{\psi}_{lmp1}} S_{lmp1} \cos 2M - \frac{g_{lp-1}}{\dot{\psi}_{lmp-1}} S_{lmp-1} \cos 2M \right] \right\}$$

$$ae\Delta M \sin M = A \left\{ -\frac{g_{lp1}}{\dot{\psi}_{lmp1}} S^*_{lmp1} \sin M - \frac{g_{lp-1}}{\dot{\psi}_{lmp-1}} S^*_{lmp-1} \sin M \right.$$
$$\left. + e \left[\left(2(l + 1) - g_{lpo} - 3(l - 2p)\frac{n}{\dot{\psi}_{lmpo}} \right) \frac{1}{\dot{\psi}_{lmpo}} S^*_{lmpo} \sin M \right] \right\}$$

Now, recalling the expressions of S_{lmpq} and S^*_{lmpq}, and noting that $\psi_{lmpq} \pm sM = \psi_{lmp,q \pm s}$, we form :

$$S_{lmpq} \cos sM = \frac{1}{2}(S_{lmp,q+s} + S_{lmp,q-s})$$

$$S_{lmpq} \sin sM = \frac{1}{2}(S^{\bullet}_{lmp,q+s} - S^{\bullet}_{lmp,q-s})$$

$$S^{\bullet}_{lmpq} \cos sM = \frac{1}{2}(S^{\bullet}_{lmp,q+s} + S^{\bullet}_{lmp,q-s}) \tag{69}$$

$$S^{\bullet}_{lmpq} \sin sM = \frac{1}{2}(S_{lmp,q-s} - S_{lmp,q+s})$$

Collecting the terms of order zero, we get :

$$\Delta r^{o}_{lmp} = A\left[\frac{2l-4p}{\dot{\psi}_{lmpo}} - \frac{g_{lp1}}{\dot{\psi}_{lmp1}} + \frac{g_{lp-1}}{\dot{\psi}_{lmp-1}}\right]S_{lmpo} \tag{70}$$

(four terms, with S_{lmp2} and S_{lmp-2}, have cancelled out...).

When $l - 2p = o$ and $m = o$, the first term is actually zero (the factor $l - 2p$ cancels the term before integration), and there remains a constant term :

$$\Delta r^{o}_{2p,o,p} = A\left[-\frac{g_{2p,p,1}}{\dot{\psi}_{2p,o,p,1}} + \frac{g_{2p,p,-1}}{\dot{\psi}_{2p,o,p,-1}}\right]\overline{C}_{2p,o}$$

This formula also shows that semi-major axis perturbations with frequencies $\dot{\psi}_{lmpo} = (l-2p)(\dot{\omega}+\dot{M}) + m(\dot{\Omega}-\dot{\theta})$ produce radial perturbations at the same frequencies ; it also implies that perturbations on a at any other frequency ($q \neq o$) produce much smaller radial perturbations. It is also seen that the terms in Δe and ΔM which yield the largest radial perturbations have the frequencies $\dot{\psi}_{lmp\pm1}$ and produce terms at the same (previous) frequencies $\dot{\psi}_{lmpo}$. A major consequence of this is that the long period perturbations on e and M result in short period radial perturbations. For example, if $m = o$ and $l - 2p = 1$, we have perturbations on e and M with frequency $\dot{\omega}$, due to the odd degree zonals ; radially, they induce a perturbation with frequency $\dot{\omega}+\dot{M}$ (once per revolution). Due to the usually large amplitude of such long period perturbations on e and M, the short period radial perturbation on r is also quite large.

The terms of order 1 give a more complicated result. After some algebra, one finds :

$$\Delta r^{1}_{lmp} = Ae\ \left[\left(\frac{C^{+1}_{1}}{\dot{\psi}_{lmp1}} + \frac{C^{+1}_{o}}{\dot{\psi}_{lmpo}} + \frac{C^{+1}_{2}}{\dot{\psi}_{lmp2}} + \frac{C^{+1}_{-1}}{\dot{\psi}_{lmp-1}}\right)S_{lmp1}\right.$$

$$+ \left(\frac{C^{-1}_{-1}}{\dot{\psi}_{lmp-1}} + \frac{C^{-1}_{o}}{\dot{\psi}_{lmpo}} + \frac{C^{-1}_{-2}}{\dot{\psi}_{lmp-2}} + \frac{C^{-1}_{1}}{\dot{\psi}_{lmp1}}\right)S_{lmp-1}$$

$$\left(\frac{C^{+3}_{1}}{\dot{\psi}_{lmp1}} + \frac{C^{+3}_{2}}{\dot{\psi}_{lmp2}}\right)S_{lmp3} + \left.\left(\frac{C^{-3}_{-1}}{\dot{\psi}_{lmp-1}} + \frac{C^{-3}_{-2}}{\dot{\psi}_{lmp-2}}\right)S_{lmp-3}\right] \tag{71}$$

Formula (70) is in agreement with previous works of other authors. In (71), we have :

$$C^{+1}_{1} = (3l-4p+1)(l-2p+3/2)$$

$$C_o^{+1} = \frac{3}{2}(p-l) - 1 + \frac{1}{4}(4p-3l)(l-4p) + \frac{3}{2}(l-2p)n/\dot{\psi}_{lmpo}$$

$$C_2^{+1} = \frac{1}{2}\left[(p-l)\left(2l-3p+\frac{5}{2}\right) - \frac{1}{4}(l-2p+2)^2\right]$$

$$C_{-1}^{+1} = \frac{1}{4}(4p-l+1)$$

$$C_{-1}^{-1} = (4p-l+1)(l-2p-3/2)$$

$$C_o^{-1} = \frac{3}{2}p + 1 - \frac{1}{4}(4p-3l)(l-4p) - \frac{3}{2}(l-2p)n/\dot{\psi}_{lmpo}$$

$$C_{-2}^{-1} = -\frac{1}{2}\left[p\left(3p-l+\frac{5}{2}\right) + \frac{1}{4}(l-2p-2)^2\right]$$

$$C_1^{-1} = -\frac{1}{4}(3l-4p+1)$$

These are the terms found by Rosborough, with (it seems) a missing factor (1/2) for $C_2^{+1} C_{-1}^{+1}, C_1^{-1}$ and a factor (- 1/2) for C_{-2}^{-1}.

This author forgot the four other terms, which are :

$$C_1^{+3} = -g_{lp1}/2$$

$$C_2^{+3} = -g_{lp2}/2$$

$$C_{-1}^{-3} = g_{lp-1}/2$$

$$C_{-2}^{-3} = -g_{lp-2}/2$$

It is interesting to quote another form of formula (70), derived by Wagner (1985). Making the substitution of indices $(l, m, p) \rightarrow (m, k, l)$ already encountered in (66), denoting $\dot{\psi}_{lmpo} = \dot{\psi}_{lkm} = \dot{\psi}_o$ (then $\dot{\psi}_{lmp\pm1} = \dot{\psi}_o \pm n$) and then $\dot{\psi}_o/n = \beta_{km} = k(1+\dot{\omega}/n) + m(\dot{\Omega}-\dot{\theta})/n$, it is easy to transform (70) into :

$$\Delta r^o = \sum_{m=o}^{L} \sum_{k=-L}^{+L} \sum_{\substack{l=\max(m,2,k) \\ l-k: \text{even}}}^{L} na\left(\frac{R}{a}\right)^l \overline{F}_{lm,q-k/2}(l) \frac{\beta_{km}(l+1)-2k}{\beta_{km}(\beta_{km}^2-1)} S_{lm,q-k/2,o} \tag{72}$$

3.2 GENERAL FORMULATION OF THE RADIAL PERTURBATIONS

We start from (17) for r and we follow Rosborough :

$$r = a \sum_{s=o}^{\infty} H_s \cos sM$$

with :

$$H_o = 1 + e^2/2$$

$$H_s = -\frac{2e}{s^2}\frac{d}{de}[J_s(se)] , \quad \text{for} \quad s > o$$

We define $H'_s = dH_s/de$, and find easily :

$$\Delta r = \Delta a\left(\sum_{s=o}^{\infty} H_s \cos sM\right) + a\Delta e\left(\sum_{s=o}^{\infty} H'_s \cos sM\right) - a\Delta M\left(\sum_{s=o}^{\infty} sH_s \sin sM\right) \tag{73}$$

We then apply (64) and, for a particular set (l, m, p, q) we find :

$$\Delta r_{lmpq} = \Delta a_{lmpq} \sum_s H_s \cos sM$$

$$+ a\Delta e_{lmpq} \sum_s H'_s \cos sM$$

$$- a\Delta M_{lmpq} \sum_s sH_s \sin sM \tag{74}$$

Looking at this series term by term, we have for any combination (l, m, p, q, s) :
$$\Delta r_{lmpqs} = \Delta a_{lmpq} H_s \cos sM + a\Delta e_{lmpq} H'_s \cos sM - a\Delta M_{lmpq} sH_s \sin sM \tag{75}$$

Since $J_s(x) = o(x^s)$, H_s is of order e^s, that is the perturbations decrease with s increasing. Replacing Δa_{lmpq} by $C^a_{lmpq} S_{lmpq}$, etc... we get an expression with products $S_{lmpq} \cos sM$, $S_{lmpq} \sin sM$, $S^*_{lmpq} \sin sM$, which we transform by (69), hence :

$$\Delta r_{lmpqs} = \frac{1}{2}(C^a_{lmpq} H_s + a C^e_{lmpq} H'_s + a C^M_{lmpq} sH_s) S_{lmp(q+s)}$$

$$+ \frac{1}{2}(C^a_{lmpq} H_s + a C^e_{lmpq} H'_s - a C^M_{lmpq} sH_s) S_{lmp(q-s)}$$

If the range of s is changed from o to $+\infty$ to be $-\infty$ to $+\infty$ and the function $\tilde{H}_s$ is defined as :

$$\tilde{H}_o = H_o$$

$$\tilde{H}_s = H_s/2 , \tilde{H}_{-s} = \tilde{H}_s , (s = 1,2,... + \infty),$$

then we can write in a compact form :

$$\Delta r_{lmpqs} = C_{lmpqs} S_{lmp(q+s)}$$

where :

$$C_{lmpqs} = C^a_{lmpq} \tilde{H}_s + a C^e_{lmpq} \tilde{H}'_s + a C^M_{lmpq} s\tilde{H}_s$$

The total radial perturbation is written :

$$\Delta r = \sum_{l=2m=o}^{\infty} \sum_{p=o}^{l} \sum_{q=-\infty}^{l} \sum_{s=-\infty}^{+\infty} \sum^{+\infty} \Delta r_{lmpqs}$$

We can make the following changes of indices :

$q' = q + s$: q' range is $-\infty, +\infty$

$s' = q$: s' range is $-\infty, +\infty$

and then rename q' as being q and s' as being s, and we obtain :

$$\Delta r = \sum_{l=2}^{\bar{}} \sum_{m=o}^{l} \sum_{p=o}^{l} \sum_{q=-\infty}^{+\infty} \Delta r_{lmpq} \tag{76}$$

where :

$$\Delta r_{lmpq} = C_{lmpq} S_{lmpq} \tag{77}$$

and :

$$C_{lmpq} = \sum_{s=-\infty}^{+\infty} C_{lmps(q-s)}$$

$$= \sum_{s=-\infty}^{+\infty} C_{lmps} \bar{H}_{q-s} + a\, C_{lmps} \bar{H}'_{q-s} + a(q-s) C_{lmps}^{M} \bar{H}_{q-s} \tag{78}$$

This form shows that the radial perturbation amplitude at a given frequency (l, m, p, q given) depends on the perturbations on a, e, M at that frequency ($s = q$ in the summation) and also of an infinite number of different frequencies (other values of s).

4 TEMPORAL CHARACTERISTICS OF THE RADIAL PERTURBATIONS AND ERRORS ON A SATELLITE ORBIT

4.1 FREQUENCY SPECTRUM

We start from (77) and want to identify all terms of different frequency. From the form of S_{lmpq} and ψ_{lmpq}, we infer that we must distinguish between the zonal ($m = o$) and non-zonal terms ($m > o$). We found also simpler to start from a formula where the frequencies are indeed identified by three indices k, q, m (cf. formula (66) for $\mathcal{R}$) : $\psi_{kqm} = k(\omega + M) + qM + m(\Omega - \theta)$.

From this, it is clear that :

- when $m = o$: terms of all different frequencies are obtained for :
  . $k = o$: $+ q$ and $- q$ but $q \neq o$ ($q = o$ gives a secular term)
  . $k \neq o$: (k, q) and $(- k, - q)$ since $\psi_{-k,-q,o} = -\psi_{kqo}$
- when $m > o$: all terms with different (k, q)'s generate different frequencies.

Hence, writing that we have a model truncated à $l = L$ and that q is limited to $| q | \leq Q$:

$$\Delta r = \sum_{m=o}^{L} \sum_{k=-L}^{+L} \sum_{q=-Q}^{+Q} \sum_{l=l_{min}}^{L} C_{lm,q-k/2,q} S_{lm,q-k/2,q} \tag{79}$$

with $l_{min} = \max(2, |k|, m)$ and $l - k$ being always even in the summation, we have :

- for the zonal terms :

   - when $k = o$:

$$\sum_{q=1}^{Q}\left[\sum_{l=2}^{L}(C_{lo,l/2,q} + C_{lo,l/2,-q})\overline{C}_{lo}\right]\cos qM$$

   - when $k > o$:

$$\sum_{q=-Q}^{+Q}\left[\sum_{l}(C_{lo,q-k/2,q} + (-1)^l C_{lo,q+k/2,-q})\overline{C}_{lo}\right]\binom{\cos}{\sin}_{k\,:\,odd}^{k\,:\,even}\psi_{kqo}$$

These two cases can be compacted in :

$$\Delta r_{(m=o)} = \sum_{q=1}^{Q}\left[\sum_{j}(C_{2j,o,j,q} + C_{2j,o,j,-q})\overline{C}_{2j,o}\right]\cos\psi_{oqo}$$

$$+ \sum_{k=1}^{L}\sum_{q=-Q}^{Q}\left[\sum_{j}(C_{k+2j,o,j,q} + (-1)^k C_{k+2j,o,k+j,-q})\overline{C}_{k+2j,o}\right]\binom{\cos}{\sin}_{k\,:\,odd}^{k\,:\,even}\psi_{kqo} \tag{80}$$

This has been derived by setting $l - k = 2j$, and we have j running from j_{min} to j_{max} :

$$j_{min} = \max(o, 1 - [k/2])$$

$$j_{max} = [L - k]/2$$

- for the tesseral harmonics using the same transformation of indices, we find :

$$\Delta r_{(m>o)} = \sum_{k=-L}^{+L}\sum_{q=-Q}^{+Q}\left[\left(\sum_{j=j_{min}}^{j_{max}} C_{k+2j,m,j,q}\check{C}_{k+2j,m}\right)\cos\psi_{kqm}\right.$$

$$\left. + \left(\sum_{j=j_{min}}^{j_{max}} C_{k+2j,m,j,q}S_{k+2j,m}\right)\sin\psi_{kqm}\right] \tag{81}$$

where we now have :

$$j_{min} = \max(o, 1 - [k/2], -k, [m - k]/2)$$

$$j_{max} = [L - k]/2, \text{ as before.}$$

By letting the indices run as indicated, that is for $m = o : k = o$ to L, $q = -Q$ to Q ($q \neq o$) ; and for $m > o : k = -L$ to $+L$, $q = -Q$ to $+Q$, we obtain all terms of different frequencies.

The amplitudes are obtained by :

$$m = o, k = o \quad : \quad | \sum_j [(C_{2j,o,j,q} + C_{2j,o,j,-q}) \overline{C}_{2j,o}] | \quad ; \quad q = 1 \quad \text{to} \quad Q$$

$$m = o, o < k \le L \quad : \quad | \sum_j [C_{k+2j,o,j,q} + (-1)^k C_{k+2j,o,k+j,-q}] \overline{C}_{k+2j,o} | \quad ; \quad q = -Q \quad \text{to} \quad +Q$$

$$m > o, -L \le k \le L \quad : \quad \left[\left(\sum_j C_{k+2j,m,j,q} \overline{C}_{k+2j,m} \right)^2 + \left(\sum_j C_{k+2j,m,j,q} \overline{S}_{k+2j,m} \right)^2 \right]^{1/2} \quad ; \quad q = -Q \text{ to} +Q$$

As an example, figure 13 shows the spectrum of the Topex-Poseidon radial orbit perturbations based on Rapp 1991 model truncated at degree and order 70.

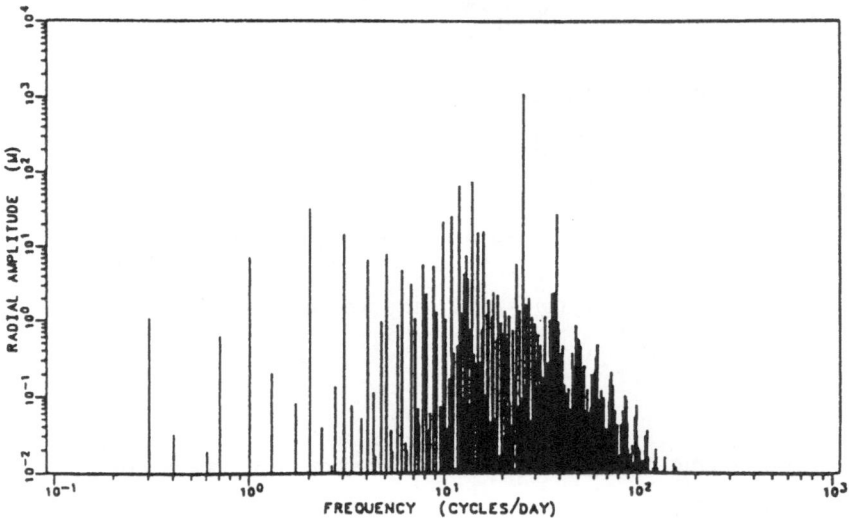

Fig. 13. Frequency spectrum of the Topex-Poseidon radial orbit perturbations

4.2 RADIAL PERTURBATIONS BY COEFFICIENT, BY ORDER, BY DEGREE

We will derive the r.m.s. perturbations for : each pair of coefficient $(\overline{C}_{lm}, \overline{S}_{lm})$, then for all coefficients of a given order m, finally for all coefficients of a given degree. Depending on the case, we will use one form or another, such as (76), or (80)-(81), of the radial perturbations, which is most suited to identify the different frequencies in order to take properly the r.m.s. of the ad'hoc terms.

4.2.1 For a pair of coefficients

We start from (76), for l and m fixed, that is :

$$\Delta r_{lm} = \sum_{p=o}^{l} \sum_{q=-Q}^{Q} C_{lmpq} S_{lmpq}$$

Then :

$$< \Delta r_{lm}^2 > = \sum_{p=o}^{l} \sum_{j=o}^{l} \sum_{q=-Q}^{Q} \sum_{s=-Q}^{Q} C_{lmpq} C_{lmjs} < S_{lmpq} S_{lmjs} >$$

Recalling the form of S_{lmpq}, it is clear that the means <...> are zero unless $\psi_{lmpq} = \pm\psi_{lmjs}$. This condition is satisfied when :

* $m = o : j = p, s = q$ and $j = l - p, s = -q$; that is :

$$< S_{lopq}^2 > = \overline{C}_{lo}^2 < \binom{\cos^2}{\sin^2} \psi_{lopq} > = \frac{1}{2}\overline{C}_{lo}^2$$

where <...> is taken over the smallest common multiple of all encountered periods.

$$< S_{lopq} S_{lo,l-p,-q} > = \overline{C}_{lo}^2 \binom{\cos}{\sin} \psi_{lopq} \binom{\cos}{\sin} \psi_{lo,l-p,-q} \text{ in distinguishing between } l \text{ even (cos...) and } l \text{ odd}$$

(sin...) ; hence <...> $= -(1)^l \overline{C}_{lo}^2/2$ in this case.

Therefore, for a zonal term :

$$< \Delta r_{lo}^2 > = \frac{1}{2}\overline{C}_{lo}^2 \sum_{p=o}^{l} \sum_{q=-Q}^{+Q} [C_{lopq}^2 + (-1)^l C_{lopq} C_{lo,l-p,-q}] \tag{82}$$

* $m > o$: we simply need $j = p, s = q$.

Since $< S_{lmpq}^2 > = \frac{1}{2}(\overline{C}_{lm}^2 + \overline{S}_{lm}^2)$ we have :

$$< \Delta r_{lm}^2 > = \frac{1}{2}(\overline{C}_{lm}^2 + \overline{S}_{lm}^2) \sum_{p=o}^{l} \sum_{q=-Q}^{+Q} C_{lmpq}^2 \tag{83}$$

The r.m.s. follows by taking the square root of (82) or (83).

4.2.2 For a given order m

We already identified precisely the indices yielding different frequencies.

* $m = o$: we start from (80), square Δr_o and take the average ; hence :

$$< \Delta r_{(m=o)}^2 > = \frac{1}{2}\sum_{q=1}^{Q}\left[\sum_{j}(C_{2j,o,j,q} + C_{2j,o,j,-q})\overline{C}_{2j,o}\right]^2$$

$$+ \frac{1}{2}\sum_{k=1}^{L}\sum_{q=-Q}^{Q}\left[\sum_{j}(C_{k+2j,o,j,q} + (-1)^k C_{k+2j,o,k+j,-q})\overline{C}_{k+2j,o}\right]^2 \tag{84}$$

* $m > o$: starting from (81), we readily find :

$$< \Delta r^2_{(m>o)} > = \frac{1}{2} \sum_{k=-L}^{L} \sum_{q=-Q}^{Q} \left[\left(\sum_{j} C_{k+2j,m,j,q} C_{k+2j,m} \right)^2 \right.$$
$$\left. + \left(\sum_{j} C_{k+2j,m,j,q} S_{k+2j,m} \right)^2 \right] \tag{85}$$

In (84) and (85), the range of index j is as prescribed in (80) and (81). From these, it is easy to find the full field perturbations in summing over all orders (since frequencies of all terms of different orders are all different), that is :

$$r.m.s.(\Delta r) = \left[\sum_{m=o}^{L} < \Delta r^2_{(m)} > \right]^{1/2} \tag{86}$$

4.2.3 For a given degree *l*

We here start from (76), for l fixed, that is :

$$\Delta r_{(l)} = \sum_{m=o}^{l} \sum_{p=o}^{l} \sum_{q=-Q}^{+Q} C_{lmpq} S_{lmpq}$$

Then :

$$< \Delta r^2_{(l)} > = \sum_{m=o}^{l} \sum_{k=o}^{l} \sum_{p=o}^{l} \sum_{j=o}^{l} \sum_{q=-Q}^{Q} \sum_{s=-Q}^{Q} C_{lmpq} C_{lkjs} < S_{lmpq} S_{lkjs} >$$

If $m \neq k$, frequencies are necessarily different and $<...>$ is zero. Hence we are left with :

$$< \Delta r^2_{(l)} > = \sum_{m=o}^{l} \sum_{p=o}^{l} \sum_{j=o}^{l} \sum_{q=-Q}^{Q} \sum_{s=-Q}^{Q} C_{lmpq} C_{lmjs} < S_{lmpq} S_{lmjs} >,$$

which is nothing but :

$$< \Delta r^2_{(l)} > = \sum_{m=o}^{l} < \Delta r^2_{lm} > \tag{87}$$

Therefore it suffices to add the terms of (82) and (83) for l fixed.

As examples, fig. 14 and 15 give the radial perturbations by order and by degree which are expected for Topex (model truncated at degree and order 50).

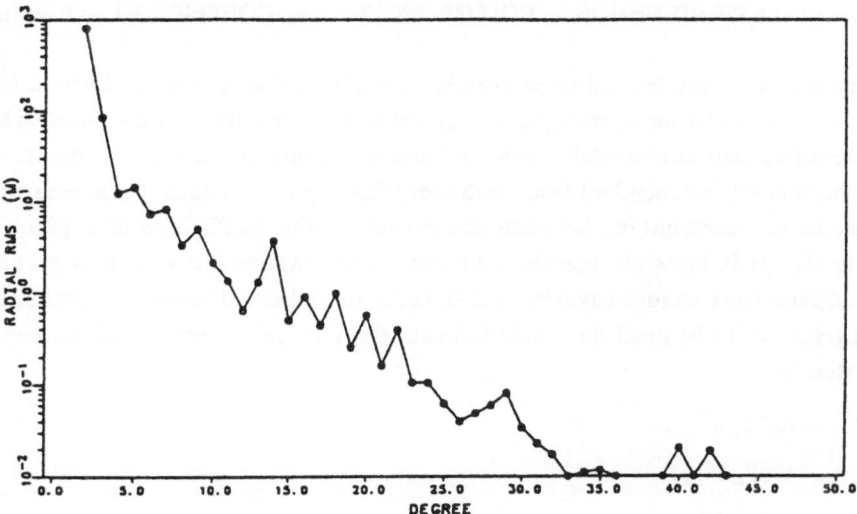

Fig. 14. r.m.s. of the Topex-Poseidon radial orbit perturbations by degree

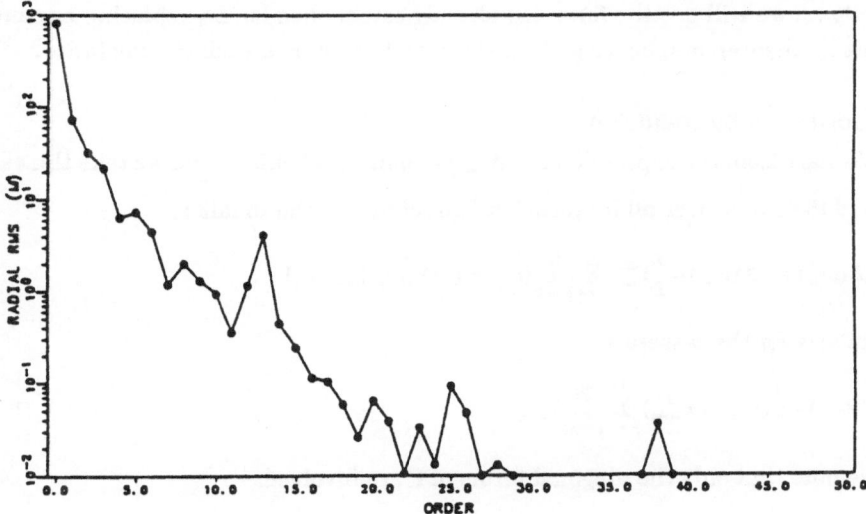

Fig. 15. r.m.s. of the Topex-Poseidon radial orbit perturbations by order

4.3 THE RADIAL ORBIT ERRORS FROM GEOPOTENTIAL COEFFICIENT COVARIANCES

These errors may be evaluated grossly by applying the previously derived formulas for the radial perturbations (by coefficient, by order, etc...) to difference coefficients between two different geopotential models. This is a too crude approximation for : (I) the models are to-day not so much independent from each other (having a lot of data sets in common) ; (II) this ignores the correlations between the errors on the coefficients of a given model. Consequently, it is better to use the statistical informations inherent to a given global solution, that is the variance-covariance matrix Γ of the spherical harmonics. The subscripts and superscripts to be used for Γ will indicate the particular element of the covariance matrix that is :

$$\Gamma^{CC}_{lml'm'} = E[\delta\overline{C}_{lm}\delta\overline{C}_{l'm'}]$$

$$\Gamma^{SS}_{lml'm'} = E[\delta\overline{S}_{lm}\delta\overline{S}_{l'm'}]$$

$$\Gamma^{CS}_{lml'm'} = E[\delta\overline{C}_{lm}\delta\overline{S}_{l'm'}] \tag{88}$$

$$\Gamma^{SC}_{lml'm'} = E[\delta\overline{S}_{lm}\delta\overline{C}_{l'm'}]$$

where E is the expectation operator, the $\delta\overline{C}$'s and $\delta\overline{S}$'s are the errors on the coefficients of the model. The Γ matrix is symmetrical, that is $\Gamma^{CC}_{lml'm'} = \Gamma^{CC}_{l'm'lm}$, $\Gamma^{SS}_{lml'm'} = \Gamma^{SS}_{l'm'lm}$, $\Gamma^{CS}_{lml'm'} = \Gamma^{SC}_{l'm'lm}$. In what follows, we will use the formulas already established for Δr, replacing harmonic coefficients by their errors, hence getting the radial error δr in each circumstance.

4.3.1 Radial error by coefficient

We start from the expression of $< \Delta r^2_{lm} >$ of paragraph 4.2.1., and we take the expected value and the r.m.s. over all frequencies. Therefore, for the zonals :

$$< E(\delta r^2_{lo}) > = \sigma^2(\delta r_{lo}) = \frac{1}{2}\Gamma^{CC}_{lolo} \sum_{p=o}^{l} \sum_{q=-Q}^{+Q} [G^2_{lopq} + (-1)^l G_{lopq} G_{l,o,l-p,-q}] \tag{89}$$

and similarly for the tesserals :

$$\sigma^2(\delta r_{lm}) = \frac{1}{2}(\Gamma^{CC}_{lmlm} + \Gamma^{SS}_{lmlm}) \sum_{p=o}^{l} \sum_{q=-Q}^{+Q} G^2_{lmpq} \tag{90}$$

We note that only the diagonal terms of Γ are involved.

4.3.2 Radial error by order

Starting from (84) and (85), we find by expanding the squares and then taking the expectation :

. for the zonals :

$$\sigma^2[\delta r_{(m=o)}] = \frac{1}{2} \sum_{q=1}^{Q} \left[\sum_j \sum_h (C_{2j,o,j,q} + C_{2j,o,j,-q}).(C_{2h,o,h,q} + C_{2h,o,h,-q}) \Gamma^{cc}_{2j,o,2h,o} \right]$$

$$+ \frac{1}{2} \sum_{k=1}^{L} \sum_{q=-Q}^{Q} \left[\sum_j \sum_h (C_{k+2j,o,j,q} + (-1)^k C_{k+2j,o,k+j,-q}). \right.$$

$$(C_{k+2h,o,h,q} + (-1)^k C_{k+2h,o,k+h,-q}) \Gamma^{cc}_{k+2j,o,k+2h,o} \qquad (91)$$

for the tesserals :

$$\sigma^2[\delta r_{(m>o)}] = \frac{1}{2} \sum_{k=-L}^{L} \sum_{q=-Q}^{Q} \left[\sum_j \sum_h C_{k+2j,m,j,q} C_{k+2h,m,h,q} (\Gamma^{cc}_{k+2j,m,k+2h,m} + \Gamma^{ss}_{k+2j,m,k+2h,m}) \right] \qquad (92)$$

To obtain this formula, we have used the expression of $(\hat{C}_{lm}, \hat{S}_{lm})$ in terms of $\overline{C}_{lm}, \overline{S}_{lm}$, and the fact that, if k - m is even we find $\Gamma^{cc}_{...} + \Gamma^{ss}_{...}$, and, if k - m is odd, we find $\Gamma^{ss}_{...} + \Gamma^{cc}_{...}$ (with the same subscripts), that is the same sum of two terms. We remark that off-diagonal terms of Γ are needed, but only between C's or between S's...

In (91) and (92), j and h run as in the formulas (84) and (85) from which they originated. Figure 16 shows an example of such radial error for the satellites ERS1 and TOPEX-PO-SEIDON, computed from the covariances of the GRIM4-C2 model.

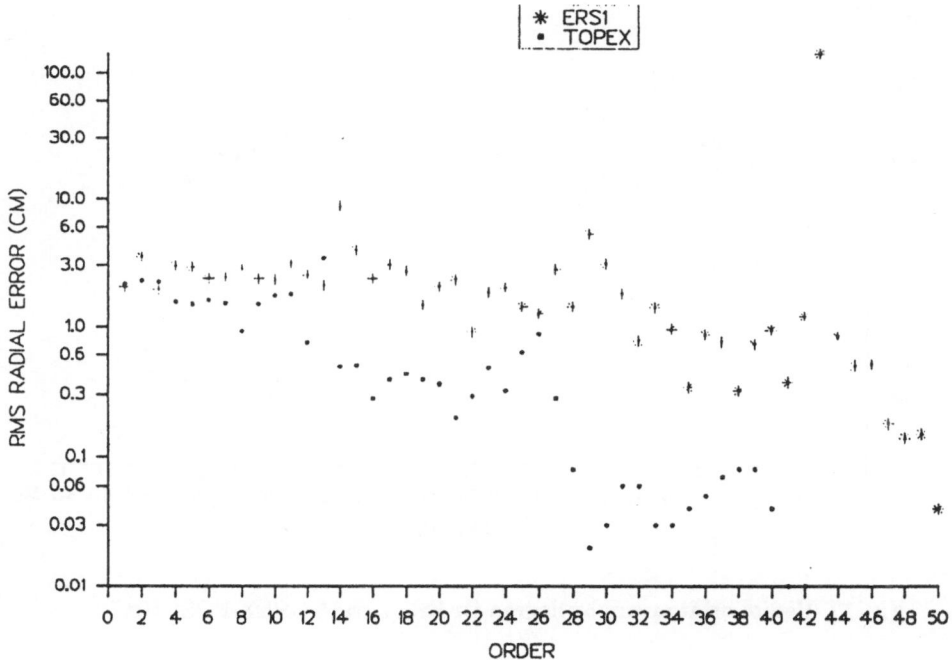

Fig. 16. Radial orbit errors by order for ERS1 and TOPEX-POSEIDON

4.3.3 Radial error by degree

From (87) and (89), (90), we easily find :

$$\sigma^2 <\Delta r_{(l)}> = \sum_{m=o}^{l} \sigma^2 <\delta r_{lm}>$$

$$=\frac{1}{2}\Gamma_{lolo}^{cc} \sum_{p=o}^{l} \sum_{q=-Q}^{Q} [C_{lopq}^2 + (-1)^l C_{lopq} C_{l,o,l-p,-q}]$$

$$+\frac{1}{2}\sum_{m>o}^{l} (\Gamma_{lmlm}^{cc}+\Gamma_{lmlm}^{ss}) \sum_{p=o}^{l} \sum_{q=-Q}^{Q} C_{lmpq}^2 \qquad (93)$$

In this formula, only diagonal elements of Γ are needed. Figure 17 shows an example of this computation for ERS1 and TOPEX-POSEIDON (same model covariances as in fig. 16).

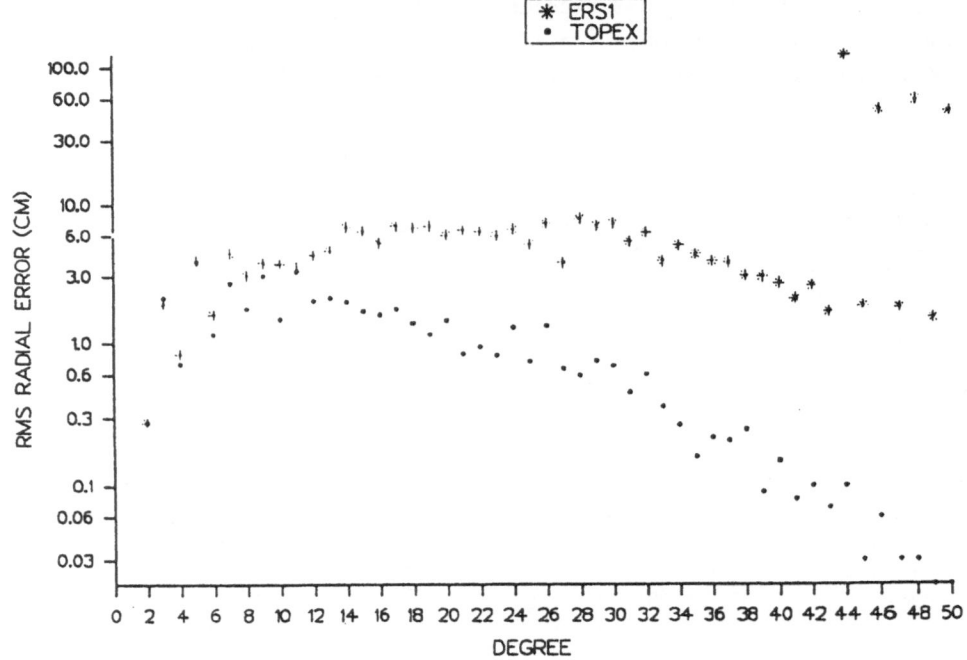

Fig. 17. Radial orbit errors by degree for ERS1 and TOPEX-POSEIDON

4.3.4 Total radial orbit error ; two approaches

The full field radial orbit error may first be evaluated, for instance by summing up all terms obtained for various orders, that is by writing :

$$\sigma^2(\delta r) = \sum_{m=o}^{L} \sigma^2 < \delta r_{(m)} > \tag{94}$$

where individual contributions are computed from (91) and (92). This gives a single number, whereas the actual behaviour of δr with time in general presents significant variability around this value. Also, the way the r.m.s. is computed cancels out all correlations between C's and S's coefficients (since the sine and cosine functions are always orthogonal).

Hence it is of great interest to study δr itself instead of its r.m.s. value, for instance to map δr as a function of time, starting from its general expression as given by (76) and (77) ; we here write :

$$\delta r = \sum_{m=o}^{L} \sum_{l=\max(2,m)}^{L(m)} (u_{lm}\delta\overline{C}_{lm} + v_{lm}\delta\overline{S}_{lm}) \tag{95}$$

where we have introduced $L(m)$ as it is usual in most geopotential models (maximum degree per order), and with :

$$u_{lm} = \sum_{p=o}^{l} \sum_{q=-Q}^{Q} C_{lmpq} \begin{pmatrix} \cos \\ \sin \end{pmatrix}_{l-m \,:\, \mathrm{odd}}^{l-m \,:\, \mathrm{even}} \psi_{lmpq} \tag{96}$$

$$v_{lm} = \sum_{p=o}^{l} \sum_{q=-Q}^{Q} C_{lmpq} \begin{pmatrix} \sin \\ -\cos \end{pmatrix}_{l-m \,:\, \mathrm{odd}}^{l-m \,:\, \mathrm{even}} \psi_{lmpq} \tag{97}$$

We introduce the vectors :

$$W = [u_{20},\ldots u_{L0}, u_{21}, v_{21}, u_{22}, v_{22}, \ldots u_{lm}, v_{lm} \ldots u_{LL} v_{LL}]^+$$

and $\quad \delta K = [\delta\overline{C}_{20}, \ldots, \delta\overline{C}_{L0}, \delta\overline{C}_{21}, \delta\overline{S}_{21}, \delta\overline{C}_{22}, \delta\overline{S}_{22} \ldots \delta\overline{C}_{lm}, \delta\overline{S}_{lm} \ldots \delta\overline{C}_{LL}, \delta\overline{S}_{LL}]^T$

and rewrite δr as :

$$\delta r = W^+ \delta K$$

Then :

$$E(\delta r^2) = W^+ E(\delta K . \delta K^+) W$$

and finally, from the definition of Γ :

$$\sigma(\delta r) = [W^+ \quad \Gamma W]^{1/2} \tag{98}$$

This expression must be evaluated all along the orbit, for instance over one day with a small time step so as to get the shortest period variations, and over several days (for instance up to the repeat period if it exists) with a large stepsize to get an insight in other time variations.

The analytic expression (94) provides a quick means for examining a priori the expected radial accuracy as a function of orbital geometry and of given covariances of a geopotential model. For instance, we can examine how $\sigma^2(\delta r)$ changes at a particular altitude for varying inclination, which can be a determining factor for selecting it. An example is given on figure 18 at the altitude of the ERS1 satellite.

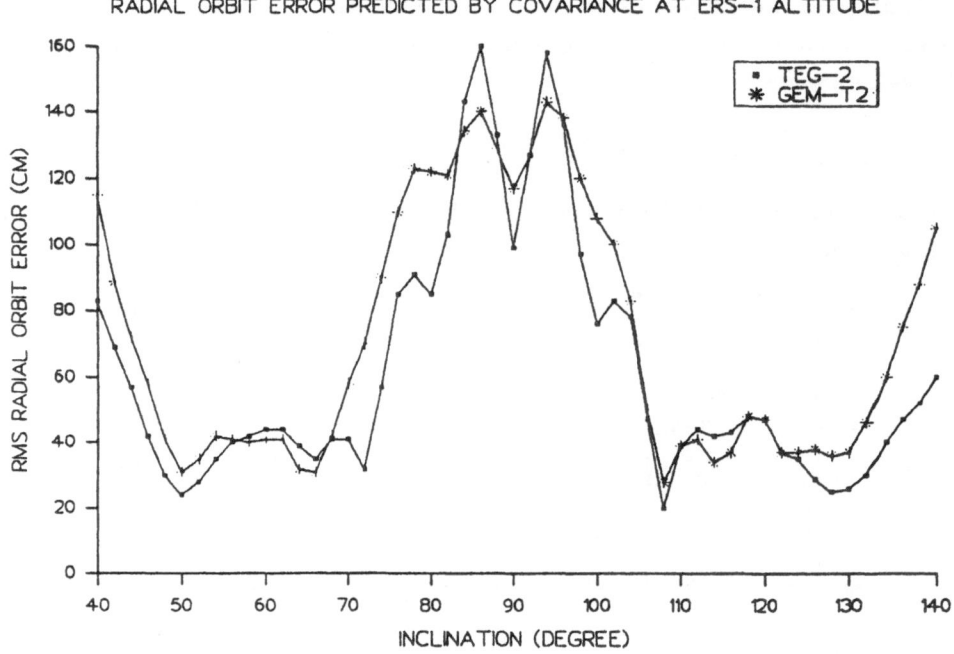

Fig. 18. The total radial orbit error at the altitude of ERS1, as a function of inclination, (based on the covariances of two-different models : GEMT2 and TEG2)

5 SPATIAL REPRESENTATION AND CHARACTERISTICS OF THE RADIAL ORBIT PERTURBATIONS AND ERRORS

The preceding results are important to understand the nature of the radial perturbation and its time variations. Equally important for altimetric satellite missions is how this perturbation and its error are manifested in the Earth fixed reference system, if there exist geographically correlated errors (which a cross-over minimisations technique will not be able to remove), etc...

We will make this transformation here, starting from the temporal solution expressed as a power series in eccentricity. We will limit ourselves to the first term of this series (the order zero in eccentricity), which is the only one which can be represented explicitly as a function of the satellite's latitude and longitude. The character of the higher order terms is such that they can be represented only partially in the body-fixed system ; the remaining part is time dependent without a strict relation to geographic location (primarily due to the circulation of the perigee). For small eccentricities, the zero order solution is quite sufficient. In other cases, it at least gives the main contribution and allows interesting mission analysis studies. We therefore start from formula (70) and follow Rosborough's derivation.

5.1 RADIAL PERTURBATIONS IN THE SPATIAL DOMAIN

We write $\Delta r \approx \Delta r^\circ$, with :

$$\Delta r^\circ = \sum_{l=2}^{\bar{\infty}} \sum_{m=o}^{l} \sum_{p=o}^{l} A_{lmp} S_{lmpo} \tag{99}$$

where :

$$A_{lmp} = na\left(\frac{R}{a}\right)^l \bar{F}_{lmp}(I)\left[\frac{2(l-2p)}{\dot{\psi}_{lmpo}} + \frac{4p-3l-1}{2\dot{\psi}_{lmp1}} + \frac{4p-l+1}{2\dot{\psi}_{lmp-1}}\right] \tag{100}$$

and, as before :

$$S_{lmpo} = \begin{pmatrix}\bar{C}_{lm} \\ -\bar{S}_{lm}\end{pmatrix}_{l-m:\,odd}^{l-m:\,even} \cos\psi_{lmpo} + \begin{pmatrix}\bar{S}_{lm} \\ \bar{C}_{lm}\end{pmatrix}_{l-m:\,odd}^{l-m:\,even} \sin\psi_{lmpo}$$

$$\psi_{lmpo} = (l-2p)(\omega+M) + m(\Omega-\theta)$$

Assuming that A_{lmp}, evaluated with the mean elements and rates, are constant coefficients, the task is simply to transform S_{lmpo} in terms of ϕ, λ, the latitude and longitude of the satellite in the Earth fixed system (cf. figure 8). In other words the transformation is a kind of inverse of the Kaula's transformation dealt with in paragraph 2.3.1. We are going to do it using elementary algebra.

We write $\Omega - \theta = \lambda - (\alpha - \Omega)$, where α is the right ascension of the satellite (in the Σ system). We transform $\cos\{[(l-2p)(\omega+M) - m(\alpha-\Omega)] + m\lambda\}$ in terms of $\cos m\lambda$ and $\sin m\lambda$, and of $\cos[(l-2p)(\omega+M) - m(\alpha-\Omega)]$ and $\sin[(l-2p)(\omega+M) - m(\alpha-\Omega)]$, and similarly for $\sin\{[...]+m\lambda\}$. Then :

$$S_{lmpo} = \begin{pmatrix}\cos\beta \\ \sin\beta\end{pmatrix}_{l-m:\,odd}^{l-m:\,even}(\bar{C}_{lm}\cos m\lambda + \bar{S}_{lm}\sin m\lambda) + \begin{pmatrix}-\sin\beta \\ \cos\beta\end{pmatrix}_{l-m:\,odd}^{l-m:\,even}(\bar{C}_{lm}\sin m\lambda - \bar{S}_{lm}\cos m\lambda)$$

with : $\beta = (l-2p)(\omega+M) - m(\alpha-\Omega)$.

We now need to transform the sines and cosines of multiples of arguments into power series of sines and cosines of these arguments, by using (when $N \geq o$) :

$$\cos Nx = \sum_{k=o}^{[N/2]} \binom{N}{2k} (-1)^k \cos^{N-2k} x \sin^{2k} x$$

$$\sin Nx = \sum_{k=o}^{[(N-1)/2]} \binom{N}{2k+1} (-1)^k \cos^{N-2k-1} x \sin^{2k+1} x$$

The argument x will successively be $\alpha - \Omega$ with $N = m$, and $\omega + M$ with $N = l - 2p$. When $N < o$ we use $\sin(Nx) = -\sin(-Nx) = (N/|N|)\sin(|N|x)$; this occurs when $p > l/2$ in the case of $\omega + M$ only.

In the spherical triangle NHS of figure 8, we now write :

$$\sin \phi = \sin I \sin(\omega + v)$$

$$\approx \sin I \sin(\omega + M)$$

to zero order in eccentricity, and :

$$\cos(\omega + v) = \cos \phi \cos(\alpha - \Omega) \approx \cos(\omega + M)$$

$$\sin(\omega + v)\cos I = \cos \phi \sin(\alpha - \Omega) \approx \sin(\omega + M)\cos I$$

Given the latitude ϕ, $\sin(\omega + M)$ can be determined uniquely by $\sin \phi / \sin I$. But $\cos(\omega + M)$, being written $\pm[1 - \sin^2(\omega + M)]^{1/2} = (-1)^\sigma (\sin^2 I - \sin^2 \phi)^{1/2}/\sin I$, must have its sign specified without ambiguity. It is geometrically obvious that it must be (+) for the satellite being on its ascending track, and (-) on the descending track. Hence we introduce $\sigma = o$ on the ascending track and $\sigma = 1$ on the descending one. We find after some elementary algebra :

$$\cos[(l - 2p)(\omega + M)] = (-1)^{\sigma H} Y_H^c / \sin^H I$$

$$\sin[(l - 2p)(\omega + M)] = (-1)^{\sigma(H-1)} \frac{l - 2p}{H} Y_H^s / \sin^H I$$

$$\cos m(\alpha - \Omega) = (-1)^{\sigma m} Z_m^c / (\cos^m \phi \sin^m I)$$

$$\sin m(\alpha - \Omega) = (-1)^{\sigma(m-1)} Z_m^s / (\cos^m \phi \sin^m I)$$

where H stands for $|l - 2p|$ (for the $\sin(l - 2p)x$ formula...), and with :

$$Y_H^c = \sum_{k=o}^{[H/2]} \binom{H}{2k} (-1)^k (\sin^2 I - \sin^2 \phi)^{H/2-k} \sin^{2k} \phi$$

$$Y_H^s = \sum_{k=o}^{[(H-1)/2]} \binom{H}{2k+1} (-1)^k (\sin^2 I - \sin^2 \phi)^{(H-1)/2-k} \sin^{2k+1} \phi$$

$$Z_m^c = \sum_{k=o}^{[m/2]} \binom{m}{2k} (-1)^k \cos^{2k} I (\sin^2 I - \sin^2 \phi)^{m/2-k} \sin^{2k} \phi \qquad (101)$$

$$Z'_m = \sum_{k=o}^{[(m-1)/2]} \binom{m}{2k+1} (-1)^k \cos^{2k+1} I (\sin^2 I - \sin^2 \phi)^{(m-1)/2-k} \sin^{2k+1} \phi$$

From these, and since $(-1)^{o(H+m)} = (-1)^{o(I-m)}$:

$$\cos\beta = (-1)^{o(I-m)} \Phi^c_{Imp}$$

$$\sin\beta = (-1)^{o(I-m-1)} \Phi^s_{Imp}$$

with :

$$\Phi^c_{Imp} = \{Y^c_H Z^c_m + [(l-2p)/H] Y^s_H Z'_m\}/(\cos^m \phi \sin^{m+H} I)$$

$$\Phi^s_{Imp} = \{[(l-2p)/H] Y^s_H Z^c_m - Y^c_H Z'_m\}/(\cos^m \phi \sin^{m+H} I)$$

(when $l-2p = o$, the factor $(l-2p)/H$ must be made equal to zero, for the term from which it originates is $\sin[(l-2p)(\omega+M)] \equiv o$).

We rewrite now S_{Impo} in separating the cases l - m : even and l - m : odd.

. l - m : even

$$S_{Impo} = (-1)^{o(I-m)} \Phi^c_{Imp} [\overline{C}_{lm} \cos m\lambda + \overline{S}_{lm} \sin m\lambda] + (-1)^{o(I-m-1)} (-\Phi^s_{Imp}) [\overline{C}_{lm} \sin m\lambda - \overline{S}_{lm} \cos m\lambda]$$

$$= \Phi^c_{Imp} [\ldots + \ldots] - (-1)^o \Phi^s_{Imp} [\ldots - \ldots]$$

. l - m : odd

$$S_{Impo} = (-1)^{o(I-m-1)} \Phi^s_{Imp} [\overline{C}_{lm} \cos m\lambda + \overline{S}_{lm} \sin m\lambda] + (-1)^{o(I-m)} \Phi^c_{Imp} [\overline{C}_{lm} \sin m\lambda - \overline{S}_{lm} \cos m\lambda]$$

$$= \Phi^s_{Imp} [\ldots + \ldots] + (-1)^o \Phi^c_{Imp} [\ldots - \ldots]$$

We now define :

$$\check{\Phi}^c_{Imp} = \Phi^c_{Imp} \text{ and } \check{\Phi}^s_{Imp} = -\Phi^s_{Imp} \text{ when } l - m \text{ is even}$$

$$\check{\Phi}^c_{Imp} = \Phi^s_{Imp} \text{ and } \check{\Phi}^s_{Imp} = \Phi^c_{Imp} \text{ when } l - m \text{ is odd.}$$

We note that we can perform the summation on p separately ; let us have :

$$M_{lm} = \sum_{p=o}^{l} A_{Imp} \check{\Phi}^c_{Imp}$$

$$V_{lm} = \sum_{p=o}^{l} A_{Imp} \check{\Phi}^s_{Imp}$$

and finally :

$$\Delta r = \sum_{l=2}^{\bar{}} \sum_{m=o}^{l} M_{lm}(\overline{C}_{lm}\cos m\lambda + \overline{S}_{lm}\sin m\lambda)$$

$$+ (-1)^{\sigma} \sum_{l=2}^{\bar{}} \sum_{m=o}^{l} V_{lm}(\overline{C}_{lm}\sin m\lambda - \overline{S}_{lm}\cos m\lambda) \tag{102}$$

This is the searched form since M_{lm} and V_{lm} are functions of the satellite latitude (and of the mean elements a and I). It shows that the first part is a mean regional orbit perturbation, the geographically correlated part, of which the error is unobservable when examining cross-over differences ; the second part has its sign which depends on the ascending (+) and descending (-) portions of the orbit, it characterizes the variability of the perturbation and its error is partly recoverable.

5.2 RADIAL PERTURBATIONS BY COEFFICIENT, BY DEGREE AND BY ORDER

In the following, we will need to separate the mean geographical radial perturbation :

$$\Delta\bar{r} = \sum_l \sum_m M_{lm}(\overline{C}_{lm}\cos m\lambda + \overline{S}_{lm}\sin m\lambda) \tag{103}$$

from the variability around the mean :

$$\Delta r^{\bullet} = \sum_l \sum_m V_{lm}(\overline{C}_{lm}\sin m\lambda - \overline{S}_{lm}\cos m\lambda) \tag{104}$$

which is to be multiplied by $(-1)^{\sigma}$, with $\sigma = o$ for ascending passes and $\sigma = 1$ for descending passes (the notations M_{lm} , V_{lm} help reminding that the M's are for the mean part and that the V's are for its variability).

To characterize Δr , $\Delta\bar{r}$ or $\Delta r^{\bullet}$, one can compute their r.m.s. values. However, due to the non linear coupling of the longitude and latitude (via the orbital motion) and due to the complexity of the M's and V's functions, it is not possible to find an analytical representation of the r.m.s. Instead, the three functions can be evaluated along the satellite ground track and the r.m.s. values are then numerically computed. Although it is easier to perform, it may not be as accurate to compute Δr , $\Delta\bar{r}$, $\Delta r^{\bullet}$ directly on a regular grid in (ϕ,λ) - of course limited to $-I \le \phi \le I$ because of the Y's and Z's functions defined by (101), because the satellite may actually never pass through a given point of the grid..., however, if one knows from a preliminary study of the coverage that the Earth's surface is to be densely covered, this simpler procedure proves to be accurate enough.

Mapping the total radial perturbation Δr requires to separate the cases of ascending and descending tracks. Representation of $\Delta\bar{r}$ and $\Delta r^{\bullet}$ can be done in one map for each function.

Let us now make a few general remarks on these functions. We will denote by Δr_{lm} , $\Delta\bar{r}_{lm}$, $\Delta r_{lm}^{\bullet}$ individual terms, which may be perturbations or errors if potential coefficients are replaced by their uncertainties.

```
  2   0.0  2.2  1.0
  3   0.0   .3   .9   .8
  4   0.0  2.3   .4  1.2  1.1
  5   0.0   .5  2.6   .9  1.1  1.0
  6   0.0  1.0   .6  1.0   .8  1.0   .9
  7   0.0   .1  1.7   .8  1.2  1.0  1.1  1.0
  8   0.0  3.5   .7  1.5   .9  1.1  1.0  1.0   .9
  9   0.0   .3  1.2   .7  1.0   .9  1.0   .9  1.0   .9
 10   0.0  3.0   .5  1.3   .9  1.1  1.0  1.0  1.0  1.0  1.0
 11   0.0   .3  2.5   .8  1.3   .9  1.0  1.0  1.0  1.0  1.0   .8
 12   0.0  2.4   .5  1.2   .8  1.0   .9  1.0  1.0  1.0  1.0  1.0  1.0
 13   0.0  1.1  1.9  1.0  1.2  1.0  1.1  1.0  1.0  1.0  1.0   .9  1.0  1.0
 14   0.0  4.1   .6  1.6   .9  1.1  1.0  1.0  1.0  1.0  1.0  1.0  1.0  1.0  1.0
 15   0.0   .2  1.6   .6  1.1   .8  1.0   .9  1.0  1.0  1.0  1.0  1.0  1.0  1.0  1.0
D16   0.0  3.3   .8  1.4  1.0  1.1  1.0  1.0  1.0  1.0  1.0  1.0  1.0  1.0  1.0  1.0  1.0
e17   0.0   .3  2.7   .7  1.2   .9  1.0  1.0  1.0  1.0  1.0  1.0  1.0  1.0  1.0  1.0  1.0  1.0
g18   0.0  2.9   .5  1.3   .8  1.1   .9  1.0  1.0  1.0  1.0  1.0  1.0  1.0  1.0  1.0  1.0  1.0
r19   0.0   .4  2.0   .8  1.3  1.0  1.1  1.0  1.0  1.0  1.0  1.0  1.0  1.0  1.0  1.0  1.0  1.0
e20   0.0        .6        .8        .9  1.0       1.0  1.0  1.0  1.0  1.0  1.0  1.0  1.0  1.0
e21   0.0   .2        .6  1.2   .8       1.0  1.0            1.0  1.0  1.0  1.0  1.0  1.0  1.0
 22   0.0  3.5   .6  1.5   .9  1.1  1.0  1.1  1.0  1.0  1.0  1.0  1.0  1.0       1.0  1.0  1.0
 23   0.0   .3        .7            1.0                           1.0  1.0            1.0
 24   0.0             1.4       1.1       1.0  1.0       1.0  1.0  1.0  1.0  1.0  1.0  1.0
 25   0.0   .3  2.1       1.3   .9       1.0            1.0  1.0            1.0
 26   0.0                               1.0  1.0  1.0  1.0  1.0            1.0
 27   0.0            1.9                           1.0
 28   0.0                               1.0       1.0  1.0  1.0
 29   0.0   .3  1.6                          1.0  1.0  1.0  1.0
 30   0.0                          1.0  1.0  1.0  1.0
 31   0.0   .3
 32   0.0                                    1.0
 33   0.0
 34   0.0                          1.0  1.0
 35   0.0   .3
 36   0.0

Order  0    1    2    3    4    5    6    7    8    9   10   11   12   13   14   15   16   17


 18   1.0
 19   1.0  1.0
 20   1.0  1.0
 21   1.0  1.0  1.0  1.0
 22   1.0  1.0  1.0  1.0
 23             1.0  1.0  1.0
D24   1.0  1.0       1.0
e25             1.0  1.0  1.0  1.0  1.0
g26   1.0
r27                       1.0  1.0  1.0  1.0
e28        1.0                 1.0  1.0
e29                       1.0  1.0
 30                            1.0  1.0  1.0
 31                            1.0  1.0  1.0  1.0
 32                            1.0
 33
 34
 35                       1.0  1.0
 36

Order 18   19   20   21   22   23   24   25   26   27   28   29   30   31   32   33   34   35   36
```

Fig. 19. Ratios of the r.m.s. of the variable part $\Delta r_{lm}^{\centerdot}$ of the radial orbital error to the r.m.s. of the geographical mean $\Delta \bar{r}_{lm}$, for each coefficient pair (based on the GEM-10B geopotential model - from Rosborough, 1986).

The r.m.s. values for a given pair of coefficients are easy to evaluate, and it is informative to compute the ratios r.m.s. $(\Delta r_{lm}^{\centerdot})/\text{r.m.s.}(\Delta \bar{r}_{lm}) = \rho_{lm}$ which show the change in geographic behaviour of the radial perturbation or error as a function of coefficient order (generally). In the case of TOPEX-POSEIDON (fig. 19), it is found that the magnitude of ρ_{lm} is a function of whether $l\text{-}m$ is even or odd, for the long-wavelength terms.

It is noted that $\rho_{lo} = o$, which is obvious since the variability due to the zonals, $\Delta r_{lo}^{\cdot}$, is always zero, while $\Delta \bar{r}_{lo} = M_{lo}\overline{C}_{lo}$. Hence, residual orbit errors due to mismodeling of the zonals cannot be removed by altimetric crossing arc techniques, and they entirely map onto the errors on the determined sea surface.

Looking at the individual terms due to the tesseral harmonics, we can write :

$$\Delta \bar{r}_{lm} = M_{lm}\bar{J}_{lm}\cos m(\lambda - \lambda_{lm})$$

$$\Delta r_{lm}^{\cdot} = \pm V_{lm}\bar{J}_{lm}\sin m(\lambda - \lambda_{lm})$$

where $\bar{J}_{lm}\cos m\lambda_{lm} = \overline{C}_{lm}$ and $\bar{J}_{lm}\sin m\lambda_{lm} = \overline{S}_{lm}$. This shows that the two quantities are 90 degrees out of phase in longitude. Hence, $\Delta r_{lm}^{\cdot}$ is minimum (and even equal to zero) where $\Delta \bar{r}_{lm}$ is maximum, and vice versa.

The r.m.s. values for a given degree are simply computed as :

$$\sigma(\overline{\Delta r_l}) = r.m.s.\left\{\sum_{m=o}^{l} \Delta \bar{r}_{lm}\right\} \tag{105}$$

$$\sigma(\Delta r_l^{\cdot}) = r.m.s.\left\{\sum_{m=o}^{l} \Delta r_{lm}^{\cdot}\right\} \tag{106}$$

$$\sigma(\Delta r_l) = r.m.s.\left\{\sum_{m=o}^{l} [\Delta \bar{r}_{lm} + (-1)^{\sigma}\Delta r_{lm}^{\cdot}]\right\} \tag{107}$$

Equations (105) and (106) may be evaluated directly on a grid. Two grids are generated (for $\sigma = o, 1$) in the case of (107), or this quantity has to be computed along the orbit as said before (and the result must be very close to what was obtained via the temporal representation (cf. figure 17).

Similarly, by order m, we have :

$$\sigma(\Delta \bar{r}_m) = r.m.s.\left\{\sum_{l=m}^{L} \Delta \bar{r}_{lm}\right\} \tag{108}$$

$$\sigma(\Delta r_m^{\cdot}) = r.m.s.\left\{\sum_{l=m}^{L} \Delta r_{lm}^{\cdot}\right\} \tag{109}$$

$$\sigma(\Delta r_m) = r.m.s.\left\{\sum_{l=m}^{L} \Delta \bar{r}_{lm} + (-1)^{\sigma}\Delta r_{lm}^{\cdot}\right\} \tag{110}$$

and these quantities are evaluated as above. We already noticed that $\Delta r_o^{\cdot} = o$. In the case of the tesserals, an interesting feature is that :

$$\Delta \bar{r}_m = \left(\sum_{l=m}^{L} M_{lm}\overline{C}_{lm}\right)\cos m\lambda + \left(\sum_{l=m}^{L} M_{lm}\overline{S}_{lm}\right)\sin m\lambda$$

$$\Delta r_m^* = \left(\sum_{l=m}^{L} V_{lm} \overline{C}_{lm} \right) \sin m\lambda + \left(\sum_{l=m}^{L} V_{lm} \overline{S}_{lm} \right) \cos m\lambda$$

which means that the two surfaces defined by these equations have $2\,m$ zeros in longitude for any given latitude.

Finally, the full field radial perturbations can be evaluated by taking the r.m.s. of equations (102), (103) and (104) under the conditions previously defined.

5.3 RADIAL ORBIT ERRORS IN THE SPACE DOMAIN BASED ON GEOPOTENTIAL CO-VARIANCES

As we did it in paragraph 4.3, we are going to look at the radial orbit errors mapped in the Earth fixed system, and evaluated from the covariance matrix Γ of a global geopotential model. We will of course distinguish between the mean part of the error and its variability.

5.3.1 Radial error by coefficient

We start from :

$$\delta \overline{r}_{lm} = M_{lm} (\delta \overline{C}_{lm} \cos m\lambda + \delta \overline{S}_{lm} \sin m\lambda)$$

$$\delta r_{lm}^* = V_{lm} (\delta \overline{C}_{lm} \sin m\lambda - \delta \overline{S}_{lm} \cos m\lambda)$$

$$\delta r_{lm} = [M_{lm} \cos m\lambda + (-1)^\sigma V_{lm} \sin m\lambda] \delta \overline{C}_{lm} + [M_{lm} \sin m\lambda - (-1)^\sigma V_{lm} \cos m\lambda] + \delta \overline{S}_{lm}$$

Squaring these expressions and taking the expected value, we find :

$$\sigma^2(\delta \overline{r}_{lm}) = M_{lm}^2 [\Gamma_{lmlm}^{CC} \cos^2 m\lambda + \Gamma_{lmlm}^{SS} \sin^2 m\lambda + 2\Gamma_{lmlm}^{CS} \cos m\lambda \sin m\lambda] \tag{111}$$

$$\sigma^2(\delta r_{lm}^*) = V_{lm}^2 [\Gamma_{lmlm}^{CC} \sin^2 m\lambda + \Gamma_{lmlm}^{SS} \cos^2 m\lambda - 2\Gamma_{lmlm}^{CS} \cos m\lambda \sin m\lambda] \tag{112}$$

and :

$$\sigma^2(\delta r_{lm}) = c_{lm}^2 \Gamma_{lmlm}^{CC} + s_{lm}^2 \Gamma_{lmlm}^{SS} + 2 c_{lm} s_{lm} \Gamma_{lmlm}^{CS} \tag{113}$$

where we have defined :

$$c_{lm} = M_{lm} \cos m\lambda + (-1)^\sigma V_{lm} \sin m\lambda$$

$$s_{lm} = M_{lm} \sin m\lambda - (-1)^\sigma V_{lm} \cos m\lambda$$

$\sigma^2(\delta \overline{r}_{lm})$ and $\sigma^2(\delta r_{lm}^*)$ can be computed over a grid for any λ, ϕ (between $-I$ and $+I$) ; an r.m.s. value over this spatial domain can then be derived numerically. In the case of $\sigma^2(\delta r_{lm})$ it has to be computed along the orbit (because of $(-1)^\sigma$).

5.3.2 Radial error by degree

We write :

$$\delta \overline{r}_l = \sum_{m=o}^{l} M_{lm}(\delta \overline{C}_{lm} \cos m\lambda + \delta \overline{S}_{lm} \sin m\lambda)$$

Squaring, taking the expected value, and noting that $M_{lm}M_{lk}\Gamma_{lmlk}^{CS}\cos m\lambda \sin k\lambda$ is identical to $M_{lk}M_{lm}\Gamma_{lklm}^{SC}\cos k\lambda \sin m\lambda$, we obtain :

$$\sigma^2(\delta \overline{r}_l) = \sum_{m=o}^{l} \sum_{k=o}^{l} M_{lm}M_{lk}(\Gamma_{lmlk}^{CC}\cos m\lambda \cos k\lambda$$

$$+ \Gamma_{lmlk}^{SS}\sin m\lambda \sin k\lambda + 2\Gamma_{lmlk}^{CS}\cos m\lambda \sin k\lambda) \tag{114}$$

Similarly :

$$\sigma^2(\delta r_l^{\bullet}) = \sum_{m=o}^{l} \sum_{k=o}^{l} V_{lm}V_{lk}(\Gamma_{lmlk}^{CC}\sin m\lambda \sin k\lambda$$

$$+ \Gamma_{lmlk}^{SS}\cos m\lambda \cos k\lambda - 2\Gamma_{lmlk}^{CS}\sin m\lambda \cos k\lambda) \tag{115}$$

For δr_l, we first use the form :

$$\delta r_l = \sum_{m=o}^{l} (c_{lm}\delta \overline{C}_{lm} + s_{lm}\delta \overline{S}_{lm})$$

from which

$$\sigma^2(\delta r_l) = \sum_{m=o}^{l} \sum_{k=o}^{l} [\Gamma_{lmlk}^{CC}c_{lm}c_{lk} + \Gamma_{lmlk}^{SS}s_{lm}s_{lk}$$

$$+ 2\Gamma_{lmlk}^{CS}c_{lm}s_{lk}] \tag{116}$$

The mean variances can be computed on a grid or along the trajectory and provide global insight on these perturbations.

5.3.3 Radial error by order

For the zonal terms, since $\delta r_{lm}^{\bullet} \equiv o$, we have $\delta r_{(m=o)} = \delta \overline{r}_{(m=o)}$, and then :

$$\sigma^2[\delta r_{(m=o)}] = E\left\{\left(\sum_{l=2}^{L} M_{lo}\delta \overline{C}_{lo}\right)^2\right\} = E\left\{\sum_l \sum_k M_{lo}M_{ko}\delta \overline{C}_{lo}\delta \overline{C}_{ko}\right\}$$

$$= \sum_{l=2}^{L} \sum_{k=2}^{L} M_{lo}M_{ko}\Gamma_{loko}^{CC} \tag{117}$$

For the non zonal terms, we easily find :

$$\sigma^2[\delta \overline{r}_{(m>o)}] = \sum_{l=m}^{L} \sum_{k=m}^{L} M_{lm}M_{km}(\Gamma_{lmkm}^{CC}\cos^2 m\lambda$$

$$+ \Gamma_{lmkm}^{SS}\sin^2 m\lambda + 2\Gamma_{lmkm}^{CS}\cos m\lambda \sin m\lambda) \tag{118}$$

and :

$$\sigma^2[\delta r^{\bullet}_{(m>o)}] = \sum_{l=m}^{L}\sum_{k=m}^{L} V_{lm}V_{km}(\Gamma^{CC}_{lmkm}\sin^2 m\lambda$$

$$+ \Gamma^{SS}_{lmkm}\cos^2 m\lambda - 2\Gamma^{CS}_{lmkm}\cos m\lambda \sin m\lambda) \tag{119}$$

For the total radial error of order m, we again use the form :

$$\delta r_{(m>o)} = \sum_{l=m}^{L} (c_{lm}\delta\overline{C}_{lm} + s_{lm}\delta\overline{S}_{lm})$$

from which we obtain :

$$\sigma^2[\delta r^{\bullet}_{(m>o)}] = \sum_{l=m}^{L}\sum_{k=m}^{L} [\Gamma^{CC}_{lmkm}c_{lm}c_{km} + \Gamma^{SS}_{lmkm}s_{lm}s_{km} + 2\Gamma^{CS}_{lmkm}c_{lm}s_{km}] \tag{120}$$

5.3.4 Full field radial error

We square $\delta\overline{r}$, as given by (103) with the coefficients replaced by their errors :

$$\delta\overline{r}^2 = \sum_{l=2}^{L}\sum_{m=o}^{l}\sum_{j=2}^{L}\sum_{k=o}^{l} [M_{lm}M_{jk}(\delta\overline{C}_{lm}\cos m\lambda + \delta\overline{S}_{lm}\sin m\lambda)(\delta\overline{C}_{jk}\cos k\lambda + \delta\overline{S}_{jk}\sin k\lambda)]$$

from which we obtain the mean geographical error :

$$\sigma^2(\delta\overline{r}) = \sum_{l=2}^{L}\sum_{j=2}^{L}\sum_{m=o}^{l}\sum_{k=o}^{l} M_{lm}M_{jk}(\Gamma^{CC}_{lmjk}\cos m\lambda \cos k\lambda$$

$$+ \Gamma^{SS}_{lmjk}\sin m\lambda \sin k\lambda + 2\Gamma^{CS}_{lmjk}\cos m\lambda \sin k\lambda) \tag{121}$$

and the variability :

$$\sigma^2(\delta r^{\bullet}) = \sum_{l=2}^{L}\sum_{j=2}^{L}\sum_{m=o}^{l}\sum_{k=o}^{l} V_{lm}V_{jk}(\Gamma^{CC}_{lmjk}\sin m\lambda \sin k\lambda$$

$$+ \Gamma^{SS}_{lmjk}\cos m\lambda \cos k\lambda - 2\Gamma^{CS}_{lmjk}\sin m\lambda \cos k\lambda) \tag{122}$$

The full error itself is computed via the form with the c_{lm} and s_{lm} coefficients, and we find :

$$\sigma^2(\delta r) = \sum_{l=2}^{L}\sum_{j=2}^{L}\sum_{m=o}^{l}\sum_{k=o}^{l} [\Gamma^{CC}_{lmjk}c_{lm}c_{jk} + \Gamma^{SS}_{lmjk}s_{lm}s_{jk}$$

$$+ 2\Gamma^{CS}_{lmjk}c_{lm}s_{jk}] \tag{123}$$

5.3.5 Practical Implementation and example

As we already said it, if we know that the Earth's surface is to be densely enough covered by the satellite ground tracks (that is with a resolution compatible with the phenomena we want to study), it is sufficient to separately compute the mean (geographically correlated) error $\delta\bar{r}$ and the variability $\delta r^{\cdot}$ around it over a regular grid. The maximum error on the ascending tracks may then be approximated by $\sigma(\delta\bar{r})+\sigma(\delta r^{\cdot})$, and over the descending tracks by $\sigma(\delta\bar{r})-\sigma(\delta r^{\cdot})$, ignoring the correlations introduced in $\sigma(\delta\bar{r}\pm\delta r^{\cdot})$.

The practical evaluation of the formulas for the variances σ^2 derived from geopotential covariances Γ, is rendered difficult because the matrix Γ usually does not fit in central memory of a computer. We may however remark the following :

- at a given (ϕ,λ), the variances by coefficient may be evaluated easily one by one by running through the Γ matrix (assumed stored on disc) only once, all necessary coefficients M_{lm}, V_{lm} having been computed and stored in central core,

- the variance by order, or by degree, requires having only one row, or one column, of the Γ matrix at a time,

- the numerical evaluation of formulas (121), (122), (123) may be done as follows :
  . for (121) and (122), $\{M_{lm}, V_{lm}\}$ being computed for a given latitude, we define the vectors :

$$\mathcal{M} = [\dots M_{lm}\cos m\lambda, M_{lm}\sin m\lambda \dots]_{l,m}^{+}$$

$$\mathcal{V} = [\dots V_{lm}\sin m\lambda, -V_{lm}\cos m\lambda \dots]_{l,m}^{+}$$

Then :

$$\sigma^2(\delta\bar{r}) = \mathcal{M}^{+} \ \Gamma\mathcal{M}$$

$$\sigma^2(\delta r^{\cdot}) = \mathcal{V}^{+} \ \Gamma\mathcal{V}$$

. for (123), we must additionally define, for each longitude :

$$\mathcal{T} = [\dots c_{lm}, s_{lm} \dots]_{l,m}^{+}$$

then :

$$\sigma^2(\delta r) = \mathcal{T}^{+} \ \Gamma\mathcal{T}$$

The advantage is that fast methods, already developed for computing the errors on the geoid associated with geopotential covariances (cf. figure 12), can be applied to the evaluation of these linear operations repeated over a regular grid.

As an example, figures 20 and 21 show $\sigma(\delta r)$ for ascending and descending tracks respectively, and figures 22, 23 show $\sigma(\delta\bar{r})$ and $\sigma(\delta r^{\cdot})$ in the case of ERS1, from the GRIM4-S1 global gravity field model. It appears, especially on figures 20-21, that the errors are dominated by long wavelength components.

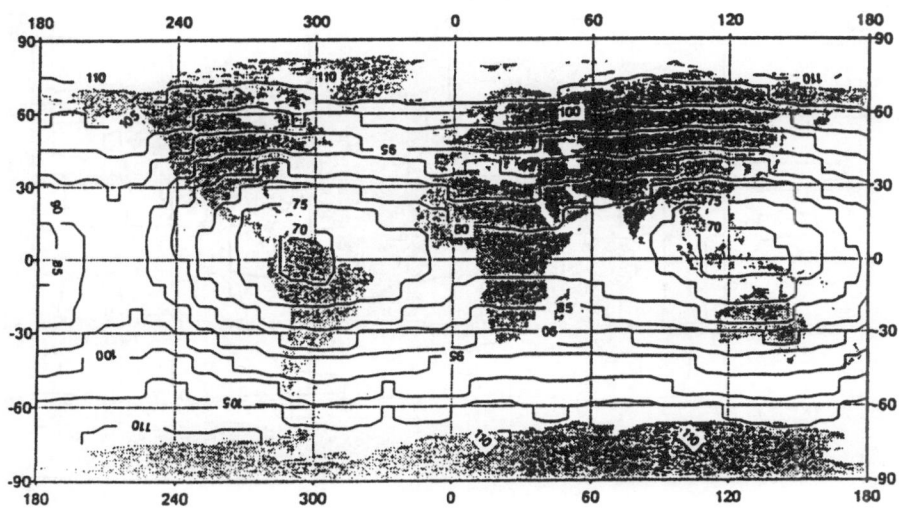

Fig. 20. Geographical distribution of ERS1 predicted total radial orbit error (ascending tracks) using calibrated GRIM4-S1 variance-covariance matrix (unit : centimeter, contour interval : 5 cm).

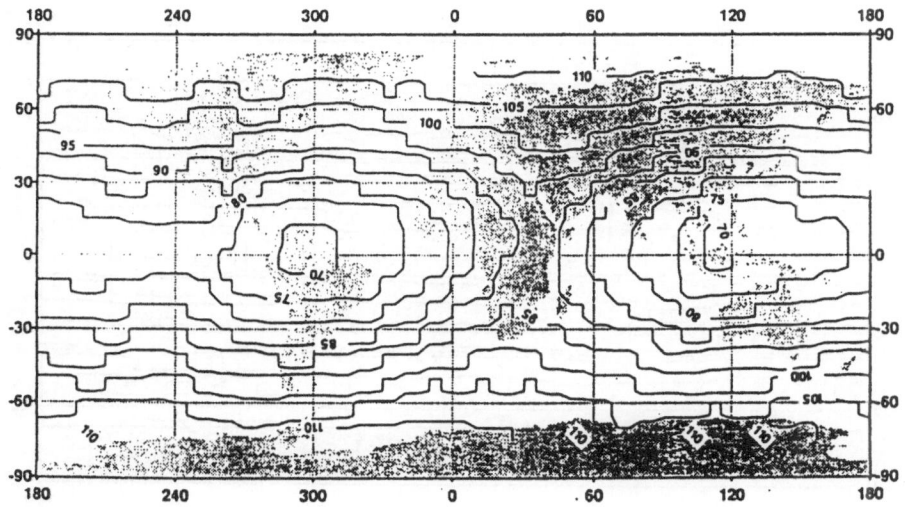

Fig. 21. Geographical distribution of ERS1 predicted total radial orbit error (descending tracks) using calibrated GRIM4-S1 variance-covariance matrix (unit : centimeter, contour interval : 5 cm).

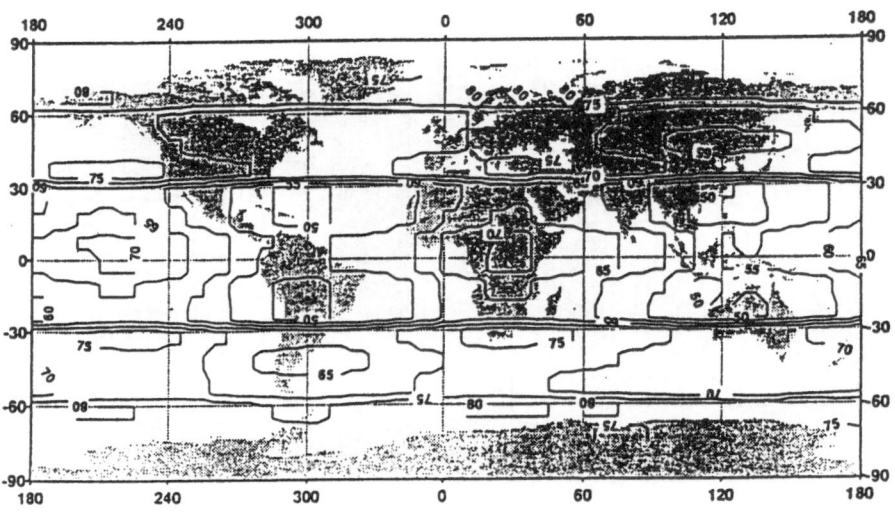

Fig. 22. Geographical distribution of ERS1 predicted geographically correlated part of radial orbit error using calibrated GRIM4-S1 variance-covariance matrix (unit : centimeter, contour interval : 5 cm).

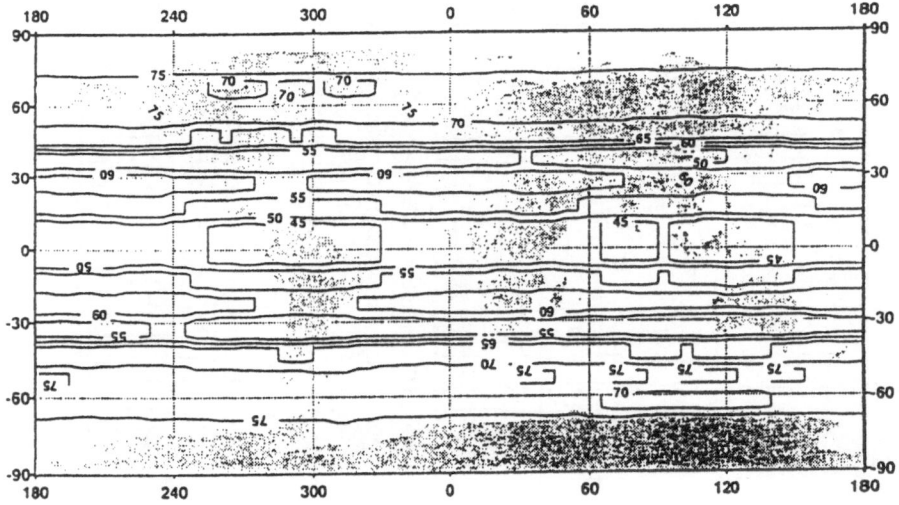

Fig. 23. Geographical distribution of ERS1 predicted variable part of radial orbit error using calibrated GRIM4-S1 variance-covariance matrix (unit : centimeter, contour interval : 5 cm).

6 CONCLUSIONS IN SATELLITE ALTIMETRY

We have shown how geopotential orbital perturbations are mapped on the radial component of a satellite position, in the time domain first and in a quite exact manner (but still in the framework of a first approximation to the analytical solution of the equations of motion), then in the space domain with respect to the Earth's surface for the terms independent of the eccentricity (hence a good enough approximation in the case of quasi-circular orbits such as those of altimeter satellites). This second approach has the advantage of depicting quite precisely the two parts of the radial perturbation, and of the radial orbit error to expect when determining an orbit with a given geopotential model with given covariances. There exists a part which is geographically correlated, that is which is the same at a given point on Earth and if the satellite moves in the same direction (ascending or descending). Around this mean, there is a variation which sign depends on the direction of motion ; it is this variation which is directly recoverable from cross-over techniques (though partly, because of other error sources). Both parts and the total error can be computed from the variance-covariance matrix of a global gravity field model and represented over the Earth's surface.

A software (PERADOR) has been written in Fortran by the author, which performs most computations treated here. Its main characteristics are :

- fast computation of the normalized inclination functions (by recursive relations) and of the Hansen functions (by Fourier transform) up to high degree, order, and (p, q) indices.

- optimized algorithms for all formulas involving the covariance matrix.

- extensive use of the vectorization capabilities of the computer compiler (CDC Cyber 2000 V).

- complete generality as concerns the planet or natural satellite around which the spacecraft is in orbit (the software is to be used for the Mars Observer mission for which one has a laser altimeter on board the orbiter to map the topography of Mars...).

In practice, the remaining radial orbit error which is due to mismodeling of the geopotential (and other forces) and of errors on the orbital elements after the determination of the trajectory has been performed, can be represented by taking into account the errors in the adjusted initial elements of the arc, δr_i, and the errors in the geopotential perturbations induced by the first ones and also due to gravity harmonics errors. It is necessary to include δr_i because they exist independently of geopotential model errors, and are caused by tracking data errors and other force parameters uncertainties.

δr_l may be evaluated from equation (13) truncated at e^2 and differentiated, leading to :

$$\delta r_l = (1 - e\cos M)\delta a_o + a(e - \cos M - e\cos 2M)\delta e_o + ae\sin M \delta M_o \tag{124}$$

This relation, valid for a keplerian orbit, may be used here with all elements (a, e, M) replaced by their mean values (including the account of the secularly perturbed mean anomaly - mainly by $\overline{C}_{20}$), and with δa_o, δe_o, δM_o being the errors on the initial elements of the arc.

To evaluate the two other components of the error, we start from (76) and (77). First, the part coming from the gravity model errors has already been expressed and writes as :

$$\delta r_g = \sum_{lmpq} C_{lmpq}(\delta \mathcal{C}_{lm}\cos\psi_{lmpq} + \delta \mathcal{S}_{lm}\sin\psi_{lmpq}) \tag{125}$$

where $\delta \mathcal{C}_{lm}$, $\delta \mathcal{S}_{lm}$ are defined in terms of the errors $(\delta \overline{C}_{lm}, \delta \overline{S}_{lm})$ on the gravity harmonics themselves.

The part of δr, coming from the total gravity perturbations Δr, and induced by initial condition errors are :

$$\delta r_{g_i} = \sum_{lmpq} \{\delta C_{lmpq}(\mathcal{C}_{lm}\cos\psi_{lmpq} + \mathcal{S}_{lm}\sin\psi_{lmpq})$$
$$+ C_{lmpq}[\mathcal{C}_{lm}(\cos\psi_{lmpq} - \cos\psi_{lmpq}^{true}) + \mathcal{S}_{lm}(\sin\psi_{lmpq} - \sin\psi_{lmpq}^{true})]\}$$

Here, we have avoided differentiating $\cos\psi_{lmpq}$ and $\sin\psi_{lmpq}$ which would have introduced $-\sin\psi_{lmpq}\delta\psi_{lmpq}$ and $\cos\psi_{lmpq}\delta\psi_{lmpq}$, with $\delta\psi_{lmpq}$ being a secular error caused by $\delta\dot{\Omega}, \delta\dot{\omega}, \delta\dot{M}$ (due to $\delta a_o, \delta e_o, \delta I_o$) - although we will do that in a moment but with care. The reason is that all terms in the expression of Δr are periodic and that the error is bounded - which would not be the case if we would at this point introduce the secular error terms.

Now, ψ_{lmpq}^{true} being the true value, we expand the differences of sines and cosines, and we obtain (assuming $(\psi + \psi^{true})/2 \approx \psi$ for any l, m, p, q) :

$$\delta r_{g_i} = \sum_{lmpq} \{\delta C_{lmpq}\mathcal{C}_{lm} + C_{lmpq}[\delta \mathcal{C}_{lm} + 2\mathcal{S}_{lm}\sin(\delta\psi_{lmpq}/2)]\}\cos\psi_{lmpq}$$
$$+ \{\delta C_{lmpq}\mathcal{S}_{lm} + C_{lmpq}[\delta \mathcal{S}_{lm} - 2\mathcal{C}_{lm}\sin(\delta\psi_{lmpq}/2)]\}\sin\psi_{lmpq} \tag{126}$$

In this formula, δC_{lmpq} is computed from $\delta a_o, \delta e_o, \delta I_o$, and ($t = o$ being the beginning of the arc) :

$$\delta\psi_{lmpq} = \delta\psi_{lmpq}^o + \delta\dot{\psi}_{lmpq}^o t$$

$$\delta\dot{\psi}_{lmpq}^o = (l - 2p)(\delta\dot{\omega} + \delta\dot{M}) + m\delta\dot{\Omega} + q\delta\dot{M}$$

$$\delta\psi^{\circ}_{l,m,p,q} = (l-2p)(\delta\omega_{o} + \delta M_{o}) + m\delta\Omega_{o} + q\,\delta M_{o}$$

$\delta\dot{\Omega}, \delta\dot{\omega}, \delta\dot{M}$ may be evaluated by differentiating (60) hence including the error δn_{o} on the mean motion, due to δa_{o}.... These quantities being usually rather small, $\sin(\delta\psi_{l,m,p,q}/2)$ is a very long period term which modulates the amplitude of $\delta r_{s_{l}}$. From what was said in paragraph 3.1, the major gravity perturbations occur at the frequency $\dot{\omega}+\dot{M}$ and are due to the odd degree zonals, hence the induced error too. Therefore, the main term of $\delta r_{s_{l}}$ has the form :

$$\delta r_{s_{l}} \approx \left[A + A^{\bullet}\sin\frac{(\delta\dot{\omega}+\delta\dot{M})}{2}t \right]\cos(\omega+M)$$

$$+ \left[A' + A^{\bullet'}\sin\frac{(\delta\dot{\omega}+\delta\dot{M})}{2}t \right]\sin(\omega+M)$$

Over a short time (usually a few days with respect to the period $4\pi/(\delta\dot{\omega}+\delta\dot{M})...$), we may (now) linearize the sine term in the brackets and we get something of the form :

$$\delta r_{s_{l}} \approx (A+Bt)\cos(\omega+M) + (A'+B't)\sin(\omega+M) \tag{127}$$

Under the same approximation (main terms, of order o in e), collecting the results of (124), (125) and (127), we obtain :

$$\delta r \approx C + (C_{1}+Bt)\cos(\omega+M) + (C_{2}+B't)\sin(\omega+M) \tag{128}$$

The real shape of the radial error is as shown on figure 24 (it is obtained from a simulation with two different geopotential models and after orbit adjustment using one of them). The bow tie, or butterfly shape is due to the initial conditions fit which tends to reduce, on average, the effects of the unmodeled linear terms in (128) : they are minimum in the middle of the arc.

One implication of this is that repeat arc differences (cf. paragraph 2.3) in satellite altimetry may be analysed with a function such as (128), where the C, C_{1}, C_{2}, B, B' coefficients are adjusted from the observations. Another implication is that cross-over minimisation techniques should work best with such a function, which is practically proven to be efficient, too, for absorbing other unmodeled effects.

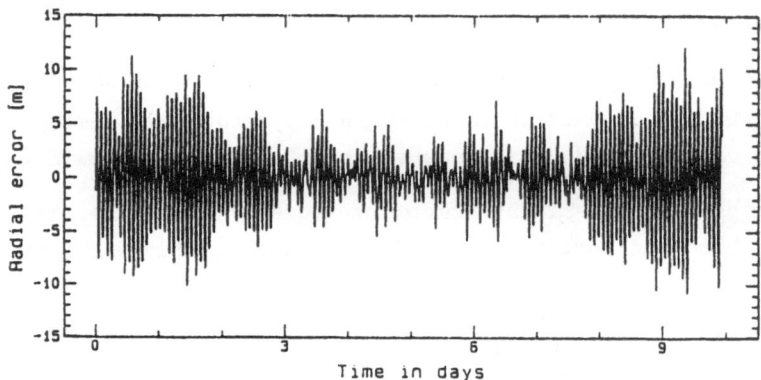

Fig. 24. Radial orbit error for Topex-Poseidon from a simulation experiment using two different geopotential models (from Schrama, 1989).

7 REFERENCES

Balmino, G., 1978, Quelques problèmes de rotation dans l'espace. Applications en géodésie, en dynamique des solides et en planétologie. Technical Report, GB/NS/R.7911/CT/GRGS.

CNES-DARA, 1991, A New Earth Gravity Field Model in support of ERS1 and SPOT2, Final Report on GRIM4-S1/C1, DGFI/GRGS.

Engelis, T., 1987, Radial Orbit Error Reduction and Sea Surface Topography Determination using Satellite Altimetry, Report 377, Dept. of Geodetic Science and Surveying, the Ohio State University.

Kaula, W.M., 1966, Theory of Satellite Geodesy, Blaisdell Pub. Co.

Rapp, R.H., Wang Y.M., and N.K. Pavlis, 1991, The Ohio State 1991 Geopotential and Sea Surface Topography Harmonic Coefficient Models, O.S.U., Rep. 410.

Reigber, Ch., Balmino G., Schwintzer P., Barth W., Massmann F.H., Raimondo J.C., Gerstl M., Bode A., Li H., Biancale, R., Moynot B., Lemoine J.M., Marty J.C., Barlier F., and Y. Boudon, 1992, GRIM 4 Earth Gravity Field Models in Support of ERS1 and SPOT2, Symposium G3, XXth IUGG, Vienna (Austria).

Rosborough, G.W., 1986, Satellite Orbit Perturbations due to the Geopotential Rep., CSR-86-1, Austin, Texas.

Schrama, E.J.O., 1989, The Role of Orbit Errors in Processing Satellite Altimeter Data, Publication n° 33, Neth. Geodetic Comm.

Wagner, C.A., 1985, Radial Variations of a Satellite Orbit Due to Gravitational Errors : Implications for Satellite Altimetry, J.G.R., Vol. 90, n° B4, pp. 3027-3036.

Theory of Geodetic B.V.P.s Applied to the Analysis of Altimetric Data

Fernando Sansò

Politecnico di Milano
Dipartimento di Ingegneria Idraulica,
Ambientale e del Rilevamento
Piazza Leonardo da Vinci 32,
20133 Milano, Italy

318

1. Introduction

It is a neverending task of geodesy to approximate more and more the anomalous
potential T of the earth's gravity field. This is done by several means, for
instance by disseminating the earth's surface with different kinds of geometric
observations as well as of measurements of different functionals of the function T;
ever since the sixties, analogous measurements can be performed by linking points on
the earth's surface with satellites, which are flying inside the domain of
harmonicity of T and more are planned to be performed purely at satellites' level.
Just as examples we can mention:

1) on the earth's surface S; theodolite observations, EDM measurements, spirit
levelling, absolute and relative gravity measurements, gradiometry, INS observations,
etc;

2) between points on S and satellites; distance measurements (SRL), doppler
observations, phase measurements (GPS), radar-altimetry, etc;

3) at satellites' level; accelerometry, differential accelerometry (gradiometry),
satellite to satellite tracking, etc.

The data of type 1), after suitable adjustment can be usefully combined to provide ,
for large areas of the earth's surface, a fairly dense coverage with some homogeneous
type of functional of T.

We can make a few examples:

Example 1.1

On continental areas, by means of classical geodetic observations we are able to give
the horizontal location of points P on the earth surface, namely the ϕ, λ
coordinates of the projection P_0 of P on the ellipsoid E; at P we can also
measure the gravity modulus g; moreover by combining gravity with spirit levelling we
come to know the gravity potential W and P.

In this example therefore we picture the situation by saying that we assume to
know, all over land areas, the four quantities ϕ_p, λ_p, g_p, W_p.

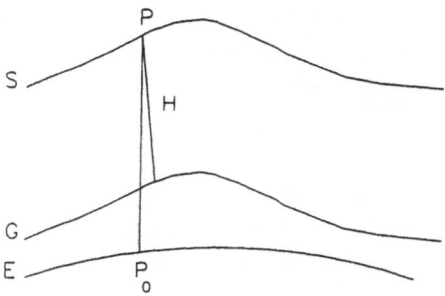

Fig. 1.1

Example 1.2

This is the same as the example 1.1 but applied to some closed seas areas (e.g. the Mediterranean, the Baltic sea, etc.).

Here let's assume that it is possible to model completely the time varying part of the sea surface and that by ship measurements we come to know the gravity value on it; if the horizontal position of the ship is known (e.g. by radar control) and if we can assume that the stationary sea surface is in fact an equipotential surface we arrive to claim that we know again ϕ, λ, W, g on this area.

This statement would be realistic, was it not for the hypothesis that the stationary sea surface is an equipotential surface, since this can be grossly false in many situations.

With data like those presented in the two examples above, we can, and it has been done in fact, formulate some boundary value problem (B.V.P.) and think of solving it in order to find what is unknown in the specific case; in both example 1.1 and example 1.2 the unknown is the field of geometric heights h_p of the points P on E, in fact once this is known, the potential W, or better T (remember that $W = U + T$, U being the normal potential) can be continued in space by solving the corresponding Dirichlet problem.

The big drawback of this approach was that not only part of the land areas of the world could not enter into this process, because either they were and still are unsurveyed or the data on them are not available for political reasons, but also most of the oceanic surface was and still is in the same situation since we don't have gravity measurements on it.

This situation has changed drastically ever since observations of type 2), particularly with GPS and radar altimeter techniques, have been available.

In fact if we can assume that the satellite's orbit is perfectly known, i.e. that we have done a good work to correct for orbit errors, then the observations can be taken as giving direct information on the ground points related to them.

Several examples can be built in this case:

Example 1.3

On land, we could assume that gravity measurements are taken together with GPS, so that we know at the same point P the quantities ϕ,λ,h,g; in this way, contrary to the previous two examples, the shape of the boundary becomes a datum of the problem and is not any more an unknown.

Example 1.4

On closed seas we perform ship gravimetry together with GPS observations so that we are very much in the same situation as in the previous example.

Example 1.5

On seas and oceans we use radar-altimetric measurements, call them H_A, (cfr. Fig. 1.2), which, when the position of the satellite A is perfectly known (i.e. we know ϕ,λ of the projection A_0 of A onto the ellipsoid, and the ellipsoidal altitude h_A), can supply ϕ,λ of $P^{(1)}$, as they are the same of A, and also

$$h_P = h_A - H_A \quad .$$

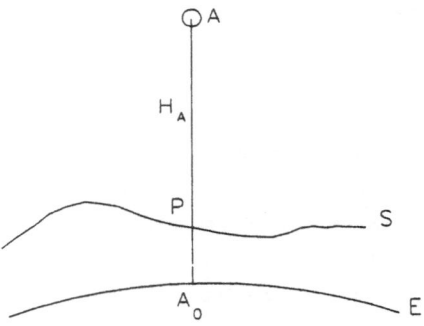

Fig. 1.2

If furthermore we can assume that we have been able to remove all the departures of the sea surface from an equipotential surface, then we can add the fourth information

$$W_P = U_P + T_P = \text{const.} \quad ,$$

---

[1] In reality the radar beam is reflected along the perpendicular to the sea surface S which can be inclined, say, less than 10^{-4} rad, what means for a satellite, as high as 1000 Km, a horizontal shift of ~ 100 m; this in turn would imply a variation of h_P of at most 1 cm.

which is necessary to set up a corresponding B.V.P.

Remark 1.1

There has been a long debate among geodesists whether data, like these we are considering here, should be used to set up a B.V.P. or whether they should be treated as discrete from the very beginning.

This is essentially a matter of density of data and of required resolution of the unknown potential, as discussed in Sansò [1]; be whatever it is, the analysis of the underlying B.V.P. has been recognized as essential in order to understand (or even to define) in which sense an approximate method is in fact "approximate".

By combining in different ways the known quantities at the earth surface, i.e. at the boundary of harmonicity of T, several types of B.V.P.s have been formulated and can be used in order to achieve a full knowledge of T throughout the space and, in most cases, also to arrive to know the shape of the boundary itself, which is, in most cases, unknown in one way or another.

Remark 1.2

Let us underline here that knowing T in the space is essential everytime we like to compute, even on the surface itself, derivates of T pointing toward the exterior of the earth; this is the case for instance when we assume to know T on the ocean, but we want to derive gravity anomalies on it.

We can organize the various B.V.P.'s formulated in the history of geodesy as shown in Table 1.1, which is also known as the hierarchy of gravimetric problems

Problem	Land		Ocean	
	known	unknown	known	unknown
1	ϕ, Λ, g, W	$\underline{r}_P$	Φ, Λ, g, W	$\underline{r}_P$
2	ϕ, λ, g, W	h_P	ϕ, λ, g, W	h_P
3	ϕ, λ, g, W	h_P	ϕ, λ, h, W	(g_P)
4	ϕ, λ, g, W	h_P	ϕ, λ, h, g	W_P
5	ϕ, λ, h, g	W_P	ϕ, λ, h, g	W_P

Pbl-1:   is known as the classical Molodensky's problem and is valuable mainly for historical and theoretical reasons ($\underline{r}_P$ is the position vector of P).

Pbl-2:   is known as the scalar Molodensky's problem and is currently used in many contexts.

Pbl-3,4: are known as the altimetry-gravimetry problem I and II respectively and will be discussed in detail in these lecture notes.

Pbl-5:   is known as a fixed boundary gravimetric B.V.P. and is increasing in importance, since we got the availability of GPS observations.

We stress that in all problems from 1 through 5 the full spatial knowledge of W is one, if not the main, target.

Problem 1 to 4 are free boundary B.V.P., while only 5 is a fixed boundary B.V.P..

In problems 1,2,5 always the same types of data are used to set up the corresponding B.V.P., irrespectively of land and ocean areas; on the contrary in problems 3 and 4 the types of data change from land to ocean, so that these can be classifed as mixed B.V.P.s.

Remark 1.3

Apart from their theoretical relevance, there are two main uses of B.V.P.s in geodesy; a global one and a local one.

In the global approach we apply the B.V.P. theory in order to build an approximation of T maybe not with the highest resolution, but valid all over the earth.

In the "local" solution, which is somehow a monster from the point of view of potential theory, what we do in reality is to subtract from our data, distributed only on a local area of the earth, the corresponding quantities computed from a global model, accounting for their long wavelength features; in this way it is known that for statistical reasons the data from remote zones will have little influence on the solution in our local area, so that we can apply the proper Green functions to derive a good approximation, even sometimes working in a tangent plane approximation. Naturally in this case the solution will have validity only for the particular area analyzed and even in space we should limit ourselves to the neighbouring layer.

To conclude this paragraph let's say that what we look at in the sequel of these notes is to solve the problem: what knowledge of the gravity potential can we achieve from altimetric data, when the oceanographic phenomena are assumed to be known?

This will be done here mainly by studying the altimetry-gravimetry problems I and II.

2. The reduction of the original B.V.P.s to linearized spherical problems.

The aim of this paragraph is to formulate the two altimetry- gravimetry (AG) problems, to linearize them and finally to reduce them to a spherical form.

This is done mainly for two reasons; first because the original problems are too difficult even to be analyzed theoretically, second because also a numerical treatment is much more easily done basically in a spherical approximation and by using a spherical boundary and then applying some corrections for perturbation effects.

Remark 2.1

In this respect it should be borne in mind that most numerical techniques in geodesy rely on suitably simplifying the geometry of the boundary , i.e. by computing perturbative effects with respect to a simple geometrical surface. This not only because many times we have to solve a free B.V.P. but also because of the extreme complication of the true earth's surface which is even more rugged than the gravity field.

At first let us formulate and linearize the two AG problems; in this process we shall consider as first order quantities only the anomalous potential T as well as the height anomaly ζ_p defined as the separation between P,Q along the same ellipsoidal normal, such that

$$W_p = U_Q \qquad\qquad (2.1)$$

(cfr. Fig. 2.1); we note that should P be lying on the geoid, Q would fall on the ellipsoid and the height anomaly ζ would become the same as the geoid undulation, also called N in geodetic literature.

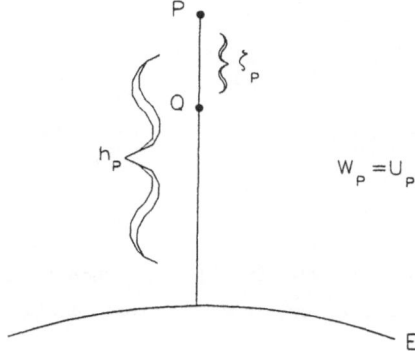

Fig. 2.1

Since both problems, AG I and AG II, share the same B.V.P. on land areas, we start by linearizing that part.
So, assume to know ϕ,λ,W,g at points P corresponding to land areas; then, using the traditional notation $\underline{\gamma} = \nabla U$,

$$W(P) = U(Q) \cong U(P) + \gamma(Q) \, \zeta_p \qquad\qquad (2.2)$$

and on the same time

$$W(P) = U(P) + T(P) \quad, \qquad\qquad (2.3)$$

so that, by comparison, we derive

$$T(P) \cong \gamma(Q) \, \zeta_P \quad , \tag{2.4}$$

which after moving P down to Q without changing $T(P) \cong T(Q)$, reads

$$\zeta_P = \frac{T(Q)}{\gamma(Q)} \quad , \tag{2.5}$$

known also as Bruns' relation.

Moreover

$$g(P) = \left| \nabla U(P) + \nabla T(P) \right| \cong \gamma(P) + \frac{\gamma(P)}{\gamma(P)} \cdot \nabla T(P) \quad ; \tag{2.6}$$

now $-\underline{\gamma}/\gamma$ differs from the unit vector $\underline{\nu}$, normal to the ellipsoid, for less than 10^{-5} at 6000 m of height so that we can easily substitute (2.6) with

$$g(P) \cong \gamma(Q) + \frac{\partial \gamma(Q)}{\partial h} \, \zeta_P - \frac{\partial T}{\partial h} \quad . \tag{2.7}$$

Introducing the gravity anomaly

$$\Delta g(Q) = g(P) - \gamma(Q) \quad , \tag{2.8}$$

and substituting (2.5), we find then

$$\Delta g(Q) \cong - \frac{\partial T}{\partial h} + \frac{\partial \gamma / \partial h}{\gamma} \, T \quad ; \tag{2.9}$$

the boundary equation (2.9) has to hold at all Q, when we let P to range on the continental part of the earth surface, i.e. on that particular surface which is called the telluroid.

We now can differentiate the two formulations:

AG I)   here on oceanic areas we assume that the ocean surface, after suitable corrections coincides with the geoid, so that

$$W_P = W_0 \quad ; \tag{2.10}$$

on the other hand by (adjusted) altimetric observations we know also h_P, which actually coincides with N_P.

By using (2.4), then, we find that

$$T(Q) \cong \gamma(Q) \, N_P \tag{2.11}$$

is known on these areas. Here the point Q is the orthogonal projection of P on the ellipsoid, so that (2.11) has to hold on that part of E which corresponds to seas.

Summarizing we can say that if we split the ellipsoid into two regions S and L corresponding respectively to sea and land areas, we can formulate for T the linearized AG I problem (Ω being the space outside the earth surface)

$$
\begin{cases}
\Delta T = 0 & \text{in } \Omega \\
T = \gamma N & \text{on S, } h=0 \\
-\dfrac{\partial T}{\partial h} - bT = \Delta g & \text{on L, } h=h_Q \ (W_P = U_Q) \\
T \to 0 & \text{for } |\underline{x}| \to \infty
\end{cases}
\tag{2.12}
$$

$$
\left(b = -\frac{\partial}{\partial h} \log \gamma(Q)\right)
$$

Remark 2.1

The drawback in this approach is that, as we know, the stationary sea surface is not in reality adapting to an equipotential surface but it rather deviates from it by an amount t, the sea surface topography, which is of great interest in oceanography because it is related to oceanic circulation; hence what we "measure" really is

$$
h_P = N_P + t_P
\tag{2.13}
$$

and introducing this quantity into (2.11) we will get a biased solution.

Fortunately it is known that for obvious physical reasons (t cannot be too rugged!) the spectral power of t is mostly in the long wavelenght part, so that if we can achieve (as we plan) by independent satellite measurements a good estimate of T in the same spectral region, we can reasonably hope to discriminate the two signals;

AG II) in this case we assume to know both h_P and g_P at the sea surface, the first obtained by altimetry, the second by marine gravimetry.

Whence we can go back to (2.6) and after introducing the so-called true anomaly

$$
\delta g(P) = g(P) - \gamma(P)
\tag{2.14}
$$

we can write directly

$$
\delta g(P) = -\frac{\partial T}{\partial h} \ ;
\tag{2.15}
$$

this equation has to hold for all P on the sea surface $h = h_P$.

So we come to the formulation

$$
\begin{cases}
\Delta T = 0 & \text{in } \Omega \\[2mm]
-\dfrac{\partial T}{\partial h} = \delta g & \text{on } S, \ h=h_p \\[2mm]
-\dfrac{\partial T}{\partial h} - bT = \Delta g & \text{on } L, \ h=h_p \\[2mm]
T \to 0 & \text{for } |\underline{x}| \to \infty
\end{cases}
\tag{2.16}
$$

Remark 2.2

Since marine gravimetry is available only on relatively small seas (e.g. in closed seas like the Mediterranean, or in coastal regions) we find that this problem can be applied only for the sake of "local" solutions (as discussed in Remark 1.3); here however it is particularly useful because it is exactly in those areas that we can expect a more irregular behaviour of t; whence by computing a solution from (2.16) one can find

$$
t = h - N = h - \frac{T}{\gamma} \ .
\tag{2.17}
$$

Remark 2.3

While in AG I the oceanic part of the B.V.P. is given directly on the ellipsoid, in AG II it has to hold at points P of the actual oceanic surface; however since this is separated from the ellipsoid by a distance of the order of the geoid undulation, we can straightforwardly shift the second of (2.16) to E, neglecting second order changes.

We get in this case

$$
-\frac{\partial T}{\partial h} = \delta g(Q) \qquad \text{on } S, \ h=0 \ .
$$

We notice however that the same reasoning doesn't apply to the land part of both AG I and AG II, because this would mean a shift which could be as large as several kilometers in mountainous areas. This shift is by no means a first order quantity as it can be up to 10^2 times bigger than N.

For the moment then we must think of solving a B.V.P. which for land areas has a boundary as complicated as the telluroid, which in turn mirrors quite strictly the behaviour of the actual surface of the earth.

There are several strategies to cope with this problem:

1) the use of Molodensky's series, which essentially works out a sequence of terms obtained by applying to the preceding a suitable singular integral operator, computed on the ellipsoid;

2) the so-called downward analytical continuation method, which is deemed to be equivalent to Molodensky's series, and it is in fact so at least formally;

3) a direct finite element method of which we have seen recently some example; this is however very heavy on the computational side.

We make here a proposal which is essentially equivalent to Molodensky's theory at least at the first order in the highest function h_Q; namely assume that a known term f is given on the known surface S for the B.V.P.

$$Bu \Big|_{h=h_Q} = f(P) \quad . \tag{2.18}$$

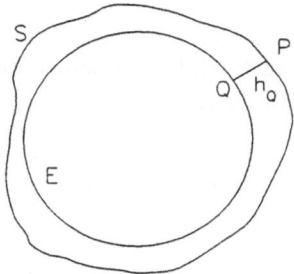

Fig. 2.2

We then define a pull-back of f on E as another function $\tilde{f}(Q)$ such that

$$\tilde{f}(Q) = f(P) \quad . \tag{2.19}$$

Note that $\tilde{f}$ has numerically the same values of f but it is a different function for it attains those values at different points.

Now let us define a function u_0 which is in fact the solution of the B.V.P.

$$Bu_0 \Big|_{h=0} = \tilde{f} \quad , \tag{2.20}$$

defined as we see on the ellipsoid[2].

If we proceed to compute

---

[2] It is important to understand that in (2.20) the operator B, acting on E, has to be suitably defined as a smooth continuation of the "true" operator B, defined on S; for instance if B is the identity on S (Dirichlet problem) then it is also the identity on E; if the projection P,Q of Fig. 2.2 is biunivocal, then any $\partial/\partial h$ on S can go into a $\partial/\partial h$ on E and any functional of the normal potential on S can go into the corresponding functional computed for h=0.

$$Bu_0 \Big|_{h=h_P}$$

we naturally won't find $f(P)$ since this equals $\tilde{f}(Q)$ and we don't expect Bu to be constant along the ellipsoidal normal.

So we have a discrepancy between the sought u and u_0 which is revealed by the residual on S

$$f - Bu_0 \Big|_S = \delta f_1 \quad . \qquad (2.21)$$

Naturally we are not able to find a v such that

$$Bv \Big|_S = \delta f_1 \quad ,$$

for otherwise we would have solved the original problem from the beginning; so we pull-back δf_1 and we content ourselves with solving

$$B\delta u_1 \Big|_{h=0} = \delta \tilde{f}_1 \quad . \qquad (2.22)$$

Then we compute again

$$f - B(u_0 + \delta u_1) \Big|_S = \delta f_2 \quad . \qquad (2.23)$$

and we iterate.

There is no direct proof of the convergence of the method, although there are hints in this direction; we shall only state here that if it converges then it converges to the right limit.

The precise meaning of this statement can be found in the Appendix where also an interesting example with spherical boundaries is fully worked out; this is particularly useful for didactic reasons.

The meaning of the above discussion is essentially that the two B.V.P.s, AG I and AG II can be formulated directly on the ellipsoid as this is the basic step for the proposed iterative procedure.

In principle the same reasoning could be applied to perform the transition from the ellipsoid to some sphere; this however would be too rough an approximation.

On the other hand the following reasoning applies: if we introduce ellipsoidal coordinates (u,θ,λ) related to cartesian through the transformation[3]

---

[3] We use only in this and in the following few pages the symbol u as ellipsoidal coordinate, not to be confused with the meaning of a harmonic function.

$$\begin{cases} x = \sqrt{u^2 + E^2} \ \sin\theta \ \cos\lambda \\ \\ y = \sqrt{u^2 + E^2} \ \sin\theta \ \sin\lambda \\ \\ z = u \ \cos\theta \end{cases} \tag{2.24}$$

$(E^2 = a^2 - b^2; \quad a,b = \text{semimajor and semiminor axes})$

we can solve the Laplace equation outside the ellipsoid $\{u \ge b\}$ by separation of variables (cfr. Heiskanen-Moritz [2]) thus finding the following general representation of a harmonic function

$$v(u,\theta,\lambda) = \sum_{0}^{\infty} {}_{n} \sum_{-n}^{n} {}_{m} \ v_{nm} \ \frac{Q_{nm}\left(i \ \dfrac{u}{E}\right)}{Q_{nm}\left(i \ \dfrac{b}{E}\right)} \ Y_{nm}(\theta,\lambda) \quad ; \tag{2.25}$$

where $Y_{nm}(\theta,\lambda)$ are the fully normalized spherical harmonics, while $Q_{nm}(ix)$ are the Legendre functions of second kind, defined in the interval $x \ge 1$.

By a direct inspection of the formula of Q_{nm}, or by a perturbation computation of (2.25) using as first order quantities the couple $e^2 = \dfrac{a^2 - b^2}{a^2}$, $\dfrac{h}{Q}$ (cfr. Sansò - Sona [3]), one finds that, R attaining any value between a and b,

$$\begin{cases} \dfrac{u}{b} \cong \dfrac{R + h}{R} \\ \\ \dfrac{Q_{nm}\left(i \ \dfrac{u}{E}\right)}{Q_{nm}\left(i \ \dfrac{b}{E}\right)} \cong \left(\dfrac{R}{R + h}\right)^{n+1} \end{cases} \tag{2.26}$$

With the approximation (2.26) and introducing just a new triplet of coordinates, with the same angles θ,λ and $r = R + h$ instead of u, we get from (2.25)

$$v(r,\theta,\lambda) \cong \sum_{0}^{\infty} {}_{n} \sum_{-n}^{n} {}_{m} \ v_{nm} \ \left(\frac{R}{r}\right)^{n+1} \ Y_{nm}(\theta,\lambda) \tag{2.27}$$

which is the very well known formula for a function of "spherical coordinates" (r,θ,λ), harmonic outside the "sphere" $\{r \ge R\}$.

Neglected terms are of a relative order of 10^{-5}, showing that already a purely spherical computation can go very close to the correct solution; naturally, when needed, one can push the computation introducing second order terms, while it seems unnecessary to go further for any practical purpose.

Remark 2.4

It is important to understand here that the coordinate r we are using, i.e.

$$r = R + h \ ,$$ (2.28)

is not the radial distance of the point P from the origin, as well as θ implicitely defined through (2.24), is not the spherical colatitude, nor the geographic colatitude ($\pi/2 - \phi$); only λ in the same in the three cases.

If we would switch from the representation (2.27) to another one in terms of spherical or geographical coordinates, we would find correcting terms of the first order in e^2 too.

Now let us come to the transformation of the boundary operators involved in AG I, AG II.

Due to the relation (2.28) we can write directly

$$\frac{\partial}{\partial h} = \frac{\partial}{\partial r}$$ (2.29)

in our formulas.

Moreover it is easy to see that on the earth's ellipsoid

$$b = \frac{2}{R} \left(1 - \varepsilon \sin^2\phi \right) \ ,$$ (2.30)

with ε being some parts in 10^{-3}, depending also on the particular R chosen.

With this is mind we can always write the land part of the B.V.P.s (2.12) and (2.16) in the form

$$- \frac{\partial}{\partial r} T - \frac{2}{R} T = \Delta g - \frac{2\varepsilon}{R} \sin^2\phi \ T$$ (2.31)

and consider again a perturbative solution by iterating on the right hand side small term.

In this way we are finally reconducted to consider as basic problems the following spherical formulations: find T harmonic in the spherical domain $r \geq R$ and regular (i.e. tending to 0) at infinity, such that

(AG I) $\begin{cases} T = \gamma N = \delta w & \text{on S} \quad (r = R) \\ \\ - \frac{\partial T}{\partial r} - \frac{2}{R} T = \Delta g & \text{on L} \quad (r = R) \end{cases}$ (2.32)

or

$$(AG\ II) \begin{cases} -\dfrac{\partial T}{\partial r} = \delta g & \text{on } S \quad (r = R) \\[3mm] -\dfrac{\partial T}{\partial r} - \dfrac{2}{R}\,T = \Delta g & \text{on } L \quad (r = R) \end{cases} \qquad . \tag{2.33}$$

Before closing the paragraph we must still introduce one further modification of (2.32), (2.33) which we will justify here from the point of view of the physical modelling of the measurements, although the decision of introducing it is not extraneous to the much better analytical properties of the modified problems.

The point here is that we have assumed in both AG I, AG II to be able to reconstruct the height of the stationary sea surface on the ellipsoid; from this height we derive then either $\delta w = \gamma N$ or δg.

This surface is given once we assume that the orbit of the altimetric satellite is known in a certain reference frame to which the earth's ellipsoid is referred too.

However the orbit of the satellite contains a significant radial orbital error which is reduced by a so-called cross over adjustment; in this step however we have a certain degree of arbitrariness in that the adjusted heights are well known in relative sense, whereas there is a bias parameters or vertical shift which remains unknown or poorly determined.

This vertical shift is reflected in the determination of δw or δg also by a constant (with good approximation) variation of such quantities. Since in this way we decrease the information supplied on our problems, we must look for some other information, to substitute it; as such we can use, and in fact is used, the value of GM which is very well known from satellite tracking only.

This value can be used to assess the value of the normal potential on the ellipsoid in such a way that

$$T = W - U \underset{r\to\infty}{\sim} O(1/r^2) \quad ; \tag{2.34}$$

this relation is equivalent in fact to the statement that the same quantity of mass M generates W and U.

Since we are using a spherical approximation, (2.34) is translated into the equivalent condition

$$\frac{1}{4\pi} \int T\, d\sigma = 0 \quad . \tag{2.35}$$

Summarizing we modify (2.32), (2.33) as follows:

$$(AG\ I) \begin{cases} T = \delta w + a & \text{on } S \\[2mm] -\dfrac{\partial T}{\partial r} - \dfrac{2}{R} T = \Delta g & \text{on } L \\[2mm] \dfrac{1}{4\pi} \displaystyle\int T\, d\sigma = 0 \end{cases} \tag{2.36}$$

and

$$(AG\ II) \begin{cases} -\dfrac{\partial T}{\partial r} = \delta g + a & \text{on } S \\[2mm] -\dfrac{\partial T}{\partial r} - \dfrac{2}{R} T = \Delta g & \text{on } L \\[2mm] \dfrac{1}{4\pi} \displaystyle\int T\, d\sigma = 0 \end{cases} \tag{2.37}$$

and these will be the formulations for which we shall perform our further analyses in the next paragraphs. This analysis will be the bases of the practical use of the two B.V.P. in both the contexts of the global and of the local determination of the potential T.

Remark 2.5

It is to be underlined that in our formulations (2.36), (2.37) we don't make any particular discussion for the first degree harmonics T_{1m} (m= -1,0,1) as it is done for instance for the Stokes problem, which corresponds to the B.V.P.

$$-\frac{\partial T}{\partial r} - \frac{2}{R} T = \Delta g \tag{2.38}$$

over the whole sphere $\{r = R\}$.

These degrees of freedom which are simply proportional to the coordinates of the barycentre, are in fact not determined in (2.38) since

$$\left(-\frac{\partial T}{\partial r} - \frac{2}{R} \right) \left(\frac{R}{r} \right)^2 Y_{1m} \equiv 0 \quad ;$$

while the uniqueness of the solution of (2.36), (2.37) will show that they are indeed determined in AG I and AG II.

3. Some recalls of mathematics

In the following paragraph we will make use of some classical concepts of the analysis of abstract equations in Hilbert spaces; we therefore will summarize hereafter the main definitions and theorems needed, for the ease of the reader.

We want to stress that we will give our definitions so as to be directly applicable to functions defined on the unit sphere, thus loosing generality, but gaining insight in what we will really need in the sequel[4].

Def. 3.1: a (real) Hilbert space H is a linear space endowed with a (real) scalar product (,)$_H$ and closed with respect to the corresponding norm

$$\|f\|^2 = \sqrt{(f,f)} \tag{3.1}$$

Example 3.1

The linear space of real functions of (ϕ,λ) (i.e. defined on the unit sphere) which are square integrable with respect to the standard measure

$$\frac{1}{4\pi} \, d\sigma = \frac{1}{4\pi} \, \cos \phi \, d\phi \, d\lambda$$

is a Hilbert space, denoted $L^2(\sigma)$, with scalar product

$$(f,g)_{L^2(\sigma)} = \frac{1}{4\pi} \int f(P) \, g(P) \, d\sigma_P \quad . \tag{3.2}$$

For reasons which will be clear after we will denote the L^2 product also simply as $(f,g)_0$.

Th. 3.1: for any product the Schwarz inequality holds

$$|(f,g)| \le \|f\| \cdot \|g\| \tag{3.3}$$

Def. 3.2: the dual of a Hilbert space H, denoted as H^*, is the linear space of linear, bounded functionals on H endowed with the norm

$$\|\phi\|_{H^*} = \sup_{\|f\|=1} |\phi(f)| \quad . \tag{3.4}$$

Th. 3.2 (the "representation" or Riesz theorem): H^* is isometrically equivalent to H, in the sense that given g ∈ H, defining $\phi_g(f) = (g,f)_H$ we see that $\phi_g \in H^*$ and

$$\|\phi_g\|_{H^*} = \|g\|_H \quad ; \tag{3.5}$$

---

[4] To be short we will use the notation Def for definition, Th for Theorem.

viceversa, given $\phi \in H^*$ there is a $g_\phi \in H$ (the "representer" of ϕ in H) such that

$$\phi(f) = (g_\phi, f)_H \quad ,$$ (3.6)

moreover

$$\|g_\phi\|_H = \|\phi\|_{H^*} \quad .$$ (3.7)

Remark 3.1

It is important to realize that the above theorem states the existence of one possible isometric representation of H^* in H, but there can be more representations indeed, as we shall see. By the way Th. 3.2 proves that H^* has the structure of a Hilbert space too.

Example 3.2

If $\Delta g(P)$ is a field of gravity anomalies and $\Delta g \in L^2(\sigma)$, then the block average over a block B

$$\phi(\Delta g) = \Delta\bar{g} = \frac{1}{\mu(B)} \int_B \Delta g \, d\sigma$$ (3.8)

is a bounded functional since

$$\left| \int_B \Delta g \, d\sigma \right|^2 \leq \mu(B) \int_B \Delta g^2 \, d\sigma \leq \mu(B) \, 4\pi \, \|\Delta g\|^2 \quad .$$ (3.9)

The representer of ϕ in $L^2(\sigma)$ is the characteristic function of B, $\chi_B(P)$ ($\chi_B = 1$ if $P \in B$, $\chi_B = 0$ otherwise), or, more precisely

$$g_0 = \frac{4\pi}{\mu(B)} \chi_B(P) \quad .$$

On the contrary the evaluation of the corresponding disturbing potential at a point P, computed by the Stokes integral,

$$T(P) = \frac{1}{4\pi} \int_0 S(\psi_{PQ}) \, \Delta g(Q) \, d\sigma_Q$$ (3.10)

$(\psi_{PQ}$ = spherical distance between P,Q

$$S(\psi) = t^{-1} - 6t + 1 - (1-2t^2)[5 - 3\log(t + t^2)]$$
$$t = \sin \psi/2)$$

is not a bounded functional because $S(\psi)$ is not a square integrable function

$$\frac{1}{4\pi} \int S^2(\psi_{PQ}) \, d\sigma_Q = + \infty \quad .$$

Def. 3.3: the "span" of a sequence $\{f_n, n=1,2,\ldots\}$ also denoted by Span $\{f_n, n=1,2,\ldots\}$, is the minimum closed subspace containing all the f_n; this is also generated by closing the space of finite linear combinations, $\sum_1^N \lambda_i f_i$, under the norm of H.

Corollary: the orthogonal complement of Span $\{f_n\}$ is the subspace of $g \in H$ such that
$$(g, f_n)_H = 0 \quad , \quad \forall n \quad . \tag{3.11}$$

Def. 3.4: a sequence $\{f_n\}$ is complete in H if
$$H = \text{Span } \{f_n\} \quad ; \tag{3.12}$$
in this case it is also said that $\{f_n\}$ is a base for H.

Corollary: a sequence $\{f_n\}$ is complete in H if
$$g \in H, \quad (g, f_n) = 0, \quad \forall n \Rightarrow g \equiv 0 \quad . \tag{3.13}$$

Def. 3.5: a sequence is orthogonal if
$$(f_n, f_m) = 0 \quad , \qquad n \neq m \tag{3.14}$$

it is normalized if
$$\|f_n\| = 1 \qquad \forall n \quad ; \tag{3.15}$$

if $\{f_n\}$ is both orthogonal and normalized it is said to be orthonormal (ON).

Def. 3.6: a sequence is an orthonormal base (ONB) of H if it is ON and complete.

Th. 3.3 (Fourier rapresentation): for any ONB $\{f_n\}$ and any $f \in H$ the following identities holds

$$f = \sum_{1}^{+\infty} {}_n (f,f_n) \, f_n \tag{3.16}$$

$$(f,g) = \sum_{1}^{+\infty} {}_n (f,f_n)(g,f_n) \tag{3.17}$$

$$\|f\|^2 = \sum_{1}^{+\infty} {}_n (f,f_n)^2 \tag{3.18}$$

Remark 3.2

Th. 3.3 is the basis for a calculus in H which is a direct generalization of the vector calculus in euclidean spaces; the great difference is in that any formula in H which includes a series, must be checked to be convergent.

Example 3.3

The sequence of spherical harmonics $\{Y_{nm}(P)\}$, where

$$Y_{nm}(P) = \bar{P}_{nm}(\sin \phi) \begin{cases} \cos m\lambda & n \geq m \geq 0 \\ \sin |m|\lambda & -n \leq m \leq 0 \end{cases} \tag{3.19}$$

is an ONB in $L^2(\sigma)$.

To verify that it is normalized is just a lengthy exercise; that it is orthogonal is a consequence of the well known formula

$$\Delta_\sigma Y_{nm} = -n(n+1) \, Y_{nm} \tag{3.20}$$

stating that Y_{nm} is an eigen-function of the Laplace-Beltrami operator

$$\Delta_\sigma = \nabla_\sigma \cdot \nabla_\sigma = \frac{\partial^2}{\partial\phi^2} + \mathrm{tg} \, \phi \, \frac{\partial}{\partial\phi} + \frac{\partial}{\cos^2 \phi} \, \frac{\partial^2}{\partial\lambda^2} \quad . \tag{3.21}$$

In fact, owing to the spherical Green's identity

$$0 = \frac{1}{4\pi} \int \left\{ Y_{nm} \, \Delta_\sigma \, Y_{jk} - Y_{jk} \, \Delta_\sigma \, Y_{nm} \right\} d\sigma =$$

$$= \{j(j+1) - n(n+1)\} \frac{1}{4\pi} \int Y_{nm} \, Y_{jk} \, d\sigma \quad ,$$

we find that Y_{nm} is orthogonal to Y_{jk} for $n \neq j$; when $n = j$, the integral in $d\lambda$, along the parallels, is always zero unless m=k, due to the well known orthogonality of $\sin m\lambda$, $\cos m\lambda$.

That it is complete, is a deeper result, of which we just sketch a simple proof. We want to show that if $f \in L^2$ is such that

$$\frac{1}{4\pi} \int f \, Y_{nm} \, d\sigma = 0 \qquad\qquad \forall \, n,m \quad, \tag{3.22}$$

then f has to be zero. Let us consider, in space, the potential u generated by a single layer of density f, namely

$$u(P) = \frac{1}{4\pi} \int \frac{f(Q)}{r_{PQ}} \, d\sigma \quad . \tag{3.23}$$

By using the classical developments (note that $r_Q = 1$)

$$\frac{1}{r_{PQ}} = \sum_0^\infty {}_n \, r_P^n \, P_n(\cos \phi_{PQ}) \qquad\qquad r_P < 1$$

$$\frac{1}{r_{PQ}} = \sum_0^\infty {}_n \, \frac{1}{r_P^{n+1}} \, P_n(\cos \phi_{PQ}) \qquad\qquad r_P > 0$$

and the summation rule (cfr. Heiskanen-Moritz [2])

$$P_n(\cos \phi_{PQ}) = \frac{1}{2n+1} \sum_{-n}^n {}_m \, Y_{nm}(P) \, Y_{nm}(Q) \quad,$$

we see that if f satisfies (3.22) then $u(P) \equiv 0$ in both $r_P > 1$ and $r_P < 1$. Since on the other hand the jump relation

$$\frac{\partial u}{\partial r}\bigg|_{r^+ = 1} - \frac{\partial u}{\partial r}\bigg|_{r^- = 1} = f$$

has to hold, we find that also $f(P) \equiv 0$ on the unit sphere, as it was to be proved. Sometimes we come to formulate a problem by saying that we would like to have a development of the form

$$f = \sum_1^\infty {}_n \, c_n \, f_n \tag{3.24}$$

with f given in H and $\{f_n\}$ a sequence which is not orthonormal, although it is known to be complete because condition (3.13) is satisfied. This problem is considerably more difficult and it has to do with the existence of a so-called biorthogonal sequence $\{h_n\}$.

<u>Def. 3.7</u>: given $\{f_n\}$ the sequence $\{h_n\}$ is said to be biorthogonal to $\{f_n\}$ if $(f_n, h_k) = \delta_{nk}$. $\qquad$ (3.25)

<u>Corollary</u>: if $\{f_n\}$ admits a biorthogonal sequence then both the following representations hold (and are unique)

$$f = \sum_1^\infty {}_n \ (f, \ h_n) \ f_n \qquad\qquad (3.26)$$

$$f = \sum_1^\infty {}_n \ (f, \ f_n) \ h_n \qquad\qquad (3.27)$$

Remark 3.3

The existence of a biorthogonal sequence <u>can never rely</u> on the fact that $\{f_n\}$ are finitely linearly independent, i.e. that for finite N

$$\sum_1^N {}_n \ \lambda_n \ f_n = 0 \quad \Rightarrow \quad \lambda_n = 0 \ ; \qquad\qquad (3.28)$$

the condition (3.28) is not sufficient as the following counterexample shows.

Example 3.4

The sequence $\{\sin n\theta, \cos n\theta\}$ is non orthogonal but complete in $L^2(0\pi)$; in fact if $f \in L^2(0\pi)$, by setting

$$\bar{f}(\theta) = \begin{cases} f(\theta) & 0 < \theta \leq \pi \\ 0 & \pi < \theta \leq 2\pi \end{cases}$$

we can extend it to the whole circle, and it turns out that

$$\int f \begin{Bmatrix} \sin n\theta \\ \cos n\theta \end{Bmatrix} d\theta = 0 \ , \quad \forall \ n \ \rightarrow \ \bar{f} \equiv 0 \ \rightarrow \ f(\theta) = 0 \ \text{ in } \ 0 < \theta \leq \pi \ .$$

On the other hand $\{1\}$ and $\{\sin \theta\}$ are not orthogonal in 0π, as

$$\int_0^\pi \sin \theta \ d\theta = 2 \ .$$

Moreover the above sequence is finitely linearly independent because any trigonometric polynomial which is zero on any interval, is identically zero everywhere. For this sequence however there can be no biorthogonal partner, because the subsequence $\{\sin 2n\theta, \cos 2n\theta\}$ is indeed orthogonal and complete in $L^2(0\pi)$. It follows that if we try to compute $\{c_n, b_n\}$ from a truncated representation

$$f(\theta) = \sum_0^N {}_n a_n \cos n\theta + b_n \sin n\theta \quad ,$$

by some least squares method, the determination will become more and more unstable for $N \to \infty$, because in the limit there is no uniqueness of such coefficients

Remark 3.4

The example 3.4 can be transferred to the spherical analysis of functions defined on a part only of the sphere, σ_0, in terms of spherical harmonics.

The sequence $\{Y_{nm}\}$ is indeed complete in $L^2(\sigma_0)$, non-orthogonal there and no biorthogonal sequence exists, because there is no unique representation of the form

$$f = \sum_0^\infty {}_n \sum_{-n}^n {}_m f_{nm} Y_{nm}(P) \quad , \qquad P \in \sigma_0 \quad .$$

Def. 3.8: given a linear bounded operator A in H, we define the adjoint $A^\bullet$ as that operator for which the identity

$$(g,Af) = (A^\bullet g,f) \qquad\qquad (3.29)$$

holds $\forall f,g \in H$.

Remark 3.5

When A is bounded, $A^\bullet$ turns out to be bounded too, with the same norm.

When A is unbounded (3.29) has to hold $\forall f \in D_A$ (domain of A in H), $\forall g \in D_{A^\bullet}$ (domain of $A^\bullet$ in H).

Def. 3.10: an operator K is compact if it trasforms every bounded sequence $\{f_n\}$ into a sequence $\{Kf_n\}$ with at least an accumulation point.

Th. 3.4: a compact selfadjoint operator A is characterized by having a sequence $\{\lambda_n\}$ of real eigenvalues, a sequence $\{f_n\}$ of eigenfunctions which are orthogonal one another and furthermore

$$\lim_{n\to\infty} \lambda_n = 0 \quad ; \qquad\qquad (3.30)$$

moreover if $\lambda_n \neq 0 \; \forall n$, the inverse A^{-1} exists and it is an unbounded selfadjoint operator.

Finally if A is even only compact, the following property, known as Fredholm alternative, holds: if the equation

$$f - Af = 0 \qquad\qquad (3.31)$$

admits only the trivial solution f=0, then the equation

$$f - Af = g \qquad (3.32)$$

admits a solution for every $g \in H$. If (3.31) admits non-null solutions $\{f\}$, then the same is true for the adjoint equation $w - A^*w = 0$ and (3.32) admits (non unique) solutions only if g is orthogonal to all such $\{w\}$.

Example 3.5

Let H be a space of harmonic functions f in the exterior of the unit sphere, such that $f \big|_\sigma \in L^2(\sigma)$; in this space, with norm $\| f\big|_\sigma \|_{L^2(\sigma)}$, the following operators are unbounded

$$- r \frac{\partial}{\partial r} g \bigg|_\sigma$$

$$- \Delta_\sigma f \bigg|_\sigma$$

$$\left(- r \frac{\partial}{\partial r} - 2 \right) f \bigg|_\sigma$$

because they all admit $\{Y_{nm}\}$ as eigenfunctions, with eigenvalues respectively $(n+1)$, $n(n+1)$, $(n-1)$.

The first operator is invertible (the inverse of $- \frac{\partial}{\partial r}$ being the Hotine integral operator H), the other two are invertible only on subspaces, since they admit zero eigenvalues.

By the way the conditional inverse of the third operator, is the Stokes' integral operator S.

Both H and S are selfadjoint compact operators in $L^2(\sigma)$.

Def. 3.11: the Sobolev spaces H^λ $(\lambda > 0)$ are defined as subsets of $L^2(\sigma)$, costituted by functions

$$f = \sum_{n,m} f_{nm} Y_{nm} \quad , \qquad (3.33)$$

such that

$$\|f\|_\lambda^2 = \sum_{n,m} (1+n)^{2\lambda} f_{nm}^2 < + \infty \quad . \qquad (3.34)$$

the following inclusion is obvious

$$H^\mu \supset H^\lambda \quad , \qquad \mu < \lambda \tag{3.35}$$

because

$$\|f\|_\mu^2 \leq \|f\|_\lambda^2 \quad . \tag{3.36}$$

Remark 3.6

The embedding (3.35) is compact, in the sense that if $\{f_n\}$ is bounded in H^λ, then the same sequence seen in H^μ ($\mu < \lambda$) becomes compact.

In fact we note that, calling

$$D = -r\frac{\partial}{\partial r} \quad , \tag{3.37}$$

owing to the discussion in Example 3.5, D is selfadjoint, unbounded and positive definite (i.e. with positive eigenvalues), so we have

$$f \in H^\lambda \quad \longleftrightarrow \quad D^{\lambda-\mu} f \in H^\mu$$

and furthermore this correspondence is an isometry since

$$\|f\|_\lambda^2 = \sum (1+n)^{2\lambda} f_{nm}^2 = \|D^{\lambda-\mu} f\|_\mu^2 = \sum (1+n)^{2\mu} (1+n)^{2\lambda-2\mu} f_{nm}^2 \quad .$$

So if

$$\|f_n\|_\lambda^2 < C$$

Then also

$$\|D^{\lambda-\mu} f_n\|_\mu^2 < C \quad ,$$

so that

$$f_n = D^{\mu-\lambda} (D^{\lambda-\mu} f_n) \tag{3.38}$$

is the transform in H^μ of the sequence $D^{\lambda-\mu} f_n$, which is bounded in H^μ, through the operator $D^{\mu-\lambda}$ (with eigenvalues $(1+n)^{-(\lambda-\mu)}$) which is obviously compact in H^μ, as it was to be proved.

Remark 3.7

Assume that ϕ is a linear bounded functional in H^λ, then according to Riesz theorem, there is a $g \in H^\lambda$ such that

$$\phi(f) = (g,f)_\lambda = \sum g_{nm} f_{nm} (1+n)^{2\lambda} \tag{3.39}$$

If we write, formally,

$$\begin{cases} h = D^{2\lambda}g \quad, \\ h_{nm} = (1+n)^{2\lambda} g_{nm} \quad, \end{cases} \tag{3.40}$$

we see that (3.39) takes the form

$$(g,f)_\lambda = \sum h_{nm} f_{nm} = (h,f)_0 \tag{3.41}$$

The series in (3.41) is still convergent and, by applying suitably the Schwarz inequality we see that

$$|(h,f)_0|^2 \le \sum h_{nm}^2 (1+n)^{-2\lambda} \sum f_{nm}^2 (1+n)^{2\lambda} = \sum g_{nm}^2 (1+n)^{2\lambda} \sum f_{nm}^2 (1+n)^{2\lambda} \tag{3.42}$$

What (3.42) says is that if we take h as a member of a space with norm

$$\sum h_{nm}^2 (1+n)^{-2\lambda} \quad,$$

then we can perform a generalized $L^2(\sigma)$ coupling between h and f since this last belongs to H^λ

Def. 3.12: we define the Sobolev spaces $H^{-\lambda}$ ($\lambda > 0$) in such a way that $h \in H^{-\lambda}$ if

$$\sum h_{nm}^2 (1+n)^{-2\lambda} < + \infty \quad; \tag{3.43}$$

the members of this space, $h = \sum h_{nm} Y_{nm}$, need not to be functions, but they are rather distributions which can be coupled with H^λ functions f by computing

$$(h,f)_0 = \sum h_{nm} f_{nm} = \frac{1}{4\pi} \int hf \, d\sigma \tag{3.44}$$

Remark 3.8

With the above definition we have found another representation of linear bounded functionals on H^λ; so we have the following inclusions

$$\ldots\; H^{-\lambda} \supset H^{-\mu} \supset \ldots\; L^2 \ldots \supset H^{\mu} \supset H^{\lambda} \ldots \qquad (3.45)$$

with the further identification

$$H^{-\lambda} = (H^{\lambda})^{\bullet} \quad,$$

with respect to the representation (3.44).

Example 3.6

Let us take the interesting case $\lambda = 1/2$; then $(H^{1/2})^{\bullet} = H^{-1/2}$ and

$$h \in H^{-1/2} \longleftrightarrow h = Dg \quad, \quad g \in H^{1/2}$$

Furthermore (remember $r=1$ on the unit sphere)

$$(h,f)_0 = (Dg,f)_0 = -\frac{1}{4\pi} \int \left(\frac{\partial}{\partial r}\, g \right) f \; d\sigma$$

and, choosing $g=f$, we get

$$(Df,f)_0 = (D^{1/2}f, D^{1/2}f)_0 = \|f\|_{1/2}^2 = -\frac{1}{4\pi} \int \left(\frac{\partial}{\partial r}\, f \right) f \; d\sigma \quad. \qquad (3.46)$$

This is indeed a very classical norm, which through the first Green identity, can be identified with the Dirichlet norm because

$$-\frac{1}{4\pi} \int_{\sigma} \left(\frac{\partial}{\partial r}\, f \right) f \; d\sigma = \frac{1}{4\pi} \int_{(r>1)} |\nabla f|^2 \; r^2 \; dr \; d\sigma \qquad (3.47)$$

Since this will be very much needed, we want to give a meaning, for $h \in H^{-\lambda}$, to the statement that we have, on a part only of the sphere, σ_0,

$$h = 0 \qquad (\text{on } \sigma_0) \quad. \qquad (3.48)$$

Since h is defined through the coupling (3.44), we must use it to give meaning to (3.48). We use then the following definition.

Def. 3.13: Fixing σ_0 on σ and calling σ_c the complement of σ_0 in σ, we first define $\overset{\circ}{H}{}^{\lambda}(\sigma_0)$ as the subspace of $H^{\lambda}(\sigma)$ of those f that satisfy

$$f\bigg|_{\sigma_c} = 0 \qquad (f \in H^\lambda) \ . \tag{3.49}$$

Then we say that, for $h \in H^{-\lambda}$,

$$h\bigg|_{\sigma_0} \equiv 0 \tag{3.50}$$

if

$$(h,f)_0 = 0 \qquad \forall f \in \overset{\circ}{H}{}^\lambda(\sigma_0) \ . \tag{3.51}$$

Remark 3.9

It comes natural to the mind to ask whether (3.49) in fact creates a constraint on f, so that (3.51) also assumes a particular meaning; in fact for instance if $f \in L^2(\sigma)$ the fact that $f \equiv 0$ on σ_c doesn't put any constraint on f on σ_0. It turns out that the situation is like the one in $L^2(\sigma)$ for H^λ $(0 < \lambda < 1/2)$, because even if you take a function with a sharp jump it can still belong to those spaces; while for H^λ $(\lambda > 1/2)$ you can't have a sharply discontinuous f.

Just to grasp the idea let us make the one-dimensional example of the step function (cfr. Fig. 3.1).

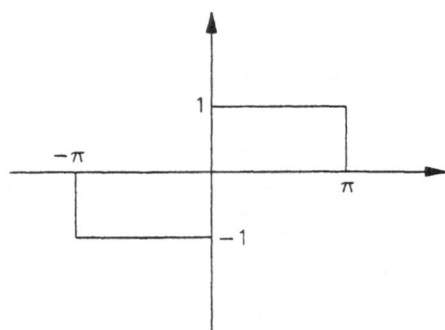

Fig. 3.1

This function has a pure sine Fourier series with coefficients

$$f = \sum C_n \frac{1}{\sqrt{2\pi}} \sin n\theta$$

$$C_n = \begin{cases} 0 & n \text{ even} \\ \dfrac{2\sqrt{2}}{\sqrt{\pi}\, n} & n \text{ odd} \end{cases}$$

Therefore

$$\|f\|_\lambda^2 = \sum_k \, k \, \frac{8}{\pi} \, \frac{1}{(2k+1)^{2-2\lambda}} \quad ;$$

as we see then

$$\|f\|_\lambda^2 \ < \ +\infty \qquad\qquad \lambda < 1/2$$

$$\|f\|_\lambda^2 \ = \ +\infty \qquad\qquad \lambda \geq 1/2 \quad .$$

4. The analysis of AG I

In this paragraph we shall analyze the problem AG I, i.e. we shall determine conditions sufficient for the existence, uniqueness and good behaviour of the solution of (2.36).
In particular we aim at proving the following basic theorem.

<u>Th. 4.1</u>: assume δw_S to be the restriction to S of a function $\delta w \in H^{1/2}(\sigma)$ and Δg_L to be the restriction to L of a $\Delta g \in H^{-1/2}(\sigma)$, then defining

$$\bar{H}^{1/2}(\sigma) = \left\{ T; \ T \in H^{1/2}(\sigma), \ \int T \, d\sigma = 0 \right\} , \qquad\qquad (4.1)$$

there is one and only one solution (T,a), with $T \in \bar{H}^{1/2}(\sigma)$ of the problem (AG I)

$$\begin{cases} T \ \Big|_S = \delta w_S + a \\[2mm] -\dfrac{\partial T}{\partial r} - \dfrac{2}{R} \, T \ \Big|_L = \Delta g \quad . \\[2mm] \int T \, d\sigma = 0 \end{cases} \qquad\qquad (4.2)$$

In order to prove the Th. 4.1 we must first study a simplified case, namely (AG 0)

$$\begin{cases} T \ \Big|_S = \delta w_S \\[2mm] -\dfrac{\partial T}{\partial r} \ \Big|_L = f_L \end{cases} \qquad\qquad (4.3)$$

with the purpose of proving the

<u>Th. 4.2</u>: let $\delta w_S = \delta w \big|_S$ with $\delta w \in H^{1/2}(\sigma)$ and $f_L = f \big|_L$ with $f \in H^{-1/2}(\sigma)$, then there is one and only one $T \in H^{1/2}(\sigma)$ satisfying (4.3).

To prove this theorem we first define two subsets, $U \subset H^{1/2}(\sigma)$ and $V \subset H^{-1/2}(\sigma)$ as follows

$$U = \left\{ u \in H^{1/2}(\sigma); \qquad u \mid_S = \delta w_S \right\} \tag{4.4}$$

$$V = \left\{ v \in H^{-1/2}(\sigma); \qquad v \mid_L = f_L \right\} \tag{4.5}$$

U and V are non empty sets as $\delta w \in U$, $f \in V$; further on they are closed in their respective spaces. We prove this for U as the other proof is similar. Let $\{u_n\} \in U$ be a Cauchy sequence in $H^{1/2}$; then there is $\bar{u} \in H^{1/2}$ such that $u_n - \bar{u} \to 0$ in $H^{1/2}$. On the other hand let $\{\phi\}$ be any element of $H^{-1/2}$ such that $\phi \mid_L = 0$. Then

$$\frac{1}{4\pi} \int_\sigma (u_n - \delta w) \, \phi \, d\sigma \equiv 0 \quad , \qquad \forall \, n, \; \forall \, \phi \tag{4.6}$$

because $u_n - \delta w \mid_S = 0$ and $\phi \mid_L = 0$. Then we must also have, by the continuity of the functional $\phi \in H^{-1/2}$ with respect to $(u_n - \delta w)$,

$$\frac{1}{4\pi} \int_\sigma (\bar{u} - \delta w) \, \phi \, d\sigma \equiv 0 \quad , \qquad \forall \, \phi \in H^{-1/2} \left(\phi \mid_L = 0 \right) \quad . \tag{4.7}$$

The relation (4.7), implies

$$\bar{u} - \delta w \mid_S = 0 \quad , \tag{4.8}$$

as it was to be proved.
We now form a subset of $H^{-1/2}$ defined as follows

$$Z = \left\{ z = \frac{\partial u}{\partial r} - v; \quad u \in U, \; v \in V \right\} \quad ; \tag{4.9}$$

that $Z \subset H^{-1/2}$ follows from the fact that when $u \in H^{1/2}$ then $\frac{\partial u}{\partial r} \in H^{-1/2}$. Z is not empty. Z is closed, as if $\{z_n\} \in Z$ is a Cauchy sequence in $H^{-1/2}$ then

$$z_n - z_m = \frac{\partial}{\partial r} (u_n - u_m) - (v_n - v_m) \quad ; \tag{4.10}$$

if we can prove that $\frac{\partial}{\partial r} (u_n - u_m)$ is orthogonal to $(v_n - v_m)$ in $H^{-1/2}$ then $z_n - z_m \to 0$ implies also

$$\left\| \frac{\partial}{\partial r} (u_n - u_m) \right\|_{-1/2} \to 0 \quad , \quad \left\| (v_n - v_m) \right\|_{-1/2} \to 0 \quad . \tag{4.11}$$

The first of (4.11) implies that

$$\left\| u_n - u_m \right\|_{1/2} \to 0 \quad ,$$

so that there are $\bar{u} \in H^{1/2}$, $\bar{v} \in H^{-1/2}$ such that $u_n \to \bar{u}$ in $H^{1/2}$ $\left(\frac{\partial u_n}{\partial r} \to \frac{\partial \bar{u}}{\partial r} \text{ in } \right.$ $H^{-1/2} \left. \right)$ and $v_n \to \bar{v}$ in $H^{-1/2}$; but then $z_n \to \bar{z} = \frac{\partial \bar{u}}{\partial r} - \bar{v}$, with $\bar{u} \in U$ and $\bar{v} \in V$ so that $\bar{z} \in Z$ too. It remains to be proved that

$$< \frac{\partial}{\partial r} (u_n - u_m) , v_n - v_m >_{-1/2} = 0 \tag{4.12}$$

if $u_n, u_m \in U$ and $v_n, v_m \in V$.

In fact let us remember that if

$$H = \left(- \frac{\partial}{\partial r} \right)^{-1} \tag{4.13}$$

is the Hotine operator, from $H^{-1/2}$ to $H^{1/2}$ we can define

$$< f, g >_{-1/2} = \frac{1}{4\pi} \int (H f) g \, d\sigma \quad . \tag{4.14}$$

So, exploiting (4.13), we get

$$< \frac{\partial}{\partial r} (u_n - u_m) , v_n - v_m >_{-1/2} =$$

$$= \frac{1}{4\pi} \int (u_n - u_m) (v_n - v_m) \, d\sigma \quad . \tag{4.15}$$

On the other hand

$$u_n, u_m \in U \quad \to \quad u_n - u_m \Big|_S = 0$$

$$v_n, v_m \in V \quad \to \quad v_n - v_m \Big|_L = 0 \quad ,$$

so that necessarily

$$\int (u_n - u_m) (v_n - v_m) \, d\sigma \equiv 0 \quad . \tag{4.16}$$

Relations (4.16), (4.15) imply (4.12) and the closedness of Z is proved.
Let us notice that essentially Z can be viewed as the subspace of $H^{-1/2}$ functions
y such that

$$y = \left(\frac{\partial}{\partial r} u - v \right)$$

$$u \Big|_S = 0 \tag{4.17}$$

$$v \Big|_L = 0 \quad ,$$

shifted from the origin e.g. by the element $\tilde{z} = \frac{\partial \delta w}{\partial r} - f$; i.e. Z is a closed linear
manifold in $H^{-1/2}$. Therefore there is one and only one element in Z which is of
minimum norm $\bar{z} = \frac{\partial \bar{u}}{\partial r} - \bar{v}$ ($\bar{u} \in U$, $\bar{v} \in V$); this element is characterized by the
orthogonality

$$< \bar{z}, y >_{-1/2} = < \bar{z}, \frac{\partial u}{\partial r} - v >_{-1/2} = 0 \tag{4.18}$$

for every u $\left(u \Big|_S = 0 \right)$ and v $\left(v \Big|_L = 0 \right)$.

From (choosing v=0)

$$0 \equiv <\bar{z}, \frac{\partial u}{\partial r} >_{-1/2} = \frac{1}{4\pi} \int \bar{z} \, u \, d\sigma \tag{4.19}$$

for every u $\left(u \Big|_S = 0 \right)$, we find

$$\bar{z} \Big|_L = 0 \quad . \tag{4.20}$$

From (choosing u=0)

$$0 \equiv <\bar{z}, v >_{-1/2} = \frac{1}{4\pi} \int (H\bar{z}) \, v \, d\sigma$$

for every v $\left(v \Big|_L = 0 \right)$, we find

$$H\bar{z} \Big|_L = 0 \quad . \tag{4.21}$$

But (4.20) and (4.21) together imply that

$$\|\bar{z}\|^2_{-1/2} = \frac{1}{4\pi} \int \bar{z} \, (H\bar{z}) \, d\sigma = 0 \quad,$$

i.e. $\bar{z} = 0$. This is turn means that

$$\frac{\partial \bar{u}}{\partial r} - \bar{v} = 0 \quad, \tag{4.22}$$

so that $\bar{u}$ is a function $H^{1/2}$ with

$$\bar{u} \Big|_S = \delta w_S \quad, \tag{4.23}$$

because $\bar{u} \in \bar{U}$ and with

$$\frac{\partial \bar{u}}{\partial r} \Big|_L = \bar{v} \Big|_L = f_L \quad, \tag{4.24}$$

because $\bar{v} \in V$.

So we have proved the existence, part of Th. 4.2. The uniqueness follows from the observation that if $T \in H^{1/2}$ and

$$T \Big|_S = 0$$

$$\frac{\partial T}{\partial r} \Big|_L = 0 \quad,$$

then also

$$\|T\|^2_{1/2} = -\frac{1}{4\pi} \int T \, \frac{\partial T}{\partial r} \, d\sigma \equiv 0 \quad,$$

i.e. $T = 0$.

Remark 4.1

From Th. 4.2 it follows that there are two operators R_S and R_L acting respectively on $H^{1/2}$ and $H^{-1/2}$ such that, given $\delta w \in H^{1/2}$, $f \in H^{-1/2}$, give rise to

$$\bar{u} = R_S \, \delta w + R_L f \quad ; \tag{4.25}$$

in particular since R_L is a bounded operator from $H^{-1/2}$ to $H^{1/2}$ and since $H^{1/2} \subset H^{-1/2}$, the embedding being compact, we see that R_L restricted to $H^{1/2}$ is in fact a compact operator.

We can come now to problem (4.2) which we rewrite in the form

$$
\begin{cases}
T \big|_S = \delta w_S + a \\[2mm]
-\dfrac{\partial T}{\partial r}\Big|_L = \Delta g + \dfrac{2}{R} T \\[2mm]
\displaystyle\int T \, d\sigma = 0
\end{cases}
\tag{4.26}
$$

By making use of the Remark 4.1, we transform (4.26) into

$$
T = R_S(\delta w + a) + R_L\left(\Delta g + \frac{2}{R} T\right) = F + a\,\tau + \frac{2}{R} R_L T
\tag{4.27}
$$

where

$$
\tau = R_S(1) \in H^{1/2}
$$

and

$$
F = R_S\,\delta w + R_L\Delta g \in H^{1/2} \quad .
$$

If we impose the condition $\int T \, d\sigma = 0$ we find

$$
\int_S (\delta w + a) \, d\sigma + \int_L T \, d\sigma = 0
\tag{4.28}
$$

which allows to eliminate a from (4.27), thus arriving at the unique equation in T

$$
T = \tilde{F} + \frac{2}{R} R_L T - \tau \frac{1}{\mu_S} \int_L T \, d\sigma
\tag{4.29}
$$

where

$$
\tilde{F} = F - \tau \frac{1}{\mu_S} \int_L \delta w_S \, d\sigma \quad .
$$

To study (4.29) we can exploit the compactness of R_L, recalled in remark 4.1, as well as the compactness of the second operator

$$
T \to \tau \left(\frac{1}{\mu_S} \int_S T \, d\sigma \right) \quad ,
$$

which has a 1-dimensional range.

In view of this compactness we need only to prove the uniqueness of this solution, or

equivalently of the homogeneous form of (4.26), in order to get existence[5].
We then suppose that

$$
\begin{cases}
T \Big|_{S} = a \\[2ex]
-\dfrac{\partial T}{\partial r}\Big|_{L} = \dfrac{2}{R}\, T \\[2ex]
\displaystyle\int T\, d\sigma = 0
\end{cases}
\tag{4.30}
$$

and we must prove that $T = 0$, $a = 0$.
To this aim we first notice that the third of (4.30) implies

$$
\int \frac{\partial T}{\partial r}\, d\sigma = 0 \quad,
$$

so that, also using (4.30), the following chain of equalities holds

$$
\int_{S} \frac{\partial T}{\partial r}\, d\sigma = -\int_{L} \frac{\partial T}{\partial r} = \frac{2}{R}\int_{L} T\, d\sigma = -\frac{2}{R}\int_{S} T\, d\sigma \quad.
\tag{4.31}
$$

We then have

$$
\begin{aligned}
\|T\|_{1/2}^{2} &= -\frac{1}{4\pi}\int \frac{\partial T}{\partial r}\, T\, d\sigma = -\frac{a}{4\pi}\int_{S} \frac{\partial T}{\partial r}\, d\sigma + \frac{2}{R}\frac{1}{4\pi}\int_{L} T^{2}\, d\sigma = \\[2ex]
&= \frac{2}{R}\frac{a}{4\pi}\int_{S} T\, d\sigma + \frac{2}{R}\frac{1}{4\pi}\int_{L} T^{2}\, d\sigma = \\[2ex]
&= \frac{2}{R}\frac{1}{4\pi}\int_{S} T^{2}\, d\sigma + \frac{2}{R}\frac{1}{4\pi}\int_{L} T^{2}\, d\sigma = \frac{2}{R}\,\|T\|_{0}^{2}
\end{aligned}
\tag{4.32}
$$

If we express (4.32) in terms of spherical harmonics, noting that $T_{00} = 0$, we get

$$
\frac{1}{R}\sum_{1}^{\infty}{}_{l}\sum_{m}(1+1)\, T_{lm}^{2} = \frac{2}{R}\sum_{1}^{\infty}{}_{l}\sum_{m} T_{lm}^{2} \quad;
\tag{4.33}
$$

since the summation on l starts from $l=1$ we see that it must be

$$
T_{lm} = 0 \qquad \forall\; l > 1 \quad.
\tag{4.34}
$$

---

[5] In reality we should study the uniqueness of the adjoint of (4.29), however as we observed $(I-K)x = 0$ admits the only solution $x = 0$, when K is compact, then the same is true for $(I-K^{*})x = 0$.

But then

$$T = \sum_m T_{1m} Y_{1m}$$

so that, for the first of (4.30),

$$\sum_m T_{1m} Y_{1m} - a \equiv 0 \qquad P \in S \; ; \tag{4.35}$$

recalling the Remark 3.3, since (4.35) involves only finitely many harmonics, it implies

$$T_{1m} = 0 \; , \qquad a = 0$$

as far as $\mu_s > 0$, as it is.
This proves Th. 4.2.

Remark 4.2

Crucial for our proof was that the condition $T_{00} = 0$ has been imposed, without which it is not any more true that

$$\|T\|^2_{1/2} - \|T\|^2_0 \geq 0$$

and uniqueness is not garanteed in general.
So the modification produced by introducing the unknown a and the condition $T_{00} = 0$ makes the solution unconditionally unique.

Remark 4.3

If the data δw_s, Δg in (4.2) become more regular, then also the solution becomes such; however the data themselves must satisfy a further condition to garantee the existence, because if $T \in H^\lambda$ ($\lambda > 1/2$) in both S and L areas, then we can compute T along the coast coming from the sea area as well as from the land (in fact it happens that $T|_{coast} \in H^{\lambda-1/2}$), so that, so to speak, along this line we must impose that the two limits agree if we want the solution to belong to H^λ over all σ.
In this sense it seems that $\delta w \in H^{1/2}$, $\Delta g \in H^{-1/2}$ are somehow natural conditions for this problem.

5. The analysis of AG II

Let us first remember the formulation of AG II, i.e. find T and a such that

$$
\begin{cases}
\left. -\dfrac{\partial T}{\partial r} \right|_S = \delta g_S + a \\[2ex]
\left. -\dfrac{\partial T}{\partial r} - \dfrac{2}{R} T \right|_L = \Delta g_L \\[2ex]
\displaystyle\int T \, d\sigma = 0
\end{cases}
\tag{5.1}
$$

We can first formally write the two conditions on S and L in one equation; in fact introducing the characteristic function of S

$$
\chi_S(P) = \begin{cases} 1 & P \in S \\[1ex] 0 & P \notin S \end{cases}
\tag{5.2}
$$

and

$$
\chi_L(P) = 1 - \chi_S(P) \quad ,
\tag{5.3}
$$

we can just sum the two equations to get

$$
-\frac{\partial T}{\partial r} - \frac{2}{R} \chi_L T = f + a \chi_S
\tag{5.4}
$$

with

$$
f = \chi_S \, \delta g + \chi_L \, \Delta g
\tag{5.5}
$$

Remark 5.1

If we want to prescribe some regularity of f in term of Sobolev spaces, we must take into account the Remark 3.9. In fact if $f \in H^\lambda$, we have $\chi_L f$, $\chi_S f \in H^\lambda$ too, only if $\lambda < 1/2$; furthermore for reasons that will be cleared later we also need to restrict λ so that $\lambda > -1/2$; whence we shall assume that

$$
-1/2 < \lambda < 1/2 \quad .
\tag{5.6}
$$

We want then to prove the following.

Th. 5.1: if $f \in H^\lambda(\sigma)$ $(-1/2 < \lambda < 1/2)$, then (5.1) admits one and only one solution (T,a) with $T \in H^{\lambda+1}(\sigma)$.

To prove that we start by trying to write a unique equation in the unknown T and see that Fredholm's alternative holds; it turns out that this can be more easily done here than in the previous paragraph.

In fact let us start with the third of (5.1); as we know this is equivalent to

$$\int \frac{\partial T}{\partial r} \, d\sigma = 0 \quad , \tag{5.7}$$

which can be written

$$\int_S (\delta g + a) d\sigma + \int_L (\Delta g + \frac{2}{R} T) d\sigma = 0 \quad . \tag{5.8}$$

Solving with respect to a, we get

$$a = - \frac{1}{\mu_S} \int f \, d\sigma - \frac{2}{R} \frac{1}{\mu_S} \int \chi_L \, T \, d\sigma \tag{5.9}$$

Then we go back to (5.4) and multiply this equation by the Hotine's operator.

$$H = \left(- \frac{\partial}{\partial r} \right)^{-1} \quad ; \tag{5.10}$$

after a rearrangement and using (5.9), we find

$$T = \frac{2}{R} H \left(\chi_L T \right) - \eta \frac{2}{R} \frac{1}{\mu_S} \int \chi_L \, T \, d\sigma + h \tag{5.11}$$

where

$$\eta = H \chi_S \tag{5.12}$$

$$h = Hf - \eta \frac{1}{\mu_S} \int f \, d\sigma \quad . \tag{5.13}$$

Now H is an integral operator with kernel

$$H(P,Q) = \sum_{n,m} \frac{R}{n + 1} Y_{nm}(P) Y_{nm}(Q) \quad , \tag{5.14}$$

so that it can be viewed as a bounded operator (an isometry as a matter of fact) between $H^\lambda \to H^{\lambda+1}$ or as a compact operator in H^λ, for every λ.
On the other hand we know that $\chi_S, \chi_L \in H^{1/2-\varepsilon}$ for any small $\varepsilon > 0$ and also that

$$\|\chi_L T\|_{1/2-\varepsilon} \leq C \|T\|_{1/2-\varepsilon} \quad . \tag{5.15}$$

So the following conclusions can be drawn: chosen a conveniently small ε, such that for a fixed λ satisfying (5.6),

$$1/2 - \varepsilon > \lambda \quad ; \tag{5.16}$$

then

$$\eta = H \chi_s \in H^{3/2-\varepsilon} \subset H^{1+\lambda} \quad ; \tag{5.17}$$

moreover, noting that by (5.6) $1 + \lambda > 1/2$,

$$T \in H^{1+\lambda} \to \chi_L T \in H^{1/2-\varepsilon}$$

and

$$H \chi_L T \in H^{3/2-\varepsilon} \subset H^{\lambda+1} \quad . \tag{5.18}$$

The embedding $H^{3/2-\varepsilon} \subset H^{\lambda+1}$ being compact, we see that (5.18) proves that $H \chi_L$ is a compact operator in $H^{\lambda+1}$.

Furthermore, owing to (5.17) and since

$$\left| \frac{1}{4\pi} \int \chi_L T \, d\sigma \right| \leq \|T\|_0 \leq \|T\|_{1+\lambda} \quad ,$$

we see that the finite rank operator

$$\eta \, \frac{2}{R} \, \frac{1}{\mu_s} \int \chi_L T \, d\sigma \tag{5.19}$$

is a compact operator too in $H^{\lambda+1}$, as it has a one dimensional range. Moreover, since $f \in H^\lambda$ by hypothesis, we also have $h \in H^{\lambda+1}$.

Therefore the equation (5.11) has the form

$$T = K T + h \tag{5.20}$$

with $h \in H^{\lambda+1}$ and K compact in $H^{\lambda+1}$; for such an equation the Fredholm's alternative holds and the existence of a solution is guaranteed once uniqueness is proved.

So we are reduced to prove that

$$\begin{cases} - \dfrac{\partial T}{\partial r} \Big|_s = a \\[2mm] - \dfrac{\partial T}{\partial r} - \dfrac{2}{R} T \Big|_L = 0 \\[2mm] \displaystyle\int T \, d\sigma = \left(- \int \dfrac{\partial T}{\partial r} \, d\sigma \right) = 0 \end{cases} \tag{5.21}$$

implies $T = 0$, $a = 0$.

We shall prove the uniqueness in two steps; first we assume $\lambda=0$ and then we prove it for every $-\frac{1}{2} < \lambda < \frac{1}{2}$.

If $\lambda = 0$ we consider the problem (5.21) with $T \in H^1$ and we manage to compute its norm.

To this aim we first observe that

$$\int_S \frac{\partial T}{\partial r} \, d\sigma = - \int_L \frac{\partial T}{\partial r} \, d\sigma = \frac{2}{R} \int_L T \, d\sigma = - \frac{2}{R} \int_S T \, d\sigma \; ; \qquad (5.22)$$

then

$$\|T\|_1^2 = \frac{1}{4\pi} \int \left(\frac{\partial T}{\partial r} \right)^2 d\sigma = \frac{1}{4\pi} \int_S \left(\frac{\partial T}{\partial r} \right)^2 d\sigma + \frac{1}{4\pi} \int_L \left(\frac{\partial T}{\partial r} \right)^2 d\sigma =$$

$$= -\frac{a}{4\pi} \int_S \frac{\partial T}{\partial r} \, d\sigma - \frac{2}{4\pi R} \int_L T \frac{\partial T}{\partial r} \, d\sigma =$$

$$= \frac{2}{R} \frac{a}{4\pi} \int_S T \, d\sigma - \frac{2}{R} \frac{1}{4\pi} \int_L T \frac{\partial T}{\partial r} \, d\sigma =$$

$$= -\frac{2}{R} \frac{1}{4\pi} \int_S T \frac{\partial T}{\partial r} \, d\sigma - \frac{2}{R} \frac{1}{4\pi} \int_L T \frac{\partial T}{\partial r} \, d\sigma =$$

$$= \frac{2}{R} \left[-\frac{1}{4\pi} \int_S T \frac{\partial T}{\partial r} \, d\sigma \right] = \frac{2}{R} \|T\|_{1/2}^2 \qquad (5.23)$$

If we write (5.23) in terms of harmonic components, recalling also that the spectrum of

$$-\frac{\partial T}{\partial r} \quad \text{is} \quad \left(\frac{n+1}{R} \right) \quad ,$$

we get

$$\frac{1}{R^2} \sum_1^\infty {}_n \sum_m (n+1)^2 \, T_{nm}^2 = \frac{2}{R^2} \sum_1^\infty {}_m (n+1) \, T_{nm}^2 \quad . \qquad (5.24)$$

From (5.24) again we immediately derive

$$T_{nm} = 0 \; , \qquad \forall \, n > 1, \quad \forall \, m \; ,$$

so that the solution of (5.21) reduces to

$$T = \sum_m T_{1m} Y_{1m} \quad . \qquad (5.25)$$

Inserting (5.25) into the first of (5.21), we obtain

$$\frac{2}{R} \sum_m T_{1m} Y_{1m} - a = 0 \quad , \qquad P \in S \tag{5.26}$$

Showing that

$$T_{1m} = 0 \; , \qquad a = 0 \tag{5.27}$$

because (5.26) holds on a set of non zero measure $(\mu_s > 0)$.
With (5.27) the uniqueness has been proved when $T \in H^1$, $(\lambda = 0)$.

If $-\frac{1}{2} < \lambda < \frac{1}{2}$, i.e. when $T \in H^\gamma \left(\frac{1}{2} < \gamma < \frac{3}{2} \right)$, we still have to show that (5.21), or its equivalent form (5.20) or (5.21), imply $T = 0$, $a = 0$. If $1 \le \gamma < \frac{3}{2}$ then it is also $T \in H^1$ and the uniqueness holds for sure.
On the other hand if $T \in H^\gamma \left(\frac{1}{2} < \gamma < 1 \right)$, then $\chi_L T \in H^{1/2-\varepsilon}$ so that, if $\varepsilon < 1/2$,

$$H \chi_L T \in H^{3/2-\varepsilon} \subset H^1 \quad . \tag{5.28}$$

Furthermore it is also $\chi_s \in H^{1/2-\varepsilon}$ and

$$\eta \in H^{3/2-\varepsilon} \subset H^1 \quad . \tag{5.29}$$

Then if T is a solution of the homogeneous equation

$$T = \frac{2}{R} H(\chi_L T) - \eta \frac{2}{R} \frac{1}{\mu_s} \int \chi_L T \, d\sigma \quad , \tag{5.30}$$

and $T \in H^\gamma \left(\frac{1}{2} < \gamma < 1 \right)$ it is also $T \in H^{3/2-\varepsilon}$, since this is true for the right hand side according to (5.28), (5.29). Therefore $T \in H^1$ too and indeed the uniqueness $(T = 0)$ has to hold.
According to our discussion, the existence of the solution is then guaranteed in H^γ $\left(\frac{1}{2} < \gamma < \frac{3}{2} \right)$ and Th. 5.1 is proved.

6. Applications

We want to show how the mathematical theory presented in the above paragraphs could be used to construct numerical solutions of some important geodetic problems.
What we will show here is not the main strategy currently adopted to achieve the solutions, although both proposals are already present in the geodetic literature; they are just, in the author's opinion, the most natural ways to apply the theory

developed up to now.

a) The application of AG I to the construction of global models

By a global model of the anomalous potential we mean here a representation of T in terms of a truncated series of a spherical harmonics up to some degree N_{max}

$$T_M = \sum_{2 \ n}^{N_{max}} \sum_{-n \ m}^{n} T_{nm} \ Y_{nm}(\phi, \lambda) \quad ; \tag{6.1}$$

so what we need, to construct such a model, are the coefficients T_{nm}.

These can be estimated from the data actually provided in AG I, i.e. δw_s, Δg_L, following different strategies: what is usually done is to transform δw_s into Δg_s by solving locally a so called inverse Stokes problem and then the B.V.P. approach is applied taking Δg as a uniform datum on the surface of the earth.

What we propose here is a direct approach.

In fact let us summarize what we did in § 4 in the following way: defining the boundary operator B as

$$BT = T\chi_S + \left(-\frac{\partial}{\partial r} - \frac{2}{R} \right) T \ \chi_L = \sum T_{nm} \ B \ Y_{nm} =$$

$$= \sum_{n,m} T_{nm} \left\{ \chi_S + \frac{n-1}{R} \chi_L \right\} Y_{nm} \quad , \tag{6.2}$$

we have solved the problem

$$BT = f = \delta w_s \ \chi_S + \Delta g_L \ \chi_L \quad ,$$

showing essentially that the Fredholm's alternative holds for B and that $BT=0$ implies $T=0$.

Now if instead of the whole T we want for instance only one coefficient T_{jk} we see that the problem is just to find that Z_{jk} for which

$$< Z_{jk}, \ BT > = \sum T_{nm} < Z_{jk}, \ BY_{nm} > \equiv T_{jk} \tag{6.3}$$

i.e. such that

$$< Z_{jk}, \ BY_{nm} > = \frac{1}{4\pi} \int Z_{jk} \left\{ \chi_S + \frac{n-1}{R} \chi_L \right\} Y_{nm} \ d\sigma = \delta_{jn} \delta_{mk} \quad . \tag{6.4}$$

In other words what we need is the biorthogonal sequence $\{Z_{jk}\}$ of the non orthogonal sequence $\{BY_{nm}\}$.

But does this sequence exist? It does exactly because our original problem was solvable. In fact let us rewrite (6.4) in the form

$$<B^* Z_{jk}, Y_{nm}> = \delta_{jn}\delta_{mk} \quad ; \tag{6.5}$$

we see that the sought solution is also the solution of the equation

$$B^* Z_{jk} = Y_{jk} \tag{6.6}$$

The question is whether the equation

$$B^* v = w \tag{6.7}$$

has always solution. The answer is that this is true because B has been proved to be invertible with bounded inverse so that, at least formally, the inverse of B^* also exists and is given by

$$(B^*)^{-1} = (B^{-1})^* \quad . \tag{6.8}$$

We can also verify directly that the solution of (6.7) is unique, i.e.

$$B^* u = 0 \rightarrow u = 0$$

In fact we have shown that the equation

$$BT = f$$

has always a solution for every f so that if

$$0 = B^* v$$

then, $\forall u$

$$0 = < B^* v, u > = < v, Bu > \quad , \tag{6.9}$$

and it is sufficient to choose u such that Bu = v, to see that (6.9) implies v = 0.

This proves that if the B.V.P.

$$BT = f$$

enjoys the Fredholm alternative and admits always a solution in a certain space H, this solution being unique, then if $\{Y_{nm}\}$ is an ONB of H automatically there exists a sequence $\{Z_{jk}\}$ biorthogonal to $\{BY_{nm}\}$.

To find practically this sequence, equations (6.4) have to be suitably discretized and solved; this problem is actually under analysis from the numerical point of view.

b) The application of AG II to local problems

A "local" B.V.P. is a B.V.P. of the general form

$$BT = f \tag{6.10}$$

to which, first of all, a "reference model" equation

$$BT_M = f_M \tag{6.11}$$

has been subtracted. In this way in fact we arrive at an equation of the form

$$B\delta T = \delta f \tag{6.12}$$

where it is known that the values attained by δf at points far away from the area where we want to compute a solution δT, are influencing very little this solution. In practice if we have a resolving kernel $S(P,Q)$ for (6.12), so that

$$\delta T(P) = \int S(P,Q)\ \delta f(Q)\ d\sigma \tag{6.13}$$

we can come to the conclusion that, at the price of commiting insignificant errors, we can extend the integral (6.13) only to an area slightly larger than the area where we want to compute δT, instead of covering the whole sphere.

Now let us imagine that we have δg data on a sea area and Δg on the neighbouring land area, covering together say a rectangle D (cfr. Fig. 6.1) and we want to compute T in a smaller area A.

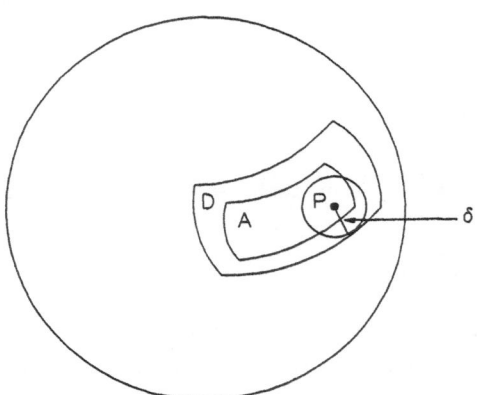

Fig. 6.1 -

To this aim it is necessary to determine first a (spherical) radius δ and a model T_M such that if we put

$$f = \delta g\, \chi_S + \Delta g\, \chi_L \quad,$$

$$f_M = \delta g_M\, \chi_S + \Delta g_M\, \chi_L \quad,$$

$$\delta f = f - f_M \quad,$$

and we take the Hotine kernel

$$H(\psi_{PQ}) = \frac{1}{t} - \ln\left(1 + \frac{1}{t}\right) \quad, \tag{6.14}$$

$$t = \operatorname{sen}\frac{\phi}{2}$$

then the following approximation

$$\int_\sigma H(\psi_{PQ})\, \delta f(Q)\, d\sigma \cong \int_C H(\psi_{PQ})\, \delta f(Q)\, d\sigma \tag{6.15}$$

$(C = C(P,\delta) = $ cap of centre P and radius δ)

is acceptable for every P in A.
Under the above conditions we see that by applying (6.15) for every P in A the function

$$F(P) = \int H(\psi_{PQ})\, \delta f\, d\sigma \tag{6.16}$$

can be computed with sufficient approximation by using only the data available in D. We can then go back to the AG II, B.V.P. and recall that it could be expressed in the integral form

$$T = \frac{2}{R}\int_L H(\psi_{PQ})\, T(Q)\, d\sigma + \int H(\psi_{PQ})\, f(Q)\, d\sigma \quad; \tag{6.17}$$

after subtracting the corresponding model equation, (6.17) becomes

$$\delta T = \frac{2}{R}\int_L H(\psi)\, \delta T\, d\sigma + F(P) \tag{6.18}$$

with F(P) given by (6.16).

Now since also δT is conveniently small, the land area outside the data rectangle D will not contribute significantly to the integral in (6.18) so that this can be transformed into

$$\delta T(P) = \frac{2}{R} \int_{L \cap D} H(\psi_{PQ}) \, \delta T(Q) \, d\sigma + F(P) \quad .$$

Now if we take in (6.19) P ∈ L∩D only, we can read this as an integral equation for functions defined in (L ∩ D). This equation is easily solved for instance by discretizing the integral and solving algebrically the corresponding linear system; numerical experiments in this sense have shown that the resulting system is very well conditioned (cfr. Stock-Sansò [4]).

Once δT has been computed in L ∩ D, it can be substituted back into the right hand side of (6.19) and δT can be computed all over A.

Remark 6.1

The above approach can be meaningfully applied only if highly accurate marine gravity data are available, namely in the range of ± 0.1 mgal.

In fact after subtracting a global model the residual δT has a small order of magnitude, say δT ~ γ·3 m; therefore the term $\frac{2}{R}$ δT has values ranging around 1 mgal; so that the difference between δg and Δg can really be perceived only if the measuring noise is significantly smaller than 1 mgal.

7. Conclusions

In these lecture notes the two altimetry-gravimetry problems have been analyzed showing that they enjoy reasonable mathematical properties, namely that unique solutions exist under reasonable regularity conditions on data.

In order to get unconditional uniqueness one has to modify slightly the original formulation; without this modification there exist theorems of existence and uniqueness if the surface covered by land is sufficiently small (cfr. Sacerdote-Sansò [5]).

This analysis can be taken as the basis to attack numerically the solution of the relevant B.V.P.s, in particular the method of constructing biorthogonal sequences to {BY_{nm}} deserves more attention and work because it seems to be the most "natural" tool to construct global models from the actual data, without going through the solution of improperly posed problems.

Naturally the trend of geodesy today is to achieve more and more data sets and to improve their accuracy, so that new estimation problems have arisen where data are even more than those strictly necessary to determine a solution; so a mathematical apparatus has to be developed to treat such overdetermined problems.

Although they are essentially out of the scope of these lecture notes, we won't fail to mention at least two of them which seem to be of interest in this context. As a first overdetermined problem we could consider one in which the sea surface is covered by gravity data and on the same time it is considered as an equipotential surface; in other terms we assume that both T and δg are known on S.

As a second overdetermined problem we can take AG I, to which a data set of second derivates of T is added on a sphere at 200 Km height $\left(e.g. \ \dfrac{\partial^2 T}{\partial r^2} \right)$ where hopefully a gradiometric mission (Aristoteles) will fly sooner or later.

It has to be noted that these additive data sets could be used to strengthen the solution, but they could also be used to improve our model, e.g. introducing a sea surface topography as a new unknown.

But this will be maybe the item of still another International School to come in the future.

Appendix

Assume we have two surfaces S and E like in Fig. A.1 and that we want to solve the B.V.P.

$$Bu \ \Big|_S \ = f \ , \tag{A.1}$$

with u a regular harmonic potential in Ω i.e. in the domain exterior to S.

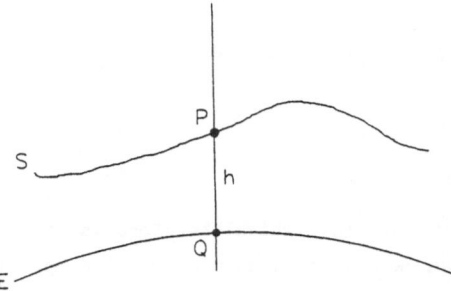

Fig. A.1

We shall make the hypothesis that (A.1) has one and only one solution when $f \in H_o$

(a Hilbert space with norm $| \ |$); by the very definition we can say that the solution u belongs to the Hilbert Space H_1 endowed with the graph norm

$$\|u\| = |B \ u| \tag{A.2}$$

and that B is an isometry between H_0 and H_1.

Example A.1
Assume S to be sufficiently smooth (e.g. to satisfy a cone con- dition) and take $H_0 = L^2(S)$, $B = I$; then clearly $H_1 = H_0 = L^2(S)$.

Example A.2
Take for instance the case that $H_0 \equiv H^{-1/2}(S)$ and $B = \partial/\partial\nu$, with the normal ν pointing to the exterior of S, i.e. in Ω.
Since $f \in H^{-1/2}(S)$ means essentially that we can give a finite value to the form

$$\phi(u) = \int_S fu \ dS \tag{A.3}$$

for any $u \in H^{1/2}(S)$ and since $H^{1/2}(S)$ can be taken as the subspace of $L^2(S)$ such that the norm

$$\|u\| = -\int (\partial_\nu u) \ u \ dS \tag{A.4}$$

is finite, we see that by applying Riesz theorem, the identity

$$\phi(u) = \int_S fu \ dS = -\int (\partial_\nu v) \ u \ dS \tag{A.5}$$

has to hold $\forall \ u \in H^{1/2}(S)$ and for a suitable v; it follows that $(-v)$ is the solution of

$$\partial_\nu(-v) = f$$

and it is proven to belong to $H^{1/2}(S)$ which therefore is the sought space H_1.
Now let us assume that we can extend in a natural way the operator B from S to E ; for instance if S is in one to one correspondence with E through the orthogonal projection on this surface, we can take values of functions defined at P and transport them to define new functions at Q (the pull-back operation that we shall indicate with a tilda $\tilde{f}(Q) = f(P)$); we can also take directional derivatives at P

and transport them parallely to Q. In this way a $\partial/\partial h$ in P will go into $\partial/\partial h$ in Q.

At least the case of particular interest that B = I, corresponding to the Dirichlet problem, can be easily and unambiguously treated.

Assume now to construct the following iterative scheme

$$
\begin{array}{lcl}
\mathbf{S} & & \mathbf{E} \\
f & \rightarrow & \tilde{f} = Bu_0 \\
& & \downarrow \\
Bu_0\big|_S & \leftarrow & u_0 \\
\downarrow & & \\
\delta f_1 = f - Bu_0\big|_S & \rightarrow & \delta\tilde{f}_1 = B\,\delta u_1 \\
& & \downarrow \\
B\delta u_1\big|_S & \leftarrow & \delta u_1 \\
\downarrow & & \\
\delta f_1 = f - B(u_0 + \delta u_1) & \rightarrow & \delta\tilde{f}_2 = B\,\delta u_2
\end{array}
$$

As we see, we go from left to right by pulling back and from right to left by the upward harmonic continuation from E to S ; the first operator transports identically the numerical values of functions defined on S so that any kind of "local" norm on S can be transported as well on E: the second operator, we call it U, is continuous (and even compact), for all types of choices of B we can be interested in.

Now let us assume that

$$\bar{u} = u_0 + \sum_1^\infty {}_1 \; \delta u_1 \tag{A.6}$$

is convergent in H_1.

We want to prove then, that

Th. A.1: $\bar{u}$ is the solution of the B.V.P.

$$B\bar{u} = f$$

and therefore it coincides with u.

In fact if (A.6) converges in H_1 , then

$$f - B\left(u_0 + \sum \delta u_1\right) = f - B\bar{u} \quad , \tag{A.7}$$

the equality holding in H_0.

Assume (A.7) to be different from zero.

Since, setting

$$u_n = u_0 + \sum_1^n {}_i \; \delta u_i \quad ,$$

$$B \; \delta u_{n+1} \; \Big|_E = \tilde{f} - B^{\sim} u_n \; \Big|_S \tag{A.8}$$

we also have

$$B \; \delta u_{n+1} \; \Big|_S = U\!\left(\tilde{f} - B^{\sim} u_n \; \Big|_S \right) \tag{A.9}$$

so that by the continuity of U

$$B \; \delta u_{n+1} \; \Big|_S \to U\!\left(\tilde{f} - B^{\sim} u \; \Big|_S \right) \neq 0 \quad , \tag{A.10}$$

but this contradicts the hypothesis that $\sum \delta u_i$ converges in H_1, i.e. $\sum B \; \delta u_i$ converges in H_0.

Hence $f - B\bar{u}$ has to be zero, as it was to be proved.

Remark A.1

It is important to understand that the only convergence we can expect from the above approach is at best "on S", but not certainly in the layer between E and S ; in fact such a convergence can happen only if the sought solution u has in fact a backward harmonic continuation and this can happen only under very restrictive conditions.

This situation is well illustrated in the following example.

Example A.3

We assume here that both S and E are spheres and that the ratio between the radius is

$$q = \frac{R_E}{R_S} < 1 \quad .$$

Moreover we assume that the operator B is diagonal and isotropic on spherical harmonics, i.e.

$$B \left(\frac{R}{r} \right)^{n+1} Y_{nm}(\sigma) = b_n(r) \left(\frac{R}{r} \right)^{n+1} Y_{nm}(\sigma) \quad . \tag{A.11}$$

This already covers several examples of interest in geodesy; in particular to cope with our hypothesis of invertibility we have to assume that

$$b_n \neq 0 \quad , \qquad \forall \, n \tag{A.12}$$

This would exclude the most interesting case of Stokes problem, for which however we can make the ad hoc hypothesis that the Hilbert spaces H_0 , H_1 do not include first order harmonics so that (A.12) is still satisfied; naturally the known term f has to satisfy this condition too.

As for H_0 let us choose the simple $L^2(\sigma)$ norm, i.e. if f on the sphere S has the development

$$f = \sum f_{nm} Y_{nm}(\sigma) \quad ,$$

then

$$|f|^2 = \sum f_{nm}^2 = \frac{1}{4\pi} \int f^2(\sigma) \, d\sigma \quad . \tag{A.13}$$

Accordingly H_1 has to be endowed with the norm

$$\|u\|^2 = \sum u_{nm}^2 \, b_{Sn}^2 \tag{A.14}$$

if

$$u \Big|_S = \sum u_{nm} Y_{nm}(\sigma)$$

and

$$b_{Sn} = b_n(R_S) \quad . \tag{A.15}$$

Following our iterative scheme we see that, due to the strongly simplifying hypotheses, we can perform a purely spectral calculus, remembering also that pulling back a function from S to E means just to use the same harmonic coefficients, while upward continuing a function from E to S means to multiply its coefficients by

$$\left(\frac{R_E}{R_S} \right)^{n+1} = q^{n+1} \quad ;$$

another point to watch is that B, when acting on E , will multiply harmonic coefficients by

$$b_{En} = b_n(R_E) \quad .$$

(A.16)

With all that in mind, following the same symbolism used before and setting

$$\frac{b_{Sn}}{b_{En}} = p_n \quad , \quad 1 - q^{n+1} p_n = \Delta_n \quad ,$$

we get

S		E
f_{nm}	$\rightarrow$	$\tilde{f}_{nm} = f_{nm}$
		$\downarrow$
$q^{n+1} p_n f_{nm}$	$\leftarrow$	$u_{0nm} = \dfrac{f_{nm}}{b_{En}}$
$\downarrow$		
$\delta f_{1nm} = \Delta_n f_{nm}$	$\rightarrow$	$\delta \tilde{f}_{1nm} = \Delta_n f_{nm}$
		$\downarrow$
$q^{n+1} p_n \Delta_n f_{nm}$	$\leftarrow$	$\delta u_{1nm} = \dfrac{\Delta_n}{b_{En}} f_{nm}$
$\downarrow$		
$\delta f_{2nm} = \Delta_n^2 f_{nm}$	$\rightarrow$	$\delta \tilde{f}_{2nm} = \Delta_n^2 f_{nn}$

As we see, we have

$$|\delta f_1|^2 = \sum_{m,n} \Delta_n^{21} f_{nm}^2 = |f - B(u_0 + \sum_1^1 {}_k u_k)|^2 \quad ;$$

(A.17)

if the condition

$$0 < a \le p_n q \le 1$$

(A.18)

is satisfied we find that

$$0 < \Delta_n < 1$$

(A.19)

and on the same time

$$\lim_{n \to \infty} \Delta_n = 1 \quad .$$

(A.20)

Now take N such that

$$\sum_{N+1}^{\infty}{}_{n,m} \quad f_{nm}^2 < \varepsilon \quad , \tag{A.21}$$

so that for every 1

$$\sum_{N+1}^{\infty}{}_{n,m} \quad \Delta_n^{21} f_{nm}^2 < \varepsilon \quad , \tag{A.22}$$

holds too; then

$$|\delta f_1|^2 \le (1 - aq^N)^{21} \sum_0^N {}_{nm} \quad f_{nm}^2 + \varepsilon \quad , \tag{A.23}$$

so that

$$\overline{\lim_{1 \to \infty}} \quad |\delta f_1|^2 \le \varepsilon \tag{A.24}$$

and, since ε is arbitrary,

$$\lim_{1 \to \infty} |\delta f_1|^2 = 0 \tag{A.25}$$

The relation (A.25) proves an the same time that $u_0 + \sum_1^1 {}_k \delta u_k$ is indeed convergent in H_1 and that

$$\lim_{1 \to \infty} B \left(u_0 + \sum_1^1 {}_k \delta u_k \right) \Bigg|_S = f \tag{A.26}$$

in H_0 .

On the other hand it is also quite instructive to compute directly $B \left(u_0 + \sum_1^1 {}_k \delta u_k \right)$ on both S and E: in fact we find

$$(B \delta u_k)_{nm} \Bigg|_E = p_n \Delta_n^k f_{nm}$$

$$(B \delta u_k)_{nm} \Bigg|_S = p_n q^{n+1} \Delta_n^k f_{nm}$$

that is, also recalling the elementary formula $\sum_0^1 {}_k \Delta_n^k = \dfrac{1 - \Delta_n^{1+1}}{1 - \Delta_n}$,

$$B\left(u_0 + \sum_1^l k\ \delta u_k\right)\Bigg|_E = \sum_{n,m} p_n\ \frac{1 - \Delta_n^{l+1}}{p_n\ q^{n+1}}\ f_{nm}\ Y_{nm} \tag{A.27}$$

$$B\left(u_0 + \sum_1^l k\ \delta u_k\right)\Bigg|_S = \sum_{n,m} p_n q^{n+1}\ \frac{1 - \Delta_n^{l+1}}{p_n\ q^{n+1}}\ f_{nm}\ Y_{nm} \quad . \tag{A.28}$$

By direct inspection we find that, while (A.28) is certainly convergent, as it had to be, (A.27) can converge only if strong restrictive conditions are imposed on f, like

$$\sum_{n,m} q^{-n-1}\ f_{nm}^2 < +\infty \quad ,$$

essentially meaning that f can be harmonically continued from S to E.

To conclude this example, as well as this appendix, we observe that (A.18) is certainly verified in two interesting cases:

a) for the Dirichlet problem, for which

$$p_n = \frac{b_{nS}}{b_{nE}} = 1 \ < \ 1/q \quad ,$$

b) for the Stokes problem, for which assuming (n $\neq$ 1 in agreement with a previous remark)

$$B\left(\frac{R}{r}\right)^{n+1} Y_{nm} = \frac{n-1}{r}\left(\frac{R}{r}\right)^{n+1} Y_{nm} \quad ,$$

so that

$$p_n = \frac{\dfrac{n-1}{R_S}}{\dfrac{n-1}{R_E}} = \frac{R_E}{R_S} = q \ < \ 1/q \quad .$$

References

[1] Sanso' F. " Talk on the Theoretical Foundations of Physical Geodesy. "
Contribution to Geodetic Theory and Methodology. XIX General Assembly of the IUGG -
IAG- Section IV, ed. by Dept of Surveying Engineering, University of Calgary,
Vancouver, August 1987.

[2] Heiskanen W.A., Moritz H. " Physical Geodesy."
Repr. Inst. of Physical Geodesy TUG, Graz, 1981.

[3] Holota P. "The altimetriy-gravimetry boundary value problem I: linearization,
Friedrich's inequality."
Bollettino di Geodesia e Scienze Affini, vol. 42, 1983.

[4] Holota P. "The altimetry-gravimetry boundary value problem II: Weak solution, V-
elliplicity."
Bollettino di Geodesia e Scienze Affini, vol. 42, 1983.

[5] Rapp R.H. "Combination of satellite altimetric and terrestrial gravity data."
In Theory of Satellite Geodesy and Gravity Field Determination, F. Sansò and R.
Rummel (Eds.). Springer-Verlag, Berlin, Heidelberg New York, pp. 261-283.

[6] Sacerdote F., Sansò F. "A contribution to the analysis of the altimetry-
gravimetry problem."
Bull. Geod., vol. 57, 1983.

[7] Sacerdote F., Sanso' F. " Further Remarks on the Altimetry - Gravimetry
Problems."
Bull. Geod., N. 61, pp. 65-82, 1987.

[8] Sansò F. "A discussion on the altimetry-gravimetry problem."
Geodesy in Transition (dedicated to H. Moritz), University of Calgary, Alberta,
Canada, 1983.

[9] Sanso' F., Sona G. " The challenge of computing the geoid in the nineties. "
Surveys in Geophysics. (in publication)

[10] Sanso' F., Stock B. " A numerical experiment in the altimetry - gravimetry.
Problem II. "
Manuscripta Geodetica, vol. 10, N. 1, pp. 23-31, 1985.

[11] Sjoberg L. "A recurrence relation for the β_n-function."
Bull. Geod., vol. 54, 1980.

[12] Svensson S.L. "Solution of the altimetry-gravimetry problem."
Bull. Geod., vol. 57, 1983.

[13] Svensson S.L. " Pseudodifferential operators - a new approach to the boundary
problems of physical geodesy."
Manuscripta Geodaetica, vol. 8, 1983.

Use of Altimeter Data in Estimating Global Gravity Models

Richard H. Rapp

Department of Geodetic Science and Surveying
The Ohio State University
1958 Neil Avenue
Columbus, Ohio 43210-1247, U.S.A.

1. Introduction

A global gravity model is generally taken as a spherical harmonic representation of the Earth's gravitational potential. This series is given up to some maximum degree depending on the data used in the estimation of the model. The higher the spherical harmonic degree the greater, in principle, is the resolution of the model. In the analysis of the perturbations of a satellite orbit an estimate of potential coefficients can be carried out. The maximum degree to which the series expansion can be estimated depends on the height and inclination distribution of the satellites as well as the data or observational accuracy. Although expansions to degree 20 were the norm in the early 1980's today solutions from orbital analysis to degree 50 have been published (Lerch et al, 1992, Schwintzer et al., 1991).

In order to improve and extend the satellite alone models it is appropriate to combine the perturbation implied information with other measurements that can be related to the Earth's gravitational potential. One obvious measure is that of gravity anomalies, averaged into appropriate cells with appropriate correction terms. An advantage of surface gravity data is the high resolution of such data in many parts of the world. One must recognize, however, that the resolution and accuracy of surface gravity data is quite geographically dependent.

Another data type that can be used for gravity field improvements is the altimeter measurement from an altimeter satellite to the ocean surface. Although the ocean surface is not an equipotential surface it still lies close (±1 m) to the geoid, and so a measurement to the surface can be expected to imply gravitational information. Since measurements along the altimeter ground track are spatially dense one expects significant information in this direction. Since cross track spacing may be on the order of 4 to 200 km the resolution is almost always less in the cross track direction than in the along track direction.

The satellite altimeter measurement is analogous to a range measurement to a satellite. However the altimeter measurement is much more affected by a gravitational effect since the signal has not been attenuated as it has at satellite altitude. It is thus clear that satellite altimeter data provide significant information on the Earth's gravitational field, primarily in the ocean areas.

The use of surface gravity and altimeter measurements in the modeling of the Earth's gravitational potential yields a combination potential coefficient model. The model development can yield a high degree (say 360) expansion but it does not have to. This paper will discuss the various ways in which the high degree models can be estimated with particular attention paid to the role of satellite altimeter data.

2. Fundamental Gravimetric Relationships

2.1 Basic Equations

In this section we consider the definition of the gravitational potential and related quantities. This material is taken, almost directly, from Rapp and Pavlis (1990).

The spherical harmonic representation of the Earth's gravitational potential, V, will be defined as:

$$V(r, \theta, \lambda) = \frac{GM}{r}\left[1 + \sum_{n=2}^{\infty}\left(\frac{a}{r}\right)^n \sum_{m=-n}^{n} \overline{C}_{nm}^s \overline{Y}_{nm}(\theta, \lambda)\right]$$

(2.1)

where r is the geocentric distance, θ is the geocentric co-latitude and λ is the longitude. GM is the geocentric gravitational constant, while "a" (usually the equatorial radius of an adopted mean-Earth ellipsoid) is the scaling factor associated with the fully-normalized spherical (s) geopotential coefficients, $\overline{C}_{nm}^s$. In addition:

$$\overline{Y}_{nm}(\theta, \lambda) = \overline{P}_{n|m|}(\cos\theta)\begin{cases} \cos m\lambda & \text{if } m \geq 0 \\ \sin|m|\lambda & \text{if } m < 0 \end{cases}$$

(2.2)

In (2.1), $\overline{P}_{n|m|}(\cos\theta)$ are the fully-normalized associated Legendre functions of the first kind (Heiskanen and Moritz, 1967, section 1-11). Geocentricity of the coordinate system used, enforces the absence of first-degree harmonics in (2.1).

The disturbing potential, T, at a point P (r, θ, λ) is the difference between the actual gravity potential of the Earth and the "normal" potential associated with a rotating equipotential ellipsoid (Somigliana-Pizzetti normal field) at P. Based on (2.1), the spherical harmonic representation of T is:

$$T(r, \theta, \lambda) = \frac{GM}{r} \sum_{n=2}^{\infty}\left(\frac{a}{r}\right)^n \sum_{m=-n}^{n} \overline{C}_{nm}^s \overline{Y}_{nm}(\theta, \lambda)$$

(2.3)

The zero-degree term in (2.3) has been set to zero, assuming equality of the actual mass of the Earth and the mass of the reference ellipsoid. In addition the even zonal coefficients in (2.3) represent the difference between the harmonic coefficients of the actual and the normal gravitational potentials.

The geoid is an equipotential surface of the Earth's gravity field. It is loosely defined as the surface that can be closely associated with the mean ocean surface. The geoid undulation, N, is the separation between the reference ellipsoid and the geoid. The undulation is given by Brun's equation:

$$N_P = \frac{T_P}{\gamma}$$

(2.4)

where γ is an average value of normal gravity. Using (2.3) in (2.4) we can write:

$$N_P = \frac{GM}{r_P \gamma_P} \sum_{n=2}^{\infty} \left(\frac{a}{r}\right)^n \sum_{m=-n}^{n} \overline{C}_{nm} \overline{Y}_{nm}(\theta, \lambda) \tag{2.5}$$

Equation (2.5) could be inappropriate to use for the precise calculation of N for land stations. More rigorous procedures involve the computation of the height anomaly from T_P at the topographic surface and converting this value to a geoid undulation using the following relationship:

$$N - \zeta = \frac{\Delta g_B}{\gamma} H \tag{2.6}$$

where Δg_B is the Bouguer gravity anomaly and H is the orthometric height.

The surface free-air gravity anomaly, Δg, (Heiskanen and Moritz, 1967, p. 293) is defined as the difference between the magnitude of the actual gravity acceleration, at the surface point P, minus the magnitude of the normal gravity acceleration at the corresponding telluroid point Q, i.e.

$$\Delta g = \left| \vec{g}_P \right| - \left| \vec{\gamma}_Q \right| \tag{2.7}$$

The definition of the telluroid, employed here, can be found in Pavlis (1988, section 2.1.2). Assuming that the gravity potential on the geoid equals the normal gravity potential on the reference ellipsoid, it can be shown (ibid, section 2.2.1) that the fundamental boundary condition relating the gravity anomaly to the disturbing potential, takes the form:

$$\Delta g = -\left(\frac{\partial T}{\partial r}\right)_Q - \frac{2}{r_Q} T_Q + \left(\varepsilon_h + \varepsilon_\gamma + \varepsilon_P\right)_Q \tag{2.8}$$

where ε_P accounts for the difference between the gravity anomaly as defined in (2.7) and the isozenithal projection of the gravity anomaly vector (ibid, section 2.1.1), and ε_h, ε_γ account for ellipsoidal effects up to the order of the eccentricity squared. These terms, along with other systematic corrections to be applied to the observations will be discussed in more detail in the next section. For the moment, consider Δg^c to represent the gravity anomaly after application of all systematic corrections, so that it fulfills:

$$\Delta g^c = -\left(\frac{\partial T}{\partial r}\right)_Q - \frac{2}{r_Q} T_Q \tag{2.9}$$

Substitution of (2.3) into (2.9) yields

$$\Delta g^c (r, \theta, \lambda) = \frac{GM}{r^2} \sum_{n=2}^{\infty} (n - 1)\left(\frac{a}{r}\right)^n \sum_{m=-n}^{n} \overline{C}_{nm} \overline{Y}_{nm} (\theta, \lambda) \tag{2.10}$$

The gravity anomaly Δg (r, θ, λ) is not a harmonic function, but $r\Delta g$ (r, θ, λ) is, in the space outside the generating masses of the potential (the effect of the atmosphere is assumed to be taken into account in Δg). Based on (2.10), $r\Delta g$ (r, θ, λ) can be written as:

$$r\Delta g^c\left(r, \theta, \lambda\right) = a \sum_{n=2}^{\infty} \left(\frac{a}{r}\right)^{n+1} \sum_{m=-n}^{n} \overline{g}_{nm} \, \overline{Y}_{nm}\left(\theta, \lambda\right)$$

(2.11)

where

$$\overline{g}_{nm}^s = \frac{GM}{a^2}(n-1)\,\overline{C}_{nm}^s$$

(2.12)

Equation (2.11) provides the _spherical_ harmonic representation of $r\Delta\overline{g}$ (r, θ, λ), and the convergence of the infinite series in (2.11) is guaranteed in the region outside the smallest sphere enclosing all the generating masses (Brillouin sphere).

Although spherical harmonic expansions are common for satellite orbit computations, the decomposition of the gravitational potential into harmonic components, based on data residing on the physical surface of the Earth (such as the gravity anomalies), can be performed more appropriately using ellipsoidal harmonics. Such a procedure circumvents the need for approximate correction terms, such as the ones developed by Cruz (1986). In this analysis ellipsoidal harmonics are employed, following the discussions of Jekeli (1988) and Gleason (1988). The application of ellipsoidal harmonics in combination solutions has been discussed by Rapp (1989).

The system of ellipsoidal coordinates (u, δ, λ) used (Heiskanen and Moritz, 1967, section 1-19), is based on a family of confocal ellipsoids of constant linear eccentricity $E = ae$, where a is the semi-major axis and e the first eccentricity of the reference (base) ellipsoid. u is the semi-minor axis of the confocal ellipsoid passing through a given point, δ is the reduced co-latitude of the point, and λ its longitude. The Cartesian coordinates of a point are related to its spherical and ellipsoidal coordinates as follows:

$$\left. \begin{array}{l} x = r \sin\theta \cos\lambda = \sqrt{u^2+E^2} \, \sin\delta \cos\lambda \\ y = r \sin\theta \sin\lambda = \sqrt{u^2+E^2} \, \sin\delta \sin\lambda \\ z = r \cos\theta \qquad = u \cos\delta \end{array} \right\}$$

(2.13)

Jekeli (1988) has discussed the representation of harmonic functions in both spherical and ellipsoidal harmonics, and has developed closed formulas for the transformation between spherical and ellipsoidal harmonic coefficients, valid in the domain where harmonicity of the functions is ensured. Gleason (1988) has extended the studies of Jekeli and emphasized efficient algorithms, applicable for the transformation of harmonic coefficients obtained from the integration of boundary values given on spherical or ellipsoidal boundary surfaces. His developments are used in the following discussion. An arbitrary harmonic function F (u, δ, λ) can be represented in terms of ellipsoidal harmonics as (ibid, eq. (1.14)):

$$F(u,\delta,\lambda) = \sum_{n=0}^{\infty} \sum_{m=-n}^{n} \frac{\overline{S}_{n|m|}\left(\frac{u}{E}\right)}{\overline{S}_{n|m|}\left(\frac{b}{E}\right)} \cdot \overline{f}_{nm}^e \, \overline{Y}_{nm}(\delta,\lambda)$$

(2.14)

where b is the semi-minor axis of the reference ellipsoid, $\overline{Y}_{nm}$ are as defined in (2.2), but evaluated in terms of ellipsoidal coordinates (δ, λ). The renormalization functions $\overline{S}_{nm}$ are related to the associated Legendre functions of the second kind (ibid, eq. (1.10)); their introduction causes the fully-normalized ellipsoidal harmonic coefficients $\overline{\overline{f}}_{nm}^{e}$ to be real.

The function $r\Delta g_g^c (r, \theta, \lambda)$, being harmonic, can thus be represented as (ibid, eq. (2.7)):

$$r\Delta g^c (u, \delta, \lambda) = a \sum_{n=0}^{\infty} \sum_{m=-n}^{n} \frac{\overline{S}_{n|m|}\left(\frac{u}{E}\right)}{\overline{S}_{n|m|}\left(\frac{b}{E}\right)} \overline{\overline{g}}_{nm}^{e} \overline{Y}_{nm} (\delta, \lambda)$$

(2.15)

The transformation of the ellipsoidal harmonic coefficients $\overline{\overline{g}}_{nm}^{e}$ to the corresponding spherical ones $\overline{g}_{nm}^{s}$, can be performed according to (ibid, eq. (2.10)):

$$\overline{g}_{nm}^{s} = \sum_{k=0}^{s'} \frac{1}{\overline{S}_{n-2k,|m|}\left(\frac{b}{E}\right)} \cdot L_{nmk} \, \overline{\overline{g}}_{n-2k,m}^{e}$$

(2.16)

where s′ is the greatest integer less than or equal to $(n - |m|)/2$, and L_{nmk} is defined in (ibid, eq. (1.23)).

Consider now that Δg_g is analytically continued from the surface of the Earth, to the surface of the reference ellipsoid to define $\Delta g_g^E (b, \delta, \lambda)$. The particular means of this analytical continuation will be discussed in the next section. Equation (2.15), for $\Delta g_g^E (b, \delta, \lambda)$ becomes:

$$r\Delta g^E (b, \delta, \lambda) = a \sum_{n=0}^{\infty} \sum_{m=-n}^{n} \overline{\overline{g}}_{nm}^{e} \overline{Y}_{nm}(\delta, \lambda)$$

(2.17)

and the orthogonality of the surface spherical harmonics $\overline{Y}_{nm} (\delta, \lambda)$ over the unit sphere σ yields (ibid, eq. (2.11)):

$$\overline{\overline{g}}_{nm}^{e} = \frac{1}{4\pi a} \iint_{\sigma} \left[r\Delta g_g^E (b, \delta, \lambda) \right] \overline{Y}_{nm} (\delta, \lambda) d\sigma$$

(2.18)

where $d\sigma = \sin \delta d\delta d\lambda$ is the area element on the unit sphere.

In the real world situation the gravity anomalies to be analyzed are given in terms of discrete area-mean values over equiangular blocks on the reference ellipsoid. If $\Delta\lambda$ is the longitude (and latitude) extent of the blocks, and $N = 180°/\Delta\lambda°$, then discretization of (2.18) results in a numerical quadrature formula. The "area-mean" type of such a formula, as suggested by Colombo (1981) yields:

$$\overline{\overline{g}}_{nm}^{e} = \frac{1}{4\pi a} \sum_{i=0}^{N-1} \frac{1}{q_n^i} \sum_{j=0}^{2N-1} \overline{(r\Delta g)}_{ij}^{E} \int_{\delta_i}^{\delta_{i+1}} \int_{\lambda_j}^{\lambda_{j+1}} \overline{Y}_{nm} (\delta, \lambda) \sin\delta d\delta d\lambda.$$

(2.19)

In (2.19), $\overline{(r\Delta g)}_{ij}^E$ represents the average value over the block on the ellipsoid. However, since r^E (b, δ) varies slowly, for small blocksizes (e.g. $1° \times 1°$, $30' \times 30'$), $\overline{(r\Delta g)}_{ij}^E$ can be substituted by $r_i^E \overline{\Delta g}_{ij}$, where r_i^E is evaluated at the mean geodetic latitude of the block (Pavlis, 1988, Appendix A). The smoothing factors q_n^i are evaluated based on Colombo's (1981) suggestion, in terms of the Pellinen smoothing operators β_n^i as:

$$q_n^i = \begin{cases} \left(\beta_n^i\right)^2 & \text{if } 0 \le n \le N/3 \\ \beta_n^i & \text{if } N/3 < n \le N \\ 1 & \text{if } N < n \end{cases}$$

(2.20)

The computation of β_n^i is performed according to Meissl's formula by (ibid):

$$\beta_n^i = \frac{1}{1-\cos\psi_0^i} \cdot \frac{1}{2n+1}\left[P_{n-1}\left(\cos\psi_0^i\right) - P_{n+1}\left(\cos\psi_0^i\right)\right]$$

(2.21)

where ψ_0^i is the semi-aperture of a spherical cap having the same area as the equiangular block on the ith latitude belt. It is computed by (ibid, p. 85):

$$\psi_0^i = \cos^{-1}\left[\frac{\Delta\lambda}{2\pi}\left(\cos\delta_{i+1} - \cos\delta_i\right) + 1\right]$$

(2.22)

The above formulation properly accounts for the latitude dependence of ψ_0^i and q_n^i, as discussed in Katsambalos (1979).

The combination of equations (2.12), (2.16) and (2.19) yields:

$$\overline{C}_{nm}^s = \frac{1}{4\pi a\gamma} \sum_{i=0}^{N-1} r_i^E \sum_{k=0}^{s'} \frac{L_{nmk}}{\overline{S}_{n-2k,|m|}\left(\frac{b}{E}\right)} \cdot \frac{I\overline{P}_{n-2k,|m|}^i}{(n-2k-1)q_{n-2k}^i} \cdot$$

$$\cdot \sum_{j=0}^{2N-1} \overline{\Delta g}_{ij}^E \begin{cases} IC \\ IS \end{cases}_m^j \begin{array}{l} \text{if } m \ge 0 \\ \text{if } m < 0 \end{array}$$

(2.23)

where:

$$\gamma = \frac{GM}{a^2},$$

(2.24)

$$I\overline{P}_{n|m|}^i = \int_{\delta_i}^{\delta_{i+1}} \overline{P}_{n|m|}\left(\cos\delta\right)\sin\delta d\delta$$

(2.25)

and

$$\begin{cases} IC \\ IS \end{cases}_m^j = \int_{\lambda_j}^{\lambda_{j-1}} \begin{cases} \cos m\lambda \\ \sin|m|\lambda \end{cases} d\lambda$$

(2.26)

Equation (2.23) is the fundamental mathematical model that enables the computation of the unitless fully-normalized spherical harmonic coefficients $\overline{C}_{nm}^s$ of the disturbing potential, once a complete ($N \times 2N$) set of discrete area-mean values of gravity anomalies is defined on the surface of the reference ellipsoid. Such coefficients are directly comparable with corresponding estimates derived from satellite perturbation analysis, and this comparison is the basis of the adjustment procedure employed in the combination solution. In addition, equation (2.23) is used in the harmonic analysis of an adjusted set of gravity anomalies, to provide estimates of $\bar{C}_{nm}^S$ above the maximum degree of the satellite harmonics available, and up to a degree corresponding to the gravity anomaly block size.

2.2 Consideration of Systematic Effects

To obtain the gravity anomaly $\overline{\Delta g}_{ij}^E$ required in equation (2.23), a number of systematic reductions need to be applied to the surface, area-mean free-air anomaly, $\overline{\Delta g}_{ij}$, available from the gravity data sources used here. These reductions are considered here in some detail.

2.2.1 Atmospheric Correction: δg_A

To ensure harmonicity of $r\Delta g$ (r, θ, λ), in the space outside the Earth's surface, the effect of the atmosphere on Δg (r, θ, λ) has to be removed analytically (Moritz, 1980, p. 422). If in the definition of the normal potential, the mass of the reference ellipsoid includes the mass of the atmosphere (as is the case here), then the correction amounts to the <u>addition</u> of the quantity δg_A to the gravity anomaly $\overline{\Delta g}_{ij}$, where

$$\delta g_A (\text{mgal}) = 0.8658 - 9.727 \times 10^{-5} \overline{H}_{ij}(\text{m}) + 3.482 \times 10^{-9} \overline{H}_{ij}^2(\text{m}) \tag{2.27}$$

Equation (2.27) was developed by Wichiencharoen (1982), by fitting a quadratic function to the tabulated values in (IAG, 1971, p. 72). $\overline{H}_{ij}$ in (2.27) is the mean elevation above MSL of the block, so that the correction amounts to about + 0.87 mgals at sea level.

An alternative expression to compute the atmosphere correction is the following (DMA, 1987, p. 4-10):

$$\delta g_A = 0.87 e^{-0.116 h^{1.047}} \tag{2.28}$$

where h is the elevation in km and δg_A is in mgal.

2.2.2 Ellipsoidal Corrections: ε_h, ε_γ, ε_ϱ

The term ε_h represents the effect on the boundary condition of the linear term in a Taylor series expansion of $\partial/\partial h$ around its "spherical" value $\partial/\partial r$ (Pavlis, 1988, section 2.2.1). Its analytical expression is (ibid, eq. (2.58a)):

$$\varepsilon_h = e^2 \sin\theta \cos\theta \left(\frac{\partial T}{r \partial \theta} \right). \tag{2.29}$$

The term ε_γ represents the effect on the boundary condition of the linear term in a Taylor series expansion of the normal gravity γ and its gradient $\partial\gamma/\partial h$ around their "spherical" values. It is given by (ibid, eq. (2.58b)):

$$\varepsilon_\gamma = \left[6J_2 \frac{a^2}{r^3} P_2(\cos\theta) - \frac{3\omega^2 r^2}{GM} \sin^2\theta \right] T. \tag{2.30}$$

Finally, ε_p arises from the difference between the gravity anomaly as defined in equation (2.7) and the isozenithal projection of the gravity anomaly <u>vector</u>, and is given by (ibid, eq. (2.34)):

$$\varepsilon_p = \left(1 - \cos\overline{\theta}\right)\frac{\partial W}{\partial h} - \xi\frac{\partial W}{M\partial\varphi} - \eta\frac{\partial W}{N\cos\varphi\partial\lambda} \tag{2.31}$$

where W is the gravity potential of the Earth, $\overline{\theta}$ the total deflection of the vertical, (ξ, η) its meridional and prime vertical components, (M, N) the radii of curvature of the meridian and prime vertical, and φ represents geodetic latitude.

The above terms are on the order of tens of μgals and of long wavelength nature. Their effects have been studied in detail by Cruz (1986) and Pavlis (1988). Computation of integrated area-mean values of these quantities (denoted IE_h, IE_γ and IE_p respectively) was performed here, following the equations given by Pavlis (ibid, sections 2.3.4, 2.3.6) using the OSU86F (Rapp and Cruz, 1986a) geopotential model to degree 180.

2.2.3 Second order vertical gradient of the normal gravity: δg_{h^2}

The surface free-air anomaly provided in almost all current data sources is evaluated by:

$$\Delta g = g_p - \left[\gamma_{Q_0} + \left(\frac{\partial\gamma}{\partial h}\right)_{Q_0} H \right] \tag{2.32}$$

where Q_0 is the straight ellipsoidal normal projection of the surface point P and H is the orthometric height. More precisely Δg should be evaluated as:

$$\Delta g = g_p - \left[\gamma_{Q_0} + \left(\frac{\partial\gamma}{\partial h}\right)_{Q_0} H + \frac{1}{2!}\left(\frac{\partial^2\gamma}{\partial h^2}\right)_{Q_0} H^2 \right] \tag{2.33}$$

The effect of the term

$$\delta g_{h^2} = -\frac{1}{2!}\left(\frac{\partial^2\gamma}{\partial h^2}\right)_{Q_0} H^2 \approx -3\gamma_{Q_0}\left(\frac{H}{a}\right)^2 \tag{2.34}$$

has been investigated by Pavlis (1988), where it was shown that its omission can cause systematic errors in the geopotential coefficients (computed soley from terrestrial data) implying geoid undulation errors on the order of 22 cm with a maximum error of 1.80 m in the Himalaya region. Terrestrial anomalies used in this analysis should be corrected by adding δg_{h^2} computed based on the mean elevation above MSL of each block.

According to the systematic corrections (a) - (c), the anomaly $\overline{\Delta g}_{ij}^c$ is:

$$\overline{\Delta}\,g_{ij}^c = \overline{\Delta}g_{ij} + \left(\delta g_A\right)_{ij} - \left(IE_h + IE_\gamma + IE_p\right)_{ij} + \left(\delta g_{h^2}\right)_{ij} \tag{2.35}$$

2.2.4 Analytical downward continuation: g_1

Now consider the relationship between the "surface" anomaly Δg^s and the anomaly on the ellipsoid. The method of analytical downward continuation (Moritz, 1980, p. 378) is considered here. One writes a Taylor series in the following form:

$$\Delta g^s = \Delta g^E + h\,\frac{\partial \Delta g^E}{\partial h} + \frac{h^2}{2!}\,\frac{\partial^2 \Delta g^E}{\partial h^2} + .. \tag{2.36}$$

where h is the height of the surface point above the ellipsoid. In Rapp and Cruz (1986a) the partial derivatives were calculated from a high degree (n = 360) spherical harmonic expansion. Since such expansions do not reflect the high frequency information in the gradient of the anomaly the correction terms used by Rapp and Cruz (ibid) are too small by a factor of approximately three. (See Wang (1988, Table 15)).

Neglecting all but the first derivative in eq. (32) one can write the following approximation:

$$\Delta g^E = \Delta g^s + g_1 = \Delta g^s - h\,L\left(\Delta g\right) \tag{2.37}$$

where

$$L(\Delta g) = \frac{R^2}{2\pi} \iint_\sigma \frac{\Delta g - \Delta g_p}{\ell_0^3}\,d\sigma \tag{2.38}$$

where ℓ_0 is the distance between the variable point and the computation point P and R is a mean-Earth radius (e.g. 6371 km). This reduction (i.e. eq. (2.37)) is called the "gradient solution" of the downward continuation problem.

Now assume that the free-air anomalies are linearly correlated with elevation:

$$\Delta g = a + bh \tag{2.39}$$

where $b = 2\pi G\rho$ and ρ equals the crustal density. Substituting (2.39) into (2.38) and then (2.37) yields:

$$g_1 = -G\rho R^2 h_p \iint_\sigma \frac{h - h_p}{\ell_0^3}\,d\sigma \tag{2.40}$$

In planar approximation (which is sufficient for these purposes) eq. (2.40) becomes (Wang, ibid, eq. (15)):

$$g_1 = -h_p \rho G \iint\limits_{\tau} \frac{h - h_p}{\ell_0^3} \, dxdy \tag{2.41}$$

where τ is the x y plane.

The computation of g_1 for point gravity values requires a dense elevation grid. In the usual here the solution will use 30' mean gravity anomalies. Therefore, a data grid finer than 30' but, less dense than that used for point reductions, can be used. Specifically Wang (ibid) used a global set of 5' x 5' mean elevations (ETOPO5, NGDC, 1986) to calculate a global set of 5' x 5' g_1 values using a Fast Fourier Transform technique. The 5' x 5' g_1 values were then averaged to form a global set of 30' mean values. The maximum 30' g_1 term was 45 mgal with a standard deviation (with respect to a mean of 0.27 mgal) of ± 1.54 mgal. The largest g_1 values were associated with mountainous regions. However 98.7% of the g_1 values were within ± 5 mgal. The set of 30' g_1 values calculated by Wang (ibid) can be applied to the terrestrial data file in the development of the global anomaly data file to be discussed later.

$\overline{\Delta g}_{g_{ij}}^{E}$ in equation (2.23) is obtained from $\overline{\Delta g}_{ij}$ (as provided by a gravity source) by:

$$\overline{\Delta} g_{ij}^{E} = \overline{\Delta} g_{ij} + (\delta g_A)_{ij} - \left(IE_h + IE_\gamma + IE_p \right)_{ij} + \left(\delta g_{h^2} \right)_{ij} + (g_1)_{ij} \tag{2.42}$$

It should be noted that depending on the source of $\overline{\Delta g}_{ij}$, (e.g. terrestrial estimate, altimeter derived estimate, etc.) some of the reductions in (2.42) may not be applicable.

2.3 Estimation of the Geopotential Coefficients

The estimation of the complete high degree set of geopotential coefficients is performed here as a two step procedure. First, the global set of $\overline{\Delta} g_{ij}^{E}$, provides, through equation (2.23), a "terrestrial" estimate, $[\overline{C}_{nm}^s]_T$, for the harmonics that are included in the satellite-only model. In addition, equation (2.23) is used to propagate estimates of the variances of $\overline{\Delta} g_{ij}^{E}$ to $[\overline{C}_{nm}^s]_T$, and form the full covariance matrix cov $\{[\overline{C}_{nm}^s]_T\}$. Based on the satellite-derived harmonics $[\overline{C}_{nm}^s]_s$ and their associated cov $\{[\overline{C}_{nm}^s]_s\}$ and their "terrestrial" counterparts, a least-squares adjustment is used to estimate a unique set for these harmonic coefficients, essentially as a weighted average of the two estimates, under the assumption that these are uncorrelated. In addition, the least-squares adjustment provides a global set of adjusted gravity anomalies. In a second step, the set of adjusted anomalies is used in equation (2.23) to yield the higher degree coefficients, up to a maximum degree consistent with the data resolution (Nmax = 360 for 30' mean values). This general procedure was originally proposed by Kaula (1966) and has been used in previous high degree expansions (Rapp, 1981; Rapp and Cruz, 1986a).

The adjustment process may be formulated in terms of a "generalized" adjustment model (Uotila, 1986), of the form:

$$F\left(L_F^a, L_x^a\right) = 0 \tag{2.43}$$

where L_F^a corresponds to the adjusted observations (the gravity anomalies) and L_x^a to the adjusted parameters (the potential coefficients). Linearization yields:

$$B_F V_F + B_{F_x} V_x + W_F = 0 \tag{2.44}$$

where:

$$B_F = \left.\frac{\partial F}{\partial L_F^a}\right|_{\substack{L_F^a = L_F^b \\ L_x^a = L_x^b}} \tag{2.45a}$$

$$B_{Fx} = \left.\frac{\partial F}{\partial L_x^a}\right|_{\substack{L_F^a = L_F^b \\ L_x^a = L_x^b}} \tag{2.45b}$$

$$W_F = F\left(L_F^b, L_x^b\right) \tag{2.45c}$$

and L_F^b corresponds to $\overline{\Delta g}_{ij}^E$ while L_x^b to $[\overline{C}_{nm}^s]_s$. Denoting P_F the weight matrix of the anomalies (assumed diagonal) and P_x the one for the satellite-derived coefficients (i.e. $P_x^{-1} = \text{cov }([\overline{C}_{nm}^s]_s)$), a minimum variance solution is obtained by minimizing the target function:

$$\phi = V_F^T P_F V_F + V_x^T P_x V_x \tag{2.46}$$

subject to the condition (2.44). Denoting:

$$M_F = B_F P_F^{-1} B_F^T \tag{2.47}$$

the solutions for V_x and V_F are given by:

$$V_x = -\left(B_{Fx}^T M_F^{-1} B_{Fx} + P_x\right)^{-1} B_{Fx}^T M_F^{-1} W_F \tag{2.48a}$$
$$V_F = -P_F^{-1} B_F^T M_F^{-1}\left(B_{Fx} V_x + W_F\right) \tag{2.48b}$$

Adjusted observations, $[\overline{\Delta g}_{ij}^E]$, and parameters, $[\overline{C}_{nm}^s]$ are obtained from:

$$\left.\begin{aligned}\widehat{L}_F^a &= L_F^b + V_F \quad (a) \\ \widehat{L}_x^a &= L_x^b + V_x \quad (b)\end{aligned}\right\} \tag{2.49}$$

The degrees of freedom (D.F.) equals the number of equations of the form (2.43) (which here equals the number of satellite-derived coefficients), and the a-posteriori variance of unit weight is:

$$\hat{\sigma}_0^2 = \frac{\left(V_F^T P_F V_F + V_x^T P_x V_x\right)}{(D.F.)} \tag{2.50}$$

The error covariance matrix of $\hat{L}_x^a$ is:

$$\Sigma_{\hat{L}_x^a} = \sigma_0^2 \left(B_{Fx}^T M_F^{-1} B_{Fx} + P_x\right)^{-1} \tag{2.51}$$

where σ_0^2 is the a-priori variance of unit weight (taken to be unity).

In the current problem the function F is defined as:

$$F = [\overline{C}_{nm}^s]_s - [\overline{C}_{nm}^s]_T = 0 \tag{2.52}$$

and from (2.20) it is obvious that F is linear in both L_F^a and L_x^a. In addition B_{Fx} here is the identity matrix while the elements of B_F are obtained from the coefficients of $\overline{\Delta}_{g_{ij}}^E$ in (2.23), with opposite sign (due to (2.52)). Hence, (2.48a) and (2.48b) become:

$$V_x = -\left(M_F^{-1} + P_x\right)^{-1} M_F^{-1} W_F \tag{2.53a}$$

$$V_F = P_F^{-1} B_F^T P_x V_x \tag{2.53b}$$

The adjusted coefficients $[\overline{C}_{nm}^s]$ are obtained from (2.53a) and (2.49b).

The adjusted anomalies, $[\overline{\Delta}_{g_{ij}}^E]$ obtained from (2.53b) and (2.49a), are then used in (2.23) to determine the high degree coefficient set. The linearity of the model F (eq. (2.52)), ensures that the values obtained from this expansion, for the harmonic coefficients involved in the adjustment, will be identical to the adjusted coefficients implied by (2.53a) and (2.49b). To obtain accuracy estimates for the coefficients _not_ involved in the adjustment, would require evaluation of the full covariance matrix of the adjusted anomalies and then propagation of this covariance to $\overline{C}_{nm}^s$ based on (2.23). This task is computationally difficult. However, based on the observation that the diagonal dominance of $\Sigma_{\hat{L}_x^a}$ will yield a diagonally dominant covariance matrix of the adjusted anomalies, due to the orthogonality relations, (see also Pavlis (1988)), an approximate method was used to yield estimates of the accuracies of the coefficients not adjusted. We observe that the factors L_{nmk} appearing in eq. (2.23) are of the order of $(e^2)^k$, (see also Gleason (1988, eq. 1.23)), where e is the first eccentricity of the reference ellipsoid. Hence, neglecting all but the $k = 0$ terms in eq. (2.23) we have:

$$\overline{C}_{nm}^s = \frac{1}{4\pi\gamma(n-1)\,\overline{S}_{n|m|}\left(\frac{b}{E}\right)} \sum_{i=0}^{N-1} \left(\frac{r_1^E}{a}\right)^i \cdot \frac{1}{q_h^i} \cdot I\overline{P}_{n|m|}^i \cdot$$

$$\sum_{j=0}^{2N-1} \overline{\Delta g}_{ij}^E \begin{cases} IC \\ IS \end{cases}_m^j \begin{array}{l} \text{if } m \geq 0 \\ \text{if } m < 0 \end{array} \tag{2.54}$$

which implies:

$$\sigma^2_{\overline{C}_{nm}} \approx \frac{1}{\left[4\pi\gamma(n-1)\,\overline{S}_{nm}\left(\frac{b}{E}\right)\right]^2} \sum_{i=0}^{N-1}\left(\frac{r_i^E}{a}\right)^2 \cdot \frac{1}{(q_{i\hbar})^2} \cdot \left\{\mathbb{P}^i_{nm}\right\}^2 \cdot$$

$$\sum_{j=0}^{2N-1}\sigma^2_{\Delta_{u_j}^a}\left\{ \begin{array}{l} IC^2 \\ IS^2 \end{array} \right\}^j_m \quad \begin{array}{l} \text{if } m\geq 0 \\ \text{if } m<0 \end{array} \tag{2.55}$$

Equation (2.55) can be used to estimate the propagated error of the coefficients not adjusted, due to the anomaly errors. The approximations involved in eq. (2.55) are $O(e^2)$, so that this equation is considered sufficient for the estimation of the propagated coefficient errors. These values along with the diagonal elements of $\Sigma \hat{L}^a_{\underline{x}}$, define the complete set of propagated coefficient errors.

Finally, the total error of the coefficients of the high degree expansion was defined by quadratically adding to the propagated error the sampling error (Colombo, 1981). This error component arises due to the finite size of the equiangular blocks and may be computed from the following quartic expression developed by Jekeli (ibid, eq. 3.10):

$$\sigma_n^{SAMPL} = \frac{\sigma_n}{100} \cdot \left[\left(-16.19570\left(\frac{n}{N}\right) + 30.34506\right)\left(\frac{n}{N}\right) + 40.29588\right]\left(\frac{n}{N}\right)^2 \tag{2.56}$$

where:

$$\sigma_n = \left\{\sum_{m=-n}^{n}\left(\overline{C}^s_{nm}\right)^2\right\}^{1/2} \tag{2.57}$$

Colombo (ibid, pg. 78) reports that eq. (2.56) was tested in simulated cases for three block sizes, and has yielded excellent agreement with the actual sampling errors found from numerical tests.

2.4 Summary

This section has been presented to show the procedure to carry out one type of combination method that can yield a high degree potential coefficient model. This procedure involves the use of the orgonality relationships with a least squares adjustment up to the degree to which the a priori potential coefficient model is known. The coefficients above this degree are not estimated in a least squares sense. Although the use of altimeter data has not been specifically considered, we will see later the strong role such data plays in this combination procedure.

3. Satellite Altimeter Data and Orbit Improvement

3.1 Basic Model

We now turn to the role of satellite altimeter data in gravitational improvement studies. The discussion will take place in two parts. This first part describes the analysis when the altimeter measurement is considered a satellite tracking data type. The second part describes the analysis in which the altimeter data is used to defined local altimeter data is used to define local variations of the gravity signal in the ocean areas.

Let h_s be the ellipsoidal height of the satellite as defined through an orbit determination process. Let h_m be the ellipsoidal height of the sea surface which is being measured by a radar altimeter. The distance between the center of mass of the altimeter satellite and the sea surface, ρ, is then:

$$\rho = h_s - h_m \tag{3.1}$$

so that

$$h_m = h_s - \rho \tag{3.2}$$

The actual altimeter measurements is effected by factors that include: bias term, ionospheric and topographic effects, off nadir measurement, sea state effects, etc. The sea surface responds to factors that must be taken into account to reduce the sea surface height to a common baseline. These factors include: ocean tides, solid Earth tides, atmospheric pressure loading. These factors will have been discussed elsewhere in this volume. A good source of more detailed information is the workshop report on altimeter algorithms prepared by Chelton (1988).

If h_m is the sea surface height based on the measured altimeter distance and the computed orbit, the corrected sea surface height will be

$$h' = h_m - c = h_s - \rho - c \tag{3.3}$$

where we lump together all the correction terms noted above. In considering c we must recognize that this quantity can contain errors created by incorrect or inadequate environmental and geophysical models. The value of h_c in (3.3) is still subject to the orbit errors that remain in the value of h_s that is computed from the position of the satellite implied by the satellite ephemeris.

We now enter the discussion on the error in h_s and how it may be reduced through the processing of altimeter data, surface gravity data, and other satellite data. We will assume that the potential coefficient model used in the generation of the altimeter orbit has not included the altimeter data from a satellite whose orbit we with to improve. In this discussion we closely follow Denker and Rapp (1990) where that discussion is based, in part, on Engelis (1987a).

The ellipsoidal height h_s is computed from the satellite ephemerides, and it is therefore affected by errors in the determination of the initial state vector and errors in the force model parameters used for the orbit fits. Thus one can write

$$h_s = h_c + \Delta h_G + \Delta h_I \tag{3.4}$$

where h_c is the computed ellipsoidal height of the satellite, Δh_G are errors due to the modeled geopotential coefficients, and Δh_I are errors caused by initial state errors and residual effects due to air drag, solar radiation pressure, unmodeled resonant effects, and other satellite surface forces, as well as errors in the reference ellipsoid.

On the other hand, the sea surface height h can be expressed as

$$h = N_c + \Delta N_G + \Delta N_O + \zeta + \tau + (w) \tag{3.5}$$

where N_c is the geoid undulation computed from the geopotential model that was used for the orbit integration, ΔN_G are the undulation errors due to the geopotential model uncertainties, ΔN_O is the omitted higher-degree undulation that is not modeled by the geopotential model, ζ is the stationary sea surface topography (dynamic height), τ are the solid earth and ocean tides, and w are the other motions of the ocean such as wind driven waves, eddies etc., that will be neglected in the following.

Substituting (3.4) and (3.5) in (3.1), using the correction term c, and defining a new observation, the so-called residual sea surface height, by

$$\Delta h = h_c - \rho - N_c - \Delta N_O - \tau(-w) - c \tag{3.6}$$

we obtain the following observation equation:

$$\Delta h = \Delta N_G(\Delta \overline{C}_{lm}, \Delta \overline{S}_{lm}) - \Delta h_G(\Delta \overline{C}_{lm}, \Delta \overline{S}_{lm}) - \Delta h_I + \zeta \tag{3.7}$$

In (3.7), ΔN_G and Δh_G are both functions of the potential coefficient corrections $\Delta \overline{C}_{lm}, \Delta \overline{S}_{lm}$. A complete modeling of these quantities was made by Engelis [1987a], where the interested reader may find the detailed observation equations. Note that in (3.7) we assume c is known.

An appropriate representation for ζ is needed since one will try to estimate ζ as part of the modeling effort. The procedure generally used is a representation of ζ in a low degree surface spherical harmonic expansion:

$$\zeta = \sum_{l=1}^{l_{max}^{SST}} \sum_{m=0}^{l} \left(\overline{C}_{lm}^{SST} \cos m\lambda + \overline{S}_{lm}^{SST} \sin m\lambda \right) \overline{P}_{lm}(\sin \overline{\phi}) \tag{3.8}$$

where $\overline{\phi}$ and λ are geocentric latitude and longitude. Of special interest in (3.8) are the degree 1 coefficients. If one uses only satellite altimeter data as a tracking data type Wagner (1985) has argued that you can not estimate the (1, 0) coefficient. Denker and Rapp (1990) also pointed out that the (1, 0) coefficient can not be separated from the 1 cycle/revolution sine coefficient (the a_2 term in eq. (3.10)) in the Δh_I model. In addition the estimation of the degree 1 coefficients is sensitive to the accuracy of the geocentricity of the reference coordinate system as noted by Denker and Rapp (1990) and Koblinsky et al (1989).

Since ζ is not defined on land estimates of the coefficients can lead to spurious values of ζ in land areas with some impact on oceanic conclusions near coastal regions. In addition the use of spherical harmonic coefficients can imply incorrect spectral characteristics of sea surface topography. To reduce these problems Hwang (1991) developed a function set that was orthonormal over the oceans. This representation would yield almost the same spatial information as the spherical harmonic relationship but would have greater spectral integrity.

Given a function, $f(\phi, \lambda)$, defined only over the oceans we write (Hwang, 1991, eq. (6.60)):

$$f(\theta, \lambda)\Big|_{ocean} = \sum_{n=0}^{N_{max}} \sum_{m=0}^{n} \left(\alpha_{nm} O_{nm}(\theta, \lambda) + \beta_{nm} Q_{nm}(\theta, \lambda) \right) \qquad (3.9)$$

where α_{nm}, β_{nm} are the coefficient set to be estimated and $O_{nm}(\phi, \lambda)$, $Q_{nm}(\phi, \lambda)$ are orthonormal functions over the oceans. These functions are numerically determined once the "ocean" is numerically defined. Numerical results on the use of orthonormal functions for the representation of dynamic heights are discussed in Hwang (ibid) and is beyond the scope of this paper.

Colombo [1984, 1989] and Engelis [1987a] showed by rigorous modeling that Δh_I is composed largely of a constant bias, a 1-cycle/revolution (cy/rev) effect with both time dependent and independent amplitudes, and a 2-cy/rev term with a term with a time dependent amplitude. Resonant perturbations are mapped into a radial perturbation of about 1 cy/rev. In this case, 1 cy/rev effects with quadratic time dependent amplitudes need to be considered, as was pointed out by Colombo [1984]. Therefore in the context of this discussion the following formulation is adopted:

$$\begin{aligned} \Delta h_I = a_0 &+ a_1 \cos\dot{\psi}t + a_2 \sin\dot{\psi}t + a_3 \Delta t \cos\dot{\psi}t \\ &+ a_4 \Delta t \sin\dot{\psi}t + a_5 \Delta t \sin 2\dot{\psi}t + a_6 \Delta t^2 \cos\dot{\psi}t \\ &+ a_7 \Delta t^2 \sin\dot{\psi}t \end{aligned} \qquad (3.10)$$

where $\dot{\psi}$ is the frequency associated with the 1 cy/rev, t is the time from the beginning of the arc, and Δt is the time relative to the middle of the arc, which is done because residual drag and resonant errors often exhibit a signature that is minimum in the middle of the arc and grows in amplitude approaching the beginning and end of the arc (the so-called "bow-tie" pattern [Colombo, 1984; Engelis and Knudsen, 1989]).

It is appropriate to note here many investigations have been carried out to understand the appropriate representation of the radial orbit error. Most studies look at error models that resemble (3.10). However such studies do not consider the radial error terms that are associated with the potential coefficient errors that have a significant influence on the radial error. However a restricted radial error modeling may be appropriate when orbit improvement in limited geographic regions is being carried out. Examples of different approaches to modeling the radial orbit error may be found in Tai (1988), LeTraon, Boissier and Gaspar (1991), and Chelton and Schlax (1992).

Now consider the parameters that would be part of an adjustment that would incorporate satellite altimeter data to reduce the radial orbit error. We have first the potential coefficients that were used in the generation of the preliminary orbits. We then have the coefficients of the sea surface topography representation. Such coefficients could be estimated for different altimeter data or different time periods or as time dependent coefficients. Different types of solution for sea surface topography are discussed in Lerch et al (1992), Nerem and Koblinsky (1992), Rapp, Wang, Pavlis (1991) (designated RWP in future references) etc.

Other parameters of the adjustment would include the parameters of the Δh_I model used in (3.10). As written there are 8 parameters per altimeter arc to be analyzed. It should be recognized that eq. (3.10) is just one of numerous possible models that might be used. Although this model is based on analytic arguments it may be that all parameters may not be estimated due to factors such as the length of the arc being processed. Applications with (3.10) have used 5 or 6 day arcs.

One should note that the method described above is a special case of a more general orbit determination procedure. In the more general case many types of satellite data are simultaneously processed as part of the modeling effort. No a priori gravity field model is incorporated in the solution. The more general model seeks an improvement in the complete orbit, not just the radial distance. In doing this the number of parameters to be estimated will increase because such parameters can be associated with certain observation types and station coordinates that are to be recovered. The more general solution is discussed by Lerch et al (1992) for the development of the GEM-T3 potential coefficient set.

3.2 Altimeter Measurement Processing

In the procedure discussed in the previous section one incorporates altimeter measurements in the adjustment process. However it is difficult to incorporate all available data because of the voluminous amount of data. Because of this, the original data, but corrected for environmental and geophysical effects, needs to be modified to achieve a more workable number of data points. One way (Denker and Rapp (1990), RWP)) to do this is to calculate sample points. First one starts with the 1 sec residual height data that is defined by eq. (3.6). A linear fit is carried out over a 20 sec interval. An iterative outlier rejection criteria with a 3σ limit is used. The sample point was computed from the linear fit at the mid time point of the interval. In cases where the 20 sec fit would yield insufficient data in a region, RWP found it is necessary to reduce the time interval to 10 sec.

A key step in preparing the altimeter data for processing is the calculation of the ΔN_0 term in eq. (3.6). This is done by removing the undulation contribution from the maximum degree +1 of the a priori model (or more generally the potential coefficient model being estimated) to some high degree (e.g. 360) based on an existing geopotential model. Unless this step is carried out the higher frequencies can alias the estimation of the lower degree field. Since the high degree (360) model will not be perfect, and there are still effects above degree 360, there will still be residual errors that could be reduced after an improved high degree field is estimated. This basically infers a broad iterative improvement.

3.3 The Use of Surface Gravity Data

The process described in the previous sections would lead to a geopotential model that has not considered the information provided by terrestrial gravity measurements. The gravity measurement can be connected to surface gravity anomalies after a normal reference surface (ellipsoid) is defined. This anomaly could be represented in a spherical harmonic expansion by eq. (2.10). For practical purpose the point gravity anomalies are represented as a real mean values over different size cells. For applications in this section the appropriate size is a $1° \times 1°$ cell on the surface of the Earth. Before the original mean anomaly is used to develop a set of normal equations, a number of corrections must be applied to the data to make it compatible with the use of eq. (2.10) as an observation equation. These corrections can be characterized as follows:

A. Atmospheric correction

B. Ellipsoidal corrections (see eq. (2.8))

C. Second-order vertical gradient of normal gravity correction

D. Gravity formula transformation

E. Downward Continuation of the Surface Anomaly to the Ellipsoid

F. Removal of high frequency effects from the anomaly.

A detailed discussion of effects A, B and C can be found in Pavlis (1988). Effect D is discussed in RWP (p. 10) and is just a transformation from the anomaly with respect to the original gravity formula to the reference formula consistent with the adopted constants of the estimation system. Correction E is a correction that would not be applied if one chose to model the anomaly given at the surface of the Earth. Tests described by Pavlis (p. 20) in RWP indicate, however, more accurate geoid undulation results are obtained if the surface anomalies are reduced to the ellipsoid. This reduction can be computed from topographic information if certain assumptions are made as noted in Section 2.2.4.

The correction term F is one that removes the anomaly signal from the maximum degree +1 of the a priori model (or more generally the potential coefficient model being estimated) to some high degree (e.g. 360) based on an existing model. This correction is analogous to the high frequency correction made for the altimeter data sample points. In this case, however, the correction, must be computed over the cell in which the anomaly is defined. This requires the integrated evaluation (with respect to ϕ, λ) of eq. (2.10) from $N + 1$ to (usually) 360 using the same model as used in the altimeter analysis if possible. In this step two points need to be made:

1. The harmonic coefficients used to evaluate the high frequency contribution must be consistent with the anomaly data to be used in the adjustment.

2. The high frequency contribution must be evaluated at the same height to which the mean values refer (topographic surface or the ellipsoid).

3.4 The Combination Solution With Satellite Altimeter Data

We now consider how to merge together the data discussed in the two prior sections. The general normal equation will be designated N. They are formed from the observations using the design matrix (A), the weight matrix (P) and the misclosure vector (U). The general normal equation set would be:

$$N\hat{X} = U \tag{3.10}$$

where

$$N = A^TPA \; ; \; U = A^TPL_b \tag{3.11}$$

where L_b is the vector of observational residuals. The $\hat{X}$ vector contains the estimated parameters (potential coefficients, sea surface topography coefficients, arc dependent parameters, etc.). In our estimation process we use an a priori set of potential coefficients with a calibrated error covariance matrix whose inverse yields the weight matrix P_M. Let N_A and N_G be the normal equations associated with the altimeter data and the surface gravity data. Then the correction vector to the a priori parameter values would be:

$$V_X = - (N_A + N_G + P_M + P_X)^{-1}(U_A + U_G) \tag{3.12}$$

where P_X is the a priori weights on selected (e.g. sea surface topography coefficients) parameters of the adjustment. The surface gravity related vector, U_G, must be properly referenced to the a priori potential coefficient model, X_M. This can be done by writing (RWP (eq. (3.23)):

$$U_G = U - N_G(X_M - X_{ELL}) \tag{3.13}$$

where U is given by (3.11) where L_b are the corrected and reduced gravity anomaly observations and X_{ELL} are the zonal harmonic coefficients associated with the normal potential of the reference ellipsoid.

The solution vector from eq. (3.12) is added to the a priori parameter vector to obtain the adjusted parameters of the solution. The error covariance matrix of these parameters would be the matrix that is the coefficient of $(U_A + U_G)$. We then will obtain the improved geopotential model to degree N, sea surface topography coefficients, and 8 parameters for each altimeter arc used in the adjustment.

3.5 The Weighting Problem

In all these computations decisions must be made on the weights to be assigned for all measurements, as well as, the definition of any a priori weight matrices. We consider first the weighting of the altimeter data. The first step is to consider the altimeter measurement and/or observational residual. This quantity is needed when the altimeter normal equations are computed. The standard deviation used can differ from one point to the next and between altimeter missions. Approximate standard deviations are: Geos-3 ($\pm$ 20 cm); Seasat ($\pm$ 10 cm); Geosat ($\pm$ 5 cm). These values are measurement accuracies and not sea surface height accuracies.

Another consideration relates to the use of data from an exact repeat mission (ERM). In such a case one complete cycle will cover a certain geographic area in the oceans. Additional ERMs will cover the same geographic area and therefore not bring new gravity information into the picture. However the use of more than one ERM is highly desirable to reduce data noise and ocean variability effects. The incorporation of many ERMs in a solution enables one to study, as part of a general solution, time variations in the representation of the sea surface topography. However, as ERM data is added the repeated gravity information could yield a solution dominated by the altimeter data. To avoid this problem a procedure needs to be implemented that down weights the altimeter normal equations as they are used in the general solution. Numerous tests have been carried out at Ohio State to find a proper down weighting factor. Such tests have been empirical in nature with an examination of data fits, standard deviations, etc., calibration factors, to form a conclusion. We first define the down weighting factor as that quantity by which all elements in an altimeter normal equation set are to be multiplied. This factor (DW) will depend on the number of ERMs used in the solution. The analysis described in RWP started with 22 Geosat ERMs. Down weighting factors of 1/24 and 1/96 were considered with a conclusion that

1/96 gave, in an overall sence, the most reasonable results. Fortunately the overall solution is not sensitive to any reasonable changes in weights. The greatest impact is on the standard deviations of the adjusted quantities. As the downweighting factor gets smaller the standard deviations of the adjusted quantities increases. If M is the number of ERMs being used in a solution a preliminary suggestion on setting the DW is:

$$DW = 1/(4\ M) \qquad\qquad (3.14)$$

Tests need to be carried out with several DW factors to judge the best one for a given case.

The a priori weighting of some of the parameters in the adjustment is also an important aspect to be discussed. If a solution is being carried out with an a priori set of potential coefficients (such as is assumed in eq. (3.12)) the a priori error covariance matrix, properly calibrated, should be used, to compute P_M. If some coefficients have not been estimated in the model, a decision needs to be made if any a priori standard deviation should be used based on an a priori estimate of zero for the coefficient. Such a procedure is usually done for solutions based purely on satellite perturbation analysis. However such a procedure is not needed when direct altimeter measurements and surface gravity data are to be used in the solution.

In some solutions that incorporate sea surface topography coefficients as parameters it may be appropriate to assign a priori weights to the a priori estimates of zero coefficients. Such a procedure can help control the magnitude of the estimated coefficients so that unreasonably large values are not estimated. This is a problem where spherical harmonic coefficients are being estimated, in a least squares adjustment, from non-global (i.e. ocean data only) data. The a priori weighting can also reduce large magnitude estimates of sea surface topography in the land areas. Although such estimates on land have no meaning, they may influence sea surface topography behavior near the shorelines. Marsh et al (1990) computed a linear fit to a set of coefficients computed by Engelis (1987b) based on Levitus (1982) sea surface topography estimates. The function generated was:

$$\sigma_{nm} = -\ 0.1344\ n + 0.13954 \qquad\qquad (3.15)$$

where σ_{nm} is the root mean square spherical harmonic coefficient in meters. Unfortunately this model gives negative values for σ above degree 10. The solution for sea surface topography described in RWP used a priori estimates based directly on the Engelis solution to degree 10 with estimates above degree 10 based on the value at degree 10. As improved solutions are obtained they can be used to obtain more reliable a priori standard deviations of they are needed to stabilize the solution. If the orthonormal functions are used for the representation of sea surface topography the use of a priori standard deviations for the estimable coefficients would not be necessary. Hwang (1991, p. 175) did use such estimates in test solutions based on an orthonormal representation of the Levitus data.

3.6 Summary

This section has been designed to show how satellite altimeter measurements can be incorporated in a limited orbit improvement strategy where the output includes an adjusted potential coefficient model, sea surface topography and a radial orbit error model for each arc of altimeter data processed. In principle the potential coefficient model could extend to as high a degree as is consistent with the data being analyzed. For example if 1°

x 1° surface data were used and the sample points were based on 7 second data fits instead of 20 sec fits, a solution conceptually would be possible to degree 180. However there would be a significant computer impact if the solution is carried out according to the lines described here. Alternative solutions to this type of problem have been and are now being studied. Such solutions usually make assumptions that make the normal equations block diagonal so that the formation of the normal equations and their solution can be simply done.

At the present time the solutions that combine satellite data, altimetry and surface gravity data are taken to a maximum spherical harmonic degree 50. Representative solutions include GEM-T3 (Lerch et al, 1992) and the degree 50 part of the OSU91A model which is described in RWP. These solutions will be discussed in Section 5. It is clear, however, that the current modeling effort does not yield the high degree (e.g. 360) expansions that are in use today. In the next Section we start to examine the way in which higher resolution information can be computed and subsequently used for the determination of a high degree expansion.

4.0 The Estimation of Geoid Undulations and Gravity Anomalies from Satellite Altimeter Data

4.1 The Adjusted Sea Surface Heights

We now assume the improved altimeter orbits have been obtained by the process described in the last section or by the more general process described in Lerch et al. (1992) where all types of data are simultaneous processed. From these computations we can generate the adjusted sea surface heights by writing eq. (3.4):

$$h_A = h_c + \Delta h_G + \Delta h_I \tag{4.1}$$

where h_c is the sea surface height based on the original orbit, and Δh_G and Δh_I are now known terms. The application of the correction terms yields the adjusted sea surface along all tracks used on the adjustment. These data points will be typically spaced at 1 second intervals or 7 km spacing.

If the data is given on repeat tracks one has to recognize that the tracks will only repeat to approximately ± 1 km. If averaging of the many tracks is to be considered the actual ground track should be reduced to a single reference ground track taking into account sea surface height or geoid gradients. The need for gradient reduction was noted in Brenner et al (1990) with specific techniques for the computation of such gradients described in Wang and Rapp (1991). After the sea surface heights have been reduced to a common track they can be averaged to determine a mean sea surface track. An example of such procedures is described in Wang and Rapp (1992). For our further discussion we will assume that there exists a fundamental set of sea surface heights computed from a precise altimeter orbit. We recognize that the heights are not perfect but the residual orbit error remaining in the data should be small (± 15 cm).

4.2 The Estimation Procedure for Point Values

We assume that we have a set of point altimeter measurements. It is this data that is to be used to recover geoid undulations, gravity anomalies, and other pertinent information. The recovery of this information has been discussed by many authors. A representative sample include: Rapp (1979, 1986), Hwang (1989), Farelly (1991), Knudsen (1992), McAdoo and Marks (1992), Sandwell and McAdoo (1990), Haagmans (1988), etc.

The methods represented in these papers are basically those of least squares collocation and a Fast Fourier Transform (FFT) approach. There are several different procedures in which FFT methods can be used for gravity anomaly recovery. A discussion of such procedures is beyond the scope of this paper.

The production procedure for the calculation of gravity anomalies and geoid heights on a grid has been carried out by the least squares collocation method at Ohio State. The most recent set of computations is described by Bašić and Rapp (BR) (1992) where Geos-3, Seasat, and Geosat altimeter data were used with bathymetric data. The details of these computation will be briefly reviewed here.

We assume that the altimeter implied sea surface heights have been corrected, to the extent possible, for sea surface topography effects. This can be done in practice, by using the long wavelength sea surface topography models such as OSU91. One assumes that averaging over time and different types of data sets that higher frequency sea surface topography effects will be minimized.

The basic least squares collocation procedures used by Rapp (1985) and Hwang (1989) were expanded by BR (ibid) with respect to the initial remove-restore technique involving a high degree potential coefficient model. (The following discussion is taken almost directly from BR, pages 2 and 3). But in contrast to the above solutions BR expanded the remove-restore procedure by using a set of potential coefficients and bathymetric/topographic heights (see e.g. Tscherning and Forsberg, 1987, Sünkel, 1987). The simultaneous use of satellite altimetry, spherical harmonic coefficients and topography makes it possible to improve the prediction when only one data type are used. Moreover, a slight complication is introduced if the contribution from spherical harmonic expansion is subtracted since the terrain effects associated with the same wavelengths as the harmonic expansion are removed implicitly.

There are two main methods to account for the topography in LSC. The first alternative is to use the topographic-isostatic reduction of the observed gravity field parameters which formalizes the prevailing tendency of the earth's topography to be compensated at depth. But reality is that many regions are either compensated or uncompensated at deeper levels, especially deep-sea trenches and mid-oceans ridges (Forsberg, 1984).

The second alternative is to work with the so called residual terrain model (RTM) reduction using a model of the topography equal to the difference between a point topographic height define topography and a mean topographic heights formed as moving averages over blocks of the same size as those used when determining the spherical harmonic coefficients (see Forsberg and Tscherning, 1981 or Forsberg, 1984). The RTM method takes only the short wavelengths of the actual topography into account and keeps the mean values of the residual quantities close to zero with a small signal variance.

A strong similarity between RTM and isostatic reductions exists. By a special choice of the mean elevation surface a nearly complete correspondence may be obtained (Forsberg, 1984, p. 38). Due to the advantages noted above BR decided to use the residual terrain modelling (RTM) method for the remove-restore procedure.

The mathematical model for predicted point gravity anomalies $\hat{\Delta}g$ and geoid undulations $\hat{N}$ is defined through the following basic equations:

$$\hat{\Delta g} = \underline{C}_{\Delta gh}(\underline{C}_{hh} + \underline{D})^{-1}(\underline{h} - \underline{h}_{REF} - \underline{h}_{RTM}) + \Delta g_{REF} + \Delta g_{RTM} \qquad (4.2)$$

$$\hat{N} = \underline{C}_{Nh}(\underline{C}_{hh} + \underline{D})^{-1}(\underline{h} - \underline{h}_{REF} - \underline{h}_{RTM}) + N_{REF} + N_{RTM} \qquad (4.3)$$

where

$\underline{C}_{\Delta gh}$, $\underline{C}_{Nh}$	—	are the cross covariance matrices between predicted and observed quantities, including the noise of the reference field,
$\underline{C}_{hh}$	—	is the autocovariance matrix of observables,
$\underline{D}$	—	is the error covariance matrix (diagonal matrix containing the error variances of observables),
$\underline{h}$	—	is the column vector of altimeter implied geoid undulations,
$\underline{h}_{REF}$	—	is the column vector of reference undulations computed from a given set of potential coefficients up to degree and order 360,
$\underline{h}_{RTM}$	—	is the column vector containing the effects of residual terrain modelling on geoid heights,
Δg_{REF}, h_{REF}	—	are the reference gravity anomaly and geoid height values computed at the prediction points from a spherical harmonics up to degree and order 360,
Δg_{RTM}, h_{RTM}	—	are the effects of residual terrain modelling on the gravity anomaly and geoid height computed at the prediction points.

The assumption for the method of least squares collocation is that the expected values of both, observations and predicted signals are equal to zero (Moritz, 1980, p. 76). This means that equations (4.2) and (4.3) implicitly assume that the expectation of the residual data vector ($\underline{h} - \underline{h}_{REF} - \underline{h}_{RTM}$) over the area in which the predictions are carried out is zero. In both the 1985 and 1989's predictions at Ohio State (Rapp, 1985, Hwang, 1989) such an assumption was neglected causing systematic biases in some limited regions. It was seen in the case of Hwang's (1989) predictions, where a single prediction area was relatively small ($1° \times 1°$) and the zero-mean-value assumption did not hold well everywhere, systematic anomaly prediction errors occurred (Rapp and Pavlis, 1990). Hwang analyzed the influence of neglecting this assumption and found that the use of centered data is always a better procedure. Therefore BR decided to force the data residuals ($\underline{h} - \underline{h}_{REF} - \underline{h}_{RTM}$) in equations (4.2) and (4.3) to be centered by subtracting the mean residual value over the used data area from every single residual values. One should be suspect that the expectation of the residual data which includes the RTM reductions is "closer" to zero than the expectation of the residuals without this reduction.

The accuracy estimates of the predicted gravity anomalies and geoid undulations are given as:

$$M_{\hat{\Delta g}}^2 = \underline{C}_{\Delta g \Delta g} - \underline{C}_{\Delta gh}(\underline{C}_{hh} + \underline{D})^{-1}\underline{C}_{h\Delta g} \qquad (4.4)$$

$$M_{\hat{N}}^2 = \underline{C}_{NN} - \underline{C}_{Nh}(\underline{C}_{hh} + \underline{D})^{-1}\underline{C}_{hN} \qquad (4.5)$$

where

$C_{\Delta g \Delta g}$ — is the autocovariance matrix of the predicted gravity anomaly,

C_{NN} — is the autocovariance matrix of the predicted geoid undulation,

$C_{\Delta gh}$, C_{Nh} — are the cross covariance matrices between predicted and observed quantities.

The general strategy for the prediction of gravity anomalies and geoid undulations using equations (4.2) and (4.3) consists in the combination of the following three kinds of data:

— the geopotential model up to degree and order 360 for the computation of long- and mid-wavelength structures of the gravity field parameters,
— the altimeter implied geoid undulations for the recovery of mid- and short-wavelength of the gravity field parameters, and
— 5' x 5' bathymetric/topographic heights for the computation of the short-wavelength of the actual gravity field parameters.

The implementation of eq. (4.2) and (4.3) requires a careful determination of h from multiple satellite data types. Given three different types of altimeter data a cross over adjustment can be carried out to achieve an optimum consistency of the data, building in this case, on the accurate OSU Geosat orbits. In the computation being described the residual orbit error was modeled by a bias parameter only where the altimeter data was considered in a 8° x 8° area. In addition the geoid undulation implied by the altimeter data were fitted to the geoid undulations implied by the reference potential coefficient model (Rummel and Rapp, 1977). In the calculations done by Bašić and Rapp (ibid) the three kinds of altimeter data were merged to eliminate duplicate geographic coverage and tracks that would be would create improbable predictions due to their closeness. The crossover discrepancies after the adjustment in the 8° x 8° cell were typically as follows: Geos-3 (± 43 cm); Seasat (± 10 cm) and Geosat (± 4 cm).

The ocean wide predictions were carried out in 36 32° x 40° or 40° x 40° blocks. The result was a 0°.125 grid of gravity anomalies and geoid undulations with individual standard deviations. In case there was insufficient altimeter data to carry out the prediction the predicted value was set to the OSU91A (to 360) value and the standard deviation set to 99.

The next step was to carefully edit the predicted geoid undulations and gravity anomalies to delete values on land and shallow areas; values whose magnitude were unreasonable considering depth etc. After the first edit was completed, a second edit was carried out that modified some predicted geoid undulations based on geoid gradient tests carried out with Geosat along track altimeter data. The final predicted data set contained 2,312,964 gravity anomalies and 2,325,669 geoid undulations. These data points covered most of the ocean areas between ± 72° latitude.

4.3 The Estimation Procedure for Mean Values

The point values described in the above analysis are more useful in high degree expansion studies if they are converted to mean values in a cell sizes commensurate with the highest degree sought in the expansion. In this case we will seek expansions to degree 360 so that the computation of 30' x 30' mean values is appropriate. The computation of such values was discussed by Hwang (1989) taking into account the correlated predictions of the point values. Such computations were repeated by Basic and Rapp (ibid) to determine and improve set of 30' x 30' mean gravity anomalies and geoid undulations. The reliability of these values would be previously dependent on the number of 0°.125 point values that were averaged to form the mean values. Basic and Rapp (ibid) recommended that the 30' values be considered reliable if at least 13 (out of 25) points were available to form the average. This criteria would lead to 144,784 mean anomalies and 143,770 mean geoid undulations.

4.4 The Formation of a Global 30' Anomaly Data File

The method for the determination of a high degree expansion described in Section 2.3 requires an estimate of a global anomaly data set. To create such a set one must merge a number of different data sources including the following:

A. 30' x 30' mean anomalies based on terrestrial gravity measurements;

B. 30' x 30' mean anomalies based on predictions from satellite altimeter data;

C. 30' x 30' mean anomalies derived from 1° x 1° mean anomalies in area where 30' values do not exist;

D. 30' x 30' mean anomalies for those areas in which no value is available from data types A, B, or C. Such values might be derived from a satellite derived potential coefficient model. In the development of the OSU89B and OSU91A degree 360 models, the "empty areas" were computed from a long wavelength potential coefficient model (e.g. GEM-T2 to some degree) plus the potential implied by the topographic/isostatic influence of the Earth's topography. This method brings into play additional information (the topography) about the Earth. The details of the modeling procedure are described in Pavlis and Rapp (1990). The details on specific computations for the OSU89B and OSU91A model may be found in the pertinent references.

An example of the coverage map of 30' data used in the development of the OSU91A model is shown in Figure 1. This data did not include the altimeter derived anomalies by Basic and Rapp (1992), rather the anomalies derived by Hwang (1989) and modified, in some cases to correct for bias problems.

In some developments of high degree potential coefficient models it may not be necessary to develop a global data file. In such a case the existing data set derived from satellite altimeter data could be used to form a set of normal equations (corresponding to a high degree expansion) that would be used in conjunction with other detailed data. The procedure for such a solution will be discussed in a later section.

5. High Degree Expansions Using Satellite Altimeter Data

5.1 Solution Based on Section 2 Discussion

At this point we have a 30' x 30' global gravity anomaly data set. We assume that all pertinent corrections have been made to reduce these anomalies to the ellipsoid. We next assume that we have a set of a priori potential

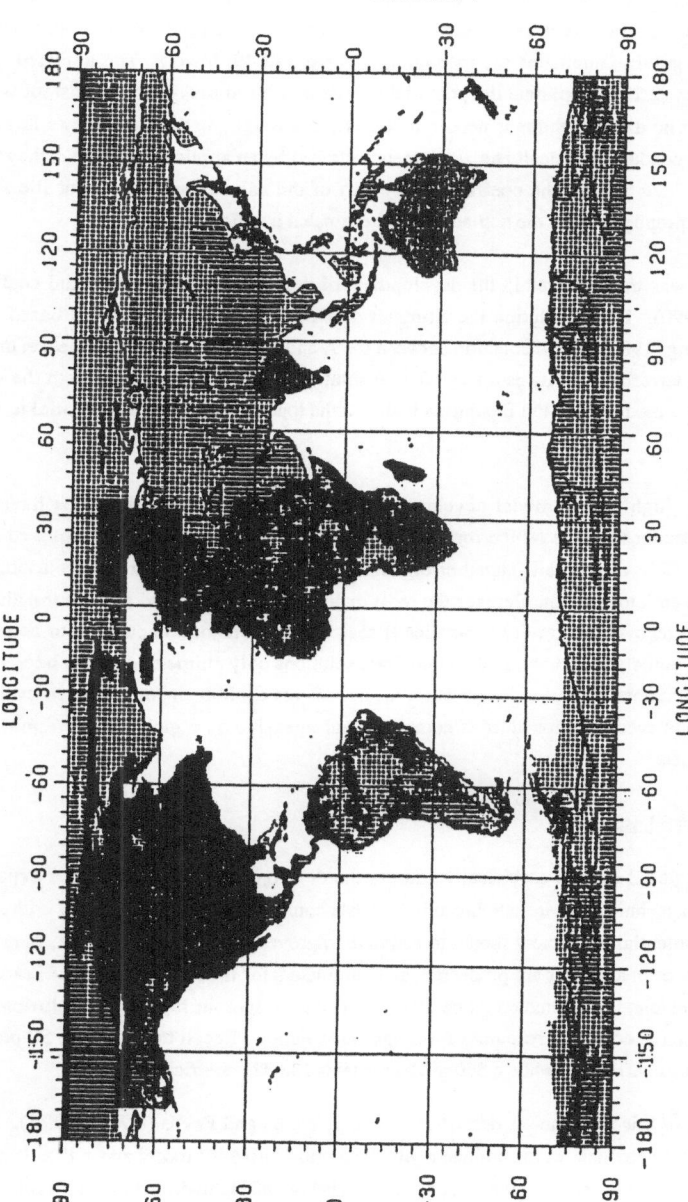

Figure 1. Identification of 30' x 30' anomalies in the merged data file used for the OSU91A model solution. The darkest striping represents the original 30' terrestrial anomalies. The darker vertical striping represents 30' terrestrial anomalies obtained from the split up of 1° terrestrial anomalies. The lightest striping represents 30' values determined by "fill-in" techniques. The blank areas designate the location of the altimeter derived 30' anomalies.

coefficients based on satellite perturbation analysis. The key point is that the a priori model should be independent of direct altimeter measurements and surface gravity data. A least squares adjustment is carried out using equation (2.53). The adjusted potential coefficients corresponding to those that exist in the a priori file are found using eq. (2.49, upper). The adjusted gravity anomalies are then computed from (2.49, lower). The adjusted gravity anomalies are then used in eq. (2.23) to estimate the potential coefficients up to the degree consistent with the anomaly block size. Note that no data weighting is needed in this step but weighting decisions play a key role in the complete process. This procedure is basically an adjustment followed by an application of the orthogonality relationships. The standard deviations of the coefficients are part of the adjustment process for the a priori coefficients followed by error propagation for the non-adjustment estimated coefficients.

The above procedure was implemented in the development of the OSU89A and B potential coefficient models (Rapp and Pavlis, 1990). In this solution the altimeter derived gravity anomalies were based on the predictions described by Hwang (1989). The distinction between the A and B solutions related to the treatment of the anomalies in areas lacking terrestrial gravity estimates. The A solution used information only from the a priori (GEM-T2) potential coefficient model while the B solution included the topographic/isostatic potential to degree 360.

Although this type of high degree model development is fairly straight forward it does have some disadvantages. The method assumes that the orbit errors in the satellite altimeter data have been removed so that the altimeter derived anomalies have no orbit related errors and the solution assumes sea surface topography effects have properly been taken into account. Perhaps the most major point is the need to process the altimeter implied sea surface heights to recover the gravity anomalies at sea. This is a major computation to achieve an ocean wide, high resolution, anomaly coverage. In the Ohio State solutions only altimeter data has been used to derive the anomalies while in a more general case terrestrial measurements could be incorporated in estimation process (Knudsen, 1992). However this procedure is computational intensive on a global scale requiring the processing of very large data sets.

5.2 Solution Based on Section 3 Discussion

The discussion in Section 3 did not specifically address the development of a high degree expansion. Instead the method described a technique where satellite altimeter data could be directly incorporated with surface gravity data and an a priori potential coefficient model to estimate improved potential coefficients, parameters modeling radial orbit error for each altimeter arc processed, and parameters for the representation of sea surface topography. Although, in principle, the method could be used to calculate a rigorous high degree solution, it has not been done because of the large computer resources for a rigorous solution. Recall that a degree 50 potential coefficient model contains 2601 coefficients while a 360 model contains 130,321 coefficients.

The latest application of this technique is described by Rapp, Wang, and Pavlis (RWP) (1991). In this solution 22 ERMs of Geosat altimeter data were combined with the a priori GEM-T2 model and a 1° x 1° surface gravity data set. The altimeter data was selected so that data was used below -60° latitude and in the Mediterranean Sea. This is noted as some solutions do not incorporate such data in their model. A total of (approximately)

819,000 sample points were used. The empty 1° anomaly cells for land areas were based on the GEM-T2 potential coefficient set to degree 9 plus topographic/isostatic effects to degree 50. The higher frequency effects (51 to 360) were removed from the altimeter sample points and the 1° x 1° surface gravity data using the coefficients of the OSU89B model. The location of the 1° x 1° terrestrial (45932) plus "fill-in" anomalies (8116) is shown in Figure 2.

The results of this solution are fully described in RWP. Several results are of interest here. First we have the adjusted set of potential coefficients to degree 50. In addition we have the parameters of the radial orbit correction terms and the sea surface topography representation to degree 10 to 15. The sea surface topography solution to degree 10 is shown in Figure 3. Note the slope in the Mediterranean Sea and the large magnitude of the topography (approaches -2 m) below -60° latitude.

The analysis that was carried out in this solution led to results on the temporal variation of the ocean surface during the time period (November 1986 to October 1987) in which the altimeter data was used. The variations found were compatible with results from other investigations and with tide gauge measurements.

But the aim of this paper is to discuss the development of high degree potential coefficient models using altimeter data. But before returning to this theme it should be clear that oceanographic effects play a significant role in the analysis.

At this point we only have a degree 50 potential coefficient model which is the best estimate based on the input data set and the weighting system used. In order to obtain a model to 360, a solution as described in Section 2 can be carried out using 30' x 30' mean anomalies globally distributed. This data set was be formed from the 30' mean anomalies derived from the 30' mean anomalies derived from Geos-3/Seasat data (Hwang, 1989) since the newer results from Basic and Rapp were not yet available. A Section 2 solution was carried out yielding a degree 360 model. The lower degree part (n ≤ 50) of the model would be inferior to that found from the direct processing of the altimeter data and solving for orbit errors, sea surface topography, etc. Therefore the final model was created by combining the 2 to 50 coefficients from the first part of the solution (the Section 3 procedure) with the higher degree (51 to 360) coefficients from the Section 2 type of solution.

A concern in this merger was that the power at degree 50 could be significantly different from the power at degree 51 since two different estimation methods were used. Fortunately this did not occur (see Figure 20 in RWP) although a discontinuity did occur in the error estimates from degree 50 to 51. The merged potential coefficient model, complete to degree 360, with standard deviations for each coefficient was called the OSU91A model. A set of sea surface topography coefficients (OSU91) complete to degree 10 are considered to be complimentary to the OSU91A potential coefficient model. A summary of the methods used in the OSU91A development is shown in Figure 4.

The method just described has, as a primary advantage, the direct incorporation of satellite altimeter data in the model estimation. This enables the incorporation of radial orbit error models and sea surface topography parameterization. The use of an extended, in time, data set enables an averaging process (in the oceans) to take

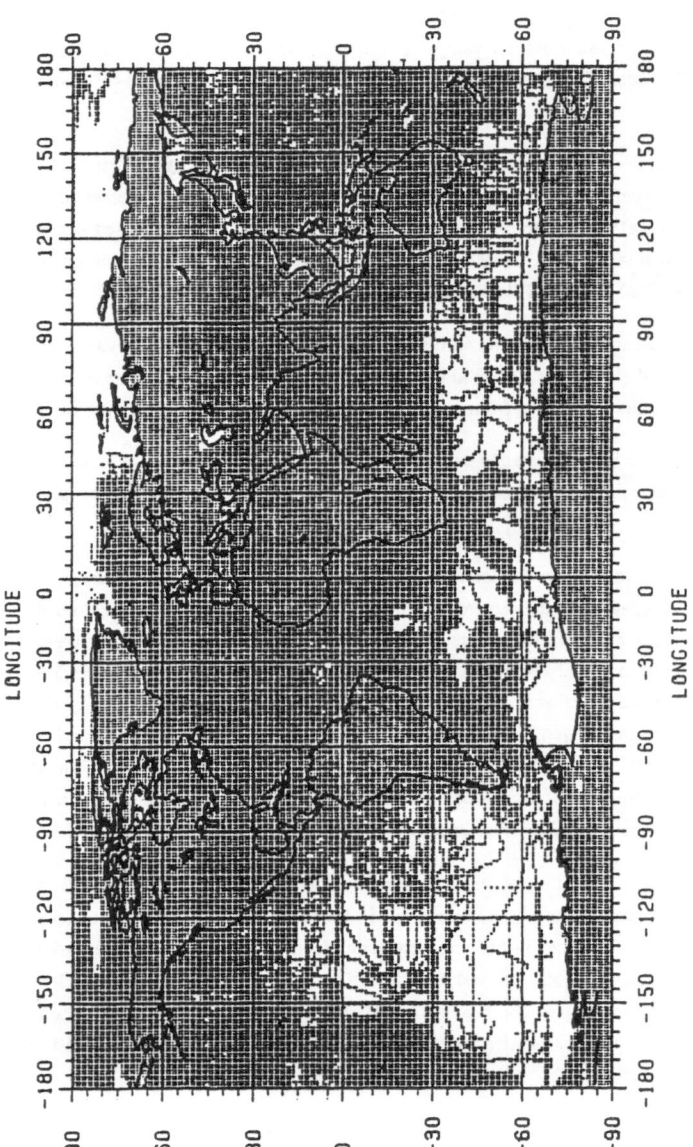

Figure 2. Geographic Distribution of the 54048 1° x 1° Mean Free-air Anomalies Used in the OSU91A Solution. "x" Identifies Values Originating From Gravity Measurements (45932) and "1" Δ^{TI} Values (8116).

403

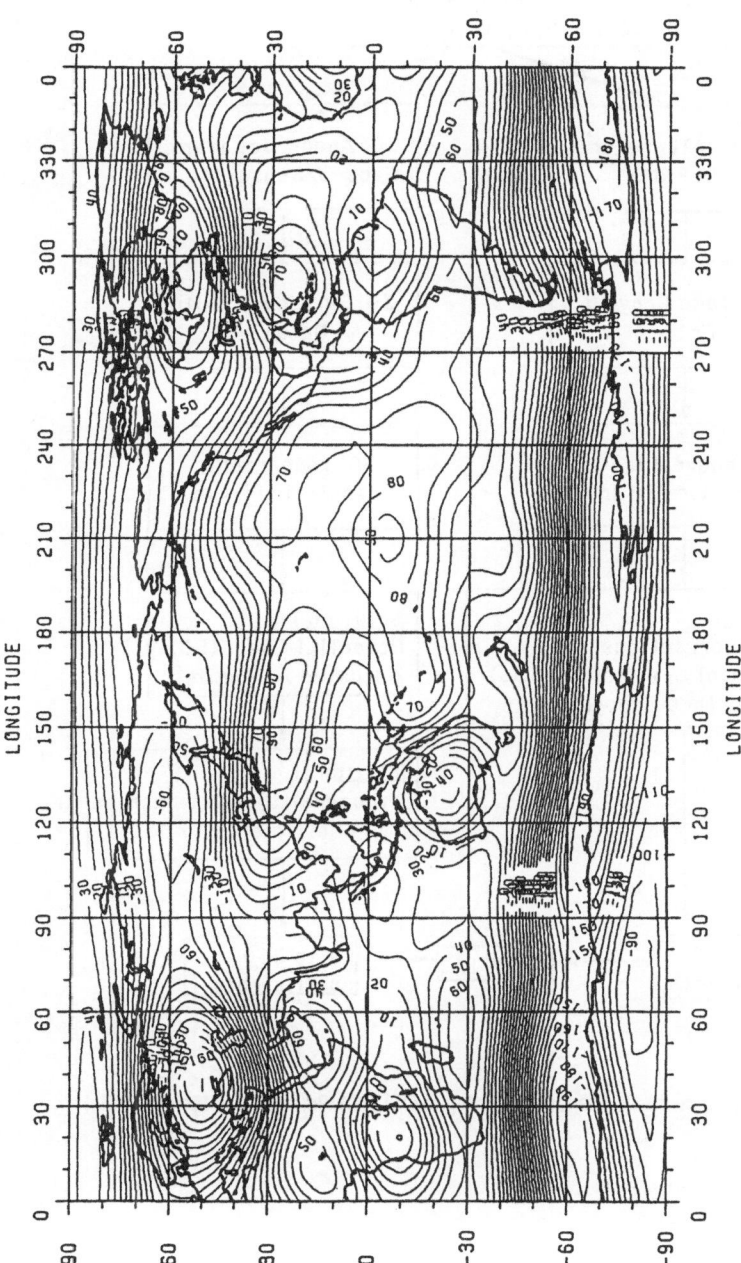

Figure 3. Sea Surface Topography Implied by Degree 10 Solution Based on One Year (November 1986 to October 1987) of Geosat Data. Contour Interval is 10 cm.

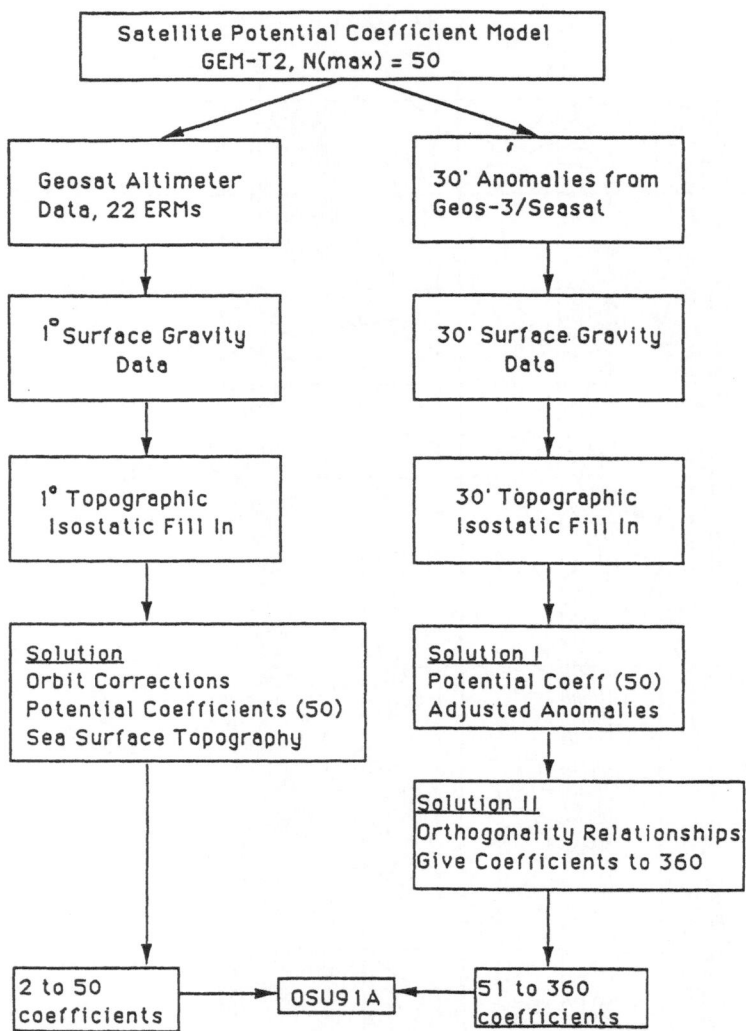

Figure 4. General Development Leading to OSU91A Potential Coefficient Model

place that reduces higher frequency sea surface topography effects. The extent that such effects (e.g. on the potential coefficient estimation) are averaged out is not clear.

A disadvantage of the method as implemented is the lack of a single solution yielding the high degree model. In addition the altimeter data used for deriving the ocean gravity anomalies was based on the older Geos-3/Seasat data. Now the Bašić/Rapp data is available for use when and if a new high degree model is estimated.

One should note that the solution described above is restrictive. The altimeter orbit error is only modeled in the radial direction. No additional satellite tracking data has been incorporated into the solution. Such solutions have been carried out with altimeter data from the three (Geos-3, Seasat, and Geosat) altimeter mission and are described in Lerch et al (1992) where the development of GEM-T3 is described. The model is complete to degree 50 and could form the basis for an OSU91A type of solution if altimeter data below -60°, in the Mediterranean Sea, and a few other areas, had not been edited out of the solution.

5.3 An Approximately Rigorous Combination Solution

The development of the high degree fields as described in the previous sections involves a least squares estimation to some low (e.g. 50) degree followed by the application of orthogonality relationships to adjusted anomalies. A more rigorous procedure would be to form normal equations from available data to estimate the coefficients to the highest degree consistent with the data resolution. Because of the large number of unknowns with a large amount of data the computational effort to carry out a completely rigorous solution is quite large. However if one makes some assumptions and approximations on the shape of the normal equations high degree solutions can be carried out. A discussion of one type of solution may be found in Wenzel (1985) where the author considered geoid undulations derived from satellite altimeter data and surface gravity data with an a priori potential coefficient model to estimate a set of potential coefficients to degree 200. Bosch (1987) described a method which studied a general and rigorous least squares approach to estimate a high degree model. This approach ordered the coefficients in such a way that the normal equations became quite patterned making a high degree solution feasible. Gruber and Bosch (1992) have described the computation of a degree 360 potential coefficient model (DGFI92A) whose estimation is dependent on the structure of the normal equations. The data used in the development of this new model were: a) the GRIM4-S2 potential coefficient set (to degree 50) and standard deviations; b) 30' x 30' and 1° x 1° terrestrial gravity anomaly data sets; c) the Geosat one year (1987) DGFI sea surface height model given on a 0°.25 x 0°.25 grid. The sea surface was converted to geoid undulations assuming a Levitus (1982) historical sea surface topography model based on oceanographic data. Normal equations were formed from the surface gravity data (after approximate corrections for atmospheric and topographic effects), and from the geoid undulation data (in the oceans). The normal equations were added to the normal equations of the GRIM4-S2 model and solved using a diagonal a priori weight on the normal equation. The solution was iterated so that the improved potential coefficient model could be used to compute gravity anomalies in areas lacking terrestrial gravity data.

The procedures used by Gruber and Bosch (ibid) are quite interesting as they lead to a near optimum solution. As presently formulated the model assumes that the altimeter orbits are correct and the sea surface topography is known. Their paper describe future developments that can reduce the assumptions they needed to make.

6.0 A Comparison of Four Potential Coefficient Models

We now compare and evaluate several geopotential models that have incorporated satellite altimeter data in their estimation. The material provided in this section is a selection of information from Rapp and Wang (1992) with the addition of material related to DGFI92A. Table 1 shows the root mean square difference over the Earth between several models taken to degree 50 only. These values have been obtained on a sphere of radius a (the equatorial radius).

Table 1

RMS Geoid Undulation Difference Implied by Selected Geopotential Models to Degree 50. Units are cm.

Model	Model		
	OSU91A	GEM-T3	DGFI92A
OSU91A	—	47	131
GEM-T3	47	—	136
GRIM4-C2	70	65	132
DGFI92A	131	136	—

The maximum differences between various models can easily reach 10 times the RMS difference. For example, Figure 30 in RWP shows the undulation differences between OSU91A and GEM-T3 can reach 4.5 m with most of the large differences on land and below -60° latitude.

The RMS geoid undulation difference between OSU91A and DGFI92A when coefficients to degree 360 are considered is ± 1.39 m. The corresponding RMS gravity anomaly difference is ± 10.7 mgal. Plots have not yet been made to show the location of the largest differences in these two models. The largest difference between the GRIM4-C2 model and OSU91A (to degree 50) is 6.3 m (at two locations in Antarctica).

The evaluation of a potential coefficient model can be done numerous ways. One way is to see the magnitude of satellite data residuals in an orbit determination process where the gravitational model is held fixed. The second procedure is to compare geoid undulations computed from the model undulations estimated from analysis independent of the development of the potential coefficient model estimation. Orbit fit analysis for several models is described in RWP (page 66). Since I do not have such results for the DGFI92A model the orbit fit results will not be reported here. Instead several types of geoid undulation comparisons will be described.

The first test is the comparison of geoid undulations derived from stations whose positions were originally determined in the NSWC 9Z-2 satellite references system using Doppler positioning techniques. Before the undulation comparisons were carried out the Doppler positions were translated to a center of mass system and properly scaled. The statistics of this comparison are shown in Table 2. The coefficients of the GEM-T3 and GRIM4-C2 model were augmented from degree 51 by the coefficients of the OSU91A model. The mean

GRIM4-C2 model were augmented from degree 51 by the coefficients of the OSU91A model. The mean difference (Doppler minus model), the standard deviation of the difference, and the number of stations used with a 4 m residual rejection criteria, are shown.

Table 2

Comparison of Geoid Undulations Implied by Doppler Positioning with Undulations from the Augmented (when necessary) Model

Model	Mean Difference	Standard Deviation	Number of Stations
OSU91A	15 cm	± 1.58 m	1802
GEM-T3	12	1.57	1788
GRIM4-C2	7	1.57	1796
DGFI92A	63	1.63	1722

The first three models listed in the table have comparable results. The DGFI92A model shows a substantial systematic difference not seen in the other models. The number of stations that were accepted with the DGFI92A were approximately 70 to 80 less than when the other models were used. If the magnitude of the residual rejection was reduced to 1.6 m the standard deviation given in Table 2 would reduce to ± 81 cm (1229 stations) for OSU91A and ± 85 cm (1045 stations) for DGFI92A. These standard deviations are very similiar to those found by Gruber and Bosch (ibid).

Another test with Doppler undulations was suggested by Tscherning a number of years ago. This test examines the correlation of the undulation residuals (actual-model) with elevation; the less correlation the better the model is the argument. To carry out the fit the residuals are grouped by elevation and the parameters of a straight line are fitted to the data. The bias and slope, as well as an RMS residual, of the line for several lines are given in Table 3. The large bias (0.356 m) and large slope (.378) for DGFI92A are out of character with the results for the other models. The reasons for these significant terms needs to be addressed. An interesting point is that the large bias and tilt terms are not present in the GRIM4-C2 model which was used as the a priori model for the DGFI92A model.

Table 3

Bias and Slope Parameters of a Line Fitted to Geoid Undulation Residuals as a Function of Elevation for a Global Station Set

Model	Bias, m	Slope, m/km	RMS Residual, m
OSU91A	.034 ± .099	.037 ± .082	.216
GEM-T3	.006 ± .105	.038 ± .087	.230
GRIM4-C2	-.013 ± .099	.028 ± .082	.215
DGFI92A	.356 ± .151	.378 ± .126	.331

The next set of comparisons is at a set of stations, in four areas, whose positions were determined from GPS observations. The standard deviation of the differences along the traverses (described in Rapp and Pavlis, 1990) are given in Table 4.

Table 4

Standard Deviation of the Geoid Undulation Differences at GPS Stations in Four Areas. Units are cm.

Model	Area			
	Europe	Canada	Australia	Tennessee
OSU91A	33	36	35	21
GEM-T3	42	36	35	23
GRIM4-C2	41	44	26	19
DGFI92A	59	21	63	27

The DGFI92A results are interesting since they are quite poor in Europe and Australia, good in Canada and somewhat poorer (than the other models) for the Tennessee networks.

Comparisons were also made at the GPS stations using undulation differences. This differencing removes some of the long wavelength error in the geopotential model. The results for these comparisons are shown in Table 5. The results are given in terms of the root mean square difference and in parts per million (ppm) of the distance between the stations.

Table 5

Relative Geoid Undulation Comparisons Based on GPS Positioning

Model	Area							
	Europe		Canada		Australia		Tennessee	
	RMS	ppm	RMS	ppm	RMS	ppm	RMS	ppm
OSU91A	23	3.6	9	6.6	22	5.3	26	3.9
GEM-T3	24	3.7	9	6.6	22	5.3	26	3.9
GRIM4-C2	24	3.8	19	8.0	23	5.4	26	3.8
DGFI92A	26	4.4	9	6.3	29	6.2	28	3.7

Table 5 indicates that most solutions are comparable although the GRIM4-C2 model does not perform well in the Canadian test. The DGFI92A model, which performed poorly in Europe and Australia now is almost comparable to the other model results. This suggests there are long wavelength errors in the DGFI92A model which are deleted when differences are taken.

The last evaluation was carried out by comparing a geoid undulation from the augmented geopotential model with a corrected sea surface height implied by Geosat altimeter data. The corrections used were those from the GEM-T2 orbit improvement process and the reduction of a sea surface height to geoid undulation using the OSU91 sea surface topography model (RWP). The comparisons were done for the one year Geosat average track described in Wang and Rapp (1992) on an ocean wide basis and in a more restricted basis excluding data below -60° latitude and in the Mediterranean Sea. The latter comparison was carried out because data deletion in the GEM-T3 and GRIM4-C3 model development had an impact on the accuracy in these comparisons. Comparisons were made at points where at least 5 ERM values were used to form the average. Table 6 gives the standard deviation of the difference between the two geoid undulation estimates and the number of residuals that exceed 1.0 m.

Table 6

Comparison of the Model and One Year Geosat Average Geoid Undulations

Model	Ocean Area		Restricted Area	
	Std. Dev.	No. ≥ 1 m	Std. Dev.	No. ≥ 1 m
OSU91A	32 cm	18,142	32	15,913
GEM-T3	45	41,182	39	24,624
GRIM4-C2	54	57,706	47	36,530
DGFI92A	68	67,609	67	60,154

Based on these comparisons one sees that the OSU91A model gives the best fit to the geoid undulations implied by the Ohio State Geosat orbits and sea surface topography model. This may be understandable since the Geosat data was used directly in the development of the OSU91A model to degree 50 and the OSU91 sea surface topography model. The GEM-T3 augmented model gives better results in the restricted area comparison than in the full ocean because the altimeter data was deleted in certain areas in the T3 development. The GRIM4-C2 model did not directly use altimeter data in its solution and so the somewhat poorer agreements are not surprising. The DGFI92A model shows the largest standard deviation (± 67 cm) of the four models tested and in the restricted area shows 65% more residuals ≥ 1.0 m than the GRIM4-C2 model.

The large standard deviation shown in Table 6 for the DGFI92 model is inconsistent with the ± 29 cm difference reported by Gruber and Bosch (1992) in comparing their altimeter input data set (corrected for sea surface topography) with the geoid undulations implied by the DGFI92A model. This difference may relate to the use of the GEM-T2 orbits for the determination of the Geosat mean sea surface used in the DGFI92A development. RWP (Table 7) reported root mean square height corrections to the GEM-T2 orbit of ± 75 cm.

Figures 5 and 6 show the location of geoid undulation residuals greater than 1 m for the OSU91A and DGFI92A model, for the average Geosat track the Ohio State orbits. The 18,142 residuals with the OSU91A model are primarily associated with areas of high frequency gravity signal, although there are some points that clearly lie n a ground track. Figure 6 also shows the 67,609 residuals from the DGFI92A solution in the high frequency areas and in some coastal regions.

7. Conclusion

The purpose of this paper is to describe how satellite altimeter data can be, and has been, used in the development of global gravity models. We have seen that there are a variety of ways to do this depending on the

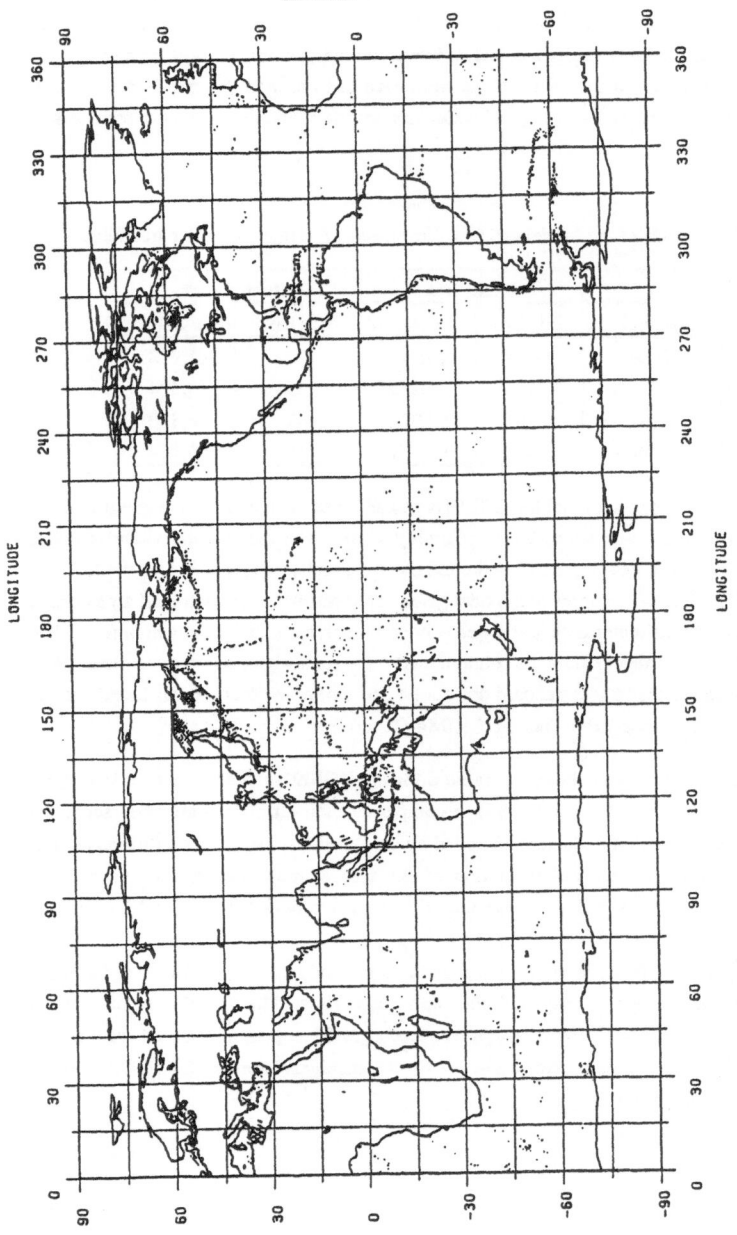

Figure 5. Location of 18,142 Points on the Geosat Track Where the Corrected Geoid Model, OSU91A, Minus the Corrected Sea Surface Height is ≥ 1.0 m.

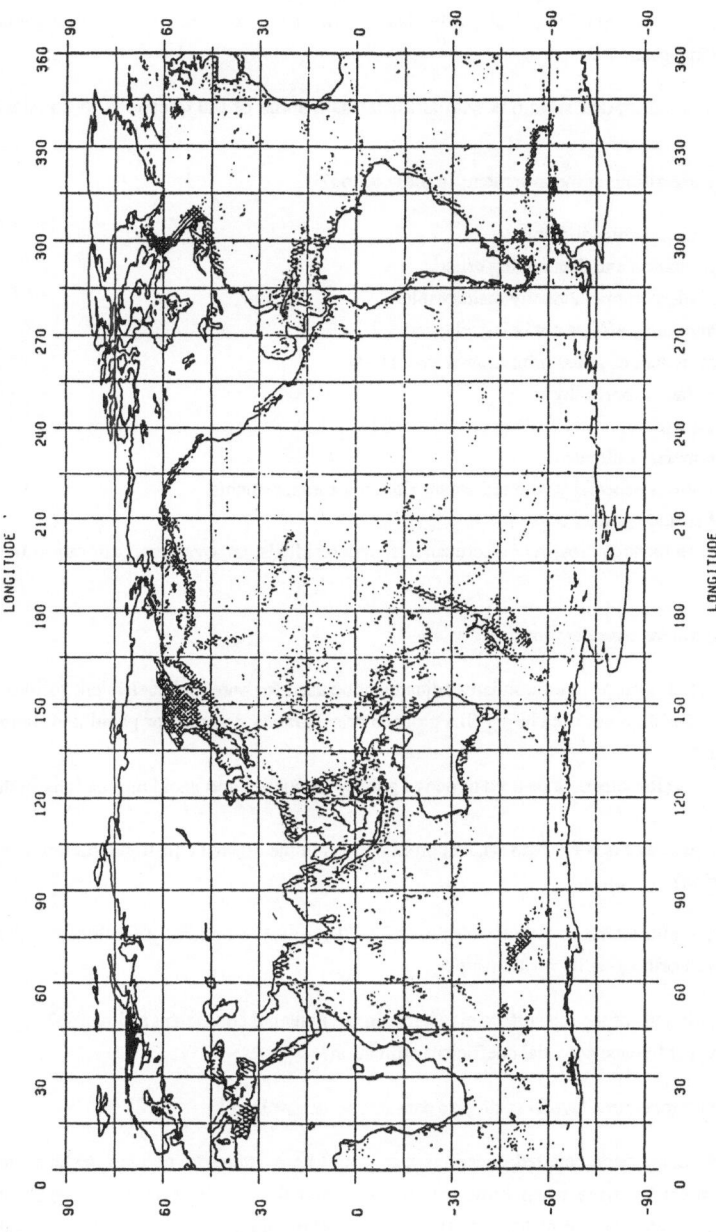

Figure 6. Location of 67,609 Points on the Geosat Track Where the Corrected Geoid Model, DGFI92A, Minus the Corrected Sea Surface Height is ≥ 1.0 m.

goals of the solution. The discussion in the paper is oriented towards the development of high (n = 360) degree potential coefficient models that incorporate surface gravity data, satellite altimeter data, other satellite tracking data or an a priori potential coefficient model.

In carrying out the combination solutions with altimeter data some of the questions to consider are as follows:

1. Do you wish to directly use altimeter measurements in the solution?

2. If yes to 1 then some pertinent consideration:
   A. Sample point determination and data editing criteria.
   B. Highest degree solution computationally manageable.
   C. Removal of high frequency effects not being adjusted.
   D. General orbit improvement or radial orbit improvement only.
   E. Treatment of sea surface topography:
      E1: Assume known
      E2: Model parameters estimated
   F. Weighting considerations especially with redundant altimeter measurements.
   G. Appropriate use of surface gravity data.
   H. Evaluation of solution through crossover discrepancies, data residuals, arc overlaps, calibration factors, etc.

3. If no to 1, then some pertinent considerations:

   A. Will orthogonality relationships or least squares adjustment be used for potential coefficient solution?
   B. If orthogonality relationships are used how will a global a priori gravity anomaly or geoid undulation database be formed?
      B1: How will satellite altimeter data be used to derive gravity anomalies and geoid undulations in the ocean areas?
   C. If a least squares adjustment is performed what approximations in the normal equation structure can lead to an acceptable solution?

4. How can the direct use of altimeter data for a low degree (50 to 70) model be optimally combined with the high degree solution with orthogonality relationship?

5. When does temporal variations of the sea surface play a role in the potential coefficient estimation?

6. How do ocean tide errors influence potential coefficient estimation?

7. Can optimum weighting procedures involving diverse data sets be determined?

Although there are many more detailed questions that would be written, the above questions show the variety of decisions that must be made when using satellite altimeter data to develop high degree potential coefficient models. The altimeter data provides a wealth of information, both gravimetric and oceanographic.

This data can be used to infer, with the help of other data, gravity field information at a variety of wavelengths, down even to 5 km if sufficient altimeter data is available. The proper treatment of such data is a continuing challenge to the scientific community.

Acknowledgment

The preparation of this paper was supported, in part, under NASA's TOPEX Altimeter Research in Ocean Circulation Mission funded through the Jet Propulsion Laboratory under contract 958121 with The Ohio State University Research Foundation. Dr. Yan Ming Wang provided most of the numerical results reported in Chapter 7.

8. References

Basíc, T. and R. Rapp, Oceanwide predictions of gravity anomalies and sea surface heights using Geos-3, Seasat and Geosat altimeter data, and ETOPO5U bathymetric data, Report No. 416, Dept. of Geodetic Science and Surveying, The Ohio State University, Columbus, 1991.

Bosch, W., High degree spherical harmonic analysis by least squares, presented at the XIX IUGG General Assembly, (DGFI, München), August 1987.

Brenner, A.C., C.J. Koblinsky and B.C. Beckley, A preliminary estimate of geoid induced variations in repeat orbit altimeter satellite observations, J. Geophys. Res., 95, 13, 3033-3040, 1991.

Chelton, D., WOCE/NASA Altimeter Algorithm Workshop, O.S. WOCE Technical Report No. 2, 70 pp., U.S. Planning Office for WOCE, College Station, TX, 1988.

Chelton, D.B. and M.G. Schlax, Spectral characteristics of time-dependent orbit errors in altimeter height measurements, manuscript submitted to J. Geophys. Res., Oceans, April 1992.

Colombo, O.L., Numerical methods for harmonic analysis on the sphere, Report No. 310, Dept. of Geodetic Science and Surveying, The Ohio State University, Columbus, 1981.

Colombo, O.L., Altimetry, orbits and tides, NASA Tech. Memo. 86180, Greenbelt, MD, 1984.

Colombo, O.L., The dynamics of global positioning system orbits and the determination of precise ephemerides, J. Geophys, Res., 94, 9167-9182, 1989.

Cruz, J., Ellipsoidal corrections to potential coefficients obtained from gravity anomaly data on the ellipsoid, Report No. 317, Dept. of Geodetic Science and Surveying, The Ohio State University, Columbus, 1986.

Denker, H. and R.H. Rapp, Geodetic and oceanographic results from the analysis of 1 year of Geosat data, J. Geophys. Res., 95, C8, 13,151-13,168, 1990.

DMA, Supplement to Department of Defense World Geodetic System 1984 Technical Report, Part I, Methods, Techniques, and Data Used in WGS84 Development, DMA, Washington, D.C., 20305-3000, 1987.

Engelis, T., Radial orbit error reduction and sea surface topography determination using satellite altimetry, Report No. 377, Dept. of Geodetic Science and Surveying, The Ohio State University, Columbus, 1987a.

Engelis, T., Spherical harmonic expansion of the Levitus sea surface topography, Report No. 385, Dept. of Geodetic Science and Surveying, The Ohio State University, Columbus, 1987b.

Farelly, B. The geodetic approximations in the conversion of geoid height to gravity anomaly by Fourier transform, Bulletin Geodesique, 65, 2, 92-101, 1991.

Forsberg, R., A study of terrain reductions, density anomalies and geophysical inversion methods in gravity field modeling, Report No. 365, Dept. of Geodetic Science and Surveying, The Ohio State University, Columbus, 1984.

Gleason, D.M., Comparing ellipsoidal corrections to the transformation between the geopotential's spherical and ellipsoidal spectrums, manusc. geod., 13, 114-129, 1988.

Gruber, T. and W. Bosch, A new 360 gravity field model, presented at the European Geophysical Society meeting, Edinburgh, April 1992.

Haagmans, R., Detailed gravity anomalies derived from Seasat altimeter data, a comparison of two alternative approaches; least squares collocation and a method based on FFT, Afstudeerscriptel, August 1988.

Heiskanen, W. and H. Moritz, Physical Geodesy, W.H. Freeman, New York, 1967.

Hwang, C., High precision gravity anomaly and sea surface height estimation from Geos-3/Seasat satellite altimeter data, Report No. 399, Dept. of Geodetic Science and Surveying, The Ohio State University, Columbus, 1989.

Hwang, C., Orthogonal functions over the oceans and applications to the determination of orbit errors, geoid and sea surface topography from satellite altimetry, Report No. 414, Dept. of Geodetic Science and Surveying, The Ohio State University, Columbus, December 1991.

International Association of Geodesy, Geodetic reference system 1967, Spec. Publ. Bull. Geod., p. 72, Paris, 1971.

Jekeli, C., The exact transformation between ellipsoidal and spherical harmonic expansions, manusc. geod., 13, 106-113, 1988.

Katsambalos, K.E., The effect of the smoothing operator on potential coefficient determinations, Report No. 287, Dept. of Geodetic Science and Surveying, The Ohio State University, Columbus, 1979.

Kaula, W.M., Tests and combinations of satellite determinations of the gravity field with gravimetry, J. Geophys. Res., 71, 5303-5314, 1966.

Knudsen, P., Estimation of sea surface topography in the Norwegian sea using gravimetry and Geosat altimetry, Bulletin Geodesique, 66, 1, 27-40, 1992.

Koblinsky, C.J., L.E. Braata, T.L. Engelis, S.M. Klosko, and R.G. Williamson, Geocenter definition in the determination of dynamic height using Geosat satellite altimetry (abstract), Eos Trans. AGU, 70, (43), 1051, 1989.

Lerch, F.J. et al., Geopotential models of the Earth from satellite tracking, altimeter and surface gravity observations: GEM-T3 and GEM-T3S, NASA Technical Memorandum 104555, Goddard Space Flight Center, Greenbelt, MD, 1992.

LeTraon, P.Y., C. Boisser, and P. Gaspar, Analysis of errors due to polynomial adjustment of altimeter profiles, J. Atmos. Oceanic Technol., 8, 385-396, 1991.

Levitus, S., Climatological Atlas of the World Ocean, NOAA Professional Paper 13, U.S. Govt. Printing Office, Washington, D.C., 1982.

Marsh, J.G. et al., Dynamic sea surface topography, gravity, and improved orbit accuracies from the direct evaluation of Seasat altimeter data, J. Geophys. Res., 95, C8, 13, 129-13, 1990.

McAdoo, D. and K. Marks, Gravity fields of the southern oceans from Geosat data, J. Geophys. Res., 97, B3, 3247-3260, 1992.

Moritz, H., Advanced Physical Geodesy, Herbert Wichmann, Karlsruhe, Germany, 1989.

National Geophysical Data Center, ETOPO5, Digital relief of the surface of the Earth, Report 86-MGG-07, Boulder, Colorado, 1986.

Nerem, R.S. and C.J. Koblinsky, The geoid and ocean circulation, in Geophysical Interpretation of the Geoid, ed. Vaníček and Christou, CRC Press, Boca Rotan, Fl., 1992.

Pavlis, N.K. and R.H. Rapp, The development of an isostatic gravitational model to degree 360 and its use in global gravity modelling, Geophys. J. Int., 100, 369-378, 1990.

Pavlis, N.K., Modeling and estimation of a low degree geopotential model from terrestrial gravity data, Report No. 386, Dept. of Geodetic Science and Surveying, The Ohio State University, Columbus, 1988.

Rapp, R.H. and J.Y. Cruz, Spherical harmonic expansions of the Earth's gravitational potential to degree 360 using 30' mean anomalies, Report No. 376, Dept. of Geodetic Science and Surveying, The Ohio State University, Columbus, 1986a.

Rapp, R.H. and N.K. Pavlis, The development and analysis of geopotential coefficient models to spherical harmonic degree 360, J. Geophys. Res., 95, B13, 21,885-21,911, 1990.

Rapp, R.H. and Y.M. Wang, Geoid undulation differences between geopotential models, Surveys in Geophysics (in press), 1992.

Rapp, R.H., Combination of satellite, altimetric and terrestrial gravity data, in Theory of Satellite Geodesy and Gravity Field Determination, edited by F. Sanso and R. Rummel, pp. 261-284, Springer-Verlag, New York, 1989.

Rapp, R.H., GEOS 3 data processing for the recovery of geoid undulations and gravity anomalies, J. Geophys. Res., 84, 3784-3792, 1979.

Rapp, R.H., Gravity anomalies and sea surface heights derived from a combined GEOS 3/Seasat Altimeter Data Set, J. Geophys. Res., 91, B5, 4867-4876, 1986.

Rapp, R.H., The Earth's gravity field to degree and order 180 using Seasat altimeter data, terrestrial gravity data, and other data, Report No. 332, Dept. of Geodetic Science and Surveying, The Ohio State University, Columbus, 1981.

Rapp, R.H., Y.M. Wang and N.K. Pavlis, The Ohio State 1991 geopotential and sea surface topography harmonic coefficient models, Report 410, Dept. of Geodetic Science and Surveying, The Ohio State University, Columbus, December 1991.

Reigber, C. et al., GRIM4-C1 and C2; Combination solutions of the global earth gravity field, presentation at the meeting of the European Geophysical Society, Wiesbaden, Germany, April 1991.

Rummel, R. and R. Rapp, Undulation and anomaly estimation using Geos-3 altimeter data without precise satellite orbits, Bulletin Geodesique, 51, 73-88, 1977.

Sandwell, D.T. and D.C. McAdoo, Marine gravity of the Southern Ocean and Antarctic margin from Geosat, J. Geophys, Res., 93, B9, 10,389-10.396, 1988.

Schwintzer, P. et al., A new earth gravity field model in support of ERS-1 and SPOT-2: GRIM4-S1/C1, Final Report to the German Space Agency (DARA) and the French Space Agency (CNES), 1991.

Suenkel, H. Gravity field determination, Gerlands Beitraege zur Geophysik 96, 54-74, Leipzig, 1987.

Tai, C.K., Accuracy assessment of widely used orbit error approximations in satellite altimetry, J. Atmos. Oceanic Technol., 6, 147-150, 1988.

Tscherning, C.C. and R. Forsberg, Geoid determinations in the Nordic countries from gravity and height data, Bulletino di Geodesia e Science Affini, 21-43, Florence, Italy, 1987.

Uotila, U.A., Notes on adjustment computations, Part I. Dept. of Geodetic Science and Surveying, The Ohio State University, Columbus, 1986.

Wagner, C.A., Radial variations of a satellite orbit due to geopotential errors: Implications for satellite altimetry, J. Geophys. Res., 90, 3027-3036, 1985.

Wang, Y.M. and R.H. Rapp, Geoid gradients for Geosat and Topex/Poseidon repeat ground tracks, Report No. 408, Dept. of Geodetic Science and Surveying, The Ohio State University, Columbus, June 1991.

Wang, Y.M. and R.H. Rapp, The determination of a one year mean sea surface height track from Geosat altimeter data and ocean variability implementations, Bulletin Geodesique, (in press), 1992.

Wang, Y.M., Downward continuation of the free-air gravity anomalies to the ellipsoid using the gradient solution, Poisson's integral and terrain correction: Numerical comparison and the computations, Report No. 393, Dept. of Geodetic Science and Surveying, The Ohio State University, Columbus, 1988.

Wenzel, H.G., Hochauflösende Kugelfunktionsmodelle für das Gravitationspotential der Erde, Wiss. Arb. 137, Fachrichtung Vermess. der Univ. Hannover, Hannover, Federal Republic of Germany, 1985.

Wichiencharoen, C., FORTRAN programs for computing geoid undulations from potential coefficients and gravity anomalies, internal report, Dept. of Geodetic Science and Surveying, The Ohio State University, Columbus, 1982.

SEMINARS

THE DIRECT ESTIMATION OF THE POTENTIAL COEFFICIENTS
BY BIORTHOGONAL SEQUENCES

M.A. Brovelli[*] and F. Migliaccio[**]
[*]Istituto Nazionale di Geofisica
[**]Dipartimento di Ingegneria Idraulica, Ambientale e del Rilevamento
International Geoid Service - Politecnico di Milano
Piazza Leonardo da Vinci, 32 - 20133 Milano - Italy

1. Introduction

The knowledge of the gravity field of the earth and its anomalies provides fundamental information for the construction of physical models of its internal structure and behaviour. Up to now, the techniques developed and the methods used to determine the parameters describing the variations of the field in the space domain have produced models having different resolutions and precisions.

The classical methods to determine global gravity models consist in combining, in the least squares sense, different systems of normal equations coming both from the perturbation analysis of different satellites arcs and from data collected on the earth surface. Therefore, the data used to estimate the geopotential coefficients are mainly:

1) on the earth surface: gravimetric and astrogeodetic (vertical deflections) measurements;

2) analysis of the perturbations of satellites arcs;

3) radaraltimetric measurements.

The first coefficients determined with this approach were obtained by W. Kaula (1966); since then several models were computed and afterwards improved by different research groups in Europe and in the USA (e.g. the French- German GRIM models, the GSFC models, the Hannover University models and the OSU models).

The different solutions are due to the fact that the various groups make use of different series of data, different procedures and different numerical methods; it has been proved (F. Migliaccio and F. Sansò, 1989) that this gives rise to relative differences between the models already amounting at 100 % at degree $l = 25$.

Besides, it has to be remembered that the classical satellite methods only provide the low frequency part of the field, while the data at ground level provide the high resolution part, thus leaving a lack of information at medium wavelengths (typically in the range of 100 to 1000 km): this makes very interesting the development of new space techniques (like Satellite to Satellite Tracking or Satellite Gravity Gradiometry) with satellites at low altitude.

The work presented in this seminar can be considered within the limits of classical methods for global models estimation.

The original part is represented by the new approach used to determine the gravity model: starting from the boundary value problem of altimetry-gravimetry, the geopotential coefficients are directly computed from the observations by means of a proper bi-orthogonal series: the idea is already present, although not presented in the same mathematical frame, in a paper by A. Mainville (1987).

The first part of this report is devoted to introducing and discussing the method adopted from the theoretical point of view, while in the second part a numerical example is presented (with data simulated from the model OSU91A), in order to test the reliability of the procedure.

2. The Altimetry-Gravimetry problems

The problems of altimetry-gravimetry are part of the integral methods allowing for the estimate of the geoid starting from geodetic measurements referring to the earth surface, considered as a boundary surface for the gravity potential.

The problem was firstly formulated and tackled by Stokes who, at the end of the 19th century, found an explicit solution on the sphere. Significant developments of this theory were subsequently due to Molodensky (M. S. Molodensky et al., 1962) by setting a more direct problem where the reference surface is the unknown earth's surface.

According to this problem, the gravity potential w and the gravity vector g are determinable (and therefore known) quantities on the earth's surface S, which is unknown; starting from these assumptions it is possible to solve for the surface S and the potential w outside it.

In recent times, the development of geodesy (in particular satellite geodesy) has given further impulse to integral methods leadings to the introduction of new problems, namely the altimetry-gravimetry problems. In fact, while satellite radaraltimetry allows to determine the shape of the oceans surface, by means of marine gravimetry techniques high precision measures of gravity or of gravity disturbances δg on sea can be obtained. One can therefore think of dividing the surface S into two parts: S_S (on sea), which is supposed to be known, and S_L (on land), which is unknown.

On S two different kinds of boundary data are given, depending on the point, which belongs either to S_S or to S_L .

Besides, two altimetry-gravimetry boundary problems are given, depending on the kind of data given on S_S .

In the first problem (F. Sansò, 1981) the boundary data are: the gravity potential w_L and the gravity vector $g_L = \nabla w$ on S_L (unknown), and the gravity potential $\bar{w}_S =$ constant on $S_S = S - S_L$ (known).

From these data one wants to determine S_L and $u = w - \frac{1}{2}\omega^2(x^2 + y^2)$ such that:

$$
\begin{cases}
\nabla^2 u = 0 & \text{outside } S \\[2mm]
u = O\left(\dfrac{1}{|x|}\right) & |x| \to \infty \\[4mm]
\left.\begin{array}{l} w = w_L \\[1mm] \nabla w = g_L \end{array}\right\} & \text{on } S_L \\[4mm]
w = \bar{w}_S & \text{on } S_S .
\end{cases}
\tag{2.1}
$$

Now if (2.1) is linearized taking as reference surface a telluroid Σ such that $\Sigma_S = S_S$ and if the gravity potential w is approximated with the normal potential U ($w = U + T$), on Σ_L the following boundary condition is obtained:

$$
(T + m\cdot\nabla T)\Big|_{\Sigma_L} = m\cdot \Delta g
\tag{2.2}
$$

where m is the iso-zenithal field defined by:

$$
m = - [\nabla\underline{\gamma}]^{-1}\cdot \underline{\gamma}
\tag{2.3}
$$

and $\underline{\gamma} = \nabla U$ is the normal gravity vector.

Instead, on Σ_S the problem is already linear and we have:

$$
T\Big|_{\Sigma_S} = \delta w_S .
\tag{2.4}
$$

Let's now introduce a simplification (spherical approximation) supposing the normal gravity vector being of the type:

$$
\underline{\gamma} = - \frac{K}{r^2}\frac{r}{r}
\tag{2.5}
$$

with:

K = constant;

$\frac{r}{r}$ = radial direction with respect to the center of mass of the telluroid.

Therefore the first problem of altimetry-gravimetry in spherical approximation can be expressed in the simplified form:

$$
\begin{cases}
\nabla^2 T = 0 & \text{outside } \Sigma \\[2mm]
T = O\!\left(\dfrac{1}{r}\right) & r \to \infty \\[2mm]
-\dfrac{2T}{r} - \dfrac{\partial T}{\partial r}\Big|_{\Sigma_L} = \Delta g_L & \text{on } \Sigma_L \\[2mm]
T\big|_{\Sigma_S} = \delta w_s & \text{on } \Sigma_S \ .
\end{cases}
\tag{2.6}
$$

In the second problem of altimetry-gravimetry (P.Holota, 1980) the boundary data are the same as in the first one (w_L and g_L) on S_L, while on S_S the known quantity is $g_S = |\nabla w|$; thus after linearizing and in spherical approximation this problem is represented by:

$$
\begin{cases}
\nabla^2 T = 0 & \text{outside } \Sigma \\[2mm]
T = O\!\left(\dfrac{1}{r}\right) & r \to \infty \\[2mm]
-\dfrac{2T}{r} - \dfrac{\partial T}{\partial r}\Big|_{\Sigma_L} = \Delta g_L & \text{on } \Sigma_L \\[2mm]
-\dfrac{\partial T}{\partial r}\Big|_{\Sigma_S} = \delta g_S & \text{on } \Sigma_S \ .
\end{cases}
\tag{2.7}
$$

The two altimetry-gravimetry problems have different applications depending on the actual available data on the surface.

Typically the second problem is used for the local estimate of the geoid (particularly in the case of islands of small or medium dimensions; for example, regarding the Mediterranean sea, Sicily and Sardinia).

Instead, the first problem can have more general applications; in fact the boundary datum on the sea surface is the anomalous potential T, which can be obtained by treating radaraltimetric observations.

As these observations are (by their nature) homogeneous and equally distributed, it is possible, as we will see, to utilize the first problem for the estimate of global models of the potential.

3. The estimate of the geopotential coefficients with the first altimetry-gravimetry problem

Let's reconsider the altimetry-gravimetry problem in the simplified form obtained in the case of a spherical approximation. We must refer to expression (2.6).

The boundary conditions can be synthesized by introducing the boundary operator B such that:

$$BT = \chi_S T + \frac{\sigma_T}{\sigma_{\Delta g}} \left(- \frac{\partial T}{\partial r} - \frac{2}{R} T \right) \chi_L \qquad (3.1)$$

where: χ_S = characteristic function of sea, that is:

$$\chi_S = \begin{cases} 1 & \text{if } P \in \Sigma_S \\ \\ 0 & \text{if } P \notin \Sigma_S \end{cases} \qquad (3.2)$$

χ_L = characteristic function of land;
R = average radius of the earth;
$\frac{\sigma_T}{\sigma_{\Delta g}}$ = regularization factor, accounting for the two different kinds of boundary data.

Expressing T in a spherical harmonic expansion:

$$T = \sum_{lm} T_{lm} Y_{lm} \qquad (3.3)$$

and applying the operator B, we obtain:

$$BT = \sum_{lm} T_{lm} BY_{lm} = \sum_{lm} T_{lm} \left\{ \chi_S + \frac{\sigma_T}{\sigma_{\Delta g}} \cdot \frac{1 - l}{R} \chi_L \right\} Y_{lm} \qquad . \qquad (3.4)$$

Let's indicate the boundary data with f:

$$f = \chi_S \cdot T_o + \frac{\sigma_T}{\sigma_{\Delta g}} \cdot \Delta g_o \cdot \chi_L \qquad (3.4)$$

where T_o and Δg_o are the observations (on sea and on land respectively) of the anomalous potential and of the gravity anomaly.
Now considering the boundary equation:

$$BT = f \qquad (3.6)$$

we want to directly estimate the geopotential coefficients T_{lm} from the observations f. This is possible, as we will see, if it exists and if it is possible to construct a sequence biorthogonal to the sequence $\{BY_{lm}\}$.
Let's then briefly introduce the biorthogonal sequences and their main properties related to our purposes.
The sequences $\{a_i\}$ and $\{b_i\}$ form a biorthogonal system in a Hilbert space $\mathcal{H}$ if:

$$\langle a_i, b_j \rangle = \delta_{ij} \quad . \tag{3.7}$$

If $\{a_i\}$ is a complete sequence in $\mathcal{H}$, also $\{b_i\}$ is complete; every element h of $\mathcal{H}$ can therefore be expressed either by:

$$h = \sum_i \langle h, a_i \rangle \, b_i \tag{3.8}$$

or by:

$$h = \sum_i \langle h, b_i \rangle \, a_i \quad . \tag{3.9}$$

Let's now consider (3.9) and let's take as space $\mathcal{H}$ the one spanned by the Y_{lm}, that is a space of functions which are square integrable on the sphere.
If we can construct a sequence $\{Z_{pq}\}$ biorthogonal to $\{BY_{lm}\}$, it will be:

$$f = \sum_{lm} \langle f, Z_{lm} \rangle \, BY_{lm} \tag{3.10}$$

and consequently:

$$T_{lm}(P) = \langle f, Z_{lm} \rangle = \frac{1}{4\pi} \int_{\sigma} f(P) \, Z_{lm} d\sigma \quad . \tag{3.11}$$

But before constructing the biorthogonal series $\{Z_{pq}\}$ its existence must be studied. This implies studying the characteristics of the operator B; in F. Sansò (1992) it is explained that the operator B has the property of Fredholm's alternative, so, if $\{Y_{lm}\}$ is an orthonormal basis in the space $\mathcal{H}$, a series biorthogonal to $\{BY_{lm}\}$ must exist.
Such a series is a solution of system:

$$\langle Z_{pq}, BY_{lm} \rangle = \delta_{pl} \delta_{qm} \quad . \tag{3.12}$$

However the problem, formulated in this way, is not solvable because there are infinite unknown elements Z_{pq}.
If we want to practically determine the sequence $\{Z_{pq}\}$ the system (3.12) must be truncated; this implies the hypothesis that T is of finite degree:

$$T = \sum_{l=0}^{l_{max}} \sum_{m=-l}^{l} T_{lm} \, Y_{lm} \quad ; \tag{3.13}$$

so we look for the biorthogonal sequence $\{Z_{pq}\}$ with $p = 1,\ldots,l_{max}$ and $-1 \leq q \leq 1$ which satisfies the condition:

$$\frac{1}{4\pi} \int_\sigma Z_{pq} BY_{lm} d\sigma = \delta_{pl} \delta_{qm} \quad . \tag{3.14}$$

In order to simplify the mathematical notation, let's also rearrange the harmonics associating to the indices l and m the index j and let's introduce the new symbols:

$$BY_{lm} = W_j$$
$$Z_{pq} = b_k \quad .$$

which take into account the previous correspondence.

In this case (3.14) simply becomes:

$$\frac{1}{4\pi} \int_\sigma b_k(P) \ W_j(P) d\sigma = \delta_{kj} \quad . \tag{3.15}$$

Now coming to the construction of the sequence, let's first of all observe that, given $\{W_j\}$, $j = 1,\ldots, L$, $\{b_k\}$ isn't univocally determined.

In fact, if we define W the space spanned by the $\{W_j\}$ and $W^\perp$ the orthogonal complement of W, if the following condition holds:

$$<b_k, W_j> = \delta_{kj} \tag{3.16}$$

it is also:

$$<b_k + w^\perp, W_j> = \delta_{kj} \qquad \forall \ w^\perp \in W^\perp \quad . \tag{3.17}$$

Then it is necessary to define a criterion in order to make the b_k unique.

A possible criterion is for example the minimum norm: from it comes that the biorthogonal sequence b_k necessarily belongs to the sub-space W. In fact if we consider any $b'_k = b_k + b_k^\perp$, with $b_k \in W$ and $b_k^\perp \in W^\perp$, condition (3.16) is still satisfied, but we have:

$$|| b' || = \int_\sigma b_k'^2 d\sigma > \int_\sigma b_k^2 \ d\sigma = || b || \quad . \tag{3.18}$$

As a consequence of what we have seen, applying the minimum norm criterion the b_k can be obtained as those linear combinations of the W_j

$$b_k = \sum_{j=1}^{L} \mu_j^k \ W_j \tag{3.19}$$

which satisfy equations (3.15), which in this case are given by:

$$\frac{1}{4\pi} \int_\sigma b_k \ W_j \, d\sigma = \frac{1}{4\pi} \int_\sigma \sum_{l=1}^{L} \mu_l^k \ W_l W_j \, d\sigma = \sum_{l=1}^{L} \frac{\mu_l^k}{4\pi} \int_\sigma W_l W_j \, d\sigma = \delta_{kj} \tag{3.20}$$

and for which holds:

$$\int_\sigma b_k^2 \, d\sigma = \min.$$

Though in this way from the analytical point of view the problem of constructing the b_k is correctly set, it hasn't anyway an easy solution; then it is better, in the determination of the biorthogonal sequence, to take into account the fact that our goal is to use the sequence in order to estimate the $\hat{T}_k$, given by the relation:

$$\hat{T}_k = \frac{1}{4\pi} \int_\sigma b_k \ f(P) \ d\sigma_P \quad,$$

or better, if we want to practically compute the coefficients, by the discretized expression:

$$\hat{T}_k = \frac{1}{4\pi} \sum_{i=1}^{M} b_k(P_i) \ \bar{f}(P_i) \ S_i \tag{3.21}$$

in which S_i are the areas, of center P_i, forming the grid used to discretize the earth's surface and M is the number of areas. In (3.21) we have introduced the notation:

$$\frac{1}{S_i} \int_{S_i} f(P) d\sigma = \bar{f}(P_i) = \bar{f}_i \quad . \tag{3.22}$$

Starting from (3.22) is then more useful to think of the b_k as represented by:

$$b_k(P) = \sum_{i=1}^{M} \lambda_i^k \ \chi_i(P) \tag{3.23}$$

with:

$$\chi_i(P) = \begin{cases} 1 & \text{if } P \in S_i \\ \\ 0 & \text{if } P \notin S_i \end{cases} \quad .$$

In this way we directly have:

$$\frac{1}{4\pi} \int_\sigma b_k \, f(P) \, d\sigma_P = \frac{1}{4\pi} \sum_{i=1}^M \lambda_i^k \int_{S_i} f(P) d\sigma = \frac{1}{4\pi} \sum_{i=1}^M \lambda_i^k \, S_i \, \overline{f}(P_i) \qquad (3.24)$$

so that the estimate of the T_k is directly expressed as a function of the observables.

Yet if we put, as it is natural, $M > L$ (that is the number of the areas, and then of the observations, larger than the number of the coefficients to be estimated) the λ_i^k coefficients are still not unique and the problem consists also in this case in applying a minimum norm criterion, that is:

$$\int_\sigma b_k^2 \, d\sigma = \frac{1}{4\pi} \sum_{i=1}^M (\lambda_i^k)^2 S_i = \sum_{i=1}^M (\lambda_i^k)^2 p_i = \min \qquad (3.25)$$

with $p_i = \dfrac{S_i}{4\pi}$.

Let's minimize the previous relation by means of the Lagrange multipliers technique, considering as conditions the biorthogonality relations, given (for k fixed and j=1,...,L) by:

$$\frac{1}{4\pi} \int_\sigma \sum_{i=1}^M \lambda_i^k \, \chi_i \, W_j d\sigma = \sum_{i=1}^M \lambda_i^k \, p_i \, \frac{1}{S_i} \int_{S_i} W_j d\sigma = \delta_{kj} \qquad (3.26)$$

Setting:

$$\frac{1}{S_i} \int_{S_i} W_j d\sigma = A_{ij} \qquad (3.27)$$

we will have:

$$\frac{1}{2} \sum_{i=1}^M (\lambda_i^k)^2 p_i - \sum_{j=1}^L \gamma_j^k \left(\sum_{i=1}^M \lambda_i^k \, p_i \, A_{ij} \right) = \min \qquad (3.28)$$

in which the γ_j^k are the Lagrange multipliers.

Minimizing (3.28) we have, $\forall$ i:

$$\lambda_i^k \, p_i - p_i \sum_{j=1}^L \gamma_j^k \, A_{ij} = 0$$

that is:

$$\lambda_i^k = \sum_{j=1}^L \gamma_j^k \, A_{ij} \qquad (3.29)$$

Now sobstituting expression (2.28) in the orthogonality conditions, the following system is obtained:

$$\sum_{i=1}^{L} \gamma_i^k \left(\sum_{i=1}^{M} p_i A_{i1} A_{ij} \right) = \delta_{kj} \tag{3.30}$$

which can be written in the compact form:

$$\Gamma \cdot N = I \tag{3.31}$$

by introducing the matrices:

$$\Gamma = [\gamma_i^k]$$

$$N = A^+ P A, \quad A = [A_{ij}] \text{ and } P = [p_i \delta_{ij}] \quad .$$

The solution of (3.31) gives the estimate of the Lagrange multipliers. This passage, involving problems related to the inversion of matrix N, in general full and of large dimensions (see § 4), can be avoided. In fact let's suppose for the moment that the Lagrange multipliers are known; the sequence biorthogonal to the $\{W_j\}$ is then given by:

$$\hat{b}_k = \sum_{i=1}^{M} \frac{1}{p_i} \sum_{i=1}^{L} \hat{\gamma}_i^k A_{i1} \quad . \tag{3.32}$$

Since the $\{b_k\}$ is biorthogonal to the $\{W_j\}$ we know that:

$$\hat{T}_k = \langle \hat{b}_k, f \rangle = \frac{1}{4\pi} \int_\sigma \sum_{i=1}^{M} \hat{\lambda}_i^k \chi_i(P) f(P) \, d\sigma = \sum_{i=1}^{M} \hat{\lambda}_i^k p_i \bar{f}(P_i) \quad ; \tag{3.33}$$

substituting in the previous expression the $\hat{\lambda}_i^k$ given by (3.29) we obtain:

$$\hat{T}_k = \sum_{i=1}^{L} \hat{\gamma}_i^k \sum_{i=1}^{M} p_i A_{i1} \bar{f}(P_i) \quad . \tag{3.34}$$

Then the geopotential coefficients are obtained from

$$T = \Gamma d \quad , \tag{3.35}$$

with $T = [T_1 \ \ldots \ T_L]$ and $d = A^+ P \bar{f}$, $\bar{f} = [\bar{f}(P_1) \ \ldots \ \bar{f}(P_M)]$, which simply means solving the system:

$\Gamma^{-1} T = d$,

and therefore:

$\mathbb{N} T = d$.   (3.36)

4. A numerical experiment

In this paragraph we shall present a numerical experiment which was performed in order to check the validity of the equations obtained in the former part of the paper.

The theoretical formulas were converted into suitable algorithms and implemented in a Fortran program running on a Unisys 2200 computer.

As this was just a first test of the theory, no real data were used. Instead, the observations were simulated according to a scheme which, although simplified, was a quite realistic one.

The earth was covered with a regular geographical grid having block size equal to $5° \times 5°$: as a consequence the coast-lines were approximated in a square shape way. The blocks were superimposed on a planisphere obtained with a Mercator cylindrical projection.

An observation was supposed to be performed at the center of each block, representing the mean value of all observation performed on that block area according to the formula (3.22).

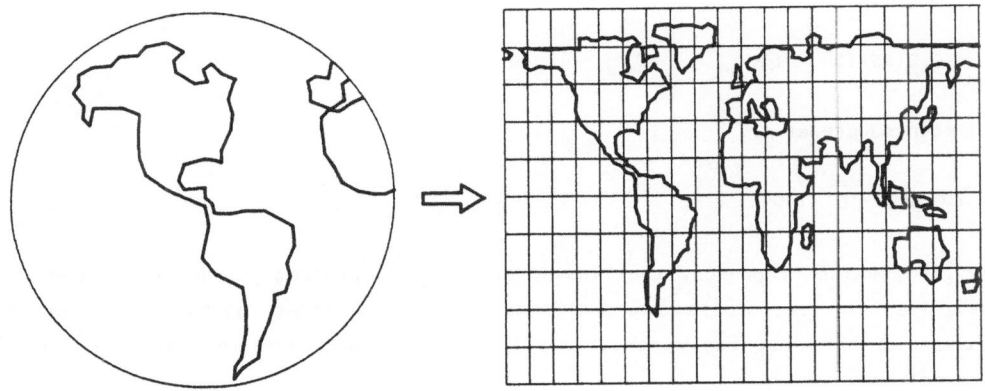

Fig. 1 - The $5° \times 5°$ grid superimposed on the planisphere.

Introducing the characteristic functions of sea and land the expression of the observation at the point P_1 can be written, for the purpose of programming, in the following way:

$$\overline{f}(P_1) = \chi_s T_o + \frac{\sigma_T}{\sigma_{\Delta g}} \cdot \Delta g_o \cdot \chi \quad . \tag{4.1}$$

The factor $\sigma_T/\sigma_{\Delta g}$ must be introduced in order to obtain a homogeneous set of data starting from the original T and Δg values.

In fact one must remember that the differences between T values at the sea level in meters and Δg values on land in mGals are of the order of 10^6. In particular, the values we simulated on sea and land (by means of software for harmonic synthesis on the sphere and using the OSU91A model with coefficients of degree 10 to 35) had the following characteristics:

$$\overline{T} = 0.2025 \cdot 10^{-7} \cdot \frac{GM}{R}$$

$$\sigma_T = 0.6034 \cdot 10^{-6} \cdot \frac{GM}{R}$$

$$\overline{\Delta g} = 0.3013 \cdot 10^{-13} \cdot \frac{GM}{R}$$

$$\sigma_{\Delta g} = 0.1284 \cdot 10^{-11} \cdot \frac{GM}{R}$$

GM = gravitational constant x earth's mass = $3.986009 \cdot 10^{14}$ m^3/s^2;
R = mean radius of the earth = 6371000 m.

The $\overline{f}(P_1)$ values obtained according to the formula (4.1) had mean value:

$$\overline{f} = 0.2145 \cdot 10^{-7} \; GM/R$$

and mean square error:

$$\sigma_f = 0.6001 \cdot 10^{-6} \; GM/R.$$

The $\overline{f}(P_1)$ values represent the "known term" of our problem. Though of course it is not a least squares problem, we shall use the same notations and terminology of least squares as they fit our purpose equally well. So much for the simulation of the observations.

Now let's turn to the algorithms for the solution of the altimetry-gravimetry problem with the use of biorthogonal sequences. What we want to achieve is the estimate of a geopotential model which, under our simplified assumptions, has coefficients ranging from degree 10 to degree 35.

The first step one has to consider is how to form the elements of the "design matrix" A. The A_{ij}, given by (3.27), represent the mean values of the harmonics up to degree and order L: in our case L = 35 (remember that we have also L_{min} = 10).

Using again the characteristic functions χ_S and χ_L, the value of W_j is given by:

$$W_j = \left\{ \chi_S + \frac{\sigma_T}{\sigma_{\Delta g}} \cdot \frac{1-1}{R} \chi_L \right\} Y_j \qquad (4.2)$$

where Y_j is the harmonic function j. Note that indexes $1, m, \alpha$ usually denoting a harmonic function have now conveyed into the single j index.

Index i refers to the S_i area on the sphere. A simple way (which of course introduces an approximation in the computations) to obtain the mean values of the harmonics is to use the Pellinen β_i factors:

$$\beta_i = \frac{2}{1 - \cos \psi} \frac{1}{2l + 1} \left[\bar{P}_{l+1}(\cos \psi) - \bar{P}_{l-1}(\cos \psi) \right] \qquad (4.3)$$

They represent the eigenvalues of a moving average operator: the field is averaged over a cap of radius ψ. The $\bar{P}_l(\cos \psi)$ are the Legendre polynomials, fully normalized. Multiplying each harmonic by the corrisponding β_j coefficient, one obtains the A_{ij} elements of the matrix A, according to the formula:

$$A_{ij} \cong P_i \left[\beta_j W_j(P_i) \right] \qquad (4.4)$$

where p_i, as already mentioned, is a weight accounting for the fact that the area of the blocks varys with the latitude.

Equation (4.4) directly allows for the calculation of the design matrix needed for the solution of the problem.

Nevertheless, one should remember what are the dimensions of that matrix: they are fixed by two parameters, the block size and the maximum degree of the model. In our case, with block size $5° x 5°$, the grid is divided into 36 x 72 = 2592 blocks represented by the corresponding 2592 rows of matrix A. Its columns correspond to the (sine and cosine) harmonics: 1196 in our case.

The memory requirement for such a matrix is 28 Mbytes (in double precision).

But luckily nearly half this storage area can be saved if the "normal matrix" N is directly computed. For larger problems this can be of great importance, as one can see from Table 1, where the dimensions of matrix A are reported, with the subsequent memory requirements (in Mbytes) for A and N.

block size	n°of rows	L	n°of coeff	memory requirements A	N
5°x 5°	2592	35	1369	28	15
3°x 3°	7200	60	3721	215	111
2°x 2°	16200	90	8281	1080	550
1°x 1°	64800	180	32761	17000	8600
0.5°x 0.5°	259200	360	130321	270000	136000

Table 1 - Memory requirements (Mbytes) for matrices A and N.

To directly compute both the normal and the known normal term is quite easy, realizing that the elements needed can be achieved summing the contributions coming from each row of matrix A: in this way only one row at a time of A has to be computed and what's more it doesn't need to be kept in core memory. The formulas allowing for such computations are the following:

$$n_{ik} = \sum_j a_{ji} \, a_{jk} \qquad (4.5)$$

$$d_i = \sum_j a_{ji} \, f_j \qquad (4.6)$$

where:

n_{ik} = element (i,k) of the normal matrix N;
d_i = element (i) of the normal term d;
a_{ji} = element (j,i) of the design matrix A;
f_j = element (j) of the known term.

For a better understanding of how these algorithms work, one may refer to Fig.2 and Fig.3.
Formulas (4.5) and (4.6) correspond, from the point of view of the software programmer, to algorithms which allow for the computation of the single contribution "j" to the element required:

$$[n_{ik}]_{\text{"j"}} = p_j \, [\beta_i \, W_i(P_j) \, \beta_k W_k(P_j)] \qquad (4.7)$$

$$[d_i]_{\text{"j"}} = p_j \, [\beta_i \, W_i(P_j)] \, \overline{f}_j \quad . \qquad (4.8)$$

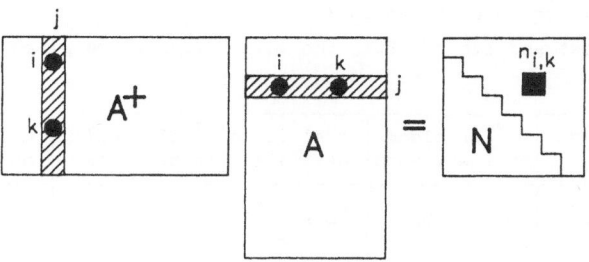

Fig. 2 - Forming the elements of N adding the contributions of the single rows of A .

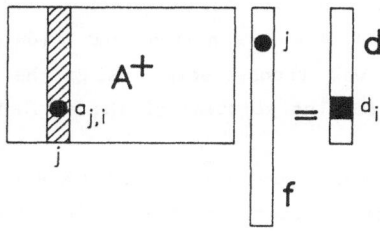

Fig. 3 - Forming the elements of d adding the contributions of the single rows of A .

As one can see, the contributions "j" come directly from the point P_j on the grid.

Before discussing the results obtained from the numerical experiment, it is worth remembering the problem of the compatibility of the grid dimensions and the maximum estimable degree of the model, which is not of secondary importance.

The data we are using represent a sampling of data on the sphere, so the whole matter is settled by the Nyquist frequency rule, stating that the Nyquist frequency must not be exceeded. For a problem on the sphere, things are more complicated than in the unidimensional case: anyway seeing it from the practical point of view, one can say that having an equiangular grid on the sphere of $2M^2$ points, the Nyquist frequency is given by the number of parallels (or rows), that are M or, to be more precise, all the coefficients of degree M can be estimated, except those of order M, as for them the Nyquist frequency is reached.

Exceeding the Nyquist frequency means giving rise to a folding of the spectrum with power from higher frequencies entering into lower frequencies and introducing an aliasing effect.

In our case, with a $5° \times 5°$ grid, M = 36. Coefficients $C_{36,36}$ and $S_{36,36}$ cannot be estimated and, in order to be far from the Nyquist frequency, we decided to deal with a model with maximum degree equal to 35.

Now coming to the test and its results, the main features are given in Table 2.

block size	n° of blocks on the sphere	model to be estimated	n° of coeffs. to be estim.	CPU time hh:mm:ss
5°x 5°	2592	L_{min} = 10 L = 35	1196	01:46:44

Table 2 - Characteristics of the test performed.

As one can see, it is a quite burdensome problem from the computational point of view, even if the model is not a large one. The data were simulated by means of a proper software for harmonic synthesis on the sphere: coefficients of model OSU91A were used. No noise was added to the Δg (on land) and T (on sea) values obtained.

A new set of coefficients was reconstructed using the Fortran program BOSALT, implementing the algorithms for the solution of the altimetry-gravimetry problem by biorthogonal sequences.

The estimated set was compared with the original OSU91A set of coefficients. First of all, relative differences were computed according to the formula:

$$d_r = \frac{C_{OSU91A} - \hat{C}}{C_{OSU91A}} \qquad (4.9)$$

where: C_{OSU91A} = coefficient of model OSU91A;
$\hat{C}$ = estimated coefficient.

The results were very good: examples are given in Fig. 4, Fig. 5 and Fig. 6. Attention must be paid to the y-axis, where relative differences are multiplied by a 10^6 factor. Both the lower and higher degrees of the model were reconstructed equally well.

Another graphic evidence of the goodness of the estimation is given in Fig. 7a) and 7b). In Fig. 7a) the histogram of relative differences calculated with equation (4.9) is shown: nearly all differences are smaller than $0.5 \cdot 10^{-6}$.

Another kind of relative differences was obtained with respect to the mean square values of the coefficients of each degree, using the following formula:

$$d'_r = \frac{C_{OSU91A} - \hat{C}}{\sqrt{\sum C^2_{OSU91A}}} \qquad . \qquad (4.10)$$

The histogram of such differences is shown in Fig. 7b), giving another proof of the precision attainable by the method described.

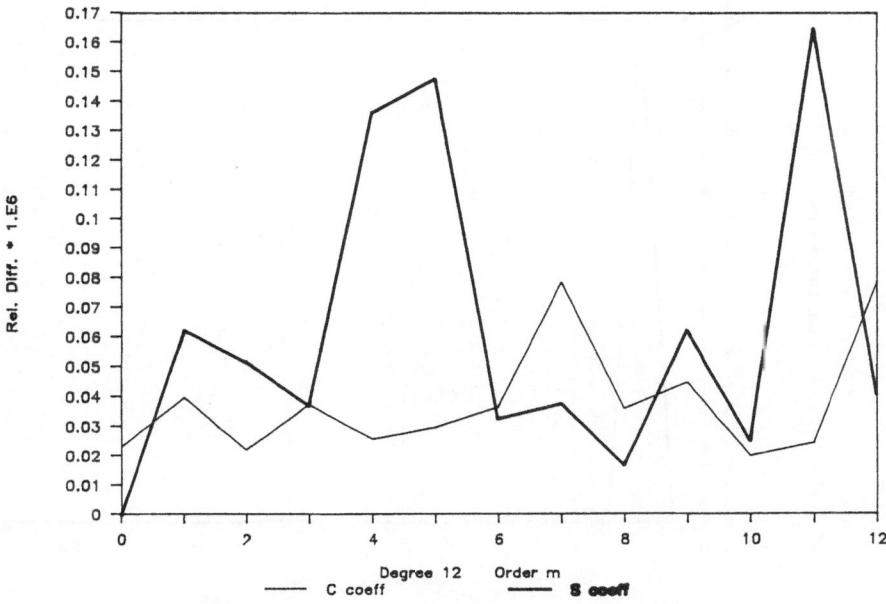

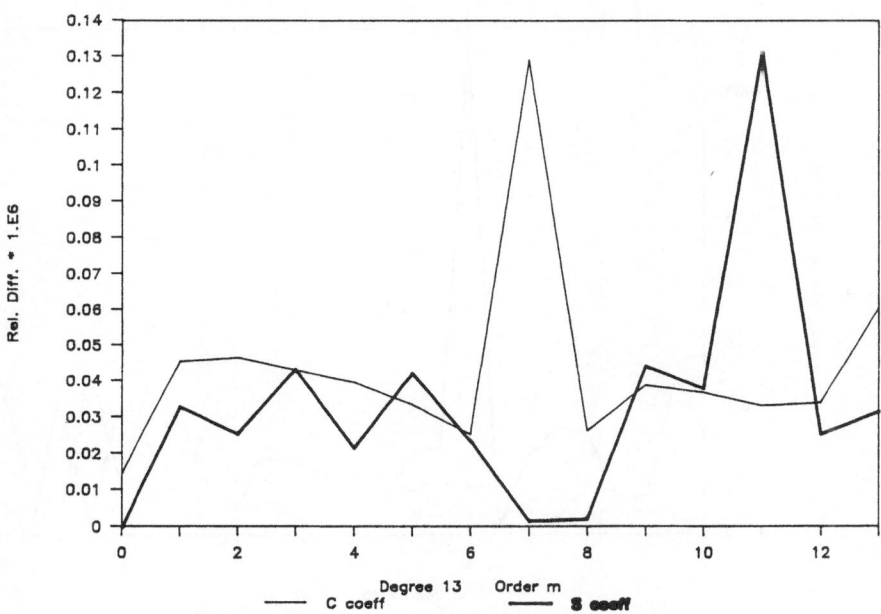

Fig. 4 a) and 4 b) - Relative differences for degrees 12 and 13.

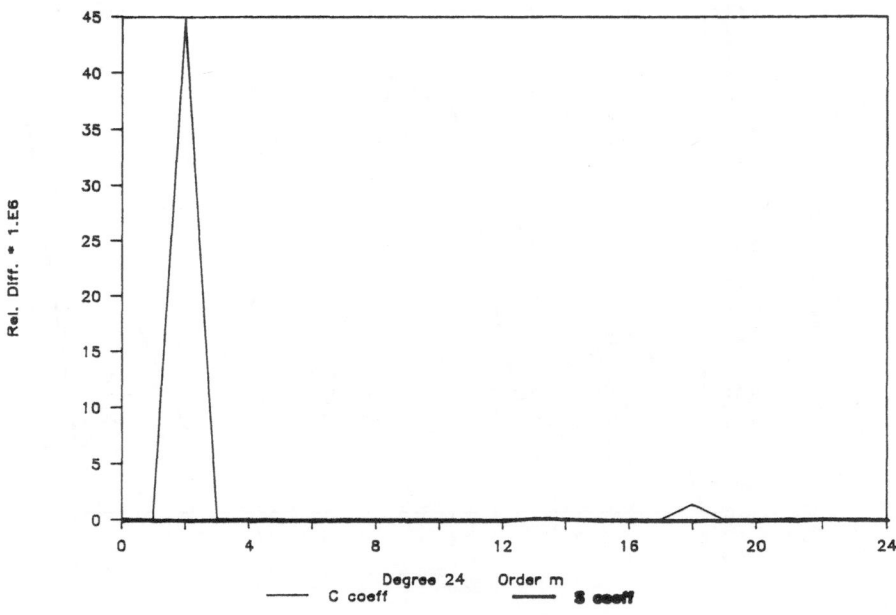

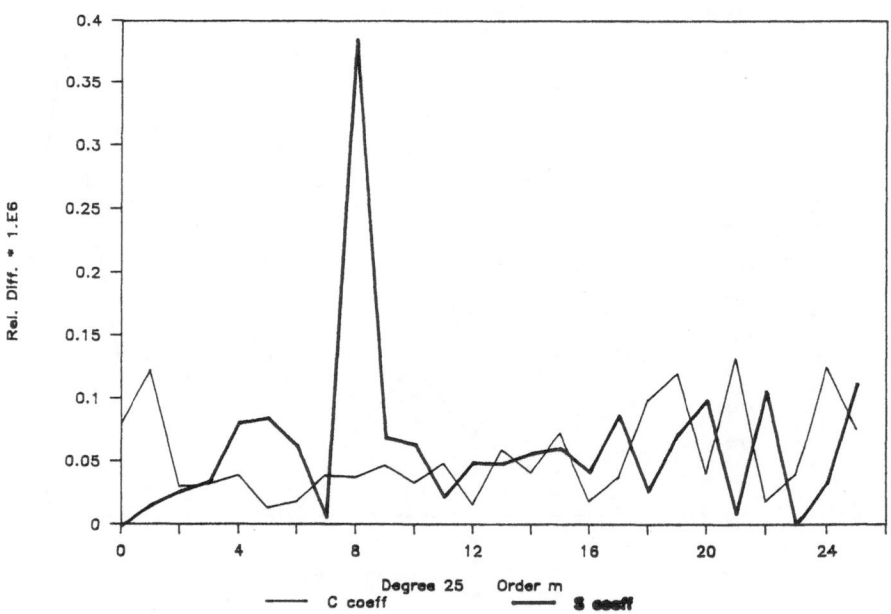

Fig. 5 a) and 5 b)- Relative differences for degrees 24 and 25.

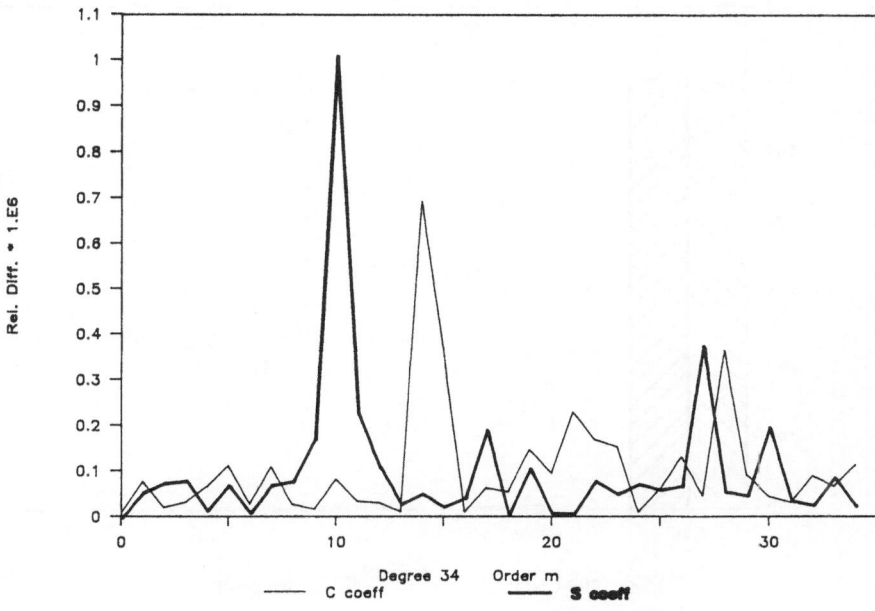

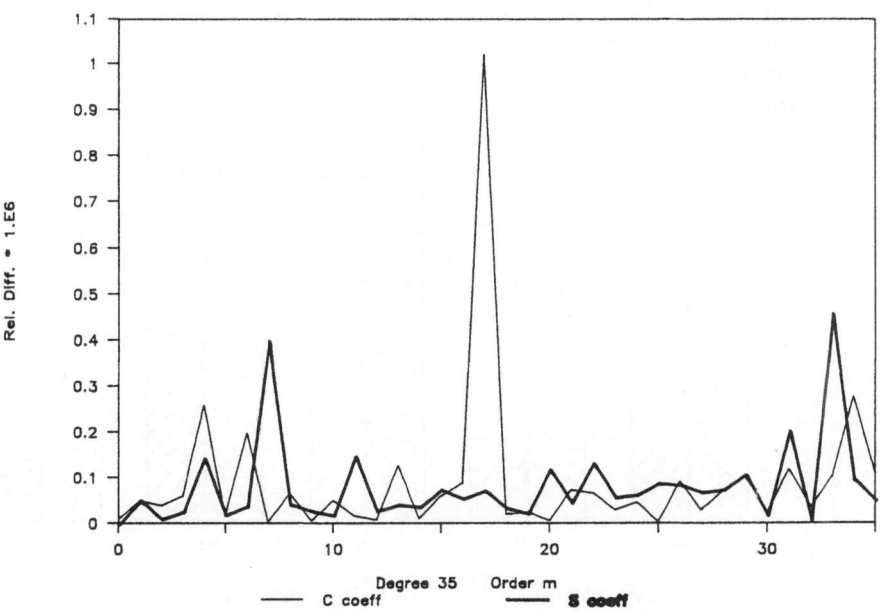

Fig. 6 a) and 6 b) - Relative differences for degrees 34 and 35.

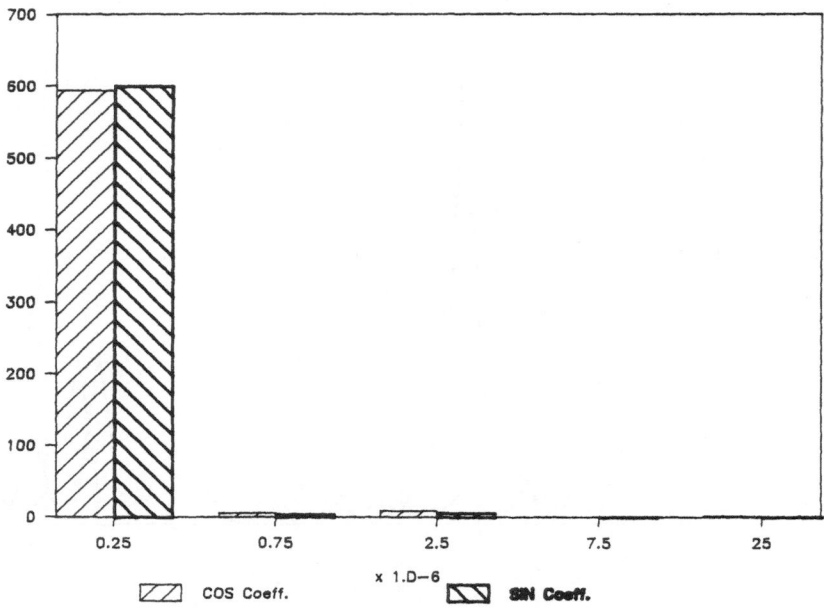

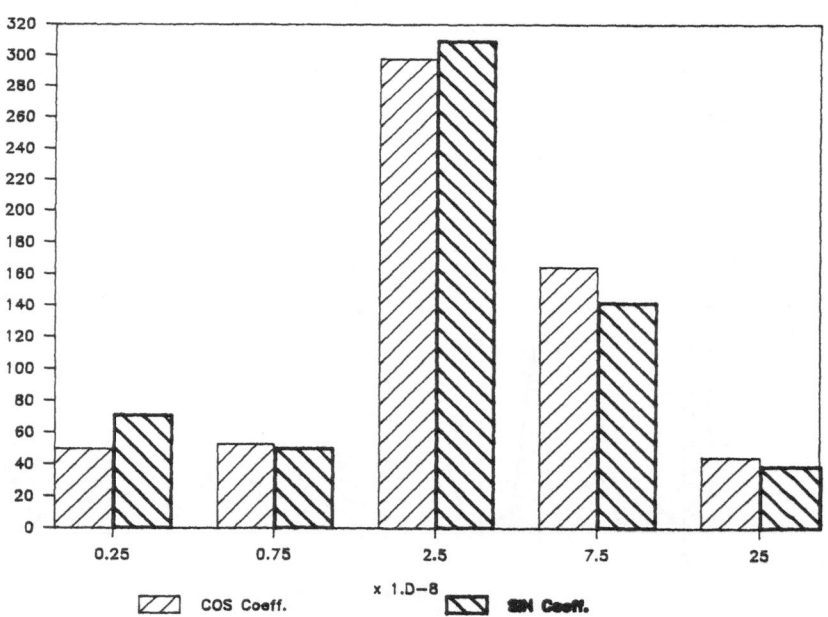

Fig. 7 a) and 7 b) - Histogram of relative differences obtained according to equations (4.9) and (4.10) respectively.

Of course this was just the very first test performed. More refined computations need to be done, in order to accurately check the proposed method. The next steps to be taken are:

- adding a noise to the data, to see how it propagates to the estimated model;
- treating real data of Δg on land and T on sea, to deal with a real estimation problem.

References

Barzaghi R., Brovelli M.A. e Sacerdote F., 1990, Altimetry-Gravimetry problem: an example, proceedings of the International Association of Geodesy Symposia, Symposium n. 104, Edimburgh, 10-11 Agosto 1989, Springer-Verlag, pp. 87-95.

Holota P., 1980, The altimetry-gravimetry boundary problem, presented at the International Scientific Conference of section 6 Intercosmos, Albena.

Kaula W.M., 1966, Theory of satellite geodesy, Blaisdell Publishing Company.

Krarup T.,1973, Letters on Molodensky's Problem, IAG Special Study Group 4.31.

Mainville A., 1987, The altimetry-gravimetry problem using orthonormal base functions, Geodetic Survey of Canada, Technical Report n. 8.

Migliaccio F. and F. Sansò, 1989, Data processing for the Aristoteles mission, Proc. Italian Workshop on the European Solid-Earth Mission Aristoteles, Trevi (Italy), 30 - 31 May.

Molodensky M.S., Eremeev V.F., Yurkin M.I., 1962, Methods for Study of the External Gravitational Field and the Figure of the Earth. Israel Pogram of Scientific Translations.

Sacerdote F. e Sansò F., 1983, A contribution to the analysis of the altimetry-gravimetry problem, Bulletin Geodesique,n. 57, pp.257-272.

Sacerdote F. e Sansò F., 1987, Further remarks on the altimetry- gravimetry problems, Bullettin Geodesique n.61, pp.65-82.

Sansò F., 1981, Recent advances in the theory of the geodetic boundary value problem, Rev. Geoph. and Space Phys., n.19, pp.437-449.

Sansò F., 1992, Theory of Geodetic B.V.P.s applied to the analysis of altimetric data, same volume.

Frozen orbits and their application in satellite altimetry

Ernst J.O. Schrama, TU Delft, Faculty of Geodesy,
Thijsseweg 11, 2629 JA Delft. The Netherlands

Summary

For several instrumental reasons, not further to be discussed in this note, there is a desire to minimize the height variations of an altimeter satellite with respect to the mean sea surface. Intuitively one would assume that this is accomplished by minimizing the mean eccentricity of an orbit. However in reality frozen orbits are chosen by which the mean eccentricity significantly differs from zero. It turns out for GEOSAT and SEASAT that it is possible to fix the mean value of the argument of perigee at approximately $90°$ with a mean eccentricity of about 10^{-3}. In this case the orbit is trapped in deep resonance, a theory originally developed for the odd $m = 0$ coefficients of the gravity field by Cook (1966). On basis of this theory we will discuss the use of frozen orbits in satellite altimetry. We will also point out that Cook's theory may be extended easily to deep resonant perturbations caused by $m \neq 0$ for orbits with repeating ground tracks. Moreover, we will address some difficulties in observing long periodic, large scale effects from satellite altimeter data.

Introduction

Consider the elements $\{a, e, I, \Omega, \omega, M\}$ in near circular orbits where the argument of perigee becomes singular and where the eccentricity parameter approaches a value of zero. In near circular orbits there are two main effects playing a role. Firstly the flattening parameter of the gravity field C_{20} causes the elements Ω, ω and M to drift linearly with respect to time. Secondly, C_{20} will cause oscillating effects resulting in radial departures of the order of 10 kilometer. Both effects may be described by means of the linear perturbations theory originally developed by Kaula (1966).

However a strict interpretation of this theory would result in a near singularity where the angular variable $\dot{\psi}_{lmpq}$ approaches $-q\dot{\omega}$ for $q = \pm 1$ remembering that $2\pi/\dot{\omega}$ equals approximately 207 days for GEOSAT and SEASAT. Formulas (3.76) in Kaula (1966) show that $\dot{\psi}_{lmpq}$ appears in the denominator which is clearly resulting in unrealistic large oscillations for all elements. In reality this doesn't happen and one would have to investigate the underlying Lagrange planetary equations.

The singular situation described in the previous paragraph is often called deep resonance. To describe perturbations in deep resonance Cook (1966) uses the so-called non singular variables $u = e \cos \omega$ and $v = e \sin \omega$. The advantage of this formulation is that there are no more singularities in the elements e and ω. Each position in the u, v plane may be translated directly into e and ω and visa versa. Physically seen u and v cannot be interpreted, mathematically seen they are a solution for a difficult problem in celestial mechanics.

An application of the use of u and v variables is to describe the secular perturbations caused by C_{20}. According to Kaula (1966) these perturbations are:

$$\left.\begin{array}{ll} e = e_0, & \dot{e} = 0 \\[2mm] \omega = \omega_0, & \dot{\omega} = \dfrac{3n_0 C_{20} a_e^2}{4(1-e^2)^2 a_0^2}(1 - 5\cos^2 I_0) \end{array}\right\}, \tag{1}$$

where $a_e = 6378137 \ [m]$ is the mean equatorial radius, $C_{20} \approx -0.00108263$ is the gravitational flattening parameter, $\mu = 3.9860044 \times 10^{14} \ [m^3/s^2]$ is the gravitational constant and $n = (\mu/a_0^3)^{1/2} \ [rad/s]$ is the mean motion. In the sequel we shall use the system of differential equations:

$$\left.\begin{array}{rcl} \dot{u} & = & \dot{e}\cos\omega - e\dot{\omega}\sin\omega \\ \dot{v} & = & \dot{e}\sin\omega + e\dot{\omega}\cos\omega \end{array}\right\}. \tag{2}$$

Substitution of $\dot{e} = 0$ and $\dot{\omega} = k$ in eq.(2) according to eq.(1) results in a compact set of first order differential equations:

$$\left.\begin{array}{rcl} \dot{u} & = & -k\,v \\ \dot{v} & = & +k\,u \end{array}\right\}. \tag{3}$$

Deep resonance

Firstly we must ask ourselves which $\{l, m, p, q\}$ combinations are causing deep resonances. To explain this we go back to the gravitational disturbing force function used by (Kaula,1966). This function expands the gravitational potential V in a Fourier series along the orbit involving inclination and eccentricity functions.

$$V = \sum_{l=0}^{\infty} \sum_{m=0}^{l} V_{lm} \tag{4}$$

$$V_{lm} = \frac{\mu a_e^l}{a^{l+1}} \sum_{p=0}^{l} \overline{F}_{lmp}(I) \sum_{q=-\infty}^{\infty} G_{lpq}(e) S(\psi_{lmpq}) \tag{5}$$

$$S(\psi_{lmpq}) = \left[\begin{array}{c} \overline{C}_{lm} \\ -\overline{S}_{lm} \end{array}\right]_{l-m:odd}^{l-m:even} \cos \psi_{lmpq} + \left[\begin{array}{c} \overline{S}_{lm} \\ \overline{C}_{lm} \end{array}\right]_{l-m:odd}^{l-m:even} \sin \psi_{lmpq} \tag{6}$$

$$\psi_{lmpq} = (l - 2p + q)(\omega + M) + m(\Omega - \theta) - q\omega \tag{7}$$

Some remarks:

- In practice V is never evaluated along the actual orbit where many other effects (e.g. drag and tides) are playing a role. Instead perturbation techniques are applied to evaluate V along an idealized secular precessing ellipse, which is also called a nominal orbit. This does not imply that a linear theory only predicts perturbations at the nominal orbit. Instead it is very well possible to predict perturbations in the neighborhood of a nominal orbit as was shown by e.g. Wagner (1985). This procedure is analogous to the treatment of a geodetic boundary value problem where the actual boundary (rugged terrain topography) is simplified to a mathematical model surface, see for instance Sansò (1992).

- Since eccentricity function contain a damping factor $e^{|q|}$ we are allowed to constrain the q index to within $[-1, 1]$.

For a linear perturbations theory Kaula (1966) uses the C_{20} secular effect so that all dependencies with respect to time are compressed in the angular variable ψ_{lmpq}. In other words:

$$\psi_{lmpq}(t) = \psi^0_{lmpq} + \dot{\psi}_{lmpq}(t - t_0) \tag{8}$$

where ψ^0_{lmpq} refers to an evaluation of ψ_{lmpq} at t_0. In this way a complicated expression for V along an actual orbit is replaced by a Fourier series along a simple nominal orbit. As mentioned before, for deep resonances an $\{l, m, p, q\}$ combination is chosen such that $\psi_{lmpq}(t) = -q\omega(t)$ which is possible for odd l and $m = 0$ potential coefficients of the gravity field (the so-called odd zonals) where $l - 2p = 1$ and $q = -1$ or where $l - 2p = -1$ and $q = +1$.

This is the original situation described by Cook (1966) where the disturbing force function becomes (see also appendix A in (Schrama,1989)):

$$V_{odd} = \frac{\mu}{a} \sum_{l=3,[2]}^{\infty} \left(\frac{a_e}{a}\right)^l \overline{F}_{l,0,(l-1)/2}(I) \, \overline{C}_{l,0}(l-1) \, e \sin \omega \tag{9}$$

It is more convenient to convert this expression in:

$$V_{odd} = Qe \sin \omega \tag{10}$$

where Q is:

$$Q = f(\mu, a, a_e, I, C_{l,0}) \tag{11}$$

for odd values of l starting at index 3. The parameter Q is independent of time and may be computed once a gravity model and a priori nominal orbit are known.

Yet there is also a second possibility for deep resonances in so-called repeat orbits where:

$$\frac{\dot{\omega} + \dot{M}}{\dot{\Omega} - \dot{\theta}} \stackrel{\text{def}}{=} \frac{\dot{\omega}_o}{\dot{\omega}_e} = \frac{N_r}{N_d} \tag{12}$$

in which N_r and N_d are integers not sharing a common prime factor. This means that the satellite moves in an orbit whereby one repeat cycle is completed after N_r revolutions after N_d so-called nodal days. (For SEASAT $N_r = 43$ and $N_d = -3$ in its last month. For GEOSAT $N_r = 244$ and $N_d = -17$ during the exact repeat mission (ERM)). In this case:

$$\dot{\psi}_{lmpq} = k' \dot{\omega}_o + m \dot{\omega}_e - q \dot{\omega} \tag{13}$$

where $k' = l - 2p + q$ so that:

$$\dot{\psi}_{lmpq} = \left[\frac{k' N_r + m N_d}{N_r} \right] \dot{\omega}_o - q \dot{\omega} \; . \tag{14}$$

It is easy to see that this may result in a deep resonant situation for resonant orders $m = \gamma N_r$ and $k' = -\gamma N_d$ where γ is an integer greater than 0. The disturbing force function may now be written as a Fourier series:

$$V_{resonant} = P e \cos \omega \; + \; Q e \sin \omega \tag{15}$$

where P and Q are time independent functions of resonant orders m of the gravity field:

$$P = f(\mu, a, a_e, I, C_{lm}, S_{lm}) \tag{16}$$

$$Q = g(\mu, a, a_e, I, C_{lm}, S_{lm}) \tag{17}$$

Some remarks:

- It is obvious that eq.(15) is more general than eq.(10). The theory described in (Cook,1966) is therefore limited to a special case where only the resonant zonal coefficients are considered. Our approach is to follow the same line of thought to other resonant orders for repeating orbits.

- We will not provide the reader with lengthy expansions of the P and Q terms that result in expressions similar to eq.(9). In our opinion this provides not much insight in the physical background of the problem and the preferred way is to rely on software that e.g. computes P and Q terms from existing gravity models.

The next step is to substitute forcing functions of the type of eq.(15) into the Lagrange planetary equations, see also (Kaula,1966). This results in a system of differential equations for $\dot{e}$ and $\dot{\omega}$ which is then translated into non singular variables. First of all the Lagrange planetary equations for $\dot{e}$ and $\dot{\omega}$ require to differentiate a forcing function with respect to the elements M, I, ω and e:

$$\dot{e} = \frac{\sqrt{1 - e^2}}{na^2 e} \left[-\frac{\partial R}{\partial \omega} + \sqrt{1 - e^2} \frac{\partial R}{\partial M} \right] \tag{18}$$

$$\dot{\omega} = \frac{\sqrt{1 - e^2}}{na^2 e} \left[+\frac{\partial R}{\partial e} - \frac{e \cos I}{(1 - e^2) \sin I} \frac{\partial R}{\partial I} \right] \tag{19}$$

In our case R is replaced by eq.(15) whereby we would like to remark that:

- The term $\dfrac{\partial R}{\partial M}$ disappears from the equation,

- The term $-\dfrac{e\cos I}{(1-e^2)\sin I}\dfrac{\partial R}{\partial I}$ is of the order of e^2 and can be ignored.

- The expression $\sqrt{1-e^2}$ is replaced by 1.

- The P and Q terms do not depend on e and ω.

Substitution of eq.(15) in (18) and (19) results in an $O(e)$ consistent system:

$$\dot{e} = \frac{1}{na^2}\left[P\sin\omega - Q\cos\omega\right] \tag{20}$$

$$\dot{\omega} = \frac{1}{na^2 e}\left[P\cos\omega + Q\sin\omega\right] \tag{21}$$

which is converted into the non singular variables u and v:

$$\dot{u} = \dot{e}\cos\omega - e\dot{\omega}\sin\omega = \frac{-Q}{na^2} \tag{22}$$

$$\dot{v} = \dot{e}\sin\omega + e\dot{\omega}\cos\omega = \frac{+P}{na^2} \tag{23}$$

This is just one of the particular cases that is added to differential equations (in the non singular elements) describing the C_{20} secular effect, see also eq.(3), resulting in the system:

$$\left.\begin{array}{ccc} \dot{u} & = & -k\,v + C_u \\ \dot{v} & = & +k\,u + C_v \end{array}\right\}. \tag{24}$$

where $C_u = -Q/(na^2)$ and $C_v = +P/(na^2)$. As mentioned before, the original problem in Cook (1966) was to analyze the effect of the odd zonal coefficients which cancels the C_v term. This results in a circle in the u, v plane about the coordinates $(0, C_u/k)$. The problem described here is slightly more complicated and the general solution of eq.(24) may be written as:

$$\left.\begin{array}{l} u(t) = A\cos(kt + \alpha) - C_v/k \\ v(t) = A\sin(kt + \alpha) + C_u/k \end{array}\right\} \tag{25}$$

with arbitrary integration constants A and α and where $k = \dot{\omega}$ due to eq.(1). This solution allows a circle in the u, v plane centered about an arbitrary point depending on the values of C_u, C_v and k.

The real world and Cook's theory

In the previous section differential equations were derived for deep resonant cases which may be caused by two sets of potential coefficients. In non repeating orbits the deep resonant set consists of the odd zonal coefficients. In orbits repeating after N_r orbital periods this set includes the resonant orders m where $m = \gamma N_r$ for integers γ where $\gamma \geq 1$.

Yet the best known deep resonance effects are coming from the odd zonal coefficients and not from resonant orders. The reason for this is twofold:

- The deep resonant forcing function is dominated by the odd zonal coefficients which are one or two orders in magnitude larger than all remaining potential coefficients in a gravity model, (in particular C_{30}).

- In some cases the length of the orbital repeat period is such that resonant orders are not sensed simply because of the characteristic damping factor $(a_e/r)^l$ in the gravitational potential.

In other words, Cook's original theory wasn't so bad after all and the Q term in eq.(15) (which is always caused by the odd zonals) dominates over the P term (which is always caused by resonant orders m). Conceptually the motion in the u, v plane is therefore very close to a circle centered at the point $(0, C_u/k)$ with some arbitrary radius A depending on the initial elements. This may result in three orbital configurations which are classified as follows, cf. (Cook,1966):

- In precessing orbits the integration constant A is greater than C_u/k. This results in a full rotation of the mean argument of perigee thereby bounding the eccentricity in between $A - C_u/k$ and $A + C_u/k$.

- In librating orbits the integration constant A is less than C_u/k. This causes the mean argument of perigee to wobble about 90° thereby restricting the mean eccentricity in between $C_u/k - A$ and $C_u/k + A$.

- In frozen orbits the integration constant A is minimized to 0. In this case the mean argument of perigee is fixed at 90° whereas the mean eccentricity is fixed at C_u/k.

Figure 1 shows the circular motion in the e, ω plane where $C_u/k = 7.9 \times 10^{-4}$ for values of A starting at 1×10^{-4} in steps of 2×10^{-4}. The numbers are derived from a simulation for SEASAT where $k = 2\pi/(-207 \times 86400)$ using the zonal coefficients of the GEM-T1 model, Marsh et al. (1986), up to degree 50. The closed curves around the coordinates $\omega = 90°$ and $e = 7.9 \times 10^{-4}$ in this figure are librating deep resonant cases. Here one could say that the orbit is trapped in deep resonance whereby each closed curve is followed in a period of $2\pi/\dot{\omega}$ seconds (the so-called apsidal period). The open wave-like curves are the precessing deep resonant cases whereby ω is unbounded.

Any analytical orbit theory should always be verified by means of numerical integration. One of the numerical results is shown in figure 2 where a librating orbit is integrated using only the first 9 zonal coefficients (Our main concern was to reduce the program execution time) of the GEM-T1 gravity model and other constants from the geodetic reference system of 1980. To obtain these results we used a 12th order Adams multi-step integrator with a step-size of 60 seconds to integrate the equations of motion 180 days in advance. The initial elements of the trajectory where chosen at $a = 7158$ km, $e = 0.75 \times 10^{-3}$, $I = 108°$, $\Omega = 0°$ $\omega = 90°$ and $f = 60°$. Furthermore to suppress short periodic oscillations the moving averages of u and v are shown after each 14th revolution. These short periodic oscillations are predominantly caused by C_{20} and result in a lemniscate resembling the numeral 8

In figure 2 the circular motion about $(0, C_u/k)$ is clearly seen. The predicted frozen point using the first 9 zonals is at $(0, 7.3 \times 10^{-4})$ which seems to coincide reasonably

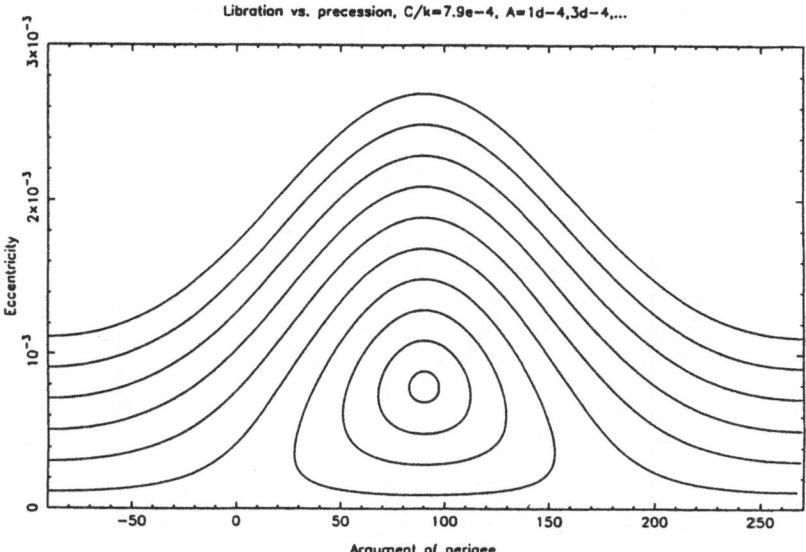

Figure 1: Libration and Precession in the e, w plane for a SEASAT type of orbit.

with the numerical results. Note that the frozen point coordinates for the first 9 zonals slightly differ from the ones computed with zonals up to degree 50 resulting in the u, v coordinates $(0, 7.9 \times 10^{-4})$.

As predicted before one does not expect too much influence on the frozen point coordinates from resonant harmonics where $m = \gamma N_r$ for integers γ greater than 1. To verify this we computed the frozen point coordinates of SEASAT ($I = 108°$, $N_r = 43, N_d = -3$) and ERS-1 (as SEASAT but at $I = 98°$) both with and without the zonal effects of GEM-T3, Nerem et al. (1991), see also table . There is indeed some small shift in the frozen point conditions but it is, as predicted by looking at the relative sizes of coefficients, rather small so that in this case the tesseral effects of order 43 may be ignored. This situation changes somewhat for deep resonant orbits that repeat e.g. once per nodal day where $N_r \leq 16$ but in most cases the argument of perigee stays fixed to within a few degrees from 90°.

Consequences for altimetry

As mentioned before, the frozen orbit technique is used in altimetry to minimize radial variations. However it turns out that altimetric orbits, which are influenced by air drag and other forces, require maintenance in order to focus at the $(0, C_u/k)$ point. These orbital maneuvers are carried out with periods of a month or so depending on the solar activity influencing atmospheric drag.

The discussion also showed that altimetric orbits may contain long periodic perturbations that could appear in results derived from altimeter data. To understand this we should mention briefly how the actual orbits are derived.

Nowadays all groups compute altimeter orbits numerically, solving the equations of

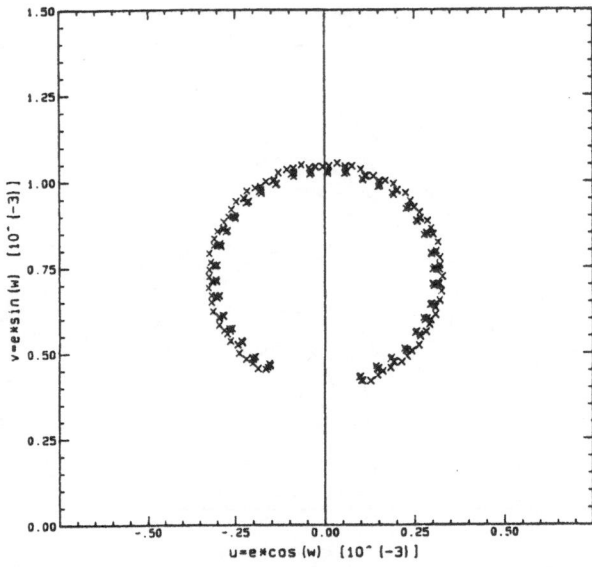

Figure 2: The 1 day mean non singular elements u and v in a librating orbit obtained by means of a numerical integration of the equations of motion using only the first 9 zonal coefficients of a gravity model.

Table 1: Some frozen points computed for the orbits of SEASAT and ERS-1 assuming that $N_r, N_d = 43, -3$ using all zonal coefficients of GEM-T3 and conditionally order 43 resonant coefficients.

Name	Tesserals included?	a [km]	I [deg]	$2\pi/k$ [days]	e [$\times 10^3$]	ω [deg]
SEASAT	No	7169.02	108	-208.64	0.8035	90.00
ERS-1	No	7152.30	98	-119.73	1.1711	90.00
SEASAT	Yes	7169.02	108	-208.64	0.8034	89.91
ERS-1	Yes	7152.30	98	-119.73	1.1710	89.94

motion and adjusting certain parameters that appear in various force models and in other models e.g. describing tracking stations coordinates and instrumental biases and drifts. Unfortunately the scope of this paper is not suitable to describe this in more detail and the reader is referred to literature.

A fact is that orbits are always computed in relatively short windows of approximately 1 to 2 weeks for the lower satellites (such as SEASAT, GEOSAT and ERS-1) whereas periods of the order of 1 month are used for the higher satellites (such as LAGEOS which is used for geodynamic research). An important argument is that the tracking residuals always increase when the 6 days orbits are pushed to e.g. 17 days (Haines (1990), private communication). Apparently it is difficult to predict accurate orbits a long time ahead and one prefers to solve again for a new set of initial state vectors that most certainly "absorb" a lot of other force parameters that are so difficult to model.

This is exactly the point where odd zonal coefficients enter into the picture. Imagine for instance that there is some problem in modelling the frozen point coordinates simply because of statistical noise in the gravity model. This causes unmodelled deep resonances that show up very close to the 1 cycle per revolution frequency in the radial orbit error spectrum, see also Wagner (1985). Over a short period of 6 days this effect is seen as the radial bow tie discussed by Colombo (1986).

The potential danger is that orbit errors are correlated in time when looking e.g. at repeat arc differences (RAD). It is true that all non-resonant radial orbit errors disappear when looking at RAD's, see Rummel (1992), but it is not necessarily true that this would be the case for the resonant radial orbit errors. The once per revolution term could be correlated over much longer times depending on the history of maneuvers. The best precaution against a mapping of deep resonances in the RAD's is therefore to avoid the area around 1 cycle per revolution in the spectrum. The relation of long periodic resonances with respect to 'annual variability studies' over 'large spatial scales' is therefore not well understood.

In contrast to processing RAD's there are techniques that envision the estimation of permanent sea surface topography (SST) and the annual and semi annual components of this topography. A well known fact, mentioned by all groups that publish results with this method, is that the initial state vector corrections are correlated (by almost 100%) to the C_{10} term of the sea surface topography field, see also (Schrama,1989). A separation of these terms is therefore only possible by adding information to the system of observation equations which is usually done by means of least squares collocation or similar techniques. The fact that errors are correlated remains and there is in general no guarantee that e.g. deep resonant orbit errors map in the C_{10} term of the SST. Also here it is justified to say that it is not well understood whether long periodic correlations of the C_{10} terms fully decorrelate from errors in the zonal coefficients and errors in the estimate of the initial state vector.

Acknowledgments

The author wishes to thank N.J. Sneeuw (TU Delft, Faculty of Geodesy), R. Zandbergen (ESOC, Darmstadt) and V. Zlotnicki (JPL) for discussions on this subject. This research is supported financially in the form of a scholarship by the Netherlands Academy of Arts and Sciences.

References

[1] Balmino G., Orbit Choice and the Theory of Radial Orbit Error for Altimetry, this issue, 1992.

[2] Colombo O.L., Notes on the Mapping of the Gravity Field Using Satellite Data, Springer Verlag series on Earth Sciences, ISBN 3-540-16809-5, 1986.

[3] Cook G.E., Perturbations of Near Circular Orbits by the Earth's Gravitational Potential, Planetary and Space Sciences, Vol. 14, pp. 433–444, 1966.

[4] Kaula W.M., Theory of Satellite Geodesy, Blaisdell Publishing Co., 1966.

[5] Marsh J.G, et al., An Improved of the Earth's Gravitational Field GEM-T1, Goddard Space Flight Center, Greenbelt Md., 1986.

[6] Nerem R.S., Lerch F.J., Putney B.H., Pavlis E.C., Marshall J.A., Earth Gravity Model Development at NASA/GSFC, Annales Geophysicae, XVI General Assembly, Wiesbaden Germany, 22–26 April 1991.

[7] Rummel R., Principle of Satellite Altimetry and elimination of radial orbit errors, this issue, 1992.

[8] Sansò F., Theory of Geodetic B.V.P.'s applied to the analysis of Altimetric Data, this issue, 1992.

[9] Schrama E.J.O., The role of orbit errors in processing of satellite altimeter data, Netherlands Geodetic commission, new series, report no. 33, 1989.

[10] Wagner C.A., Radial Variations of a Satellite Orbit Due to Gravitational Errors: Implications for Satellite Altimetry, JGR, Vol. 90, No. B4, pp. 3027–3036, 1985.

Integration of Gravity and Altimeter Data by Optimal Estimation Techniques.

Per Knudsen
Kort- og Matrikelstyrelsen
Rentemestervej 8
DK-2400 København NV

1. Introduction.

After reading the lecture notes in this volume it cannot be of any surprise that satellite altimeter data contain information about both the geoid and the sea surface topography. The stationary parts of the sea surface topography and the geoid cannot be separated unless additional information are available. Such information can be made available through a geopotential model such as a GEM or a GRIM model. Then using more or less fancy techniques the sea surface topography (including the stationary parts!) can be estimated (see e.g. lecture notes by RHR, Chapter 3). Another valuable source of information is gravity observations.

Gravity observations (both at land and sea) increase the knowledge about the gravity field and, hereby, about the geoid. Then more accurate estimates of the sea surface topography may be obtained. In this paper it is described how altimetry and gravimetry can be used simultaneously in an estimation of the gravity field and the sea surface topography by optimal estimation techniques (practically equivalent to overparameterized Gauss-Markov estimation, inverse method, or least squares collocation). It is shown how expressions for the kernel function are obtained and how they can be tuned to agree with covariance functions derived empirically from the observations. Finally, it is described how a regional solution in the Norwegian Sea can be computed using block averages of marine gravimetry and line averages of Geosat altimetry.

2. The Mathematical Model.

Especially, when different data types are combined it is important that the full signal/error content is taken into account in the mathematical model. Else fatal inconsistencies between different data types may occur, e.g. caused by an unmodelled bias in one source of data (see e.g. Arabelos *et al.*, 1987). Here altimeter data will be modelled according to the following expression:

$$h^a = N + \zeta + \epsilon \qquad (1)$$

with errors

$$\epsilon = \epsilon^o + \epsilon^{o\tau} + \epsilon^A + n \qquad (2)$$

where

N is the geoid height,
ζ is the sea surface topography,
ϵ^o is the radial orbit error,
$\epsilon^{o\tau}$ is the ocean tide residuals,
ϵ^A is additional errors (e.g. wet troposphere), and
n is the measurement noise.

The geoid height is a quantity associated with the anomalous gravity potential T. Hence, N can be expressed in terms of a linear functional (or as in this case linearized functional according to *Bruns' formula*) applied on T (γ is the normal gravity):

$$N = L_N(T) = \frac{T}{\gamma} \qquad (3)$$

At this point the important link between altimetry and gravimetry can be made, gravity anomalies are associated with T too. They are expressed as

$$\Delta g = L_{\Delta g}(T) = -\frac{\partial T}{\partial r} - 2\frac{T}{r} \qquad (4)$$

Now T is expanded into fully normalized spherical harmonic functions on the surface of a sphere with a radius R (Heiskanen & Moritz, 1967). That is

$$T = \frac{GM}{R}\sum_{i=2}^{\infty}\sum_{j=0}^{i}\left[\bar{C}_{ij}\cos j\lambda + \bar{D}_{ij}\sin j\lambda\right]\bar{P}_{ij}(\sin\phi) \qquad (5)$$

where $T=T(\phi,\lambda)=T(r=R,\phi,\lambda)$ and r,ϕ,λ are radius, latitude, and longitude respectively. C_{ij} and D_{ij} are real coefficients and P_{ij} are associated Legendre's functions of degree i and order j (bars denote that the harmonics are fully normalized). Using this spherical harmonic expansion, eq.(5), an expression for geoid heights, eq.(3), is obtained by multiplying the coefficients by $1/\gamma$. If the coefficients are multiplied by $(i-1)/R$ then an expression for gravity anomalies are obtained.

Also the sea surface topography (SST) is expanded into spherical harmonics. Ocean tides are not included in this term, but e.g. meso-scale variability exists, so the SST is a non-stationary surface. A temporal variation may be taken into account by replacing the harmonic coefficients by time dependant functions (Koblinsky *et al.*, 1991). That is

$$\zeta = \sum_{i=1}^{\infty}\sum_{j=0}^{i}\left[\bar{C}_{ij}^{\zeta}(t)\cos j\lambda + \bar{D}_{ij}^{\zeta}(t)\sin j\lambda\right]$$
$$\bar{P}_{ij}(\sin\phi)$$
$$= \Re\left(\int_{-\infty}^{\infty}e^{i\omega t}\sum_{i=1}^{\infty}\sum_{j=0}^{i}\left[\bar{C}_{ij}^{\zeta}(\omega)\cos j\lambda + \right. \qquad (6)\right.$$
$$\left.\bar{D}_{ij}^{\zeta}(\omega)\sin j\lambda\right]\bar{P}_{ij}(\sin\phi)\,d\omega\right)$$

where the time dependant functions have been expressed using their complex Fourier transforms. Hence, for each frequency ω, two sets of harmonic coefficients exist (one set associated with $\cos(\omega t)$ and one set associated

with $\sin(\omega t)$).

The stationary part of the SST may be isolated from eq.(6) (i.e. the $\omega=0$ terms) leaving an expression for the time varying parts of the SST. In practice, SST averaged over the period of measurements are determined. This surface is called the quasi-stationary SST (QSST). In any case, the stationary SST or the QSST can be expressed as

$$\zeta_s = \sum_{i=1}^{\infty}\sum_{j=0}^{i}\left[\bar{C}_{ij}^{s}\cos j\lambda + \bar{D}_{ij}^{s}\sin j\lambda\right]\bar{P}_{ij}(\sin\phi) \qquad (7)$$

The radial orbit error is modelled using a Fourier representation. For an orbit that repeats after a period T_{Arc} the orbit errors caused by geopotential errors become periodic and can be described using discrete frequencies $k\omega_a$ where $\omega_a=2\pi/T_{Arc}$. Furthermore, parts of the initial state vector errors that exhibit the so-called "bow tie" pattern centred at $t=t_m$ are taken into account (Schrama, 1989). That is

$$\epsilon^o = (t-t_m)\left(A_0^o\cos(\omega_o t) + B_0^o\sin(\omega_o t)\right)$$
$$+ \sum_{k=1}^{\infty}\left[A_k^o\cos(k\omega_a t) + B_k^o\sin(k\omega_a t)\right] \qquad (8)$$

where $\omega_o=2\pi/T_{Rev}$ is the once per revolution frequency. The initial sate vector errors are associated with the period over which the orbit has been computed (e.g. 6 days), so parameters for each orbit computation period exist.

Ocean tides occur at isolated frequencies associated with the tidal constituents (M2, S2, K1, O1, N2, P1, K2, ...). Then using eq.(6) the ocean tide residuals can be expressed as

$$\epsilon^{o\tau} = \Re\left(\sum_{k=1}^{K}e^{i\omega_k t}\sum_{i=1}^{\infty}\sum_{j=0}^{i}\left[\bar{C}_{ijk}^{o\tau}\cos j\lambda + \right.\right.$$
$$\left.\left.\bar{D}_{ijk}^{o\tau}\sin j\lambda\right]\bar{P}_{ij}(\sin\phi)\right) \qquad (9)$$

where the kth set of complex coefficients are associated with the kth constituent, which has a frequency of ω_k (lecture notes by DEC).

The additional errors, such as errors in the atmospherical corrections and sea state related biases, may be modelled using expressions similar to eq.(6) or eq.(8). Often these errors are ignored, but their effects may be considered together with the measurement noise in the error covariance matrix in a later estimation job.

Finally, it will be demonstrated how geostrophic ocean surface currents are obtained from the sea surface topography. If accelerations and friction terms are neglected and horizontal pressure gradients in the atmosphere are absent, then the components of the surface currents are obtained from the SST by

$$u = \frac{-\gamma}{fR} \frac{\partial \zeta}{\partial \phi}, \quad v = \frac{\gamma}{fR\cos\phi} \frac{\partial \zeta}{\partial \lambda} \tag{10}$$

where $f=2\omega_e\sin\phi$ is the Coriolis force coefficient. Using the spherical harmonic expansion of the SST, eq.(6), the current components are modelled as (Engelis, 1985)

$$u = -\frac{\gamma}{fR} \sum_{i=1}^{\bar{i}} \sum_{j=0}^{i} \left[\bar{C}_{ij}^{\zeta}(t)\cos j\lambda + \bar{D}_{ij}^{\zeta}(t)\sin j\lambda \right] \frac{\partial \bar{P}_{ij}(\sin\phi)}{\partial \phi} \tag{11}$$

$$v = \frac{\gamma}{fR\cos\phi} \sum_{i=1}^{\bar{i}} \sum_{j=0}^{i} j \left[-\bar{C}_{ij}^{\zeta}(t)\sin j\lambda + \bar{D}_{ij}^{\zeta}(t)\cos j\lambda \right] \bar{P}_{ij}(\sin\phi) \tag{12}$$

where

$$\frac{\partial \bar{P}_{ij}(\sin\phi)}{\partial \phi} = \bar{P}_{i(j+1)}(\sin\phi) - j\tan\phi\,\bar{P}_{ij}(\sin\phi) \tag{13}$$

3. Inversion of Altimetry.

Matrix inversion methods can be used, when the gravity field and the QSST are estimated from altimeter data. If the number of coefficients are made finite (through a truncation of the infinite series in eq.(5-9)) then a set of observations, y, can be expressed by a matrix, A, and a vector, x, holding the coefficients as

$$y = Ax + e \tag{14}$$

where e is a noise vector. The unknown parameters (i.e. the coefficients) can be estimated from a set of observations. A least squares solution is obtained by solving the following equation system

$$x = \left(A^T D^{-1} A + C_x^{-1} \right)^{-1} A^T D^{-1} y \tag{15}$$

or if the number of unknown parameters is larger than the number of observations, then

$$x = C_x A^T \left(A C_x A^T + D \right)^{-1} y \tag{16}$$

D is the error covariance matrix associated with the observations and C_x is the covariance matrix associated with the unknown parameters. Eq.(15) is used in global studies of altimetry, since a huge number of observations is used (Wagner, 1986, Engelis & Knudsen, 1989, Denker & Rapp, 1990, Marsh et al., 1990, Nerem et al., 1990). Then the number of unknowns is decreased until the size of the equation system becomes manageable. In this case it is important that the observations are low-pass filtered properly, so aliasing effects are avoided.

In local regions altimetry may be used in a detailed recovery of the high frequency parts of the gravity field and the QSST. If a resolution of 10 km is requested spherical harmonics up to degree and order 2000 are needed. That makes about 4 million unknown coefficients. Unless the normal matrix, $A^T D^{-1} A$, has certain symmetry the equation system cannot be solved using eq.(15), but if a reasonable number of observations is used, then the unknowns can be estimated using eq.(16). In fact, when eq.(16) is used a much larger number of unknowns associated with a variety of quantities can be solved for (Wunsch & Zlotnicki, 1984, Mazzega & Houry, 1989, Blanc at al., 1990, and Knudsen, 1991).

If estimations are carried out using eq.(16) then it is convenient to compress the expression and avoid a set-up of matrix A. Furthermore, the estimated coefficients are used directly to compute quantities $x = a^T x$. This is done using $C = A C_x A^T$, which is the signal covariance matrix of the observations and $C_x^T = \mathbf{a}^T C_x A^T$, where C_x is a vector of covariances between the quantity x and the observations. This results in the following expression

$$x = C_x^T (\mathbf{C} + \mathbf{D})^{-1} \mathbf{y} \qquad (17)$$

An estimate of the *a-posteriori* error covariance between two estimated quantities, x and x', is obtained using

$$\hat{c}_{xx'} = c_{xx'} - C_x^T (\mathbf{C} + \mathbf{D})^{-1} C_{x'} \qquad (18)$$

where $c_{xx'}$ is the *a-priori* (signal) covariance between x and x' (see e.g. Moritz, 1980, Tarantola, 1987).

The elements of the covariance matrices of eq.(17-18) are calculated according to the mathematical model of the observations. In this case, signals associated with the geoid and the SST are considered, eq.(1), together with the error components listed in eq.(2). If the different signal and error components are uncorrelated then the covariance values, $\mathbf{C}_{ij}$ and $\mathbf{D}_{ij}$, are obtained by adding covariances associated with each of the signal and error components. That is

$$C_{hh} = C_{NN} + C_{\zeta\zeta} \qquad (19)$$

and

$$D_{\epsilon\epsilon} = D_{\epsilon^\circ\epsilon^\circ} + D_{\epsilon^{rc}\epsilon^{rc}} + D_{\epsilon^A\epsilon^A} + D_{nn} \qquad (20)$$

Correlations between the orbit and the gravity field are not believed to be important in such a local estimation and correlations between the gravity field and the SST (Colton & Chase, 1983) are simply neglected.

The covariance values are obtained using the kernel functions. The kernel associated with the gravity field is derived using eq.(5) and some *a-priori* variances. In the finite case where the infinite series, eq.(5), has been truncated, the values are obtained through the matrix multiplication $AC_x A^T$.

An important result is obtained when the harmonic coefficients are assumed to be uncorrelated and, furthermore, that their *a-priori* variances only depend on their harmonic degree and not on their harmonic order. Then the covariance between T in the points $P(\phi, \lambda)$ and $Q(\phi', \lambda')$ is expressed as

$$K(P,Q) = \sum_{i=2}^{\infty} \sum_{j=0}^{i} \bar{\sigma}_i^{TT} [\cos j\lambda \cdot \cos j\lambda' + \sin j\lambda \sin j\lambda'] \bar{P}_{ij}(\sin\phi) \bar{P}_{ij}(\sin\phi') \quad (21)$$

$$= \sum_{i=2}^{\infty} \sigma_i^{TT} P_i(\cos\psi)$$

where $\sigma_i^{TT} = (2i+1)\bar{\sigma}_i^{TT}$ are degree variances and ψ is the spherical distance between P and Q. Hence, eq.(21) only depends on the distance between P and Q and neither on their locations nor on their azimuth (i.e. a homogeneous and isotropic kernel). Compared with the matrix product mentioned above this series has been extended to infinity, which certainly requires that the sum of the degree variances is finite. Expressions associated with geoid heights and gravity anomalies are obtained by applying the respective functionals on $K(P,Q)$, e.g. $C_{NN} = L_N(L_N(K(P,Q)))$ (more on collocation by Sansò, 1986, Tscherning, 1986). Then

$$C_{NN} = \sum_{i=2}^{\infty} \left(\frac{1}{\gamma}\right)^2 \sigma_i^{TT} P_i(\cos\psi) \qquad (22)$$

$$C_{\Delta g \Delta g} = \sum_{i=2}^{\infty} \left(\frac{i-1}{R}\right)^2 \sigma_i^{TT} P_i(\cos\psi) \qquad (23)$$

$$C_{N\Delta g} = \sum_{i=2}^{\infty} \left(\frac{i-1}{\gamma R}\right) \sigma_i^{TT} P_i(\cos\psi) \qquad (24)$$

The kernel associated with the SST, eq.(6), is believed to be expressed as

$$C_{\zeta\zeta} = \int_{-\infty}^{\infty} \cos(\omega \Delta t) \sum_{i=1}^{\infty} \bar{\sigma}_i^{\zeta\zeta}(\omega) P_i(\cos\psi) d\omega \qquad (25)$$

where $\bar{\sigma}_i^{\zeta\zeta}(\omega)$ is the temporal *a-priori* spectrum of harmonic degree i and $\cos(\omega \Delta t) = \cos(\omega t)\cos(\omega t') + \sin(\omega t)\sin(\omega t')$, $\Delta t = |t - t'|$.

If $\bar{\sigma}_i^{TT}(\omega) = \Phi(\omega)\sigma_i^{\zeta\zeta}$, where $\Phi(\omega)$ is a temporal spectrum that does not depend on harmonic degree. Then

$$C_{\zeta\zeta} \approx \int_{-\infty}^{\infty} \Phi(\omega)\cos(\omega \Delta t)d\omega$$

$$\qquad (25a)$$

$$\sum_{i=1}^{\infty} \sigma_i^{\zeta\zeta} P_i(\cos\psi)$$

where the kernel is split up into a temporal function multiplied by a spatial function. If the temporal correlations can be expressed in a similar manner as the spatial correlations except for a scale factor κ. Then

$$C_{\zeta\zeta} \approx \sum_{i=1}^{\infty} \sigma_i^{\zeta\zeta} P_i[\cos(\psi + \kappa\Delta t)] \qquad (25b)$$

Using this approximation with $\kappa=0$ and degree variances σ_i'' an expression for the QSST is obtained.

Kernels associated with ϵ^o and ϵ^{or} are found to be expressed as

$$D_{\epsilon^o\epsilon^o} = \Phi_0^o(t-t_m)(t'-t_m)\cos(\frac{2\pi}{T_{Rev}}\Delta t) \\ + \sum_{k=1}^{\infty} \Phi_k^o\cos(k\frac{2\pi}{T}\Delta t) \qquad (26)$$

$$D_{\sigma\tau\sigma\tau} = \sum_{k=1}^{K}\cos(\omega_k\Delta t)\sum_{i=1}^{\infty}\sigma_{ik}^{\sigma\tau\sigma\tau}P_i(\cos\psi) \qquad (27)$$

Kernels associated with the additional errors, ϵ^A, may be expressed using expressions like eq.(25) or eq.(26). The measurement noise of the data is assumed to be uncorrelated. Then the covariance is equal to the variance at zero lags and zero else where:

$$D_{nn} = \delta(\psi_i)n_i^2 \qquad (28)$$

Expressions associated with the geostrophic surface currents and the SST depend on the azimuth between the two points P and Q, α_{PQ}. That is (Knudsen, 1991)

$$C_{uu} = \frac{\gamma^2}{f_P f_Q}(-\cos\alpha_{PQ}\cos\alpha_{QP}C_{ll} - \\ \sin\alpha_{PQ}\sin\alpha_{QP}C_{qq}) \qquad (29)$$

$$C_{vv} = \frac{\gamma^2}{f_P f_Q}(-\sin\alpha_{PQ}\sin\alpha_{QP}C_{ll} - \\ \cos\alpha_{PQ}\cos\alpha_{QP}C_{qq}) \qquad (30)$$

$$C_{u\zeta} = \frac{-\gamma}{f_P}\cos\alpha_{PQ}C_{l\zeta} \qquad (31)$$

$$C_{v\zeta} = \frac{\gamma}{f_P}\sin\alpha_{PQ}C_{l\zeta} \qquad (32)$$

$$C_{vu} = \frac{-\gamma^2}{f_P f_Q}(-\sin\alpha_{PQ}\cos\alpha_{QP}C_{ll} + \\ \cos\alpha_{PQ}\sin\alpha_{QP}C_{qq}) \qquad (33)$$

where

$$C_{ll} = \frac{1}{R^2}\left(\cos\psi\, C'_{\zeta\zeta} - \sin^2\psi\, C''_{\zeta\zeta}\right) \qquad (34)$$

$$C_{qq} = \frac{1}{R^2}C'_{\zeta\zeta} \qquad (35)$$

$$C_{l\zeta} = \frac{-1}{R}\sin\psi\, C'_{\zeta\zeta} \qquad (36)$$

and

$$C'_{\zeta\zeta} = \partial C_{\zeta\zeta}/\partial\cos\psi \qquad (37)$$

In the next section it is described how the kernels associated with tracks of mean sea surface heights are evaluated. That is the kernel associated with the gravity field, eq.(21-24), the kernel associated with the QSST, eq.(25b) with $\kappa=0$ and σ_i'', and briefly the kernel associated with the radial orbit errors.

4. Local Empirical Covariance Functions.

When optimal estimation techniques are used it is very important that the statistical assumptions about the unknown parameters are adequately determined in order to obtain reliable results. A valuable control of the statistical assumptions is that both signal and error covariances, which are obtained using the kernels described in the previous section, agree with covariance values, which have been empirically derived from observations.

In practice, a degree variance model is used, so e.g. $\sigma_i^{TT}=f(i)$. Then the degree variance model is inserted in the kernel expression and adjusted iteratively until the kernel fits the empirical covariance function. When the degree variance model expression is chosen it is important that the decay is alright; e.g. it is known that σ_i^{TT} tend to zero somewhat faster than i^3. The decay may be estimated using power spectra derived using FFT techniques. (Note that the power spectrum will tend to zero one time faster than the degree variances.) Another valuable control is that kernels having different spectral contents (but the same degree variance model such as geoid heights and gravity anomalies) fit the respective empirical covariance functions.

A local empirical covariance function is determined from the observations located in the local region as described by Knudsen (1987a, 1988):

1) Select data so their geographical distribution becomes more homogeneous,

2) Subtract global models,

3) Wavelengths longer than the extent of the area cannot be covered, so biases and tilts should be removed. E.g. simply by using $D=a+b\phi+c\lambda$,

4) Estimate empirical covariance values in intervals of $\Delta\psi$ using:

$$C_k = M\{y_i y_j\}, \quad \psi_k - \frac{\Delta\psi}{2} < \psi_{ij} < \psi_k + \frac{\Delta\psi}{2} \quad (38)$$

where $M\{.\}$ denotes a simple arithmetic mean value, and

5) Adjust variance in order to remove the effects of measurement noise. That is $(\psi=0)$:

$$C_0 = M\{y_i^2\} - M\{n_i^2\} \quad (39)$$

The modelling of the covariance function

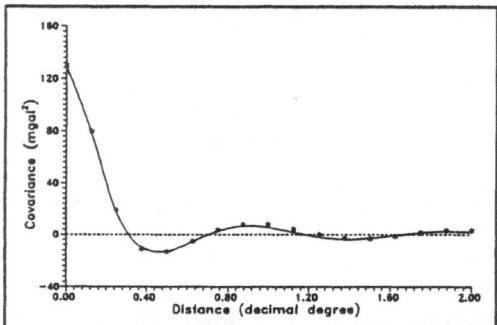

Figure 1. Empirical covariance function (*) and model (——) associated with gravity anomalies, eq.(23).

associated with the gravity field is described in Knudsen (1987a). This technique has been applied using empirical covariance values calculated from marine gravity data reduced using OSU91A to degree 360. As degree variance model a Tscherning/Rapp model (Tscherning & Rapp, 1974) was used. This expression has the advantage that the kernel can be evaluated using a closed expression instead of the infinite sum. The model is

$$\sigma_i^{TT} = \begin{cases} \epsilon_i & i=2,...,360 \\ \dfrac{A}{(i-1)(i-2)(i+4)}\left(\dfrac{R_B^2}{R^2}\right)^{i+1} & i=361,.... \end{cases}$$

$$(40)$$

where $A = 1571496$ m^4/s^4, $R_B = R - 7.4$ km were found in an adjustment. The error degree variances, ϵ_i, associated with the OSU91A model was multiplied by 0.138 (In fact the OSU89B error degree variances was used, however, above degree 50 they are equal). The covariance functions are shown in Figure 1. Here a variance of $(11.4$ mgal$)^2$ and a correlation length (which is the distance where the covariance is 50 % of the variance) of $0.16°$. The corresponding geoid height covariance function has a variance of $(0.247$ m$)^2$ and a correlation length of $0.43°$.

A determination of a covariance function model associated with the QSST may be carried out using QSST values derived from temperature/salinity data (e.g. lecture notes by CW). In Knudsen (1992a) the QSST in the Norwegian Sea has been estimated using one year of Geosat and marine gravimetry (also Zlotnicki, 1984). This was done relative to the harmonic expansion OSU89D to degree 10 (Denker & Rapp, 1990). An empirical covariance function was determined and a degree variance model chosen.

The degree variance model was constructed using 3rd degree Butterworth filters combined with an exponential factor. Hence, the spectrum of the QSST is assumed to have a decay similar to the geoid spectrum. Then the model was fitted iteratively to the empirical covariance function and, subsequently, controlled through the variances of the geostrophic ocean surface currents and

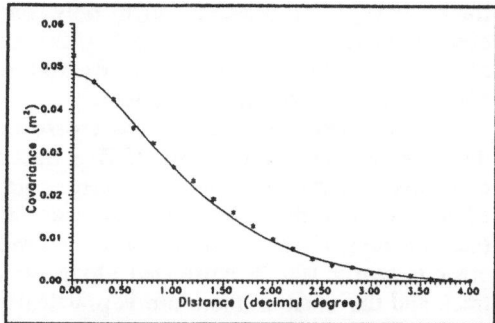

Figure 2. Empirical covariance function (*) and model (——) associated with the QSST.

the "QSST gravity equivalents". This resulted in the model:

$$\sigma_i^{ss} = b \cdot \left(\frac{k_2^3}{k_2^3 + i^3} - \frac{k_1^3}{k_1^3 + i^3} \right) \cdot s^{i+1} \cdot \left(1 - \frac{l^3}{l^3 + i^3} \right) \quad (41)$$

where $b = 6.3\ 10^{-4}$ m^2, $k_1 = 1$, $k_2 = 90$, $s = ((R-5000.0)^2/R^2)^2$, and $l = 17$. The variance and correlation length are $(0.22$ m$)^2$ and $1.11°$ respectively (see Figure 2). The variances of the current components (evaluated at $64°$ latitude) and the "QSST gravity equivalents" are $(0.22$ m/s$)^2$ and $(4.0$ mgal$)^2$.

As orbit error covariance function model an exponentially damped 1 cy/rev cosine has been used (Wunsch & Zlotnicki, 1984, Mazzega & Houry, 1989, Blanc at al., 1990, and Knudsen, 1991). In Knudsen (1992a) an expression as eq.(26) truncated at 2 cy/rev (k=488) was used. Also here the orbit error spectrum $\Phi_k^o = A_k^{o2} + B_k^{o2}$ was estimated using the results from an adjustment, where the

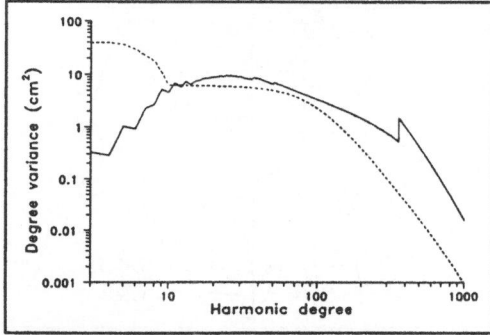

Figure 3. Degree variance models of geoid heights and QSST.

coefficients in eq. (8) were estimated from crossover differences. Such a procedure leaves the frequencies 0, ω_e, $2\omega_e$, ω_o, $\omega_o \pm \omega_e$, $2\omega_o$, $2\omega_o \pm \omega_e$, and $2\omega_o \pm 2\omega_e$ unestimated (Schrama, 1989, lecture notes by RR). These singular frequencies are important to consider, since they are parts of the geographically correlated orbit error. Therefore the use of orbit error spectra computed from the error covariance matrix of the gravity field (Tapley & Rosborough, 1985, Engelis, 1988, lecture notes by GB) needs to be tested in the near future.

The remaining covariance functions, which are associated with the time varying signals can be determined using statistical information derived from collinear and crossover analysis of satellite altimetry (e.g. Menard, 1983, Fu & Zlotnicki, 1989, Milbert et al., 1989, Tai, 1989, Knudsen, 1992b, and lecture notes by VZ). Again, it is important that the models fit the empirical covariance functions.

A final remark on the determination of covariance function models is that the long wavelength parts of the signals systematically have been removed in the empirical covariance function determination process. The effects of these parts of the signals should be restored in the degree variance models. In this case it was done rather empirically by 1) a re-scaling of ϵ_l in eq.(40) so a geoid signal of 0.25 m is added, 2) use eq.(41) with l=0 and replace the first 10 values with error degree variances of OSU89D, and 3) modify the singular frequencies in the crossover adjustment, so an (geographically correlated) orbit error of 0.20 m is added.

The degree variances of the geoid height and QSST covariance function models are shown in Figure 3. An important result is that both the geoid <u>and</u> the QSST contributes to the mean sea surface at harmonic degrees up to 150-200 (i.e. wavelengths longer than about 200 km). Above harmonic degree 150-200 the geoid dominates the mean sea surface.

5. On Redundancy and Data Selection.

As mentioned in Section 3 the number of equations corresponds to the number of observations in optimal estimation techniques. Hence, the number of observations have to be limited. Furthermore, redundant observations will cause singularities in the equation system.

An observation becomes redundant, if the correlation between this and another observation approaches 100 %. A useful "rule of thumb" is that observations distributed with a spacing about the correlation length should be used. Then e.g. gravity anomalies should be selected closest to the nodes of a $0.16° \times 0.16°/\cos\phi$ grid covering the local area of interest. This rule does not apply to altimetry, where time varying signals exist.

Also an observation that can be described by a linear combination of the other observations is redundant. An example is three height differences between three heights: Here the third difference is redundant. Also altimeter data along collinear track segments may easily become redundant: If data along the first track are selected according to correlation length, then data along the following collinear tracks are needed within the spatial correlation length of the time varying signals only, because the stationary signals already have been recovered. If e.g. mean sea surface and orbit error are considered and, again, data along the first

tracks have been selected, then only one observation from the other collinear tracks are needed, since the orbit error is nearly 100 % correlated along each track segment.

The last example shows how crossover differences become redundant: If five tracks cross five tracks and the orbit errors are estimated from the crossover discrepancies (see Knudsen, 1987b), then (because the orbit error is nearly 100 % correlated along each track and the situation therefore is practically similar to a situation, where five collinear tracks cross five collinear tracks) 16 out of 25 height differences become redundant.

Observational noise may increase the needs for a denser distribution of data. Furthermore, redundant observations may be needed in order to decrease the effects of the noise, but still redundancy can cause numerical instabilities on a computer.

6. A Regional Solution in the Norwegian Sea.

An integration of marine gravimetry and Geosat altimetry has been tested in the Norwegian Sea (see Knudsen, 1990b, Drottning & Knudsen, 1991). Here one year of Geosat data merged with GEM-T2 orbits were used. Mean sea surface heights have been obtained along collinear tracks, so the only time varying signal that is left is the radial orbit error (Knudsen & Brovelli, 1991). Marine gravity data were selected using a

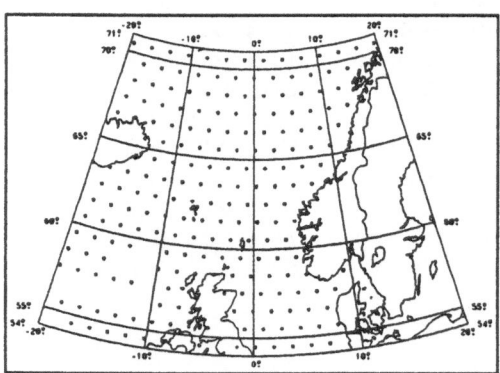

Figure 4. Locations of 1° block averages of gravity anomaly observations.

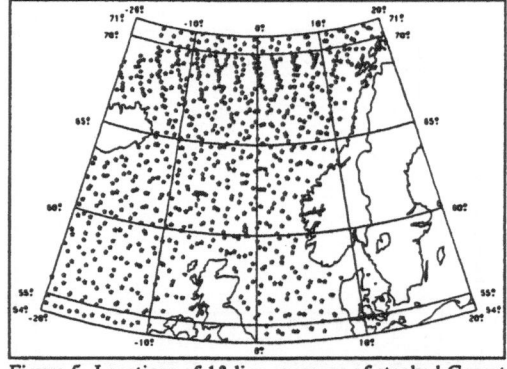

Figure 5. Locations of 1° line averages of stacked Geosat altimeter data.

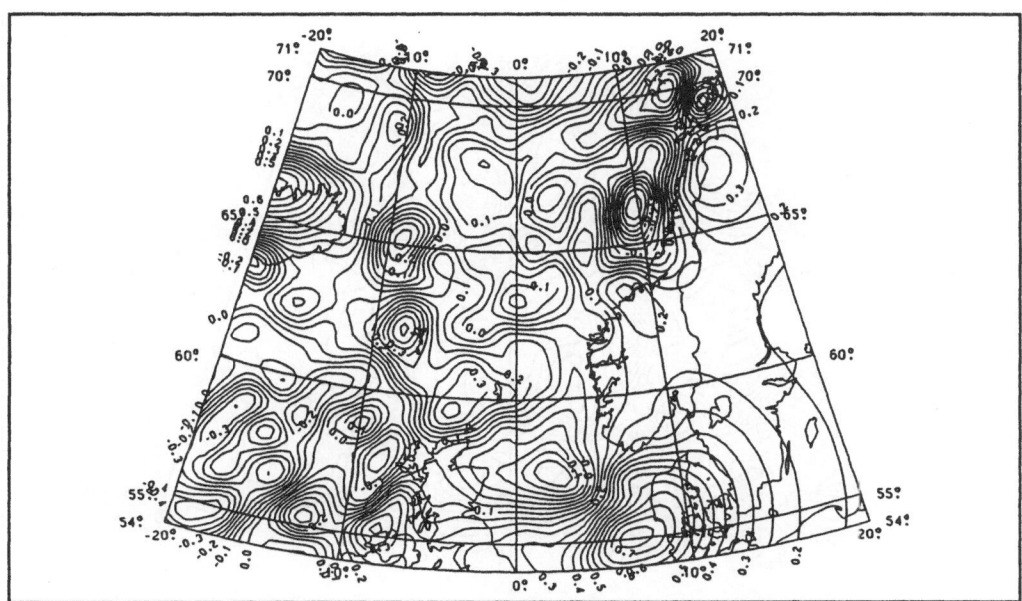

Figure 6. Estimated 1° block averages of mean sea surface heights. C.I. 0.05 m.

1/8°×1/4° grid. As mentioned in Section 4 all observation have been reduced using the OSU91A and OSU91SST models (the gravimetry has been reduced for free-air effects using the OSU91SST model; see Heck & Rummel, 1990). Finally the observations have been reduced for terrain effects of the surrounding mountains on the gravity and the geoid respectively, using the RTM technique described by Forsberg (1984).

The total number of observations is about 23,000. It is important that a solution covering the full region is computed, so the long wavelength parts of the geoid and the QSST can be recovered, but 23,000 equations cannot be solved. Therefore it was decided to use block averages of $1°×1°/\cos\phi$ and hereby recover wavelengths longer that about 222 km. However, mean values of altimetry have to be calculated along track, which result in a set of line averages.

The covariance values associated with the block averages were obtained by multiplying twice the degree variances by the Pellinen smoothing operators, β_i (lecture notes by RHR, eq.(2.21)). The covariances associated with line averages were treated differently. Such covariances can be derived analytically (such as the block average) as described by Jekeli (1989), but they are difficult to tabulate, because of their azimuthal dependency. Therefore, the line average covariances was calculated

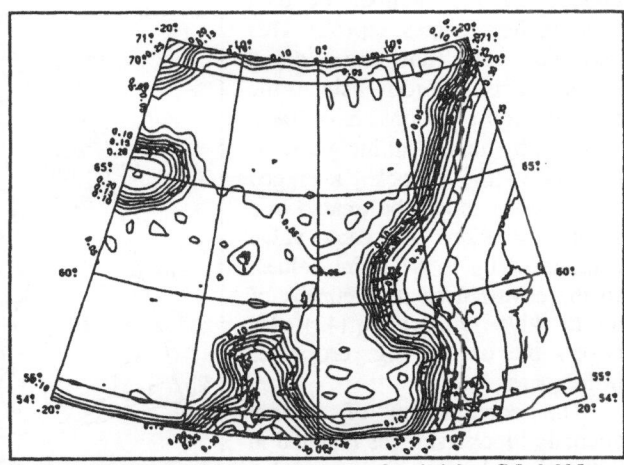

Figure 7. Errors of 1 ° block mean sea surface heights. C.I. 0.025 m.

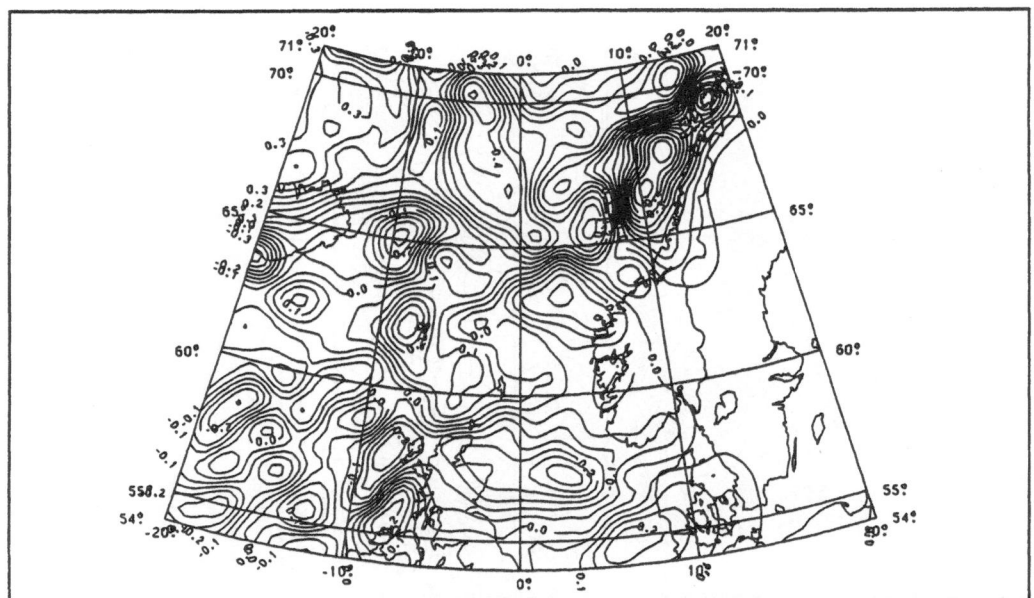

Figure 8. Estimated 1° block averages of geoid heights. C.I. 0.05 m.

numerically. In general for averaged quantities such as

$$\bar{h}_i = \sum_k w_k h_k \qquad (42)$$

the covariance is

$$C(\bar{h}_i, \bar{h}_j) = \sum_k \sum_l w_k w_l C(\psi_{kl}) \qquad (43)$$

where w_k and w_l is this case is equal to one divided by the number of data forming the averages, eq.(42). This may appear as a rude way of doing it, but it is rigorous, since the averaged observations have been obtained by numerical integrations. (In fact the block averaged observations should be treated in a similar manner, but each value consists of up to 64 point values, so the covariances may consist of up to 4096 terms.) Eq.(43) also covers the use of e.g. crossover differences.

The computation of gravity anomaly block averages resulted in 262 values having a mean value

and standard deviation of 0.35 mgal and 4.41 mgal respectively. A length of 1° was used in the computation of line averages of altimetry. The correlation between a block and a line average centred at the same point is about 80 %. This resulted in 891 values. Their mean value and standard deviation are 0.031 m and 0.208 m respectively.

Then a solution was found and 1° block averages of quantities together with error

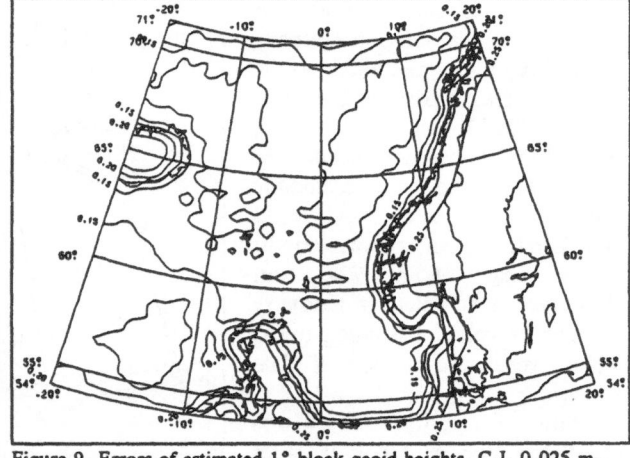

Figure 9. Errors of estimated 1° block geoid heights. C.I. 0.025 m.

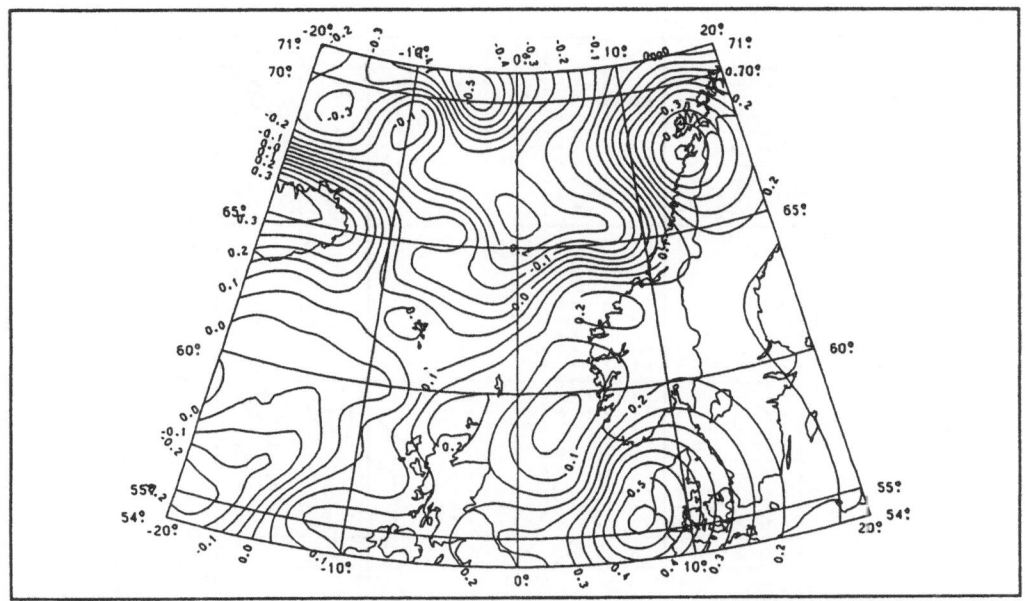

Figure 10. Estimated 1° block averages of quasi-stationary sea surface topography. C.I. 0.05 m.

estimates were calculated. The (*a-priori*) variances of 1° block averages of geoid heights, SST, and geostrophic current components at $\phi = 65°$ are $(0.263 \text{ m})^2$, $(0.304 \text{ m})^2$, and $(0.108 \text{ m/s})^2$ respectively.

The estimated block mean sea surface heights are shown in Figure 6 and the error estimates are shown in Figure 7. In the points of gravity blocks (Figure 4) these values range from -0.42 m to 0.73 m and have a mean value and a standard deviation of 0.05 m and 0.20 m respectively. In the central parts of the region the accuracy of these values is 3-5 cm (see Figure 7).

The estimated block averages of geoid heights and their errors are shown in Figure 8 and Figure 9 respectively. In the gravity block points the values range from -0.46 m to 0.59 m. Their mean value and standard deviation of 0.04 m and 0.19 m respectively. In the central parts of the region the accuracy of these values is 11-15 cm (see Figure 9). It appears that much of the variability of the mean sea

surface heights is correlated with variations in the geoid. The sea surface topography is much smoother.

The QSST averages and their errors are shown in Figure 10 and Figure 11 respectively. In the gravity block points the QSST range from -0.54 m to 0.55 m and the mean value and the standard deviation are 0.01 m and 0.19 m respectively. As the errors of the geoid estimates the accuracy of these

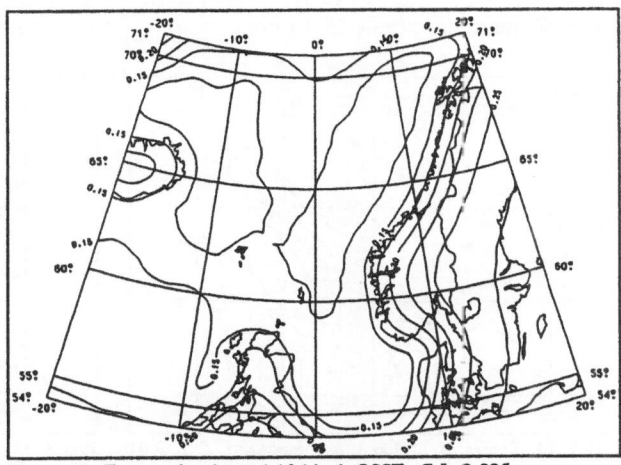

Figure 11. Errors of estimated 1° block QSST. C.I. 0.025 m.

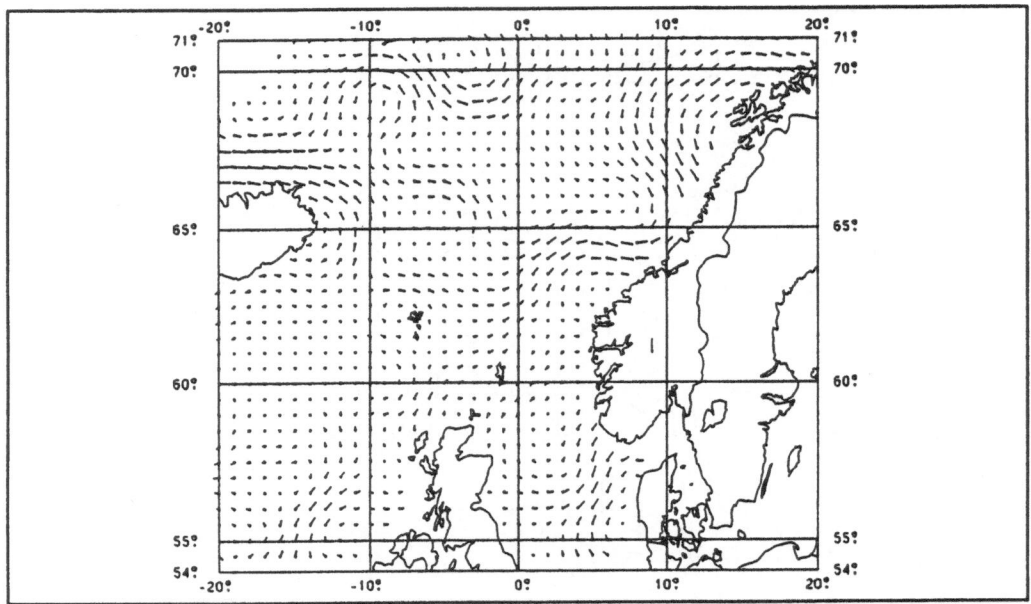

Figure 12. Vectors of estimated 1° block averages of geostrophic surface currents. A 0.2 m/s vector at 61°,9°.

values is 11-15 cm in the central parts of the region (see Figure 11). Earlier results have shown that the errors in the geoid and the errors in the QSST is up to -90 % correlated (Knudsen, 1990*b*). Hence, changes in the geoid reflect the opposide changes in the QSST, when the sum (i.e. the mean sea surface) is determined with a higher accuracy. South of Iceland and west of the UK the distribution of marine gravimetry is quite sparse, so the accuracy of QSST as well as the geoid decreases to about 15-17 cm.

The estimated 1° block averages of geostrophic surface currents are shown in Figure 12. The currents shown the Atlantic outflow into the Norwegian Sea, which is found as a northeastward current in the eastern part of the Faeroe-Shetland Channel, which continues northward off the Norwegian coast. A part of the Gulf Stream turns west south of Iceland before the Faeroe-Shetland Channel. North of Iceland the East Icelandic Current is found. Parts of this current continues southward east of Iceland, and other

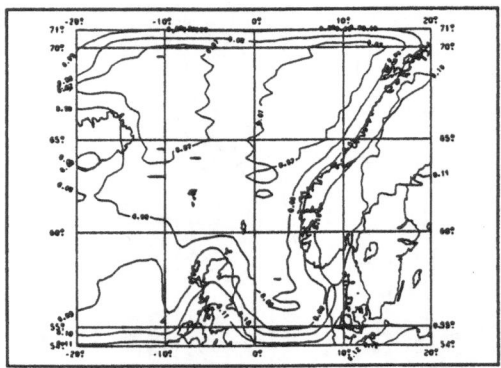

Figure 13. Errors of 1° block *u* components. C.I. 0.01 m/s.

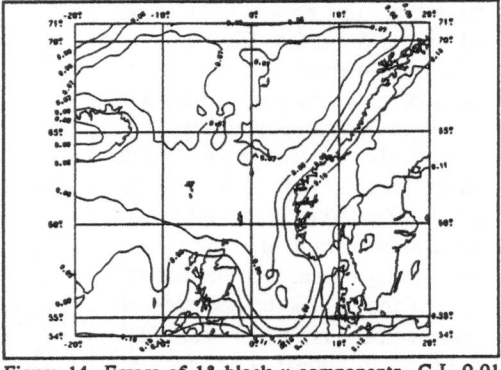

Figure 14. Errors of 1° block *v* components. C.I. 0.01 m/s.

parts continue southeast to the Faeroe Islands, where a clockwise current around the Faeroe Islands is found.

The speed of the currents, which are averaged over one year and in 1° blocks, approach 0.23 m/s north of Iceland. The accuracies of the current components in the central parts of the region are about 7-8 cm/s (see Figure 13-14).

Discussion.

In this paper it has been described how gravimetry and satellite altimetry can be merged using optimal estimation. Also the important step of determining the covariance function models is treated. Here the models were determined, so they fit empirical covariance functions. Hence, the variances and correlations have been properly represented, but many alternative expressions probably exist.

The modelling of the covariance function associated with the gravity field is well known, but developements in the modelling of the remaining covariance function could be useful. The kernel associated with the time varying SST, eq.(25), needs to be verified. For a finite set of frequencies the expression can be derived (see eq.(27)), but the extention to a continous infinite Fourier spectrum may not be trivial. Also the detemination of a covariance function assoicated with the radial orbit error needs further evaluation.

A regional solution has been computed and 1° block averages of mean surface heights, geoid heights, QSST, and geostrophic surface currents have been estimated. The results show that an estimation of the medium wavelength parts of the QSST successfully can be estimated. The estimated currents clearly displays even local characteristics about the circulation in the Norwegian Sea. More detailed information may be obtained in local areas utilizing this regional solution and point observations.

Acknowledgement. Åsmund Drottning carried out the bulk of the regional computation.

References.

Arabelos, D., P. Knudsen, and C.C. Tscherning: Covariance and Bias Treatment when Combining Gravimetry, Altimetry, and Gradiometer Data by Collocation. Proceedings of the IAG Symposia, International Association of Geodesy, Paris, 443-454, 1987.

Blanc, F., S. Houry, P. Mazzega, and J.F. Minster: A High-resolution, High-accuracy Altimeter Derived Mean Sea Surface in the Norwegian Sea. Marine Geodesy, Vol. 14, No. 1, 57-76, 1990.

Colton, M.T., and R.R.P. Chase: Interaction of the Antartic Circumpolar Current With Bottom Topography: An Investigation Using Satellite Altimetry. J. Geophys. Res., Vol. 88, No. C3, 1825-1843, 1983.

Denker, D., and R.H. Rapp: Geodetic and Oceanographic Results from the Analysis of One Year of Geosat Data. J. Geophys. Res., Vol. 95, N. C8, 13151-13168, 1990.

Drottning, Å, and P. Knudsen: An Analysis of Geosat Altimetry off Mid-Norway. Poster presented IUGG XX General Assembly, Symposium U 5, Vienna, 11-24 August, 1991.

Engelis, T.: Global Circulation from SEASAT Altimeter Data. Marine Geodesy, Vol. 9, No. 1, 45-69, 1985.

Engelis, T.: On The Simultaneous Improvement of a Satellite Orbit and Determination of Sea Surface Topography Using Altimeter Data. Manuscripta Geodaetica, Vol. 13, No. 3, 180-190, 1988.

Engelis, T., and P. Knudsen: Orbit Improvement and Determination of the Oceanic Geoid and Topography from 17 Days of Seasat Data. Manuscripta Geodaetica, Vol. 14, No. 3, 193-201, 1989.

Forsberg, R.: A Study of Terrain Reduction, Density Anomalies and Geophysical Inversion Methods in Gravity Field Modelling. Report No. 355, Department of Geodetic Science and Surveying, The Ohio State University, Columbus, 1984.

Fu, L.-L., and V. Zlotnicki: Observing Oceanic Mesoscale Eddies from Geosat Altimetry: Preliminary Results. Geophys. Res. Lett., Vol. 16, No. 5, 457-460, 1989.

Heck, B., and R. Rummel: Strategies for Solving the Vertical Datum Problem Using Terrestrial and Satellite Geodetic Data. Proceedings IAG Symposia, Edinburgh, 1989, 104: "Sea Surface Topography and the Geoid", 116-128, 1990.

Heiskanen, W. A., and Moritz, H.: Physical Geodesy, W. H. Freeman, San Francisco, 1967.

Jekeli, C.: Using Line Averages in Least Squares Collocation. Bulletin Géodésique, Vol. 63, No. 2, 203-212, 1989.

Knudsen, P.: Estimation and Modelling of the Local Empirical Covariance Function using gravity and satellite altimeter data. Bulletin Géodésique, Vol. 61, 145-160, 1987a.

Knudsen, P.: Adjustment of Satellite Altimeter Data from Cross-over Differences using Covariance Relations for the Time Varying Components represented by Gaussian Functions. Proceedings IAG Symposia, 617-628, International Association of Geodesy, Paris, 1987b.

Knudsen, P.: Determination of local empirical covariance functions from residual terrain reduced altimeter data. Reports of the Dep. of Geodetic Science and Surveying no. 395, The Ohio State University, Columbus, 1988.

Knudsen, P.: Some Accuracy Estimates of Local Geoids. In R.H. Rapp and F. Sansò: Determination of the Geoid - Present and Future. IAG Symposia 106, Springer-Verlag, 422-431, 1990a.

Knudsen, P.: An Analysis of the Quasi-stationary Sea Surface Topography in the Norwegian Sea Using Geosat Altimetry and Gravimetry. Paper presented AGU Fall Meeting, San Francisco, California, December 3-7, Abstract G12A-2 in EOS, Vol. 71, No. 43, 1990b.

Knudsen, P.: Simultaneous Estimation of the Gravity Field and Sea Surface Topography From Satellite Altimeter Data by Least Squares Collocation. Geophysical Journal International, Vol. 104, No. 2, 307-317, 1991.

Knudsen, P.: Estimation of Sea Surface Topography in the Norwegian Sea Using Gravimetry and Geosat Altimetry. Bulletin Géodésique, Vol. 66, No. 1, 27-40, 1992a.

Knudsen, P.: Separation of Residual Ocean Tide Signals in a Collinear Analysis of Geosat Altimetry. Presented EGS XVII General Assembly, G5, Edinburgh, April 6-10, 1992b.

Knudsen, P., O. Baltazar Andersen, and C.C. Tscherning: Altimetric Gravity Anomalies in the Norwegian-Greenland Sea - Preliminary Results from the ERS-1 35 days Repeat Mission. Geophys. Res. Lett., in press, July 1992.

Knudsen, P., and M. Brovelli: Collinear and Cross-over Adjustment of Geosat ERM and Seasat Altimeter Data in the Mediterranean Sea. Surveys in Geophysics, in press, 1991.

Koblinsky, C.J., R.S. Nerem, S.M. Klosko, and R.G. Williamson: Global Scale Variations in Sea Surface Topography Determined from Satellite Altimetry. Presented IUGG XX General Assembly, Symposium U 13, Vienna, 11-24 August, 1991.

Marsh, J.G., C.J. Koblinsky, F.J. Lerch, S.M. Klosko, T.V. Martin, J.W. Robbins, R.G. Williamson, and G.B. Patel: Dymanic Sea Surface Topography, Gravity, and Improved Orbit Accuracies from the Direct Evaluation of Seasat Altimeter Data. J. Geophys. Res., Vol. 95, N. C8, 13129-13150, 1990.

Mazzega, P., and S. Houry: An experiment to invert Seasat altimetry for the Mediterranean and Black Sea mean surface. Geophysical Journal, 96, 259-272, 1989.

Menard, Y.: Observations of Eddy Fields in the Northwest Atlantic and Northwest Pacific by SEASAT Altimeter Data. J. Geophys. Res., Vol. 88, No. C3, 1853-1866, 1983.

Milbert, D., B. Douglas, R. Cheney, L. Miller, and R. Agreen: Calculation of Sea Level Time Series from Noncollinear GEOSAT Altimeter Data. Marine Geodesy, Vol. 12, 287-302, 1989.

Moritz, H.: Advanced Physical Geodesy. Herbert Wichmann Verlag, Karlsruhe, 1980.

Nerem, R.S., B.D. Tapley, and C.K. Shum: Determination of the Ocean Circulation Using Geosat Altimetry. J. Geophys. Res., Vol. 95, N. C3, 3163-3179, 1990.

Sansò, F.: Statistical Methods in Physical Geodesy. In: Sünkel, H.: Mathematical and Numerical Techniques in Physical Geodesy. Lecture Notes in Earth Sciences, Vol. 7, 49-155, Springer-Verlag, 1986.

Schrama, E.J.O.: The Role of Orbit Errors in Processing of Satellite Altimeter Data. Report No. 33, Netherlands Geodetic Commission. Publications on geodesy, New series, Delft, 1989.

Tai, C.-K.: On Generating Altimetric Sea Level Time Series from Crossover Differences. Marine Geodesy, Vol. 12, 303-313, 1989.

Tapley, B.D., and G.W. Rosborough: Geographically Correlated Orbit Errors and Its Effect on Satellite Altimetry Missions. J. Geophys. Res., Vol. 90, No. C6, 11817-11831, 1985.

Tarantola, A.: Inverse Problem Theory. Elsevier Publishers B.V., ISBN 0-444-42765-1, 1987.

Tscherning, C.C.: Functional Methods for Gravity Field Approximation. In: Sünkel, H.: Mathematical and Numerical Techniques in Physical Geodesy. Lecture Notes in Earth Sciences, Vol. 7, 3-47, Springer-Verlag, 1986.

Tscherning, C.C., and P. Knudsen: Determination of bias parameters for satellite altimetry by least squares collocation. Proceedings of the International Associations of Geodesy, Hotine-Marussi Symposium on Mathematical Geodesy, Rome, June 3-6, 833-853, 1986.

Tscherning, C.C., and R.H. Rapp: Closed Covariance Expressions for Gravity Anomalies, Geoid Undulations, and Deflections of the Vertical Implied by Anomaly Degree Variances. Report no. 208, Dept. of Geodetic Science and Surveying, The Ohio State University, Columbus, 1974.

Wagner, C.A.: Accuracy Estimates of Geoid and Ocean Topography Recovered Jointly From Satellite Altimetry. J. Geophys. Res., Vol. 91, No. B1, 453- 461, 1986.

Wunsch, C., and V. Zlotnicki: The accuracy of altimetric surfaces. Geophys. J. R. astr. Soc., 78, 795-808, 1984.

Zlotnicki, V.: On the Accuracy of Gravimetric Geoids and the Recovery of Oceanographic Signals from Altimetry. Marine Geodesy, Vol. 8, 129-157, 1984.

Comparing the UK Fine Resolution Antarctic Model (FRAM) with 3-years of Geosat altimeter data.

RAYMOND C.V. FERON

Institute for Marine and Atmospheric Research, Utrecht University,
Princetonplein 5, 3584 CC Utrecht, The Netherlands

The United Kingdom Fine Resolution Antarctic Model is compared to 3 years of Geosat altimeter data. To enable a proper comparison two analysis techniques (Fourier and principal component analysis) are applied to both model results and the "real ocean" altimeter observations. In general it was found that the model succeeded in simulating important characteristics of the southern ocean. The analysis results from the very complex altimeter observations were verified, and the interpretation of Geosat data was strongly improved. The applied analysis techniques where shown to be effective in isolating ring-like phenomena and detecting possible periodic behaviour. In both the Agulhas, Brazil Malvinas, and East Australian Current regular ring formations take place, roughly every 100, 150, and 130 days respectively. The model only generated periodic ring formations in the Agulhas and East Australian Current, both with a very regular 125-130 days period. This period is clearly a model favored harmonic (1/3 year) which is however surprisingly close to the observations.

1. INTRODUCTION

The Southern Ocean is important for climate because it provides the connection between the main ocean basins. For this reason a high resolution numerical model (hereafter referred to as FRAM) has been integrated for approx. 16 years as a part of a large United Kingdom project to improve models used for climate prediction. The model performed well at mesoscale length scales ($50 < L < 300$ km). Studying variations over larger scales (> 1000 km) is not recommended given the present model results [e.g. *Webb et al.*, 1991]. The model should however be extremely useful for a combined study of spatial and temporal mesoscales in the three Southern Ocean western boundary systems (Malvinas/Brazil, Agulhas and East Australian Current). In combination with the sea level data from Geosat the FRAM results can be systematically compared to the observed variations in the sea surface over almost 3 years from Nov 1986 - Oct 1989. The model resolution ($1/2^0$ in east-west direction and $1/4^0$ in north-south) is good enough for mesoscale studies. The comparison can be used for different purposes.

Maybe the biggest advantage of the FRAM results over the altimeter data is preservation of the mean pressure (flow) field. In order to compare the altimeter relative heights with the FRAM pressure field the long term mean is subtracted. The impact of studying anomaly fields instead of absolute height (or pressure) data can be visualized in the FRAM data which enables a better understanding and possible verification of the altimeter results.

From the FRAM data the extraction of the formation frequency of rings in the Agulhas Retroflection area is straightforward because of the known mean field, while the extraction of the ring-shedding process from altimeter data alone is far from easy [e.g. *Feron et al.*, 1992]. Another way of looking at the comparison is to accept the altimeter data to be representative for the real ocean. This would enable the verification of the model results and thus answer the question how good FRAM succeeded in representing reality. Finally one can use both results for verifying the analysis methods including the data interpretation in the Southern ocean in order to be able to apply them to other regions of the world ocean such as the Gulf Stream, Kuroshio Current and Somali Current. In the next section the Fine Resolution Antarctic Model will be described, followed by the Geosat altimeter data-processing procedure (section 3). Section 4 gives an overview of the characteristic dynamics in the three Southern Ocean current systems obtained from historical and more recent observations. The analysis techniques which are used for the extraction and comparison of the typical time and spatial scales (Harmonic Analysis and Modal Decomposition) are briefly discussed in section 5. In section 6 the application of these techniques on the Geosat surface heights and the FRAM pressure fields is presented, followed by a concluding section which compares and discusses the results.

2. Short FRAM Description

The eddy resolving model is based on that of *Semtner* [1974] and *Cox* [1984]. It has 32 vertical levels and a horizontal grid spacing of $1/2^0$ in east-west direction and $1/4^0$ in north-south. The model uses viscosity coefficients of 2×10^2 m^2s^{-1} horizontally and 10^{-4} m ^{2} s^{-1} vertically. The diffusion coefficients for temperature and salinity are 10^2 m ^{2} s^{-1} horizontally and 0.5×10^{-4} m^2 s^{-1} vertically. Linear bottom friction is used, corresponding to an e-folding time of 50 days for a 4000 m ocean. The boundary condition at the northern open boundary is that proposed by *Stevens* [1990]. The model was first initialized using the *Levitus* [1982] smoothed temperature and salinity fields and zero velocity. However, this method was found to be unstable due to barotropic topographic modes trapped near Kerguelen and New Zealand. The system was re-initialized as a cold and saline motionless fluid which was dynamically relaxed to the Levitus data for the first 6 model years. The relaxation is so weak that it permits eddies, fronts and other realistic phenomena to develop. For times beyond 6 years the model runs free from relaxation terms. Annual mean winds [*Hellerman and Rosenstein*, 1983] are used for the forcing. In certain regions the model gives realistic representations of mesoscale circulation. The FRAM results allow the study of the three western boundary current systems in the Southern Ocean. Figure 1 shows the model domain. For more details on FRAM see *Webb et al.* [1991]. In this study 2 years of the FRAM results were used from day 152 model year 13 (first of June), to day 122 model year 15 (first of May) with 10-day intervals. The pressure gradients were computed from the stream function to enable a proper comparison with the altimeter sea heights. The 2 years mean can either be removed for the comparison with Geosat or conserved to observe the ring formation process in a more "realistic" view.

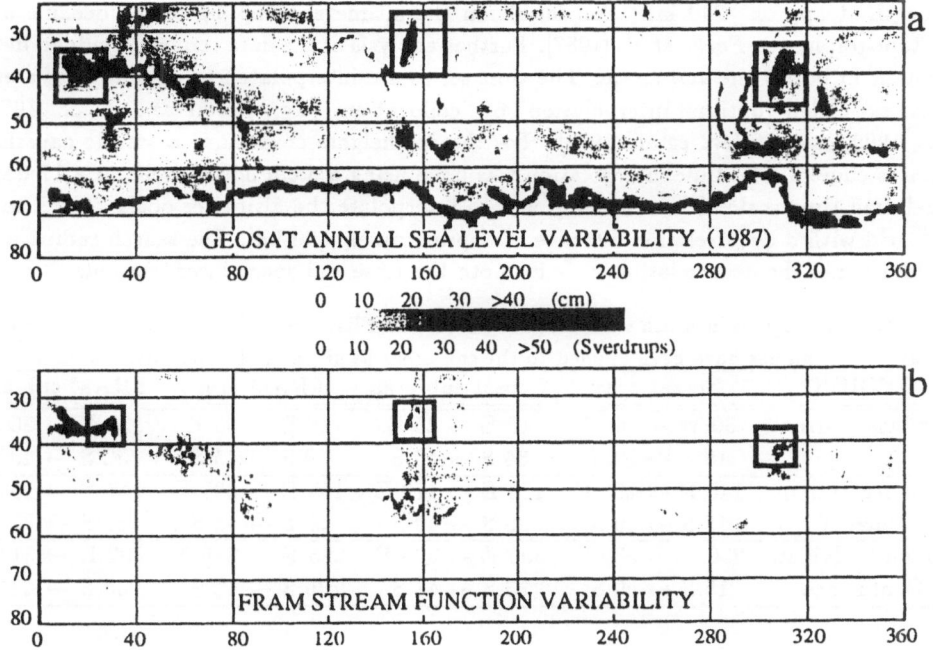

Fig. 1. (*a*) Geosat sea surface variability and (*b*) FRAM stream function variability. The small boxes show the sub areas which are studied in more detail.

3. ALTIMETER DATA PROCESSING

Geosat Exact Repeat Mission altimeter data of the period from November 1986 to September 1989 has been processed using the collinear method [e.g. *Cheney et al.*, 1983] for the area's listed in Table 1. The Geosat sub areas differ somewhat from the FRAM sub-areas because the location of the western boundary currents (and thus the locations of variability) in FRAM is close but not exactly according to reality (see also Figure 1). A sinusoidal parameterization ($2\pi/40,000$ km) to model the orbit error along tracks of approximately 6000 km was used. The orbit error model parameters were solved in a least squares sense for each pass separately. For a detailed description of the time series generation method we refer to [e.g. *Cheney et al.*, 1987, *Milbert et al.*, 1988, *Feron et al.*, 1991]. Basically all time-independent contributions are eliminated from the data by taking only differenced sea heights relative to the computed long term mean. The resolution obtained with this data is approximately 17 days in time and 120 km in space. Given these values we can roughly estimate which scales of variability can be detected and which will be aliased. Geosat's 17-days repeat period and the mid-latitude zonal and meridional resolution (≈ 120 km) sets the Nyquist period somewhere between 17 and 34 days (at crossovers) and the Nyquist wavelength at approx. 240 km. For details on the altimeter sampling of the ocean's dynamic topography see *Parke et al.* [1987]. Further analysis of the data requests gridded data equidistant in both time and space. The time series are interpolated to regular time steps in a process called "optimal interpolation" or "collocation" [e.g., *Bretherton et. al.*, 1976, *Moritz*, 1980, and *Milbert et al.* 1988]. For the sea-heights covariance a simple gaussian function is chosen. The time interpolation was done using a 5 cm noise level and a 17 days decorrelation time scale. The next step was to interpolate the altimeter observations to a $1^0 \times 1^0$ grid with a decorrelation distance of approximately 100 km. The search radius was set to four times the decorrelation scale for both the time and space interpolation.

Table 1 The ocean regions in which the Geosat altimeter and FRAM model data has been processed. The analysis techniques have been applied to the sub areas as given in the specified columns.

REGION	Geosat area	Geosat sub area	FRAM area	FRAM sub area
Agulhas Current	30 W → 90 E	10 E → 25 E	15 E → 65 E	24 E → 36 E
	15 S → 50 S	34 S → 46 S	29 S → 45 S	32 S → 40 S
East Australian	140 E → 200 E	150 E → 170 E	145 E → 295 E	147 E → 154 E
Current	15 S → 50 S	25 S → 40 S	29 S → 45 S	31 S → 38 S
Brazil Malvinas	270 E → 330 E	300 E → 320 E	295 E → 345 E	302 E → 317 E
Confluence	15 S → 50 S	33 S → 47 S	30 S → 46 S	35 S → 43 S

4. THREE SOUTHERN OCEAN CURRENT SYSTEMS

4.1. The Agulhas Current

The primary source region of the Agulhas Current is the wind-driven anticyclonic circulation cell of the Southern Indian Ocean. Secondary input from the East Madagascar Current and through the Mozambique Channel is likely but has no stationary character. The Agulhas Current follows the African continental slope. When the jet runs out of western boundary it separates and becomes a free jet with a south-westward direction. Conservation of potential vorticity and advection of planetary vorticity (β-effect) forces the current to make a large almost 180^0 anticyclonic turn which is generally referred to as the Agulhas Retroflection [e.g. *De Ruijter*, 1982]. Roughly every 100 days a large anticyclonic ring is formed in the retroflection area [*Feron et al.*, 1992]. These warm rings invade the South Atlantic subtropical gyre, thereby causing a significant contribution to the energy and fresh water flux between the Indian Ocean and the Atlantic [*e.g. Gordon and Haxby*, 1990].

4.2. The East Australian Current

The East Australian Current (EAC) flows southward along the Australian continental slope. Usually at a latitude varying from 31^0 and 33^0 this jet turns to the east and meanders towards New Zealand. The southern edge forms the Tasman Front, which separates the warm Coral Sea water from the colder Tasman Sea water in the south. Numerous observations have shown that the EAC is characterized by a quasi periodic process of formation of large anticyclonic eddies [e.g. *Nilsson and Cresswell*, 1981, and *Mulhearn*, 1987]. In these earlier studies the hypothesis was proposed that the formation of large anticyclones could be related to the "breaking" of baroclinic Rossby waves along the Tasman Front, resulting in the observed two ring-shedding events per year. The large amount of different observations (density, temperature and current measurements, bouy observations and sea-surface temperature data) make the EAC an ideal region for the verification of altimeter data and the analysis techniques which are applied to less known current systems.

4.3. The Brazil Malvinas Confluence

The Brazil Current and the Malvinas or Falkland Current converge and form a strong thermal front between 38° and 46° S [e.g., *Leckis and Gordon*, 1982]. Warm core eddies are observed more frequently than cold core eddies. *Garzoli and Garrafo*, [1989] find a 6 and 12 months period in the location of the front. Superimposed on this 6/12 months period, oscillations with periods of 1-2 months are found which can be related to instabilities of the front and/or to a north-south motion of the southern boundary of the Brazil Current at a constant latitude of separation [*Legeckis and Gordon*, 1982; *Garzoli and Garraffo*, 1989]. The wavelength of this oscillation is 400-500 km with an amplitude of 200 km [*Roden*, 1986].

5. ANALYSIS TECHNIQUES.

5.1. Harmonic Analysis

A Fast Fourier transform (FFT) is used to compute the power spectrum for each time series. In this study these spectra are used for two different purposes. The spatial average of the power at a fixed frequency gives a representable average power spectrum for the defined geographical region. The frequency of phenomena that characterize the studied area are expected to show up in the averaged power spectrum (e.g., Figure 2).

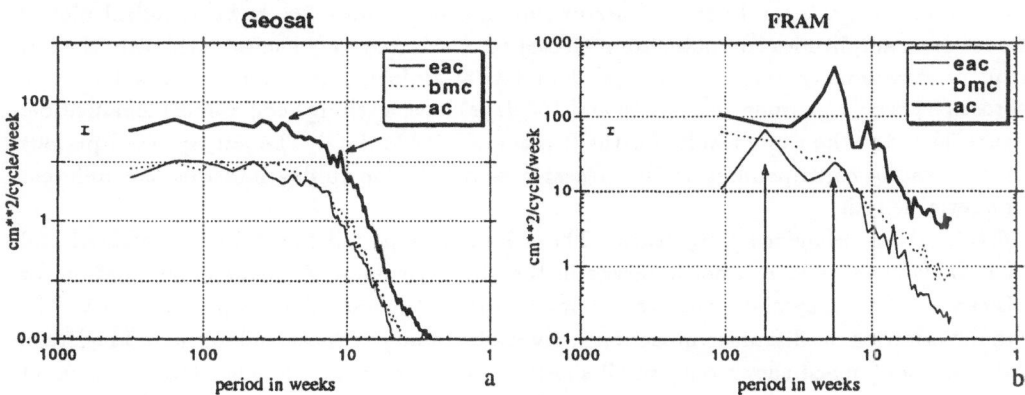

Fig. 2 (a) Geosat and (b) FRAM power spectra for the three boundary current systems as defined in Table 1.

To analyze the spatial distribution of each harmonic component separately, the amplitudes are plotted in longitude-latitude maps. The visualization of the spatial distribution of sea level amplitudes for one frequency enables insight in the local characteristics of each independent frequency. A basic problem occurs in spectral analysis if phenomena are analyzed which are not exactly harmonic, but have oscillating characteristics with a fluctuating period. The next analysis technique is the Principal component analysis which is capable of extracting the temporal and spatial behavior of statistically important phenomena.

5.2. Principal Component Analysis

Principal component analysis, which is also known as empirical orthogonal function (EOF) analysis, is an effective tool for the decomposition of variability in a few modes. All the gridded time series in the selected sub areas (Table 1) are used in the Principal component Analysis in characterizing the spatial structure. Each of these spatial structures (or modes) explains a certain amount of variance and has a time function which describes how that specific mode is changing in time. For details on the principal component analysis see e.g. *Preisendorfer*, (1988), *Kelly*, (1988). In this study the more or less conventional spatial covariance matrix is used for the analysis. The explained variance for each mode is therefore in terms of % of the total amount of temporal variance. The next section deals with the application of the analysis techniques on the Geosat and the FRAM data. Mesoscale variations from Geosat are investigated and compared to the corresponding FRAM results.

6. APPLICATION

6.1. Harmonic Analysis

As illustrated in Figure 2, the Agulhas Current is the most energetic western boundary current in the southern ocean. To investigate these power spectra in more detail the spatial distribution of the separate frequencies can be plotted. Figure 3a shows a spatial plot of the sea level amplitudes computed from Geosat for 6 selected frequencies. Evidently the low frequent variations (periods > 150 days) dominate at all locations, while the variations with mesoscale periods (periods between 50 and 150 days) have a strong geographical localization. Figure 3b shows the same results for the 2 years of FRAM data. The four selected periods are not exactly corresponding to the Geosat case due to the 10 day intervals and different time series length.

6.1.1. Brazil Malvinas Confluence. The 1-2 months period found in the infrared and XBT/CTD observations is not observed in the altimeter data. A possible explanation for this can be that the spatial scales are too small for the altimeter to be sampled correctly. The Nyquist criterion, as discussed in section 3, gives a boundary in space and in time. The Brazil Malvinas Confluence shows only small signals in the mesoscale periods. Most important frequencies from Geosat are the semi-annual, annual and longer time scales. FRAM gives basically the same preference for low frequent variations.

6.1.2. Agulhas Current. The Agulhas Retroflection area is the most energetic part of the Agulhas Current System. The interannual and seasonal period is dominating at all the locations including the retroflection area. This low frequent variability especially over the Agulhas Return Current indicates the occurrence of almost stationary or low frequent Rossby waves [e.g. Park et al. 1990]. The most intense low frequent variations (735 days) occur east of 40^{0} E (east of the south-west Indian Ridge). The mesoscale periods, associated with the formation of large anticyclonic rings [e.g. *Lutjeharms and van ballegooyen*, 1988, *Feron et al.* 1992] clearly show the location where the rings are formed. Detailed analysis of the ring formation process showed that on average every 100 days a major Agulhas Ring is formed [*Feron et al.* 1992]. This 100 days quasi-periodicity was found to have a 14 days standard deviation. Other mesoscale activity is found in the Agulhas Return Current where meanders can produce transient eddies due to baroclinic instabilities. The ring-shedding phenomenon which is also present in the FRAM data gives a clear maximum around 125 days. Hardly any seasonal signal is found in the region around Africa which is remarkable given the seasonal model-forcing.

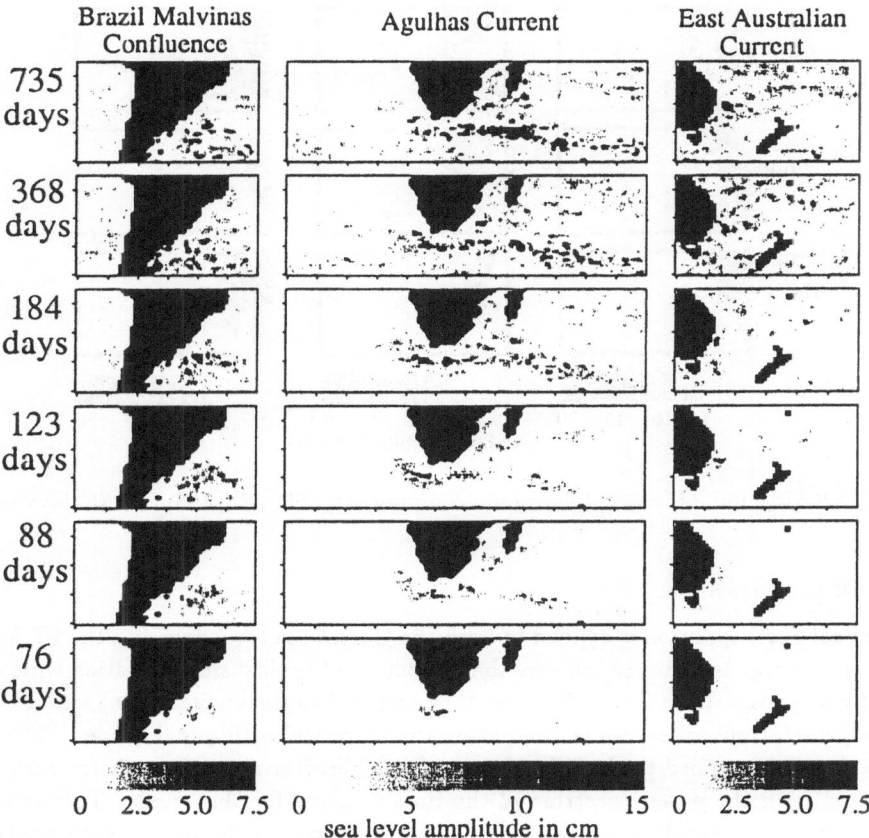

Fig. 3a Geosat plots of selected harmonic components, 735, 368, 184, 123, 88, and 76 days respectively. The images are explained in the text.

6.1.3. East Australian Current. The East Australian Current has (similar to the Brazil Malvinas Confluence) only minor variations in the mesoscale period range. Most important variations in the Geosat data are the semi-annual and lower frequency signals. FRAM gives a dominant seasonal period, which is in agreement with infrared and XBT observations. In the area where warm rings are formed, some variability with 125 day period is found besides the dominant seasonal signal.

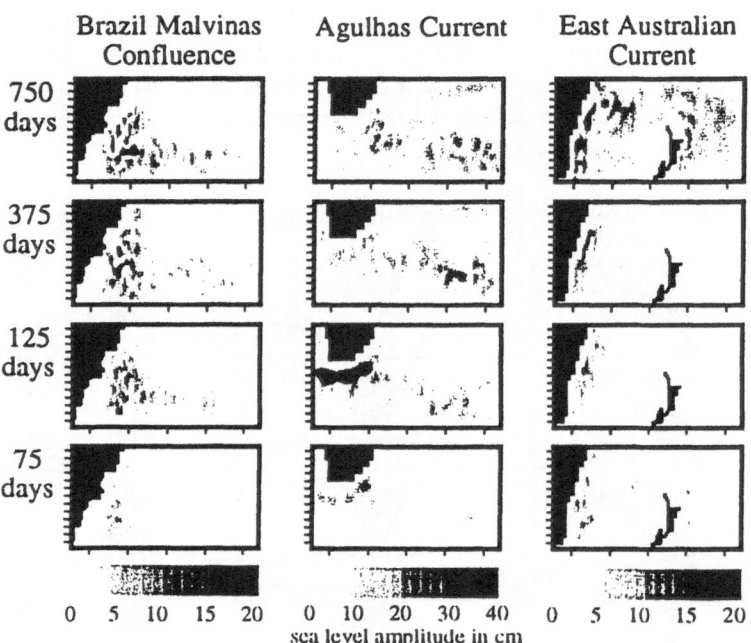

Fig. 3*b* FRAM plots of selected harmonic components, 750, 375, 125, and 75 days respectively.

6.2. Modal Decomposition.

The modal decomposition is applied to both the Geosat sea level data and the FRAM sea level pressure data. In order to improve the extraction of typical ring formation time scales the low frequent variations (periods larger than or equal to the annual cycle) were removed from the geosat time series before computing the empirical orthogonal modes. This high pass filter was not applied to the FRAM data. The characteristic time scale for each mode is estimated from the power spectrum of the time function for that mode. This spectrum is used to obtain the most important period. The variations in the Geosat data are spread over a wide range of complicated modes stressing the difficulties for a proper analysis of altimeter data [e.g., *Chelton et al.*, 1990; *Feron et al.*, 1992].

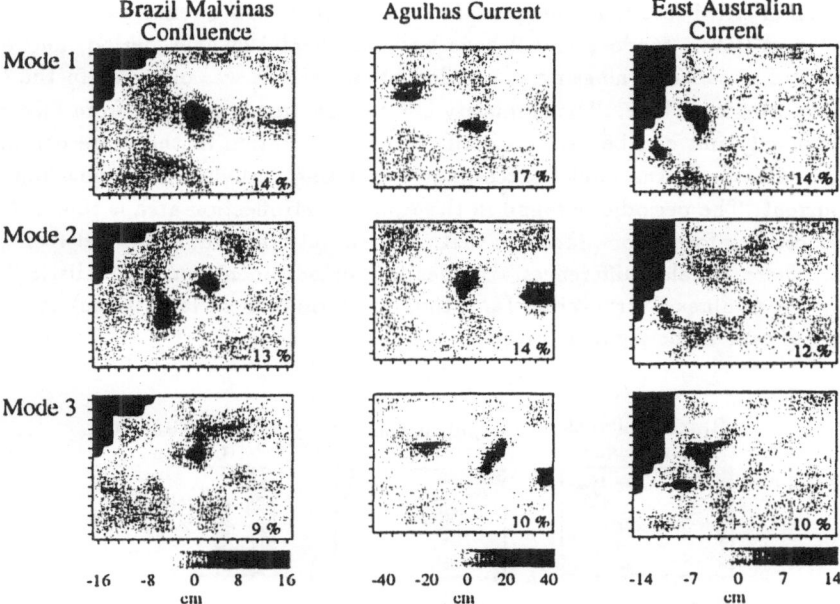

Brazil Malvinas Confluence   Agulhas Current   East Australian Current

Mode 1

Mode 2

Mode 3

-16 -8   0   8   16
cm

-40 -20 0   20 40
cm

-14 -7   0   7   14
cm

Fig. 4 (a) Geosat spatial plots of the first three PC modes. The images are explained in the text.

6.2.1. Brazil Malvinas Confluence. The first three modes from Geosat (Figure 4a) explain 14%, 13%, and 9% variance with dominant time scales of 150, 150 days, and 260 days, respectively. Evidently, Geosat indicates a 150 day periodicity in the Brazil Malvinas Confluence. Contrary to the altimeter results, FRAM gives no dominant periodic phenomenon in the Brazil Malvinas Confluence (Figure 4b). The first three modes have different time scales (> 700, 240, and 285 days respectively) which indicate no periodic ring formation. No mode is found with periods in the mesoscale range.

6.2.2. Agulhas Current. Mode 1 and 2 both have the same approximate 100 days period, where mode 2 has a 1/4 period phase shift forward in time. This indicates a translating phenomenon with a westward propagation of approximately 8 cm/s. *Feron et al.*, [1992] suggested that this propagating phenomenon was indicative for the periodic formation and translation of large Agulhas rings. The first 2 FRAM modes in the Agulhas Current (and also in the East Australian Current (Figure 4b)) show a well defined ring formation with associated south-westward propagation. The characteristics of the regular formation of rings in FRAM can be analyzed and verified because the model's mean pressure field is available and can be added to the anomaly field, something which is not possible with the Geosat sea level data. The FRAM results for the modal decomposition show a remarkably resemblance with the Geosat data. Mode 1 and 2 have both an exact 125 days period, and explain together 60% of variance. The same 1/4 period phase shift as was observed in the Geosat modes (1 and 2) occurs here. The similarity between the mode analysis from Geosat and FRAM is evident, and can be used to verify the interpretation of the more complicated Geosat data. Obviously the model succeeds in simulating periodic ring formations in the Agulhas Current. The periodicity found in the Agulhas retroflection area is thus caused by a quasi periodic ring formation process in that area. Despite the similarities between FRAM and Geosat, there are also differences. Besides the already mentioned unrealistic "exact" periodicity, the Agulhas Retroflection (and thus the formation of large rings) takes place west instead of east of the Agulhas Plateau.

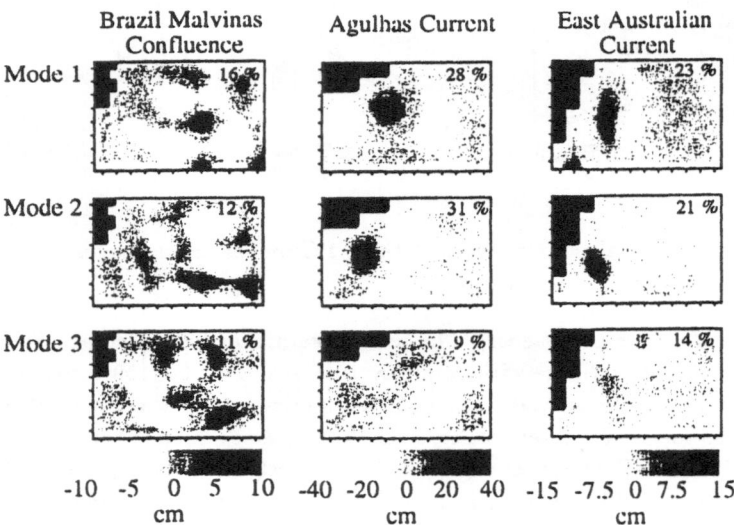

Fig. 4 (b) FRAM spatial plots of the first three PC modes. Again, the images are explained in the text.

6.2.3. East Australian Current. The first three Geosat modes explain 14%, 12%, and 10% variance with average time scales of 124 days, 124 days, and > semi-annual, respectively. Mode 1 and 2 indicate that over the three years Geosat data 8-9 rings were pinched off and translated into the Tasman Sea. This estimate is relatively high, compared to the average 1/2 year period that was found from sea surface temperature data [e.g., *Mulhearn et al.*, 1987]. A possible explanation for the different results obtained from sea surface temperature is the fact that periods of cloud cover will systematically underestimate the number of shedded rings. The first three FRAM modes explain 23%, 21%, and 14% variance, respectively. Again, the first two modes (Figure 4b) isolate a south-westward propagating phenomenon. The first two modes have a period of 125 days (the same as the FRAM Agulhas Current period) and mode 3 as a clear annual cycle. In contrast to the results in the Agulhas Retroflection, the East Australian Current model-rings are formed at the correct geographical location and surprisingly at the same frequency. The East Australian Current apparently has, similar to the Agulhas Current, also quasi-periodic ring formations. Due to the different conditions (e.g., jet separation angle, bottom topography, current velocities, continental boundaries) the warm rings west of Australia are formed less regular than the Agulhas rings and translate in a more southward than westward direction.

7. CONCLUSIONS

In general it is concluded that FRAM succeeded in simulating characteristics of the southern ocean three western boundary currents. The model gives a simplified view of mesoscale current behavior which enables the comparison with the more complex altimeter observations of the sea level. The real ocean is often characterized by variations that are simply too complex for a straightforward interpretation. The use of advanced statistical methods for the interpretation and reduction of large amounts of complex data is a direct result. The simultaneous application of these techniques on the Geosat data and FRAM results has served several purposes.

Validation of techniques for the analysis of large amounts of model output or satellite data. These methods have become increasingly important as a result of the huge amount of data coming from remotely sensed observations as well as the development of realistic numerical models.

The extraction of mesoscale phenomena such as the quasi periodic formation of large anticyclonic rings in the Agulhas Retroflection (roughly every 100 days), in the East Australian Current (roughly every 125-132 days), and in the Brazil Malvinas Confluence (roughly every 150 days), enables a better estimate of important exchange of fresh water and energy in meridional direction and between ocean basins. The Agulhas Rings transfer warm Indian Ocean water into the relative cold South Atlantic gyre. Although their importance for the large scale circulation is not yet fully understood [*Gordon*, 1985, *Gordon and Haxby*, 1990, *Rintoul*, 1991, *Gordon et al.*, 1992], the regional importance of these large coherent rings is however evident. *Lutjeharms*, [1988] reports occurrence of meridional heat transport across the Sub-tropical Convergence by a warm ring. He estimates from simultaneous hydrographical measurements and satellite infrared measurements a heat flux of 26.7 $\times 10^{17}$ Joules. Local atmospheric cooling of the upper thermocline results in an energy flux into the atmosphere. The warm rings formed east of Australia transfer momentum and warm Coral Sea water across the Tasman Front into the Tasman Sea. The Brazil Malvinas Confluence seems to have the same preference for the formation of warm (anti-cyclonic) rings that transfer energy and other properties across the confluence to the south. The altimeter data can be used to estimate the heat/energy transport by these eddy fluxes.

Another result from the combined analysis is the fact that propagating rings show up in the modal decomposition as two equally important modes with a 1/4 period phase shift. The seasonal forcing of the FRAM is effectively adsorbed in an energetic 3 per year ring formation phenomena in the Agulhas as well as the East Australian Current. The comparison with the Geosat data suggests that there could be a connection between the large scale seasonal wind forcing over the South Indian Ocean and the typical ring formation characteristics. As promising as the agreements between the "real ocean" results and the FRAM result seem, there is no clear explanation why the model generates the 125-130 days period. The fact that the Agulhas and East Australian Current both have the same 125-130 days periodic ring formation illustrates that apparently this frequency is a FRAM favored harmonic, which is however quite close to the "real ocean" period.

Acknowledgments.

Dr. D. Webb is acknowledged for giving the author the opportunity to visit IOSDL and for supplying the FRAM data, and S. Thompson for his help in the FRAM data selection and pressure computations. This investigation was partly funded through the Dutch National Research Program on Global Air Pollution and Climate Change.

REFERENCES

Bretherton, F.P., R.E. Davis, and C.B. Fandry, A technique for objective analysis and design of oceanographic experiments applied to MODE-73, *Deep Sea Res., 23*, 559-582, 1976.

Chelton, D.B., M.G. Schlax, D.L. Witter, and J.G. Richman, Geosat altimeter observations of the surface circulation of the Southern Ocean, *J. Geophys. Res., 95*, (C10), 17,877-17,903, 1990.

Cheney R.E., J.G.Marsh, and B.D. Beckley, Global mesoscale variability from collinear tracks of SEASAT altimeter data, *J. Geophys. Res., 88*, (C7), 4343-4354, 1983.

Cheney, R.E., B.C. Douglas, R.W. Agreen, L. Miller, D.L. Porter and N.S. Doyle, Geosat altimeter geophysical data record user handbook, *NOAA Techn. Memo. NOS NGS-46*, Natl. Oceanic and Atmos. Admin., Rockville, Md., 1987.

Cox, M. D., A primitive equation, 3-dimensional model of the ocean, GFDL, Ocean Gr. Tech. Rep., 1, 1984.

De Ruijter, W.P.M., Asymptotic analysis of the Agulhas and Brazil Current Systems, *J. Phys. Oceanogr., 12*, 361-373, 1982.

Dewar W., Ventilating warm rings: Theory and energetics, *J. Phys. Oceanogr., 17* , 2219-2231 , 1987.

Feron, R.C.V., W.P.M. De Ruijter, and D. Oskam, Ring-shedding in the Agulhas Current System, *J. Geophys. Res., 97*, 9467-9477, 1992.

Feron, R.C.V., M.C. Naeije, and D. Oskam, Quality estimates for ocean variability results from satellite altimetry, *Mar. Geodesy, 15*, 1-18, 1991.

Garzoli, S.L., and Z. Garraffo, Transports, frontal motions and eddies at the Brazil-Malvinas Currents Confluence, *Deep Sea Res., 36*, 681-703, 1989.

Gordon, A.L., Interocean exchange of thermocline water, *J. Geophys. Res., 91*, (C4), 5037-5046, 1986.

Gordon, A.L., J.R.E. Lutjeharms, and M.L. Gründlingh, Stratification and circulation at the Agulhas Retroflection, *Deep Sea Res., 34*, 565-599, 1987.

Gordon, A.L., and W.F. Haxby, Agulhas eddies invade the South Atlantic - Evidence from Geosat altimeter and shipboard CTD, *J. Geophys. Res., 95*, (C3), 3117-3125, 1990.

Gordon, A.L., R.F. Weiss, W.M. Smethie, Jr., and M.J. Warner, Thermocline and intermediate water communication between the South Atlantic and Indian Oceans, *J. Geophys. Res., 97*, (C5), 7223-7240, 1992.

Hellerman, S., and M. Rosenstein, Normal monthly wind stress over the world ocean with error estimates, *J. Phys. Oceanogr., 13*, 1093-1104, 1983.

Kelly, K., Comment on "Empirical orthogonal function analysis of advanced very high resolution radiometer surface temperature patterns in Santa Barbara Channel" by G.S. E. Lagerloef and R.L. Bernstein, *J. Geophys. Res., 93*, 15,753-15,754, 1988.

Legeckis, R., and A. L. Gordon, Satellite observations of the Brazil and Falklands currents-1975 to 1976 and 1978, *Deep Sea Res., 29*, 275-401, 1982.

Levitus, S., Climatological atlas of the world ocean, NOAA, Prof. Pap., 13 U.S. Dept. of Commerce, 173 pp, 1982.

Lutjeharms J.R.E., Meridional heat transport across the Sub-tropical Convergence by a warm eddy, *Nature, 331*, 251-254, 1988.

Lutjeharms, J.R.E., and R.C. van Ballegooyen, The retroflection of the Agulhas Current, *J. Phys. Oceanogr., 18*, 1570-1583, 1988.

Milbert, D., B. Douglas, R. Cheney, L. Miller, and R. Agreen, Calculation of Sea Level Time Series from non-collinear Geosat altimeter data, *Mar. Geod., 12*, 287-302, 1988.

Moritz, H., *Advanced Physical Geodesy*, Herbert Wichman Verlag, Karlsruhe, Germany, 1980.

Mulhearn, P.J., The Tasman front: a study using satellite infrared imagery, *J. Phys. Oceanogr., 17*, 1148-1155, 1987.

Nilsson, C.S., and G.R. Cresswell, The formation and evolution of East Australian Current warm-core eddies, *Prog. Oceanogr., 9*, 133-183, 1981.

Park, Y-H., Evidence of semiannual baroclinic Rossby waves south of the Indian Ocean from satellite altimetry, *C.R. Acad. Sci. Paris, 310, Ser. II*, 919-926, 1990.

Parke M.E., R.H. Steward, D.L. Farless, and D.E. Cartwright, On the choice of orbits for an altimeter satellite to study ocean circulation and tides, *J. Geophys. Res., 92*, (C11), 11693-11707, 1987.

Preisendorfer R.W. (1988) Principal Component Analysis in Meteorology and Oceanography, *Developments in Atmospheric Science 17*, Elsevier, 1988.

Rintoul, S.R., South Atlantic interbasin exchange, *J. Geophys. Res., 96*, (C2), 2675-2692, 1991.

Roden, G. I., Thermohaline fronts and baroclinic flow in the Argentine Basin during the austral spring of 1984, *J. Geophys. Res., 91*, 5075-5093, 1986.

Semtner, A. J., and R. M. Chervin, A simulation of the global ocean circulation with resolved eddies, *J. Geophys. Res., 93*, 15502-15522, 1988.

Stevens, D. P., On open boundary conditions for three-dimensional primitive equation ocean circulation models. *Geophys. and Astrophys. Fluid Dyn., 51*, 103-133, 1990.

Webb, D.J. et al. (The FRAM Group), An Eddy-resolving model of the Southern Ocean. *EOS, Vol. 72, 15*, 169,174,175, 1991.

Springer-Verlag
and the Environment

We at Springer-Verlag firmly believe that an international science publisher has a special obligation to the environment, and our corporate policies consistently reflect this conviction.

We also expect our business partners – paper mills, printers, packaging manufacturers, etc. – to commit themselves to using environmentally friendly materials and production processes.

The paper in this book is made from low- or no-chlorine pulp and is acid free, in conformance with international standards for paper permanency.

Lecture Notes in Earth Sciences